The
MICROBIAL
CHALLENGE

Jones & Bartlett Learning Titles in Biological Science

AIDS: Science and Society, Seventh Edition
Hung Fan, Ross F. Conner, & Luis P. Villarreal

AIDS: The Biological Basis, Fifth Edition
Benjamin S. Weeks & I. Edward Alcamo

Alcamo's Fundamentals of Microbiology, Body Systems Edition, Second Edition
Jeffrey C. Pommerville

Alcamo's Laboratory Fundamentals of Microbiology, Ninth Edition
Jeffrey C. Pommerville

Alcamo's Microbes and Society, Third Edition
Benjamin S. Weeks

Biochemistry
Raymond S. Ochs

Bioethics: An Introduction to the History, Methods, and Practice, Third Edition
Nancy S. Jecker, Albert R. Jonsen, & Robert A. Pearlman

Bioimaging: Current Concepts in Light and Electron Microscopy
Douglas E. Chandler & Robert W. Roberson

Biomedical Graduate School: A Planning Guide to the Admissions Process
David J. McKean & Ted R. Johnson

Biomedical Informatics: A Data User's Guide
Jules J. Berman

Botany: An Introduction to Plant Biology, Fifth Edition
James D. Mauseth

Botany: A Lab Manual
Stacy Pfluger

Case Studies for Understanding the Human Body, Second Edition
Stanton Braude, Deena Goran, & Alexander Miceli

Electron Microscopy, Second Edition
John J. Bozzola & Lonnie D. Russell

Encounters in Microbiology, Volume 1, Second Edition
Jeffrey C. Pommerville

Encounters in Microbiology, Volume 2
Jeffrey C. Pommerville

Encounters in Virology
Teri Shors

Essential Genetics: A Genomics Perspective, Sixth Edition
Daniel L. Hartl

Essentials of Molecular Biology, Fourth Edition
George M. Malacinski

Evolution: Principles and Processes
Brian K. Hall

Exploring Bioinformatics: A Project-Based Approach
Caroline St. Clair & Jonathan E. Visick

Exploring the Way Life Works: The Science of Biology
Mahlon Hoagland, Bert Dodson, & Judy Hauck

Fundamentals of Microbiology, Tenth Edition
Jeffrey C. Pommerville

Genetics: Analysis of Genes and Genomes, Eighth Edition
Daniel L. Hartl & Maryellen Ruvolo

Genetics of Populations, Fourth Edition
Philip W. Hedrick

Guide to Infectious Diseases by Body System, Second Edition
Jeffrey C. Pommerville

Human Biology, Seventh Edition
Daniel D. Chiras

Human Biology Laboratory Manual
Charles Welsh

Human Body Systems: Structure, Function, and Evironment, Second Edition
Daniel D. Chiras

Human Embryonic Stem Cells, Second Edition
Ann A. Kiessling & Scott C. Anderson

Laboratory Investigations in Molecular Biology
Steven A. Williams, Barton E. Slatko, & John R. McCarrey

Lewin's CELLS, Second Edition
Lynne Cassimeris, Vishwanath R. Lingappa, & George Plopper

Lewin's Essential GENES, Third Edition
Jocelyn E. Krebs, Elliott S. Goldstein, & Stephen T. Kilpatrick

Lewin's GENES XI
Jocelyn E. Krebs, Elliott S. Goldstein, & Stephen T. Kilpatrick

Microbial Genetics, Second Edition
Stanley R. Maloy, John E. Cronan, Jr., & David Freifelder

Molecular Biology: Genes to Proteins, Fourth Edition
Burton E. Tropp

Neoplasms: Principles of Development and Diversity
Jules J. Berman

Precancer: The Beginning and the End of Cancer
Jules J. Berman

Principles of Cell Biology
George Plopper

Principles of Modern Microbiology
Mark Wheelis

Principles of Molecular Biology
Burton E. Tropp

Science and Society: Scientific Thought and Education for the 21st Century
Peter Daempfle

Strickberger's Evolution, Fifth Edition
Brian K. Hall

Symbolic Systems Biology: Theory and Methods
M. Sriram Iyengar

20th Century Microbe Hunters
Robert I. Krasner

Understanding Viruses, Second Edition
Teri Shors

The

MICROBIAL
CHALLENGE

A PUBLIC HEALTH PERSPECTIVE

THIRD EDITION

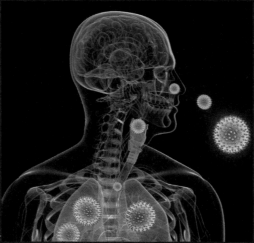

Robert I. Krasner, Ph.D., M.P.H.
Providence College

Teri Shors, Ph.D.
University of Wisconsin Oshkosh

JONES & BARTLETT
LEARNING

World Headquarters
Jones & Bartlett Learning
5 Wall Street
Burlington, MA 01803
978-443-5000
info@jblearning.com
www.jblearning.com

Jones & Bartlett Learning books and products are available through most bookstores and online booksellers. To contact Jones & Bartlett Learning directly, call 800-832-0034, fax 978-443-8000, or visit our website, www.jblearning.com.

Substantial discounts on bulk quantities of Jones & Bartlett Learning publications are available to corporations, professional associations, and other qualified organizations. For details and specific discount information, contact the special sales department at Jones & Bartlett Learning via the above contact information or send an email to specialsales@jblearning.com.

Production Credits

Chief Executive Officer: Ty Field
President: James Homer
SVP, Editor-in-Chief: Michael Johnson
SVP, Chief Marketing Officer: Alison M. Pendergast
Executive Publisher: Kevin Sullivan
Senior Acquisitions Editor: Erin O'Connor
Editorial Assistant: Rachel Isaacs
Editorial Assistant: Michelle Bradbury
Production Manager: Louis C. Bruno, Jr.
Senior Marketing Manager: Andrea DeFronzo
Production Services Manager: Colleen Lamy
Online Products Manager: Dawn Mahon Priest
VP, Manufacturing and Inventory Control:
    Therese Connell

Composition: CAE Solutions Corp.
Cover Design: Kristin Parker
Rights & Photo Research Associate: Lauren Miller
Cover Images: Background, © Roxana Gonzalez/
    ShutterStock, Inc.
    Left, © Tupungato/ShutterStock, Inc.
    Center, © Sebastian Kaulitzki/ShutterStock, Inc.
    Right, © Muriel Lasure/ShutterStock, Inc.
    Inside, left and right, © Roxana Gonzalez/
    ShutterStock, Inc.
Background Image Pages i and iii:
    © Roxana Gonzalez/ShutterStock, Inc.
Printing and Binding: Courier Companies
Cover Printing: Courier Companies

To order this product, use ISBN: 978-1-4496-7375-8

Library of Congress Cataloging-in-Publication Data
Krasner, Robert I.
    The microbial challenge / Robert I. Krasner. — 3rd ed.
        p. ; cm.
    Includes index.
    ISBN 978-1-4496-7333-8 (alk. paper)
    I. Title.
    [DNLM: 1. Communicable Diseases, Emerging. 2. Microbiological Phenomena. WA 110]
    616.9'041—dc23
                                                                    2012020607

6048

Printed in the United States of America
17 16 15 14 13   10 9 8 7 6 5 4 3 2

Authors' Dedications

From Robert I. Krasner

to Lee, with love and gratitude for 48 years of happiness and adventure. to Jon, Lisa, and Andrew, and to my grandchildren, Benjamin, Joshua, Harris, Simon, Julina, and Noah.

"Chance favors the prepared mind."
—Louis Pasteur

From Teri Shors

to the late Elaine (Motschke) Gross, my mother. *Ich vermisse dich jeden Tag.*

to John Cronn, my undergraduate microbiology mentor, colleague, and friend who opened my eyes to the invisible world of microbes and viruses.

to the hundreds of students I have taught; past and present.

"We know nothing of what will happen in the future, but by the analogy of experience."
—Abraham Lincoln

Brief Contents

Contents

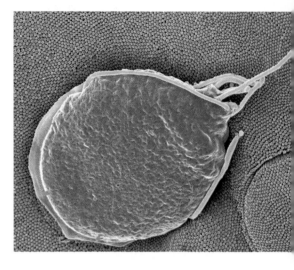

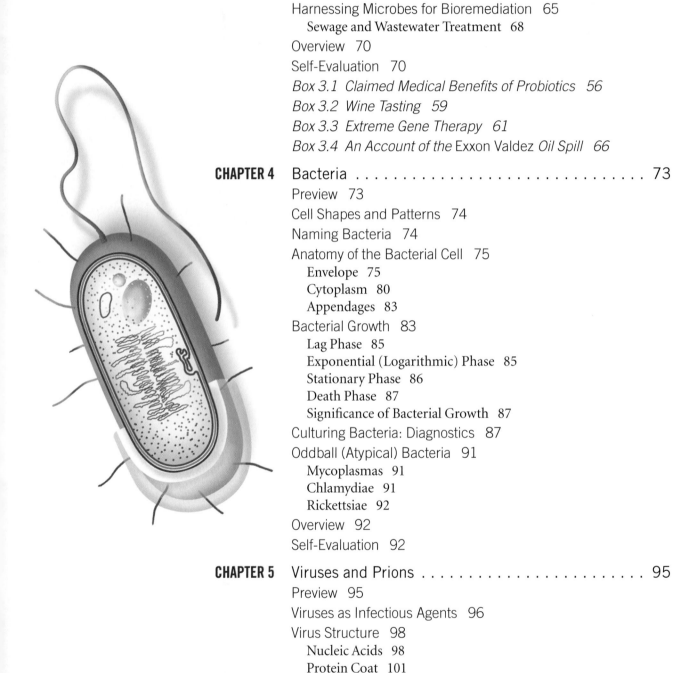

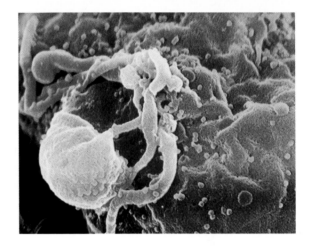

PART 2   Microbial Disease 151

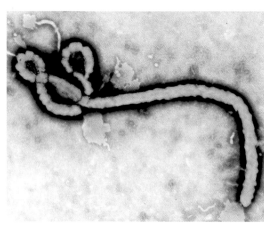

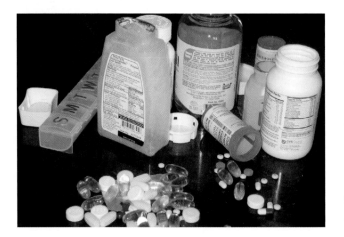

Photo Credits, Table of Contents

Page viii, © Vladimir Wrangel/ Shutterstock.; Page ix top, Courtesy of Dr. Stan Erlandsen/CDC; Page ix bottom, Courtesy of RIK; Page xi top, Courtesy of C. Goldsmith, P. Feorino, E. L. Palmer, W. R. McManus/CDC; Page xiii, © Jones & Bartlett Learning. Photographed by Christine McKeen; Page xiv, Courtesy of Cynthia Goldsmith/CDC; Page xvii, Courtesy of RIK; Page xviii, Courtesy of Jean Roy/CDC.

Preface

■ Birth and Development of *The Microbial Challenge*

After 50 years in the classroom at Providence College in Rhode Island teaching microbiology to biology majors, I decided to develop a microbiology course for nonbiology majors. Outbreaks of disease were in the news frequently, and, judging by the questions students in a nonmajors, general-biology course asked, it was apparent they, too, needed to know more about microbes and human-microbe interactions. Actually, I had been thinking about teaching a nonmajors course for a number of years, but to my surprise, there was no text available. Hence, I used handouts, assignments from the Internet, magazine and newspaper articles, and videos. I even called in a few speakers to supplement my lectures. This strategy worked, but it was cumbersome, required too many handouts, and resulted in confusion, so I decided to write my own text.

At about that time, I decided to study public health microbiology and was accepted into the Harvard School of Public Health for the 1999–2000 academic year, 41 years after I completed my Ph.D. I wanted to do this for many years, but raising a family and educating my children was my top priority. Now it was my turn! As far as I am able to determine, I am the oldest student on a full-time basis to earn the Master of Public Health (M.P.H.) degree at Harvard. My classmates were primarily young medical students who had postponed their fourth year of medical school to earn a M.P.H. before completing medical school. My Harvard studies included six weeks in a tropical disease laboratory in Brazil culminating in a two-week field trip in Manaus in the Amazon.

While at Harvard's School of Public Health, every day I passed by an inscription that reads, in several languages, "The highest attainable standard of health is one of the fundamental rights of every human being." This inscription, my studies at Harvard, and my travel experiences were major factors in the birth and in the public health perspective of my text, *The Microbial Challenge*.

Development of the third edition was a major task. I am thankful that Teri Shors, Professor in the Department of Biology and Microbiology at the University of Wisconsin Oshkosh for the past 15 years, signed on as a co-author. Her creativity, judgment, and knowledge have resulted in a text we are proud of. Her specialty is virology, and together we bring many years of teaching experience to this text.

■ Text Overview

Microbes are as much a part of our biological world as are the more familiar plants and animals. They are an extremely diverse group consisting of thousands of species, some of which are not even considered to be "alive." Most are not pathogens—disease producers—and many are necessary for the maintenance of life. A number of species have been exploited in the food industry, in genetic engineering, in environmental applications, and in many areas of research. This book focuses on the relative "handful" of microbes that produce disease in humans.

Annihilation of microbes is not a possibility, a goal, or a desirable outcome, but learning to live in harmony with microbes is realistic and necessary. All students, not only biology majors, will benefit from understanding microbes and those factors that lead to collisions between microbes and humans. As a potential parent you will deal with immunization, rashes, fevers, earaches, and sore throats that your child will develop. Further, as history has shown, epidemics and pandemics are a constant threat, and prevention is based on knowledge-based preparation. Your generation has not known a world without AIDS, and the H1N1 swine influenza remains a source of great concern throughout the world. Other noted microbial news at the time of this writing includes an outbreak of carbapenem-resistant *Klebsiella pneumoniae* at the National Institutes of Health hospital in Bethesda, Maryland, a meningitis outbreak associated with fungal contamination of pain medication at the New England Compounding Center in Massachusetts, and the discovery of three new human viruses. The new viruses include a SARS-like coronavirus from the Saudi Arabian peninsula, the Heartland virus that infected two farmers in Missouri, and a new hemorrhagic fever virus called the Bas-Congo rhabdovirus. The Bas-Congo virus killed two teenagers and infected the 32-year-old nurse attending them in central Africa. The nurse spontaneously survived and no new cases occurred.

■ Text Format

The chapters are arranged into four logical and sequential parts:

Part 1, The Challenge (Chapters 1–6)
Part 2, Microbial Diseases (Chapters 7–11)
Part 3, Meeting the Challenge (Chapters 12–14)
Part 4, Current Challenges (Chapters 15–17)

Taken as a whole, the dynamics of interaction between microbes and humans unfold. Chapter 1 considers the appearance of new, emerging, and reemerging infectious diseases and factors that contribute to their presence, including world population growth, technological advances, human behavior, and ecological disturbances. Chapter 2 introduces the array of microbes that constitute the microbial world and their distinctive properties. Lest the student think that all microbes are "bad guys and out to get us," Chapter 3 emphasizes "the other side of the coin"—the beneficial aspects of microbial life. The biology of bacteria, viruses, and prions is described in Chapters 4 and 5. Chapter 6 reviews basic genetics and describes aspects of microbial genetics with an emphasis on mechanisms of genetic exchange in bacteria.

Chapter 7 establishes that microbes do not "seek us out," but that the association between microbes and their hosts is accidental—a chance collision that may result in harm to the host. Further, the mechanisms of virulence and the stages of disease are discussed. Chapter 8 focuses on the epidemiology and cycle of microbial disease and on nosocomial (hospital-acquired) infections, an increasing worldwide problem. Chapters 9, 10, and 11 present a sample of "the challengers," namely, bacteria (Chapter 9), viruses and prions (Chapter 10), protozoa, worms, and fungi (Chapter 11), and the diseases they cause. These chapters focus on mechanisms of transmission, and each one is divided into food- and waterborne, airborne, sexually transmitted, contact, soilborne, and arthropodborne diseases.

The immune system is the topic of Chapter 12; the chapter describes the mechanisms by which molecules embedded in microbes or released (toxins) by microbes, seen by the host immune system as "foreign," are targeted for elimination. Chapter 13 considers the strategies of microbial disease control based on sanitation and clean water, immunization, and antibiotic and antiviral agents. As emphasized in Chapter 14, the burden of disease and its spread in epidemics and pandemics are most effectively prevented and controlled by partnerships at all levels ranging from the local level to the national level, to the international level, and to the private sector. Partnerships are the way to go!

Chapter 15 portrays the use of biological weapons, whether for warfare or for acts of terrorism. The history of biological weaponry is long and disturbing and remains a constant and potential threat requiring constant vigilance. Clearly, the terrorist attack in Libya on September 11, 2012, resulting in the death of the U.S. ambassador and three other Americans, serves as a reminder. Chapter 16 recognizes that plagues result from a lack of adaptation between microbes and humans and identifies AIDS, influenza, and tuberculosis as current or threatening plagues. The final chapter, Chapter 17, briefly reviews the text and emphasizes the ongoing struggle between microbes and humans.

◼ Unique Features of the Third Edition

The inside covers of *The Microbial Challenge, Third Edition* contains relevant headlines from Internet news or blogs. It is a new feature aimed to immediately "hook" to important microbiology concepts. We strive to use a writing style that is humorous (with terrible puns) and engaging and applies to the daily lives of students. The **Appendix** of the text contains a **Suggested Readings** list of books relevant to the topics of antibiotics, bacteria, fungi, history of microbiology, protozoans, prions, and viruses. The numbers of parts in this edition remain at four and are sequentially presented to assist the student in following the logic of the narrative. All of the examples that illustrate key principles in the chapters have been updated, and the art, photos, and overall design have been improved throughout the text.

The text was revised to address feedback from instructors and students. Accordingly, Chapter 6 (**Bacterial Genetics**) was added to the second edition and updated for this edition. As suggested by a reviewer, the topic of fungal diseases is a new component of the third edition. Fungal infections in the "wild" and in humans with normal immune systems or immune compromised patients were added in Chapter 11. We also added new information per reviewer suggestions to address practical microbial control in a new section titled "Disinfection and Disease Control" to Chapter 13 (**The Control of Microbial Diseases**). Topics include disinfection methods, cleaning products, soap, handwashing, and high-tech gadgets. Diagnostic bacteriology has been updated to include Analytical Profile Index 20 Enterobacteriaceae (API 20E) and rapid strep tests (Chapter 4), and molecular diagnostic procedures used to detect viral pathogens such as rapid influenza A and B tests were included in Chapter 5.

A few topics were rearranged in this edition. Immunization (vaccination) was moved from Chapter 12 (**The Immune Response**) to Chapter 13 (**Control of Microbial Diseases**) because it is a method to prevent or control microbial diseases. The biology of prions was moved from Chapter 17 (**Current Plagues** in the second edition) to Chapter 5 to become a chapter titled **Viruses and Prions**. Accordingly, prion diseases were moved to Chapter 10, **Viral and Prion Diseases**. Influenza was moved from Chapter 10 (**Viral Diseases**) to Chapter 16 (**Current Plagues**) as it is an ongoing threat to human, animal, and bird populations. Influenza was updated extensively to address the H1N1 2009 pandemic and the status of bird influenza H5N1.

We updated statistics, and we added recent disease outbreaks and topics to *The Microbial Challenge, Third Edition*. A few examples are the hantavirus cases associated with camping in Yosemite National Park, California, West Nile encephalitis, whooping cough in the United States, Ebola hemorrhagic fever in Uganda, and rare cases of fleshing eating bacterial infections that attracted attention in the news media. New topics in this edition include synthetic biology, virotherapy (to treat cancers), herd immunity, community-acquired MRSA, fecal transplants, Nipah virus, the riskiest foods regulated by the Food and Drug Administration (FDA), the triangle model of disease, and R nought or R_0 (an epidemiology used in the 2011 movie *Contagion*).

Most microbiology-based textbooks do not include helminths (worms) and understandably so. Worms are not microbes, but their presence in the body causes damage to the host resembling that which is caused by microbes. Further, the body's immune responses to helminths and microbes are similar. From a public health perspective, worm diseases play a major role in the burden of disease in developing areas of the world. This edition retains this unique feature of the text.

Partnerships in public health constitute a chapter about health organizations at all levels. We frequently mention and describe the Centers for Disease Control (CDC), the World Health Organization (WHO), and other health partners. We added information on the roles of the World Bank and Doctors Without Borders to Chapter 14, **Partnerships in the Control of Infectious Diseases**, another unique feature not found in other microbiology texts for nonmajors.

Every chapter contains at least one box feature containing applications of microbes or historical perspectives, poems, or quotations associated with the underlying principles of microbiology. For example, a box, "Microbes for

Bioremediation," in Chapter 3 includes an update on the 2010 Gulf Oil Spill. Other new boxes are

- Big and Bizarre Viruses
- Diary Entries of D. Carleton Gajdusek
- Does Life Really Beget Life?
- You Need Guts to Survive
- Oh Rats!
- *Iraqibacter*
- Where the Deer and the Antelope Play
- "Oh, Christmas Tree, Oh Christmas Tree, How I Love My Christmas Tree."
- The Armageddon Virus?
- A-Pork-Alypse Now
- Are You Ready for a Zombie Apocalypse?

The text underwent a modest facelift with new part- and chapter-opener figures, updated and new tables, and updated and new illustrations and photos. The Self-Evaluation at the end of each chapter was expanded to 10 questions for each category.

Pedagogical Features

Learning is a difficult, time consuming, and often tedious task that justifies the inclusion of strategies to help the student; therefore, we include a variety of "assists" in this third edition. Each chapter begins with a content outline and a chapter preview, allowing students to look ahead and to stay focused on the material, and each ends with a broad overview and a sampling of questions for self-evaluation. Key terms are highlighted in bold within the chapters and defined in a glossary at the back of the book. The design of the book has also been changed to make the combination of art and text more user-friendly. Numerous feature boxes with human-interest items, author's notes, and boxes containing microbiology-related information are scattered throughout the text to pique student interest, and we use humor to break the monotony of study. A number of unique photographs that I (RIK) have taken over the years depict microbial diseases described in the text. Some of them are unpleasant to look at, but they are included to let readers know that Robert I. Krasner has "been there."

Note to the Instructor

The organization of *The Microbial Challenge: A Public Health Perspective,* allows for flexibility in course design. This text, unlike many on the market today, is not intended to be encyclopedic but to allow coverage of most of the material in a one-semester course. Chapters 1 through 8 make up the core content, but even here there is room for flexibility. Chapters 9, 10, and 11 present approximately 50 diseases: some instructors assign them all, while others pick and choose representative diseases for each mode of transmission.

Dr. Krasner focuses on disease transmission and the biology of the disease, epidemiology, and diseases in the news. Students are not expected to memorize antibiotic(s) used in treatment. Dr. Shors focuses on disease transmission as well as the characteristics of microbial pathogens and the concept of selective toxicity in antibiotic therapy and tailors her course by covering pathogens that will be encountered in the daily lives of students, microbes in the news, and pathogens lurking in clinics and hospitals. Over 95% of the students in her nonmajors microbiology course are pre-nursing or other pre-professional healthcare majors.

Those diseases that might be endemic to the area where your students live would be of particular interest. Semester time restraints may dictate that a few chapters, or parts of chapters, be eliminated or assigned as self-study, depending on your own course design. This text can be easily adapted to a two-semester course by the addition of scientific papers, class-discussion and debates, digital resources, demonstrations, and "hands-on" exercises that can be performed in the classroom or in the laboratory. Newspapers and magazine articles, case studies in the primary literature, daily news broadcasts, and Internet resources including relevant podcasts and You Tube serve as excellent and timely supplements to the text.

Ancillary Materials

For the Instructor

An **Instructor's Media CD** is available to individuals who have adopted the book for their course. The CD contains a PowerPoint® Image Bank of all the figures from the book (to which Jones & Bartlett Learning holds the copyright or has permission to reproduce digitally) and PowerPoint Lecture Outlines (updated by Dr. Shors). **Answers to the Self-Evaluation Questions** (created by the authors) and a **Test Bank** updated by Fernando Monroy of Northern Arizona University are available for download as well as **Chapter Summaries**, which are downloadable from the student companion website. The **Appendix** of the textbook contains a **Suggested Readings** list of books relevant to the topics of antibiotics, bacteria, fungi, history of microbiology, protozoans, prions, and viruses.

For the Student

A student companion website, updated by Dr. Shors, has been developed exclusively for this text and is accessible at http://microbiology.jbpub.com/krasner/3e. The site includes chapter summaries, study quizzes, key-term reviews, web links, and the **Suggested Readings** list of books to enhance the understanding of microbiology concepts and topics.

20th Century Microbe Hunters by Robert I. Krasner

This book tells the story behind the achievements of thirteen 20th-century microbiologists, revealing the excitement, diligence, and often sacrifice of these eminent researchers and humanitarians. It is an engaging journey through science, history, and public health that can be used in conjunction with *The Microbial Challenge* to inspire scientific curiosity in the minds of aspiring microbiologists and other science-oriented persons.

Laboratory Component

Most likely, this text will be used in courses that do not have a scheduled laboratory session. Nevertheless, you may be able to squeeze in some exercises that can be done as a demonstration or as a "hands-on" exercise in the classroom or in the laboratory. For example, you could use antibiotic disks to demonstrate antibiotic activity; you can even use pre-streaked plates. The presence of bacteria in the environment and on and in the body can be shown by swabbing the floor, desk, a doorknob, and body parts (skin, throat, ear) and streaking the swabs on agar plates. Exercises using market-purchased yeasts are safe and inexpensive and can be used to demonstrate fermentation or disease transmission and other principles.

Note to the Student

During our many years of teaching (50 for RIK and 15 for TS), we have witnessed that the lack of frequent study, combined with the lack of organizational and time management skills are the primary cause of academic disappointment. There are many distractions college students face (e.g., social media) and more types of technology available (e.g., smartphones, PC tablets, and iPods). Cramming a few days before an exam will not earn you the best grade you can achieve. Perhaps an analogy will help illustrate what we believe to be the best strategy of action for success. Maybe in your younger years you took music lessons. If so, you would have learned quickly that the time spent practicing between lessons was at least as important as the lessons themselves; the key to improvement and accomplishment was the frequent repetition of the musical exercises assigned. And so it is in handling your college course work. Studies demonstrate that those students who review course material within 24 hours of a class lecture have higher test scores. Regular lecture attendance is imperative, but equally so is the effort spent between lectures. Best wishes to you!

Acknowledgments

Developing a textbook is a daunting task requiring a harmonious partnership between authors and a publisher. Each has particular ideas, and there is no one right way to go about producing the best text possible. Ultimately, in a spirit of compromise, a book comes together of which all involved can be proud and that will enhance the college experience of the students who read it. In the preparation and publication of this text, we have had the opportunity to work with a group of very talented and dedicated people at Jones & Bartlett Learning. Acquisition Editors Cathleen Sether and Erin O'Connor were instrumental in seeing this book through to its finished reality. Michelle Bradbury prepared the revised manuscript for production. We thank other members of the talented team: copyeditor Linda deBruyn, proofreader Jan Cocker, indexer Nancy Fulton, artists Elizabeth Morales and Electronic Publishing Services, Inc., photo researcher Lauren Miller, and the compositor, CAE Solutions Corporation.

We extend a very heartfelt thanks to Lou Bruno, a most capable and wonderful person with whom to work. Lou went beyond the call of duty as we tried to include new material up to the very last minutes of production. His work ethic is tireless. He ran production like an orchestra, addressing all of our questions and concerns. His patience, flexibility, cooperation, sense of humor, and appreciation for microbiology set the tone for a great partnership as the book went through the various stages of production. Thank you for the "dead mice," Lou! We have had some very interesting conversations that will be cherished.

We also extend gratitude to Assistant Surgeon General and Director of the National Center for Immunization and Respiratory Diseases, Dr. Anne Schuchat (Centers for Disease Control and Prevention/Office of Infectious Diseases/National Center for Immunization and Respiratory Diseases), for her willingness to create the **Foreword** for this text despite her incredible schedule. It speaks volumes of her dedication to microbiology/infectious diseases education.

A number of instructors and students who used the first and second editions of *The Microbial Challenge* provided feedback and valuable suggestions as to how the text could be improved in future editions. Thank you for your input.

Cindy Gustafson-Brown, University of California, San Diego
Cheryl Ingram-Smith, Clemson University
Mark Kainz, Ripon College
Michael R. Leonardo, Coe College
Barry Margulies, Towson University
Stacey Massulik, SUNY Onondaga Community College
Gary B. Ogden, St. Mary's University
Ofra Peled, National Louis University
Jeffrey C. Pommerville, Glendale Community College
Teri Shors, University of Wisconsin Oshkosh
Jeffrey J. Sich, Maryville University

Burton E. Tropp, Queens College, City University of New York
Robert L. Wallace, Ripon College
Stephanie A. Yarwood, Oregon State University
Brenda Zink, Northeastern Junior College

Lastly, we express our thanks and appreciation to family, friends, and colleagues that we undoubtedly have bored to death about this text during its preparation.

Robert I. Krasner, M.A., Ph.D., M.P.H.
Professor Emeritus
Department of Biology
Providence College

Teri Shors, M.S., Ph.D.
Professor
Department of Biology and Microbiology
University of Wisconsin Oshkosh

Foreword

■ **Anne Schuchat, M.D.**

Assistant Surgeon General, US Public Health Service
Director of the National Center for Immunization and Respiratory Diseases
Centers for Disease Control and Prevention, Atlanta, Georgia

We live in the microbial world, whether we are students or teachers, doctors, or dilettantes. The shower you took this morning, the runny eggs you ate for breakfast, the cat litter you changed, or the flower you picked from the garden on your way to class—any or all of these could have been the exposure that meant the difference between your unenlightened health and your miserable sick bed. Although the twentieth century saw tremendous progress in reduction of deaths from infectious diseases in the United States, microbes are generally a few steps ahead of us and their adaptation means they can evade our latest advances relatively rapidly. Some microbes, for example, the influenza virus, change so often that we have to literally make new vaccines every year to protect ourselves from them appropriately.

From time to time Hollywood has placed a microbe at the center of a fast-paced plot. It's easy to conclude that these tales are a figment of the screenwriter's wildest imagination. But whether the hemorrhagic virus at the heart of *Outbreak* or the respiratory virus conceived for *Contagion*, study of the subject will show that truth is even stranger than fiction. It is truly possible for a person to carry a scary new virus from Asia to the Americas on an airplane and launch outbreaks in multiple locations before realizing that he or she is ill. The novel coronavirus discovered in 2003 during the global epidemic of severe acute respiratory syndrome (SARS) spread to multiple continents after one person harboring the infection stayed at the same Hong Kong hotel as visitors from several other countries. As the travelers returned to their home countries, outbreaks in Canada, Vietnam, Singapore, mainland China, and several hospitals within Hong Kong resulted. The SARS coronavirus we now recognize was present in live animal markets and transmission amplified in hospitals where nurses and doctors were among the earliest victims. The SARS coronavirus turned out to have a natural reservoir in bats. Bats are also a key source of Ebola and Nipah virus and some human rabies infections.

Though Hollywood often presents an exotic destination as the "usual suspect" for the latest pandemic, this is not always the source of new infectious agents. The 2009 emergence of a novel H1N1 influenza virus took place in our own backyard. The first detected infections were found in a couple of healthy children living in southern California, and investigation revealed that earlier victims lived in our neighbor, Mexico. The major 1918 influenza pandemic that still ranks as the worst in recorded history probably originated in the midwestern United States. We cannot forget the local environment even when we are thinking globally. Because microbes can change rapidly, mutations can give them a survival advantage. Human behavior and civilizational changes have provided numerous opportunities for microbes and people, sometimes with intermediary animals or vectors, to come in contact. A single, 29-base-pair deletion in the SARS coronavirus genetic makeup is thought to have meant the difference between a virus that was poorly adapted to humans and one that could efficiently spread between people—and launched major outbreaks within hospitals and apartment building complexes.

Over the past three decades, dozens of new infectious agents have been discovered as causes of human disease. Some, like the human immunodeficiency virus, went on to cause a worldwide pandemic and killed millions of people. Other new disease outbreaks of microbial origin, like toxic shock syndrome, were brought under relatively prompt control once the exacerbating exposure (e.g., very high absorbency tampons) was removed from the market. While we need to continue to be alert and prepared to detect and respond to emerging infections, most of the world's suffering from infectious diseases derives from pathogens that have been around for centuries but which still cause illnesses that are concentrated among the poor. Limited access to clean water and sanitation means high risks of severe diarrheal and respiratory infections. Low vaccination coverage and malnutrition are a lethal combination when the measles virus is circulating. Poor control of vectors like mosquitoes in tropical climates can result in an enormous toll of misery and death from malaria and dengue virus infections. Only one major infectious scourge of humans has ever been successfully eradicated from the world—the smallpox virus—and this came about from an ambitious global partnership that searched out every last case on the planet and vaccinated in a ring around that patient to limit onward transmission. We are extremely close to achieving the same definitive result with the poliovirus. Poliomyelitis paralyzed President Franklin Delano Roosevelt when he was 39 but didn't stop him from achieving the presidency or launching a multidecade quest to develop a vaccine against polio. In 1988, the poliovirus was still paralyzing 350,000 people in 125 countries around the world—but in the past year less than a thousand cases occurred, and only three countries remain where transmission had never been interrupted. The Global Polio Eradication Initiative is trying to finish the job in Nigeria, Afghanistan, and Pakistan and send this virus to the history books.

Today, the world's poor continue to suffer disproportionately from many infectious diseases. One of the greatest advances of recent years has been public-private partnerships that have focused increased attention on preventable diseases of the poor. Philanthropists like Bill and Melinda Gates, humanitarians like Jimmy and Rosalyn Carter, celebrities like Bono, and formal public-private partnerships like the GAVI Alliance and Global Fund for AIDS, TB, and Malaria are shining a spotlight on the simple interventions that can reduce illness and death from infections among people who live in the poorer countries of the world. These efforts are saving millions of lives each year.

Wealthier settings also provide the opportunity for microbial adaptation and resulting havoc. Overuse of antibiotics has led many bacteria to become resistant to virtually all the medicines scientists and pharmaceutical companies have developed. Medical advances and technology that have permitted larger numbers of transplant recipients and longer survival for people with cancer leave these immunocompromised groups vulnerable to infections, including ones associated with the health-care environment or caused by microbes that might not harm people with stronger immune defenses. When people live or recreate further from urban centers and disrupt previously stable ecosystems, they can come into contact with vectors for diseases like Lyme and Hantavirus. Even the latest ecotourism trends can come into collision with the microbial world—for example, when survivalists swim in places people were never meant to swim, leptospirosis can follow. Natural catastrophes like earthquakes and tornadoes can disrupt the environment and cause large exposure to soil-dwelling fungi. The major Northridge, California, earthquake was followed by a large outbreak of coccidiodomycoses in 1994, and more recent tornadoes in Joplin, Missouri, have resulted in dozens of severe illnesses caused by mucormycoses. Rich or poor, adventurer or homebody, microbes have us surrounded. Of course, if you wash your hands, prepare foods properly, get all recommended vaccines, and avoid unprotected sex, you have a pretty good chance of escaping microbial assaults that could interfere with survival at least until your final exams. But becoming knowledgeable about the microbial world will also make you a better parent, cook, outdoorsperson, landscape artist, philanthropist, or Hollywood screenwriter.

And since you are still reading (and assuming you are still feeling well in the days ahead after the various incubation periods have passed), you probably did not inhale the Legionnaires bacteria in the shower this morning, or consume *Salmonella* from your breakfast, or pass toxoplasmosis on to your future baby from the cat litter you changed, or even get nicked by the sporotrichosis fungus from the thorny rose bush when you picked that flower. But you might have. Read on.

About the Authors

Robert I. Krasner

Professor Emeritus Robert I. Krasner, a member of the Department of Biology at Providence College (PC) in Rhode Island, retired after 50 years of teaching and research starting in 1958. During his tenure, he developed new courses and mentored many students in research, many of whom went on to graduate and medical schools. Dr. Krasner's courses were popular, demanding, and embellished with humor. He was recognized on several occasions for excellence in teaching. The Robert I. Krasner Teaching Award was established upon his retirement at PC to recognize outstanding graduating seniors. He is the author of *20th Century Microbe Hunters*, many scientific papers, and a contributor to other scholarly works.

Dr. Krasner's love for travel was sparked by his service as a young army medical officer in Japan. He has spent sabbaticals and leaves of absence from PC at numerous domestic and foreign institutions including Fort Detrick Army Biological Laboratories, Georgetown University School of Medicine, and those in Israel, Paris, Brazil, and London. At 69 years of age, he was accepted into the Harvard School of Public Health and earned a Master of Public Health (M.P.H.) degree and is the oldest full-time student to have accomplished this.

Dr. Krasner founded and directed the Summer Science Program for high school students at PC from 1975 to 2006; the program hosted approximately 1,000 students in its 31 years of operation. During this time he also developed and directed several grant-funded microbiology and biotechnology workshops for high school teachers.

Over the years, Dr. Krasner has presented over 60 research papers in the United States and abroad including numerous annual meetings of the American Society for Microbiology (ASM). Teaching remained his major interest, and when asked by colleagues "what (research) he was working on," his favorite reply was "students." His initiative in 1980 led to the establishment of the Division for Microbiology Educators within the ASM. He continues to lecture occasionally and enjoys gardening, pet therapy, studio art, and playing the harmonica.

In 2011, Dr. Krasner teamed with Teri Shors, a virologist at the University of Wisconsin–Oshkosh to update *The Microbial Challenge* to a third edition. Together, they bring over 65 years of teaching experience to this project.

Teri Shors

Teri Shors has been a member of the Department of Biology and Microbiology at the University of Wisconsin Oshkosh since 1997; she was promoted to the rank of Professor in 2010. Dr. Shors is a devoted teacher and researcher at the primarily undergraduate level and has been a recipient of university awards, including a distinguished teaching award and two endowed professorships. She has taught a variety of courses and laboratories and has made a strong contribution to the development of new courses in microbiology and molecular biology.

Dr. Shors' graduate and postgraduate education is virology-based and is reflected in her research. Before teaching at UW Oshkosh, she was a postdoctoral fellow in the Laboratory of Viral Diseases under the direction of Dr. Bernard Moss in the National Institute of Allergies and Infectious Diseases (NIAID) at the National Institutes of Health (NIH). While her expertise centers upon the expression of vaccinia virus genes, she is currently interested in the potential of antiviral compounds in cranberries and in other fruits and herbs used as green medicine in Hmong culture. This antiviral research has been funded by a variety of granting agencies including a prestigious Merck/AAAS award. She has mentored many students engaged in independent research projects.

Dr. Shors, as a co-author, has made major contributions to the third edition of *The Microbial Challenge*. She is also the author of *Understanding Viruses*, now in its second edition, and has recently authored *Encounters in Virology*. Additionally, Dr. Shors has contributed to and authored a variety of other texts and scientific papers.

Initiative, creativity, humor, networking, using current events and the latest technology in her courses, and leading collaborative, cross-disciplinary studies are hallmarks of Dr. Shors' talents and makes her popular among students in the classroom. She has recently developed and taught an online virology course for undergraduates.

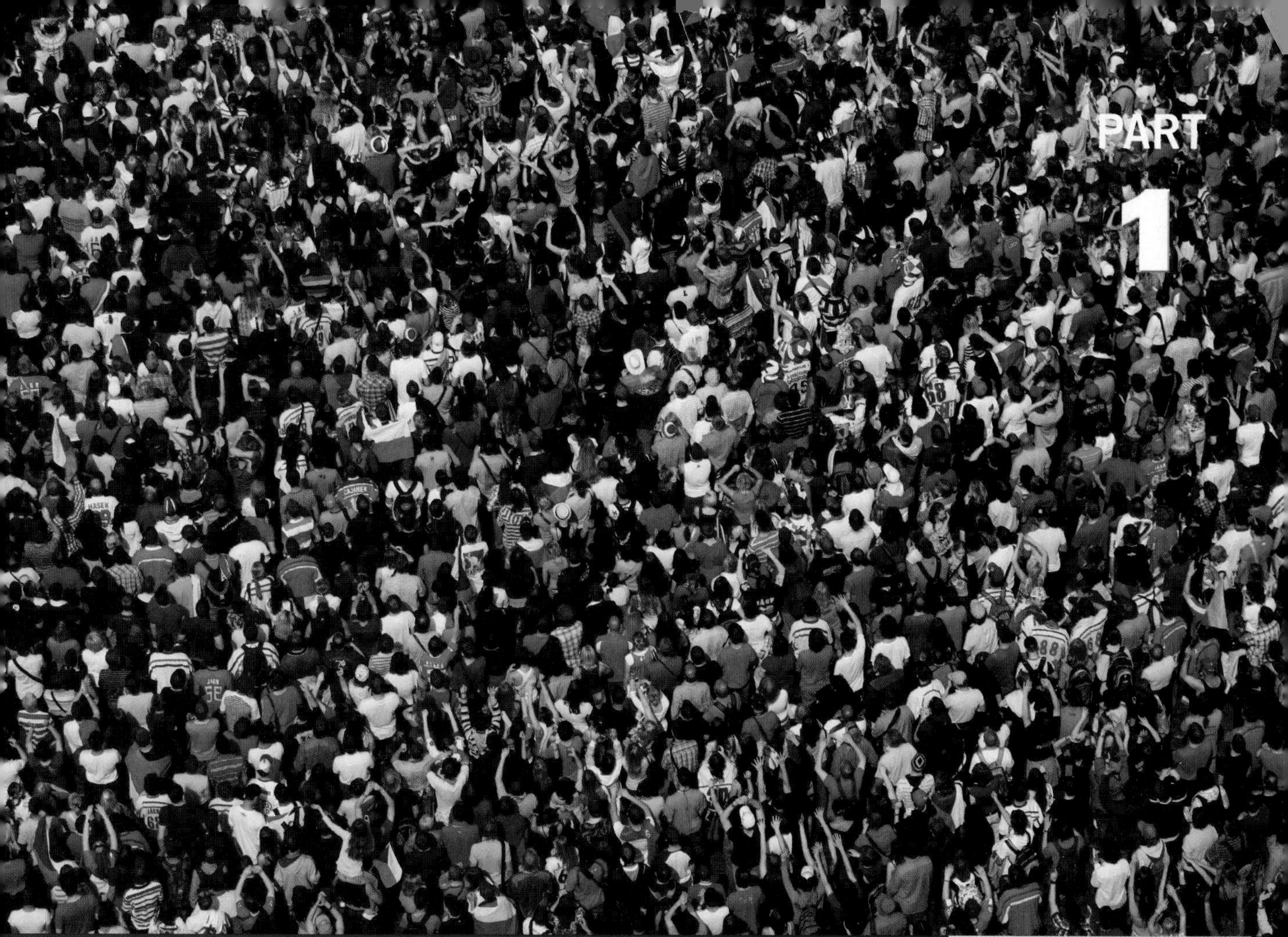

The Challenge

Photo © Vladimir Wrangel/ShutterStock, Inc.

Identifying the Challenge

If human civilization lasts, if it continues to spread, infectious diseases will increase in number in every region of the globe. Exchanges and migrations will bring the human and animal diseases of every country. The work is already well advanced; its future is assured.

—Charles Nicolle, *The Destiny of Infectious Diseases, 1932*

■ Preview

This book is about a challenge—a worldwide challenge—posed by microbes, invisible marauders that inhabit the earth, many of which cause illnesses and death. Classically, there are five distinct kinds of microbes: **bacteria**, **viruses**, **protozoans**, **fungi**, and **unicellular algae**. Most recently, **prions** can now be added to the list, bringing the total of "**infectious agents**" to six. It should be emphasized at the outset that a relatively few members of each of the groups pose the potential for infection. Most microbes are beneficial and many are essential to the cycles of nature without which higher life forms could not exist. In many cases, microbes have been harnessed for the benefit of humankind.

But this book, by intent, has a bias because its theme relates to those few microbes that are disease producers. In the language of medical microbiology, they are referred to as **pathogens** or virulent microbes. Why some of these microbial diseases now represent an increased challenge is the subject matter of this chapter and sets the stage for the remaining chapters.

The flags represent infectious disease outbreaks or hotspots.

The Challenge

The need for this book continues. Forty previously unknown infectious diseases have emerged and others have reemerged in the past two decades: AIDS, Ebola virus, *Escherichia coli*, hantavirus, West Nile virus, *Salmonella*, flesh-eating "strep," and "mad cow disease," to name only a few (FIGURE 1.1). Movies, books, and articles about microbial diseases intrigue large numbers of viewers and readers. Popular news magazine programs, including *Dateline*, *60 Minutes*, *Anderson Cooper 360°*, and *20/20*, frequently air segments relating to dangerous microbes; newspaper articles and news broadcasts appear almost daily and further alert the public to threats posed by microbes. The movie "Contagion," based on a deadly viral epidemic, debuted in September 2011, and thrilled theatergoers about a pandemic caused by an airborne **chimeric (hybrid)** influenza-Nipah-like virus.

The 1990s were especially eventful. In 1992 tuberculosis (TB), a bacterial disease almost relegated to oblivion, reemerged in New York City, resulting in almost four thousand cases; the tubercle bacillus was developing resistance to a variety of antibiotics that had once stopped the bacteria dead in their tracks. In 1993 an outbreak of cryptosporidiosis, a waterborne protozoan disease characterized by diarrhea, swept through Milwaukee, Wisconsin, causing illness in about 400 thousand people, approximately 25% of that city's population; it was the largest reported waterborne illness in U.S. history. In 1993 hantavirus, the causal agent of a potentially lethal influenza-like respiratory illness, reemerged with deadly results in New Mexico, Utah, Colorado, and Arizona. Ebola hemorrhagic fever, caused by one of the deadliest known viruses, ignited a panic in 1995 when 240 people bled to death during an outbreak that occurred in Kikwit, Zaire (now

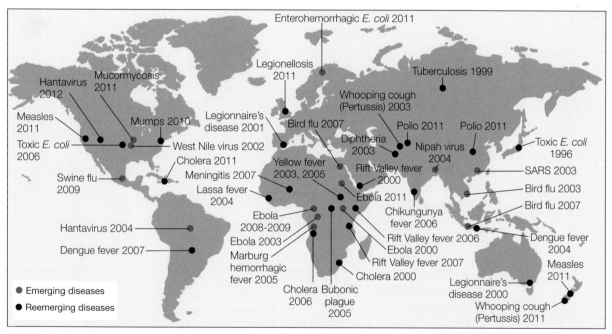

FIGURE 1.1   Emerging and reemerging diseases, 1996–2011. For more recent outbreaks, see HealthMap http://healthmap.org/en/.

the Democratic Republic of Congo). Earlier, in the winter of 1989, scientists working at a primate quarantine facility just outside Washington, DC, were terrified when Ebola virus–infected research monkeys were introduced into the facility. Richard Preston's account of the incident, *The Hot Zone,* inspired the film *Outbreak* in 1998. More recently, Ebola reemerged in the Democratic Republic of Congo in September 2007, resulting in at least 166 deaths.

Dengue fever sickened 1.2 million people in 56 countries in 1998. Reports of the bacterium *E. coli* O157:H7 frequent the news, as in 1993 when a cluster of people became ill in Seattle after eating hamburgers at a Jack in the Box fastfood restaurant. Health officials detected the organisms in hamburger patties, resulting in the recall of nearly 22 million pounds of ground beef to avert a nationwide outbreak. Can you imagine a mound of hamburger meat that size? Peanut butter contaminated with *Salmonella* caused over 425 cases of *Salmonella* infection spread over 48 states in February 2007, resulting in recalls of particular brands of the product. (Parents packing peanut butter sandwiches into their children's lunch boxes were dismayed and at a loss to find an appropriate substitute.) In the summer of 2011, a new strain of *E. coli* caused about 4,000 people to become ill and about 45 deaths; the source remains unclear. Multiple state outbreaks of Listeriosis, a serious bacterial infection, started in August 2011, and has been traced to Rocky Ford cantaloupes. It never ends! Lyme disease, severe acute respiratory syndrome (known as SARS), and avian and swine influenza are four more examples from a long list of new, emerging, and reemerging infectious diseases (TABLE 1.1). No nation can afford to be complacent regarding its vulnerability.

TABLE 1.1 New, Emerging, and Reemerging Infections

Bacterial Diseases	Protozoan Diseases
Lyme disease	Cryptosporidiosis
Ehrlichiosis	Malaria
Gastroenteritis	Babesiosis
Listeriosis	**Fungal Diseases**
Legionnaires' disease	Coccidioidomycosis
Salmonellosis	Cryptococcosis pneumonia
Tuberculosis	Pneumocystis *pneumonia*
Viral Diseases	"Bat White Nose" syndrome (bats, eastern U.S.)
Hantavirus pulmonary syndrome	Chytridiomycosis (amphibians)
Ebola hemorrhagic fever	
Dengue fever	
Rabies	
Severe acute respiratory syndrome (SARS)	
West Nile encephalitis	
HIV/AIDS	
Influenza	
Nipah encephalitis	
Measles	

The Institute of Medicine is a branch of the National Academy of Sciences that advises the government on policy matters pertaining to the health of the public. In a 1992 report, *Emerging Infections: Microbial Threats to Health in the United States,* the institute defined emerging infections as "new, reemerging or drug-resistant infections whose incidence in humans has increased within the past two decades or whose incidence threatens to increase in the near future."

Despite the tremendous progress in the latter half of the twentieth century in controlling infectious diseases, including the eradication of smallpox, the introduction of antibiotics in the 1940s, improvement in sanitation, and an increase in the diseases for which immunization is available, the battle against infectious diseases continues (BOX 1.1). David Satcher, former director of the U.S. Centers for Disease Control and Prevention (CDC) and later surgeon general of the United States, stated that "our ability to detect, contain, and prevent emerging infectious diseases is in jeopardy."

The International Red Cross warned of the danger of infectious diseases in a report published on June 6, 2000. The report spoke of "the silent tragedy" of deteriorating health services and the death of thirteen million people from preventable diseases, primarily infectious in nature, in the previous year. Further, compared with floods and earthquakes, which grab news headlines and donors' cash, the uncontrolled spread of disease steals far more lives. Over 150 million people have died of AIDS, TB, and malaria alone since 1945, compared with the more than 23 million lives lost in wars. In a recent year, 160 times more people died from AIDS, malaria, respiratory diseases, and diarrhea than were killed in that year's natural disasters, including the massive earthquakes in Turkey, floods in Venezuela, and cyclones in India. Malaria killed 781,000 in 2009, mostly children. According to the World Health Organization (WHO) *World Malaria Report, 2010,* there were 225 million cases of malaria in 2009. TB has been on the upswing in North Korea; statistics from that country reveal that 5 million of its 22 million people are infected.

Infectious diseases are the second leading cause of death worldwide, resulting in 20.7% deaths occurring worldwide each year (FIGURE 1.2). Pneumonia and diarrheal diseases are the two largest killers of children under the age of five, causing 18% and 15% of all deaths respectively in 2008. Almost ten million children under the age of five die each year, and their leading killers are infectious diseases: pneumonia, diarrhea, malaria, measles, and HIV. These figures are an underestimate, because surveillance and reporting networks are woefully deficient in many less-developed countries. The leading infectious killers in the world, according to the WHO, include bacterial, viral, protozoan, and worm diseases. Initially, one would attribute these devastating statistics to the poverty associated with developing nations, but, as surprising as it may seem, in the United States infectious diseases remain in the top ten causes of death (FIGURE 1.3). No wonder the director-general of WHO stated in a 1996 report, "We stand on the brink of a global crisis in infectious diseases. No country is safe from them. No country can any longer afford to ignore this threat." This statement remains true.

So what's the bottom line? What grade would the world now be awarded in terms of its success in coping with microbial diseases? Certainly, under

BOX 1.1 — Quotations Relating to Health and Infectious Disease

He who cures a disease may be the skillfullest, but he that prevents it is the safest physician.

—Thomas Fuller, English clergyman and historian in 1650

Not a single year passes without [which] . . . we can tell the world: here is a new disease!

—Rudolf Virchow, German doctor of medicine in 1867

The ideal way to get rid of any infectious disease would be to shoot instantly every person who comes down with it.

—H. L. Mencken, American writer in 1910

Germs come by stealth
And ruin health
So listen, pard,
Just drop a card
To a man who'll clean up your yard
And that will hit the old germs hard.

—Sinclair Lewis, *Arrowsmith*, 1925

Infectious disease is one of the few genuine adventures left in the world. The dragons are all dead and the lance grows rusty in the chimney corner. . . . About the only sporting proposition that remains unimpaired by the relentless domestication of a once free-living human species is the war against those ferocious little fellow creatures, which lurk in the dark corners and stalk us in the bodies of rats, mice, and all kinds of domestic animals; which fly and crawl with the insects, and waylay us in our food and drink and even in our love.

—Hans Zinsser, *Rats, Lice, and History*, 1935

Health is a state of complete physical, mental, and social well-being and not merely the absence of disease or infirmity. . . .

—Constitution of WHO, July 22, 1946

Everyone has the right to a standard of living adequate for the health and well-being of himself and of his family including food, clothing, housing, and medical care. . . .

—Article 25, Universal Declaration of Human Rights, adopted by the General Assembly of the United Nations, December 10, 1948

Ingenuity, knowledge, and organization alter but cannot cancel humanity's vulnerability to invasion by parasitic forms of life. Infectious diseases which antedated the emergence of humankind will last as long as humanity itself and will surely remain, as it has been hitherto, one of the fundamental parameters and determinants of human history.

—William H. McNeill, *Plagues and Peoples*, 1976

The microbe that felled one child in a distant continent yesterday can reach yours today and seed a global pandemic tomorrow. Pitted against microbial genes, we have mainly our wits.

—Joshua Lederberg, 1958 Nobel Prize winner, 1988

Pathogenic microbes can be resilient, dangerous foes. Although it is impossible to predict their individual emergence in time and place, we can be confident that new microbial diseases will emerge.

—Institute of Medicine, *Emerging Infections: Microbial Threats to Health in the United States*, 1992

It is time to strengthen our research efforts . . . so that we can unlock the mysteries behind antibiotic resistance and discover new scientific weapons in the battle to detect and control emerging infectious diseases.

—Albert Gore, former vice president of the United States, June 1996

The world may have only a decade or two to make optimal use of the many medicines presently available to stop infectious diseases. We are literally in a race against time to bring levels of infectious disease down worldwide, before the disease wears the drugs down first.

—David Heymann, Executive Director, World Health Organization's Communicable Disease Program, 2000

On a good day, we hold them at bay. On a bad day, they're winning. Our task is a lot like trying to swim against the current of a raging river.

—Michael Osterholm, former Minnesota state epidemiologist and founder of an infectious disease control company, 2000

the leadership of the United Nations, WHO, the CDC, and other organizations, the burden of infectious diseases around the world can be lessened and a higher grade achieved.

In fact, during the 1950s, 1960s, and 1970s microbial diseases appeared to be on their way out. It was a span of years heralded by optimism and progress in public health. The first polio vaccine was introduced by Jonas Salk in the 1950s and ushered in a time of successful mass vaccination campaigns. Fewer than one thousand polio cases occurred in 1967 in western Europe and North America as compared with over seventy-five thousand in 1955. It appeared that malaria would be taken off the list of diseases "of major importance," according to WHO and the Pan-American Sanitary Conference. The times were good; as the economy of nations improved, poverty decreased, and so did the burden of microbial diseases. A 1966 CDC report, in what might be considered an address on the state of the union's health, glowed with the promise of the conquest of microbial diseases. Further, William H. Stewart, surgeon general of the United States in 1967, told a gathering of health officers that it was "time to close the book on infectious diseases and shift all national efforts to chronic diseases." Stewart's optimism was echoed by health officials in other developed nations of the world.

Fortunately the advice to "close the book" was not heeded nor was the optimism in developing countries—countries that constitute a large part of the world's population. Consider that from 1980 to 1992 alone, the CDC reported a 22% increase in infectious diseases (excluding AIDS). Data presented in a 1996 article published by the *Journal of the American Medical Association* indicated a greater than 50% increase in deaths caused by microbes in the United States since 1980. Despite the tremendous strides in infectious disease control over the past century, data from the U.S. National Center for Health Statistics indicate that microbial disease remains as a leading cause of death in the United States (Figure 1.3) But the scientific community was slow to acknowledge that the bubble of antisepsis and disease control was about to burst.

Why were new diseases emerging and older ones reemerging with a vengeance, as it sometimes appeared? The 1992 Institute of Medicine's report, *Emerging Infections: Microbial Threats to Health in the United States,* warned that microbes

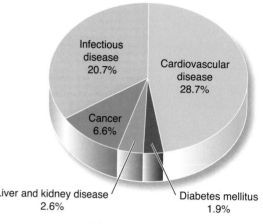

FIGURE 1.2 The five leading causes of death from disease. There were 59.4 million deaths worldwide in 2007. Cancers and cardiovascular, respiratory, and digestive diseases can also be caused by infections, and thus the percentage of deaths due to infectious diseases may be even higher than shown. *Source:* World Health Organization, *World Health Statistics,* 2008.

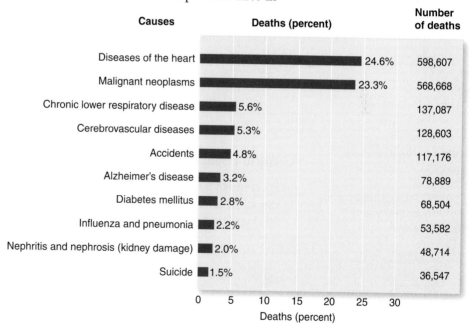

FIGURE 1.3 Leading causes of death in the United States. Data from *Deaths: Final Data for 2009, National Vital Statistics Reports,* Vol. 57, No. 4, March 16, 2011, National Center for Health Statistics/CDC.

were winning the battle and that our previous complacency and optimism had weakened our ability to counterattack. Essentially, it appeared that the choreography of adaptation between microbes and humans was beginning to come apart at the seams because of a variety of linked and overlapping factors considered below: world population growth, urbanization, ecological disturbances, technological advances, microbial evolution and adaptation, and human behavior.

■ Factors Responsible for Emerging Infections

Infections are a part of civilization and actually predate civilization. Microbes and men co-exist and share the same ecosystem. Diseases have threatened since ancient times as evidenced by numerous references in the Old Testament and other sources of antiquity; "pestilence" and "plague" speak of them. Major factors involved in the emergence are discussed in this section (TABLE 1.2).

TABLE 1.2 Factors Responsible for Emerging Diseases

World population growth	Microbial evolution and adaptation
Urbanization	Antimicrobial resistance
Ecological disturbances	Evasive strategies
Deforestation	Human behavior and attitudes
Climatic changes	Complacency
Natural disasters (drought, floods)	Migration
Technological advances	Societal factors
Air travel	
Transfusion of unsafe blood	

■ World Population Growth

Planet earth is rapidly approaching a population of seven billion people, with a growth rate at 1.092%. The United Nations estimates world population to reach 9.3 billion by 2050 and the latest long-range projection to reach 10.1 billion by 2100 (FIGURE 1.4). To add to the problem of the burgeoning population, 80% of the population is living in less-developed countries (of which 60% are tropical and subtropical areas) with a diminished capacity to cope with population increase. Several factors have been cited for the current crisis of new and emerging infectious diseases, but the population explosion is central to the issue (FIGURE 1.5).

Thomas Malthus' (1776–1834) *An Essay on the Principle of Population as It Affects the Future Improvement of Society* warned of the negative influence that unchecked population growth could have on societies,

FIGURE 1.4 World population, 1950–2050. Projections are based on an estimated annual growth rate of 1.092%. *Source:* United Nations. *World Population Prospects: The 2010 Revision.*

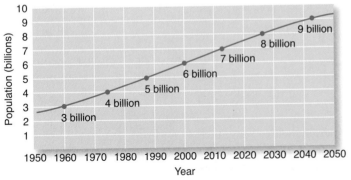

primarily because of inadequate supplies of food. (Charles Darwin's insights into evolution and the process of natural selection were strongly influenced by Malthus.) It appears that Malthus was right, as evidenced by famine in Africa and in other parts of the world over the past 100 or more years. But there is another negative consequence of overpopulation: transmission of infectious diseases. The total population of a country or a region, in itself, is not as crucial as its population density—the number of people per square kilometer in a defined area. Microbe-caused diseases can be transmitted by person-to-person contact, by **biological vectors**, including mosquitoes, ticks, lice, and flies, and by animal to human contact (**zoonotic diseases**) but whatever the mode of transmission, population density is a significant factor. Consider, for example, a classroom with fixed dimensions and assume that one person in the class has a cold, but there are only ten other students randomly spaced throughout the room; on the other hand, consider the same classroom with sixty students. Clearly, the chain of transmission is fostered in the larger population, simply because respiratory droplets are able to traverse the shorter distance from contact to contact when the population density is higher.

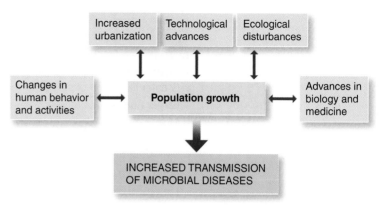

FIGURE 1.5 Population explosion: the "hub" of the problem.

The age distribution of the population is also of considerable significance in terms of the risk factors. For example, in the United States people over age sixty-five constituted 38.6% of the population in 2010. Elderly populations are more susceptible to microbial diseases, presumably because the strength of their immune system has declined. They serve as an increasing source of infection for family and community members. The point is that predictors of infectious disease need to take into account not only the total population and population density but also the demographics of age distribution in that population.

As the world population increases, there are a number of consequences that foster an increase in infectious diseases (TABLE 1.3). For example, Dhaka, the capital city of Bangladesh, has many slum areas. Bangladesh is one of the world's most populous countries, with approximately 150 million people. Population control is Bangladesh's most urgent problem, along with the attendant low per capita income. As would be expected, high levels of malnutrition exist, with much of the population getting less than one-third of the normal food intake because agricultural production has not been able to keep pace with population growth. The consequences in terms of infectious diseases, particularly diarrhea, due to poverty and poverty-related conditions are dramatic.

Urbanization

At an international conference Gerard Piel, an authoritative scientific journalist, stated that "the world's poor once huddled largely in rural areas. In the modern world they have gravitated to the cities." The following story is indicative of the depth of despair suffered by many in the world: Zaynab Begum lives in Bangladesh in a Dhaka slum, along with her husband and three children in a primitive hut, less than six square meters in size, constructed of bamboo and makeshift materials.

TABLE 1.3 Potential Effects of World Population Growth on Variables Related to Emerging and Reemerging Infections

Increased potential for person-to-person disease spread

Greater likelihood of global warming

Larger numbers of travelers

More frequent wars

Increased numbers of refugees and internally displaced persons

Increased hunger and malnutrition[a]

More crowding in urban slums

Increased numbers of people living in poverty

Inadequate potable water supply[a]

More large dam construction and irrigation projects

[a]New technologies could prevent or minimize these effects.
Reprinted from D. B. Louria, in *Emerging Infections 1*, W. M. Scheld, D. Armstrong, and J. M. Hughes (ed.), 1998, ASM Press, Washington, DC, with permission.

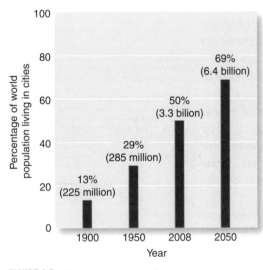

FIGURE 1.6 Progressive urbanization of our planet. Adapted from United Nations, Department of Economic and Social Affairs/Population Division, *World Urbanization Prospects: The 2009 Revision* (2010).

There is no electricity, running water, or toilet, and an open sewer runs outside the hut. Zaynab's husband is a rickshaw puller; her fate is shared by millions of others who have left their villages and migrated to the cities in search of work and a better life. It's a cruel realization that many who migrate to cities in search of a better life now live in urban poverty characterized by crowded and substandard housing lacking safe drinking water, inadequate toilets, and tenant's rights.

The trend to urbanization dates back to early in the twentieth century. In 1900 just 13% of the world's population lived in urban areas. According to the 2009 UN *World Urbanization Prospects* report, the percentage has increased to 50.1% in 2009, and is expected to climb to 68.7% by 2050 (FIGURE 1.6). The world's ten largest urban areas are listed in TABLE 1.4; about half of these megacities are in developing countries. The magnitude of the effect of urbanization on communicable diseases varies dramatically in developed and developing countries, as a function of the economy and the public health infrastructure necessary to cope with the stress of increasing population density. The challenge of maintaining acceptable standards of sanitation and hygiene is far more difficult in

TABLE 1.4 World's Ten Largest Urban Agglomerations in 2011[a]

1. Tokyo, Japan	6. Mumbai, India
2. Delhi, India	7. New York City, NY-NJ-CT (USA)
3. Seoul, South Korea	8. São Paulo, Brazil
4. Jakarta, Indonesia	9. Mexico City, Mexico
5. Manila, Philippines	10. Shanghai, China

[a]The rankings vary depending on definition of urban agglomerations and yearly estimates of current population.
Source: Demographia World Urban Areas (World Agglomerations), 7th Annual Edition, April 2011.

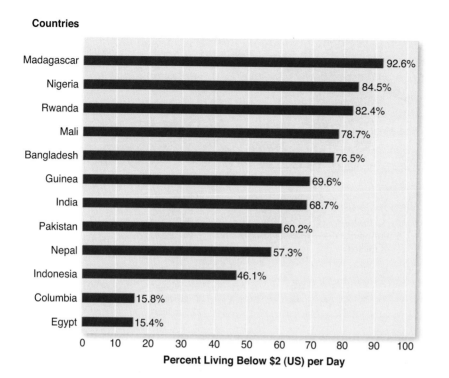

Countries

FIGURE 1.7 World's poorest countries. Poverty is especially serious where rapid population growth occurs. *Source: World Bank, 2011 World Development Indicators* (2011).

developing countries, but it is also important to keep in mind that pockets of poverty and despair also exist in the United States and other developed nations (**FIGURE 1.7**).

Urbanization frequently leads to poverty and together set up a cycle of infectious disease (**FIGURE 1.8**). In July 2000 at a meeting of the Group of 8 (G-8) countries (the top seven industrialized countries, plus Russia) in Okinawa, the nations pledged to break the vicious circle of poverty suffered by citizens of developing countries. Sub-Saharan Africa is home to 68% of the world's 33.2 million people living with HIV (**FIGURE 1.9**) because of a variety of factors, an important one of which is poverty-associated urbanization. The drain on natural resources, including safe drinking water, is excessive, whereas at the same time problems of pollution, including human waste disposal and sanitation, are magnified. Untreated human wastes are dumped by the tons into the rivers, streams, and oceans. Ultimately, slums and shantytowns develop. The United Nations estimates half of the population on the African continent—482 million people—live in slums. People live in filth and squalor (**FIGURE 1.10**). Rodent populations increase as sanitation decreases, and the cycle of disease is perpetuated. Rodents may harbor fleas, which transmit a variety of diseases, including the Black Death of fourteenth century Europe, now known simply as the Plague.

Numerous studies have concluded that city dwellers get sick more often than their rural counterparts and that people living in poverty are sick more often. Upton Sinclair's 1906 novel, *The Jungle*, portrayed the unsanitary practices and working conditions, especially for

AUTHOR'S NOTE (RIK)
I have been to Japan on a few occasions, the first being during my service as a young military officer in the U.S. Army stationed just outside of Japan. I am always impressed with its cleanliness, despite the ever-present crowds.

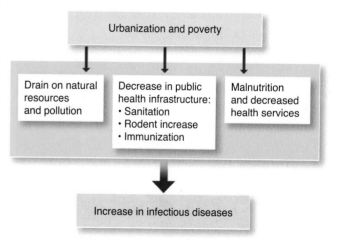

FIGURE 1.8 Relationships among poverty, urbanization, and infectious disease.

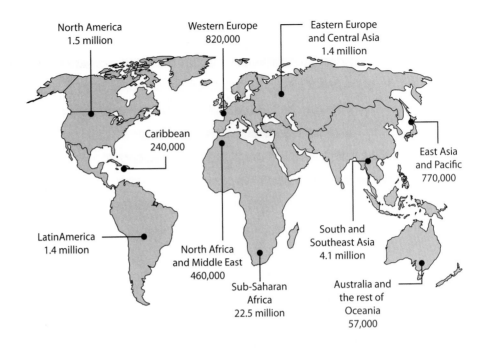

FIGURE 1.9   Global distribution of HIV in 2009. Data from UNAIDS.

North America
1.5 million

Western Europe
820,000

Eastern Europe
and Central Asia
1.4 million

Caribbean
240,000

East Asia
and Pacific
770,000

LatinAmerica
1.4 million

North Africa
and Middle East
460,000

South and
Southeast Asia
4.1 million

Sub-Saharan
Africa
22.5 million

Australia and
the rest of
Oceania
57,000

FIGURE 1.10.   Slums and shantytowns. Poverty is associated with a lack of sanitary facilities, an increase in rodent populations, a lack of safe drinking water, and other circumstances that contribute to infectious diseases. **(a)** A shack in rural Panama. Author's photo (RIK). **(b)** A shanty town in Karachi, Pakistan. © Reuters/Athar Hussain/Landov. **(c)** Individuals walking in the mud of a Nairobi slum, Kenya (October, 2011). © meunierd/ShutterStock, Inc.

(b)

(a)

(c)

The Jungle was originally published as a story in a socialist newspaper in 1906 and later republished as a novel. It chronicles immigrants working in a Chicago meatpacking factory.

There was another interesting set of statistics that a person might have gathered in Packingtown—those of the various afflictions of the workers. . . . There were the men in the pickle-rooms, for instance, where old Antanas had gotten his death; scarce a one of these that had not some spot of horror on his person. Let a man so much as scrape his finger pushing a truck in the pickle-rooms, and he might have a sore that would put him out of the world; all the joints in his fingers might be eaten by the acid, one by one. Of the butchers and floorsmen, the beef-boners and trimmers, and all those who used knives, you could scarcely find a person who had the use of his thumb; time and time again the base of it had been slashed, till it was a mere lump of flesh against which the man pressed the knife to hold it. The hands of these men would be criss-crossed with cuts, until you could no longer pretend to count them or to trace them. They would have no nails—they had worn them off pulling hides; their knuckles were swollen so that their fingers spread out like a fan. There were men who worked in the cooking-rooms, in the midst of steam and sickening odors, by artificial light; in these rooms the germs of tuberculosis might live for two years, but the supply was renewed every hour. There were the beef luggers, who carried two-hundred-pound quarters into the refrigerator cars; a fearful kind of work, that began at four o'clock in the morning, and that wore out the most powerful men in a few years. There were those who worked in the chilling-rooms, and whose special disease was rheumatism; the time-limit that a man could work in the chilling-rooms was said to be five years. There were the woolpluckers, whose hands went to pieces even sooner than the hands of the pickle-men; for the pelts of the sheep had to be painted with acid to loosen the wool, and then the pluckers had to pull out this wool with their bare hands, till the acid had eaten their fingers off. There were those who made the tins for the canned meat; and their hands, too, were a maze of cuts, and each cut represented a chance for blood poisoning.

the workers, in the Chicago meatpacking industry (**BOX 1.2**). Prevention of communicable diseases is a major component of public health and is more problematic in cities than in rural areas and wide-open spaces.

On the other hand, it does not necessarily follow that disease runs rampant in the megacities. Consider, for example, the Tokyo-Yokohama area, ranked as the world's largest urban area (Table 1.4), which is hardly poverty stricken or disease ridden. In fact, population statistics indicate that the Japanese people enjoy the longest life expectancy. The country's economy and public health infrastructure make it possible for them to cope with urbanization. By contrast, in most developing countries the crush of humanity and the tide of urbanization are overwhelming and beyond the financial resources necessary to construct sewage systems and to develop and maintain a public health infrastructure. Laura Garrett, in _The Coming Plague,_ refers to cities as "microbe magnets" and "microbe heavens." "Graveyards of mankind" is a term used by British biologist John Cairns.

Ecological Disturbances

Deforestation

Almost half of the earth's forests either no longer exist or have been damaged, possibly to the point of no return, as a result of agriculture, settlement, logging,

FIGURE 1.11 Deforestation. As people move into areas that were formerly forests, there is increased contact with animals, including insects that harbor infectious microbes. Further, the displaced animals return to neighborhoods that were once their lands in search of food. Author's photo (RIK).

and mining over the past 8,000 years. The driving force ultimately, is deliberate and financial—to make money—although wildfires and other natural phenomenon play a significant role.

Deforestation is a major factor in the eruption of emerging and reemerging diseases. Wilderness habitats serve as reservoirs for a large variety of insects and other animals that harbor infectious agents. When the village or town becomes too crowded, whether it is in a poor and developing country or in a developed country, expansion occurs into the surrounding areas for tracts of land on which to build. Generally, the first event to give notice that construction is about to take place is the whine of the chain saws signaling deforestation followed by bulldozers moving in to uproot the tree stumps (FIGURE 1.11). Every time a tree is felled or a bulldozer digs up the soil to create another shopping or housing development, microbes and other organisms are displaced. Fungi and bacteria and their spores are released into the environment and may alight and colonize on a human or animal and possibly give rise to a new or reemerging disease. Examples of outbreaks of certain fungal diseases have been reported in construction workers, particularly in the southwest. Perhaps this is what Louis Pasteur was referring to 150 years ago when he advised, "the microbe is nothing; the terrain is everything."

Deforestation favors human intrusion into the environment and fosters contact with wildlife and with insects and plays a major role in the migration of these displaced species into villages, communities, and backyards in search of food. An example is the rise of rabies in the eastern part of the United States as a result of rabies-infected raccoons foraging for food in the garbage cans of suburban and rural communities. Chagas disease is a protozoan disease carried by beetles, commonly called kissing bugs, because they bite on the face and lips where the skin is thin. They are particularly prevalent in Brazil and other areas of South America. In the early 1900s construction of the Central Railroad in Brazil was undertaken through the heavily forested tropical wilderness, a project that necessitated large-scale deforestation. You can guess the outcome—the indigenous mammals were displaced, as were the beetles that fed on them for their blood meal. Humans and their domesticated animals took up the slack and became infected, as did rodents; the latter conveyed the disease to species of beetles that inhabit housing in urban populations.

In 1998 and 1999 a new and deadly virus, named "Nipah," killed more than 100 people in Malaysia after first showing up on a pig farm. It is speculated that the pigs ate dropped fruits infected with the virus, which was then spread to farm workers. Fruit bats, also known as "flying foxes," are the world's largest bats and have been identified as the natural reservoir. In the years preceding the outbreak, massive deforestation took place and scores of fruit trees were destroyed causing the bats to forage elsewhere, including that remote pig farm surrounded by fruit trees.

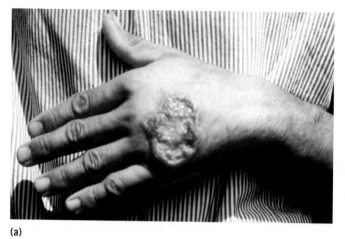

(a)

(b)

FIGURE 1.12 Leishmaniasis is a protozoan infection transmitted by infected sand flies. **(a)** A leishmaniasis skin ulcer on the hand of a Central American villager. Courtesy of Dr. D. S. Martin/CDC. **(b)** Primitive living conditions in a village in Central America. It occupies an area that was formerly a forest and is encircled by a perimeter of trees. Sand flies are poor fliers but can traverse the short distance from their forest habitat. Author's photo (RIK).

Leishmaniasis, a protozoan disease carried by infected sand flies, is a striking example of the consequence of deforestation and the emergence of urban disease. The disease, once limited to mammals of the forest, is now urban, primarily as a result of deforestation. The circumstances are similar to those described for Chagas disease (FIGURE 1.12)

Although it can be an attempt to relieve suffering and death, the intrusion of humans into ecosystems can backfire. Inadequate assessment of the public health impact can inadvertently increase microbial disease. This was the case in the construction of Egypt's one billion dollar Aswan High Dam, a ten-year project completed in 1970. The dam harnessed the uncontrolled Nile River by creating Lake Nasser. (Gamal Abdel Nasser was the Egyptian president from 1956 to 1970.) Unfortunately, the walls of the dam served as a new and convenient habitat for snails. The population of snails boomed, resulting in an increase in the incidence of schistosomiasis, a disease caused by a parasitic worm with a complicated life cycle requiring snails for its completion.

Rift valley fever, a viral hemorrhagic disease carried by mosquitoes, is another example of the downside of the Aswan High Dam project. An epidemic of this viral disease occurred close to the dam area, resulting in illness in 200 thousand people and over 500 deaths. The epidemic was the result of a thriving mosquito population in the flood lands created by the dam.

The most contemporary example of the potential consequences of humans' intrusion into the forest is that of AIDS. Most scientists agree that the origin of human AIDS is the result of the simian immunodeficiency virus (SIV) that made the species leap from infected chimpanzees and sooty mangabeys to humans (FIGURE 1.13).

Climactic Changes

What about the effect of climate on the emergence of microbial diseases? There is ample evidence that global warming and climatic changes cause ecological disturbances that affect the incidence and distribution of infectious diseases (TABLE 1.5). The twentieth century witnessed an increase in average global temperature attributed largely to the burning of fuels and forests, resulting in an

AUTHOR'S NOTE (RIK)

In the spring of 1999, I spent six weeks in Salvador, Brazil at the Institute for Tropical Medicine in completion of the requirements of the M.P.H. degree at the Harvard School of Public Health to study tropical diseases in the natural context of their host. The last week of my stay was in the Amazon and included a visit to a small village with a high incidence of leishmaniasis. It was readily apparent why this was the case; the village bordered a heavy forest (Figure 1.12). Trees had been felled to allow for the construction of primitive dwellings and yet, less than a mile away, leishmaniasis was not prevalent. The reason—sand flies are poor fliers and could not fly far from the forest. As a result of the deforestation, the village habitants and their dogs provided a blood meal for the sand flies. The old expression, "You can't see the forest because of the trees" has its virtues. Wangari Maathai died on September 25, 2011, at the age of 71. She won the Nobel Peace Prize in 2004 for her work on environmental strategies (and other accomplishments). She founded the Green Belt Movement in Kenya that resulted in the planting of 45 million trees by 900 thousand poor women who received a few shillings for the work.

FIGURE 1.13 The interspecies leap. AIDS, which originated in Africa, is presumed to have jumped the species barrier from infected chimpanzee or sooty mangabeys to humans. Other infectious diseases of humans have made a leap from animals to humans. © Jan van der Hoeven/ShutterStock, Inc.

increase in carbon dioxide and other heat-trapping greenhouse gases. The effects are seen not only in human health but also in the disruption of ecosystems and the resulting interference with food productivity. A meeting of world leaders was held in Kyoto, Japan in 1997 to develop countermeasures against the impending threat; these talks resulted in the Kyoto Protocol.

The following year, 1998, was the warmest year worldwide since 1880 when fairly accurate recordings began. The first two months were dominated by a record-breaking El Niño-influenced weather pattern, with wetter than normal conditions across much of the southern third of the United States and warmer than normal conditions across much of the northern two-thirds of the country. The increasingly high temperature exacerbated the extreme regional weather and climate anomalies associated with El Niño. That year brought into focus what scientists had long hypothesized, namely, that global warming could favor outbreaks of a variety of infectious diseases. Furthermore, seven of the eight warmest years on record have occurred since 2001.

In the case of vector-borne diseases, the vector or the microbe, or both, may be influenced by the temperature. TABLE 1.6 summarizes data on tropical diseases and indicates the likelihood of alteration in their distribution as a result of climate change. Malaria, a protozoan disease transmitted by mosquitoes, is at the top of the list; an increase in both temperature and rainfall extends habitats favorable to mosquitoes. (On the other hand, increased temperature and decreased rainfall favor the distribution of sand flies, the vectors responsible for transmission of leishmaniasis, a protozoan disease.) Estimates are that an increase in mean ambient temperature in central Africa by 2°C would extend the range of the vectors of sleeping sickness, filariasis, and leishmaniasis, allowing for these diseases of the tropics to invade marginal temperature zones. Further, higher temperatures may push malaria transmission to higher altitudes, causing epidemics, as has

TABLE 1.5 Infectious Diseases Linked to Climatic Changes

Disease	Biological Agent	Transmission
Malaria[a]	Protozoan	Mosquitoes
Rift valley fever[a]	Virus	Mosquitoes
Hantavirus[a]	Virus	Mice
Cholera[a]	Bacterium	Waterborne
E. coli infection	Bacterium	Waterborne
Cryptosporidiosis	Protozoan	Waterborne
Hepatitis	Virus	Waterborne
Leptospirosis	Bacterium	Waterborne
Lyme disease	Bacterium	Ticks
Dengue (breakbone) fever	Virus	Mosquitoes
West Nile encephalitis	Virus	Mosquitoes

[a]Directly related to El Niño.

TABLE 1.6 Status of Major Vector-Borne Diseases and Predicted Sensitivity to Climate Change

Disease	Vector	Population at Risk, in Millions[a]	Prevalence of Infection	Present Distribution	Possible Change of Distribution as a Result of Climatic Change
Malaria	Mosquito	2,400	300–350 million	Tropics, subtropics	Highly likely
Dengue	Mosquito	1,800	10–30 million	All tropical countries	Very likely
Schistosomiasis	Water snail	600	200 million	Tropics, subtropics	Very likely
Onchocerciasis	Black fly	123	17.5 million	Africa, Latin America	Very likely
Lymphatic filariasis	Mosquito	1,100	117 million	Tropics, subtropics	Likely
Yellow fever	Mosquito	450	More than 5,000 cases	Tropics, S. America, Africa	Likely
Leishmaniasis	Sand fly	350	12 million infected, 500,000 new cases per year	Asia, southern Europe, Africa, Americas	Likely
African trypanosomiasis	Tsetse fly	55	250,000–300,000	Tropical Africa	Likely

[a]Based on a world population estimate of six billion.
Adapted from Vital Climate Change Graphics. UN Environment Programme (UNEP) and GRID-Arendal 2000.

occurred in the highlands of Ethiopia and Madagascar. In Rwanda in late 1987, malaria incidence increased by 337% over the previous three-year period as a result of increases in temperature and rainfall. In the last decade the reported cases of malaria (the form of malaria with the highest fatality rate) in the North-West Frontier Province of Pakistan rose from a few hundred in 1983 to more than twenty-five thousand in 1990. This dramatic rise is attributed to unusually high temperatures at the end of the normal malaria season that extended the season. Malaria, tickborne encephalitis, and leishmaniasis (carried by sand flies) are on the upswing in Italy as a result of climate change according to an Italian environmental organization.

In hantavirus, the cause of a potentially fatal disease emerged in the Four Corners area of the United States (where New Mexico, Utah, Colorado, and Arizona meet). The disease is transmitted by deer mice, the principal animal hosts of the virus, which feed on pine kernels. Higher than normal humidity favored an abundant crop of the pine kernels, which, in turn, led to a tenfold increase in the deer mouse population between 1992 and 1993. This is an excellent example of a climatic condition triggering a chain of events resulting in the emergence of an infectious agent. It can happen again.

Diseases in which infectious agents cycle through invertebrates to complete their development are particularly sensitive to subtle climate variations compared with diseases spread from human to human. Hence, it is imperative that consideration be given to and appropriate measures be enacted regarding the influence of global warming and other climatic changes on microbes and their vectors.

Natural Disasters

Floods, hurricanes, earthquakes, drought, landslides, tsunamis, and volcanoes are environmental disturbances that place populations at risk of an increased burden of infectious diseases. In March and April of 2000 severe floods put the people of Mozambique and other southern African countries at risk for several diseases, particularly malaria and cholera. Up to 250 thousand people in Mozambique alone were endangered by these two diseases. According to a WHO press release, "The threat of a malaria epidemic in the country is increasing and will be at its most dangerous in around three to six weeks time as flood waters gradually subside, the rain stops, and warm temperatures return—ideal breeding conditions for mosquitoes. . . . Before the floods, there were between six and ten cholera cases a week; since the floods it has increased to 120 cases per week." Myanmar (formerly Burma) was hit by tropical cyclone Nargis on May 2, 2008 resulting in over 100 thousand deaths because of heavy rains and 12-foot water surges unleashed by the storm. Mudslides were triggered, contaminating wells that were a source of drinking water and blocking latrines, raising pubic health concerns as a result of a breakdown in sanitation.

The news and the photographs from Somalia are truly devastating, particularly those featuring the terribly malnourished and dehydrated. There is little doubt that most, perhaps all, of the population harbors at least one species of worms. Drought is related to increased famine and an increase in infectious diseases, as has been witnessed in eastern Africa since early 1999, placing at least sixteen million people at risk. In Ethiopia alone, about eight million people are affected. The country is the fifth poorest in the world, with an average life expectancy of fifty-seven years, the fifth lowest in the world.

On December 6, 2004, the world was shocked to witness mountains of water cascading on the northwest coast of the island of Sumatra, Indonesia triggered by an earthquake measuring 9.2 (out of 10) on the Richter scale. Natives and tourists ran to high ground for their lives, but an estimated 230 thousand didn't make it and died. Half a million people were displaced from their homes, and thousands remain unaccounted for. The rapid response and level of relief measures from the international community was unprecedented. Almost immediately, groups worked to prevent infectious disease epidemics by providing "clean" water and bed-nets, initiating a measles vaccination program, and working to prevent and treat soil-transmitted worm infections. Although gaps in the public health infrastructure of the area and in the management of catastrophic events were uncovered, no large-scale outbreaks of infectious disease occurred, and mortality from disease was lower than anticipated.

Another natural disaster occurred when Hurricanes Katrina and Rita inundated large sections of New Orleans and surrounding parishes within a month of each other, August 29 and September 24, 2005, respectively (FIGURE 1.14). The pictures on television and in the newspapers and magazines were horrific; thousands of desperate people crowded into the New Orleans Convention Center, whereas others clung atop trees and roof tops hoping for rescue from the swirling waters. Surprisingly, there were no major outbreaks of infectious disease, although there were cases of wound and gastrointestinal infection primarily due to exposure to contaminated flood waters. The major microbial

(a)

(b)

culprit was mold. As the waters receded, mold thrived and grew in the high humidity and excess moisture. Anyone exposed risked respiratory infections. Further, the CDC reported the occurrence of eighteen cases of wound-associated illness caused by two species of *Vibrio*, five of which resulted in deaths (FIGURE 1.15). These infections generally result when open wounds are exposed to warm seawater containing specific vibrios; those with weakened immune systems and the elderly are particularly at risk. More recently, in 2011, wildfires in Texas and Arizona devastated the lives of scores of people and increased the potential for microbial diseases, as did Hurricane Irene. Joplin, Missouri was hit by a tornado in 2011, which ripped through the city. In the aftermath, an unusual

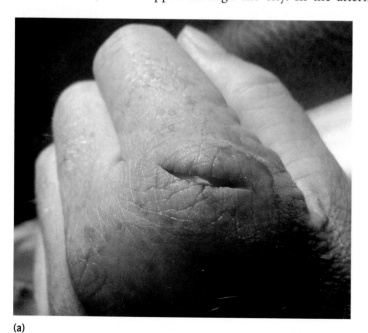

(a)

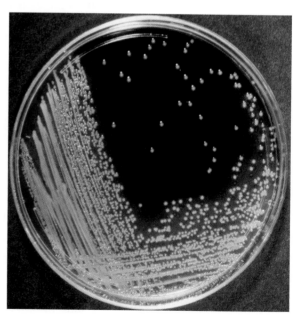

(b)

FIGURE 1.15   (a) An open wound on a hand. © Jonathan Noden-Wilkinson/ShutterStock, Inc. (b) A diagnostic culture of *Vibrio cholerae*, the cause of cholera. Courtesy of CDC.

fungal skin infection caused by *Apophysomyces trapeziformis* broke out. (Don't even try to pronounce the name. Also, be assured, the fungus is not related to trapeze artists!)

Life in less-developed countries is a struggle against poverty and disease. Approximately 1.5 billion people do not have access to safe drinking water. The lack of food and water takes its toll on the maintenance of a healthy immune system and leads to high child mortality rates. Natural catastrophic events exacerbate the potential for microbial-caused diseases.

■ Technological Advances

Human activities lead to technological advances that may pose public health risks; jet travel is an example. It has been well documented that air travel plays a significant role in the transmittance of infectious diseases from continent to continent. TABLE 1.7 lists approximate flying times from New York City to distant places. The farthest destination is Sydney, Australia, taking twenty-two hours, less than the incubation time for many microbial diseases. This means that an infectious traveler could board a jet and arrive at any world destination in less than the time it takes for that passenger to show symptoms (FIGURE 1.16). Such an incident occurred in the spring of 2000: A tourist left Tel Aviv bound for Newark International Airport, an approximately ten-hour nonstop flight, and died of bacterial meningitis approximately two hours after landing. Fortunately, there were no reports of other passengers acquiring the disease. Bacterial meningitis has an incubation period of only a few hours to about two days.

Not so lucky were thirteen travelers infected by a passenger with TB on a flight from Russia to New York. More recently, an international TB scare occurred in May 2007 when an Atlanta man previously diagnosed with an extremely drug-resistant strain of TB (XDRTB) traveled abroad. Although he did have TB, the diagnosis of XDRTB proved to be false. The good news is that none of his fellow passengers on the aircraft became infected. In these cases the best that can be done is to notify other passengers to seek medical advice.

TABLE 1.7 Jet Travel: Microbes Without Passports

Approximate Flying Time From New York City	Incubation Period for Selected Diseases
Sydney, Australia: 22 hours (1 stop)	Whooping cough: 7–10 days
Tokyo, Japan: 14 hours (nonstop)	Gonorrhea: 2–6 days
Tel Aviv, Israel: 10 hours (nonstop)	*Salmonella* food poisoning: 8–48 hours
Nairobi, Kenya: 16 hours (1 stop)	Ebola hemorrhagic fever: 4–16 days
Karachi, Pakistan: 23 hours (nonstop)	Measles: 12–32 days
Dehli, India: 22 hours (1 stop)	Chickenpox: 10–23 days
Moscow, Russia: 10 hours (nonstop)	Influenza: 24–48 hours
	Tuberculosis: weeks to years

Now consider the implications of the Airbus A380—a recently introduced and the world's largest commercial airliner—a super jumbo jet, double-decker, four engine craft with a wing span almost as big as a football field. In a three-class-seating configuration it can carry 555 passengers, but in a one-class-economy seating configuration approximately 850 can be accommodated. From an epidemiological point of view the aircraft is a nightmare—it serves as a huge potential mechanical vector capable of bringing infected people to any part of the world. Infected people can carry microbes to many different, final destinations, and in this way an epidemic can be triggered.

Microbes can be harbored and transported across borders not only in their human hosts but also in their baggage and personal items. Further, vectors harboring infectious agents can also travel; fleas can be carried in rugs transported by jet cargo from the Middle East and Asia. Public health officials inspect many items being transported from country to country and are authorized to impose quarantine in an effort to minimize the risks.

FIGURE 1.16 A flight departure board at an international airport. Jet aircraft, a major technological advance of the twentieth century, serve as vectors for microbes around the world. © Neale Cousland/ShutterStock, Inc.

The use of whole blood and blood products is a life-giving and lifesaving practice. Unfortunately, in some countries blood and blood products may be hazardous to your health. According to the WHO, most of the countries with an unsafe blood supply are developing nations, in which the chances of acquiring infectious diseases are highest. As population pressure increases, so does the demand for blood. In some countries blood is not screened and may harbor the causative agents of HIV infection, hepatitis, syphilis, malaria, and trypanosomiasis. Only recently has the blood supply in the United States been screened for the trypanosome protozoan parasite that causes Chagas disease. Although the disease is primarily spread by the bite of an infected beetle, also referred to as a "kissing bug," it can also be spread by blood transfusions and organ transplants.

To some extent advances in medical technology that make organ transplantation possible contribute to the burden of infectious diseases. Recipients are at increased risk for infection because they are on a regimen of immunosuppressive drugs to minimize organ rejection. It appears that face transplants (both partial and full) are on the map. The first face transplant was performed in France on a woman who was severely mauled by her Labrador dog. The first full facial transplant in the United States was at Brigham and Women's Hospital in March 2011. Other conditions leading to immunosuppression include AIDS, certain inherited diseases, and malnutrition. CT scans, MRI, nuclear medicine, along with other advanced technologies, although life saving in many cases contribute greatly to today's medical costs.

Prostate cancer is the second leading cause of cancer-related deaths in the United States. Transrectal ultrasound-guided biopsies of the prostate gland are common diagnostic procedures. According to the CDC, 624 thousand procedures are performed annually. On July 21, 2006, the CDC reported on four cases of infection caused by *Pseudomonas aeruginosa* after transrectal ultrasound procedures. The infections were caused by contamination of the biopsy equipment that had not been properly sterilized. The bacterial strains recovered from patients matched the strains recovered from the lumen of the biopsy needle. This is an excellent example of a **nosocomial infection** (hospital-acquired infection); these infections are described in the texts on epidemiology and the cycle of microbial disease.

Advances in food technology make it much easier to eat "on the run" by buying prepared foods in markets and frequenting fast-food restaurants. Obesity is recognized as a major health problem in the United States and elsewhere around the globe. Obesity favors the development of numerous health problems, including susceptibility to microbial diseases.

Near the Fort Myers International Airport in Florida there are many medical specialties, including ones that advertise "eyelid surgery." Who knows, someday perhaps there will be a medical practice with a large billboard advertising "colonoscopies and rectal diseases—back in and open door."

Microbial Evolution and Adaptation

The 1940s ushered in the dawn of antibiotics—agents that were rightfully called "wonder drugs." Penicillin was the first, and numerous others quickly followed; some were tailored to be effective against a broad spectrum of bacteria, whereas others were more specific. It should be emphasized that antibiotics are not effective against viruses and hence should not be prescribed for viral infections. The number of lives saved worldwide over the past fifty-five years because of antibiotic therapy is beyond estimation. An individual today who is infected with a variety of life-threatening bacteria has a fighting chance, assuming antibiotics are administered promptly, whereas an individual infected 50 or 60 years ago had little chance of recovery. The development of antibiotics was a major factor leading to the optimism of the 1970s. Many dread diseases, so it seemed, were about to become vanquished. But it turns out that the tables are turning—antibiotics are losing their punch, and increasing numbers of microbes are resistant. The expression "I'm resistant to such and such an antibiotic" has no meaning; people do not become resistant to antibiotics—their microbes do.

Emblazoned on the cover of the September 12, 1994, issue of *Time* is the headline "Revenge of the Killer Microbes" and the question "Are we losing the war against infectious diseases?" The answer to this question might be an uncomfortable "Yes" (TABLE 1.8). Resistance to antimicrobial agents is at a crisis level worldwide (FIGURE 1.17). Vancomycin, an antibiotic considered by many to be the last stronghold in certain situations, is no longer effective against many bacterial strains that responded ten years ago. Some refer to antibiotic-resistant bacteria as "super bugs." What's happening? To put it in a nutshell, the forces of natural selection are in play. Antibiotics have been grossly misused and have promoted the emergence of antibiotic-resistant organisms in a Darwinian fashion. The antibiotic-resistant strains are the result of chance mutations, and their survival is favored by the presence of antibiotics. Antibiotics are the "selecting" and not the "causing" agent.

The battle against the natural process of microbial adaptation and change, whether exhibited by resistance against antibiotics or by evasive strategies, is an

TABLE 1.8 Drug-Resistant Microbial Diseases

Bacterial Disease	Viral Disease	Protozoan Disease	Fungal Disease
Tuberculosis	HIV infection	Malaria	Aspergillosis
Gonorrhea	Hepatitis B	Visceral leishmaniasis	Candidiasis
Staphylococcal infection	Hepatitis C		Cryptococcosis
Shigellosis			
Typhoid fever			
Pneumococcal pneumonia			
Enterococcal infection			
Acinetobacter infection			

ever-present and ongoing struggle for survival. Failure to meet the challenge affords microbes the upper hand.

Insect vectors are also able to adapt to a changing environment bringing up the issue of climactic change as previously discussed. Malaria, a mosquito-borne protozoan disease, was thought to be a disease of the past, thanks to the application of the insecticide dichlorodiphenyltrichloroethane (DDT). Little did scientists realize that the forces of natural selection would again interfere as a result of the misuse of the insecticide. DDT-resistant mosquitoes emerged with a vengeance, and other vector-borne diseases shared their triumph over DDT. West Nile virus, a mosquito-borne agent, now threatens all of the contiguous United States, prompting ground and aerial spraying. Can insecticide-resistant mosquitoes carrying West Nile virus emerge?

Human Behavior and Attitudes

Complacency

How easy it is to cut corners on health-related matters when it appears that progress and improvement have taken place, leading to the false assumption that prevention and control are no longer necessary. Complacency is the belief that "it can't happen to me." The failure of people to complete their full dose of antibiotics because they are feeling better is a prime example.

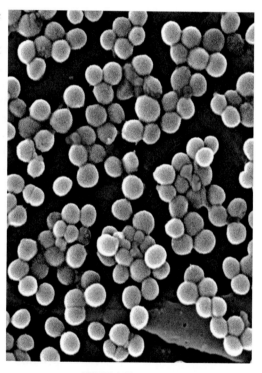

FIGURE 1.17 A scanning electron micrograph of methicillin-resistant *Staphylococcus aureus*, commonly referred to as MRSA. Courtesy of Janice Haney Carr/Jeff Hageman, M.H.S./CDC.

A dramatic example of complacency is evident in the threatened resurgence of AIDS, particularly among young gay men, because of a return to risky sexual behavior fueled by glowing reports of new drug therapies for the management of AIDS. At least five million Americans have sex and/or drug habits that put them at high risk for acquiring AIDS. The number of cases in the United States has fallen dramatically since the peak of the 1980s; the decrease is primarily attributed to safer sex habits and avoidance of dirty needles by drug abusers, but public health officials worry that the decrease in cases could cause complacency and result in an increase in the number of cases as people return to unsafe sex practices.

People have become complacent about receiving immunization shots or keeping their immunization boosters up to date. According to the CDC only 80% of two-year-olds in the United States have been given the full sequence of currently recommended immunizations, primarily because of parents' complacency and concern about safety. As an example, consider that in 2005 only about 81% of children between the ages of 19 to 35 months were fully immunized in the state of New York. This could mean trouble down the line. Past history reveals that a 10% decline in measles vaccination between 1989 and 1991 resulted in an outbreak of 55 thousand cases, several thousand hospitalizations, and 120 deaths, indicating the power of immunization. Fortunately, children entering school are required to have proof of being up to date on their immunizations; students entering college must also have proof of being fully immunized. Nevertheless, some slip through the cracks. Individuals traveling to foreign countries need to be aware that particular immunizations may be necessary against diseases prevalent in that area. For example, in 1996 tourists traveling to yellow fever areas neglected to be immunized against the disease and were responsible for infecting others with the disease upon their return to the United States and Switzerland.

Human Migration

Human migration is a major factor in the emergence and reemergence of many communicable diseases. The Population Reference Bureau estimates that in the mid-2000s about 191 million people lived outside their native countries. Populations on the move contribute to the emergence of disease beyond that resulting from voluntary urbanization fueled by a search for a better life. Population movement is frequently not a matter of choice but rather a forced movement because of wars and conflicts resulting from political upheavals. The United Nations defines **internally displaced persons** (**IDPs**) as "Persons or groups of persons who have been forced or obliged to flee or to leave their homes or places of habitual residence, in particular as a result of or in order to avoid the effects of armed conflict, situations of generalized violence, violations of human rights or natural or human-made disasters, and who have not crossed an internationally recognized State border." The term *refugee* is reserved for those who are forced under the same circumstances to cross an international border. The Office of the UN High Commissioner for Refugees estimates the number of forcibly displaced persons at nearly thirty-three million worldwide. These people carry with them their microbes and microbial vectors, resulting in an exchange with intermingling populations. Malaria is an excellent example; refugees migrating through regions where malaria is endemic can acquire the infection and disseminate the disease to other areas. Malaria is a common cause of death among refugees in numerous countries, including Thailand, Somalia, Rwanda, the Democratic Republic of the Congo, and Tanzania.

Masses of people are forced to settle in uninhabitable environments without adequate shelter, food, clean drinking water, and latrines. Personal hygiene and sanitation may be virtually nonexistent, and what few facilities are available become quickly overwhelmed. People live in filth and squalor. These camps are hotbeds for epidemics, and their potential spreads as refugees continue to flee from one area to another (FIGURE 1.18). The Darfur conflict in western Sudan is a human catastrophe and a worst-case scenario. The United Nations has estimated 200 thousand to 400 thousand have died from violence and disease and another 2.5 million have been displaced to refugee camps. The 2011 drought in Somalia, the worst in 60 years, has forced more than 1,000 people per day to migrate to refugee camps in Kenya. The camps designed for 90 thousand house over 430 thousand refugees. Try to imagine the almost overwhelming public health problems created by the surge of people, many of whom suffer from cholera and other infectious diseases. Maintenance of sanitation and provision of clean water are of the first order.

FIGURE 1.18 A refugee camp. Refugee camps are hotbeds of infection. Crowding and lack of hygiene and sanitation favor the incidence and transmission of disease. © Northfoto/ShutterStock, Inc.

Wars and civil unrest, in addition to creating refugees and displaced persons, disrupt the public health infrastructure and favor the spread of disease. The destruction of housing leads to increased human-to-human and human-to-vector contact; decline in water management programs and a lack of treatment facilities are contributory factors.

Societal Factors

In many societies, particularly in developed countries, family life and structure have changed as a result of economic growth and increased opportunities for women. In most American families both parents work, leading to an increase in child care centers. Millions of children attend day care centers, which put them at risk for a variety of intestinal parasites, diarrhea, middle ear infections, and meningitis. For example, outbreaks of shigellosis, a diarrheal disease, have caused problems in many day care centers around the country. (To a large extent, the simple act of hand washing by the staff after they change a diaper is an effective control measure.) Children convey the microbes to their family members, many of whom in turn bring their microbes to the workplace.

As longevity increases, so does the number of elderly citizens requiring nursing homes, day care centers, and assisted living environments. Like child day care centers, these facilities are potential hotbeds for the emergence and spread of communicable diseases within the resident population and the staff, visitors, and their contacts.

Food production and dietary habits also affect the spread of microbial diseases. Globalization of the food supply, centralized processing, fast-food restaurants, dining out, and take-out food are all significant. Foodborne diseases are a major public health problem in the United States. During the spring and continuing into the summer of 2008, over one thousand cases of salmonellosis occurred in the United States, presumably due to certain varieties of contaminated tomatoes. In 2011, turkey burgers, packaged bags of salad, strawberries, eggs contaminated with bacteria including *E. coli*, *Listeria* and *Salmonella* were the causes of foodborne illness. Other recent examples of foodborne outbreaks were described earlier in the chapter. In a better economy and a family structure in which both parents work, many people rely more on prepared foods to reduce household chores.

Fast-food restaurants and take-out restaurants are part of our social structure. Food has increasingly become a source of recreation. Consider, for example,

(a)

PIERCING PRIJZEN / PIERCING PRICES	
NAVEL / BELLY BUTTON	Hfl. 100,-
TONG / TONQUE	Hfl. 125,-
TEPEL / NIPPLE	Hfl. 110,-
WENKBRAUW / EYEBROW	Hfl. 90,-
NEUS / NOSE	Hfl. 95,-
OOR / EAR	Hfl. 90,-
LIP / LIP	Hfl. 95,-
NEUSSCHOT / SEPTUM	Hfl. 110,-
LABRET / LABRET	Hfl. 100,-
SCHAAMLIP / LABIA	Hfl. 150,-
CLITORUSKAPJE / CLITHOOD	Hfl. 175,-
BALZAK / SCROTUM	Hfl. 150,-
PRINCE ALBERT / P.A.	Hfl. 195,-
INCL: SIERAAD / INCL: JEWELLERY	

(b)

FIGURE 1.19 Tattooing and body piercing. Tattooing and body piercing are a risky part of popular culture. The skin is invaded, potentially resulting in serious infections because of the use of unclean instruments. **(a)** Tattoos. © PeterSVETphoto/ShutterStock, Inc. **(b)** Anything goes if the price is right (Amsterdam). Author's photo (RIK).

a typical conversation: "So, what'll we do tonight?" Answer: "Let's eat out!" This is all well and good, assuming that personal hygiene and sanitary control measures practiced by food handlers are not compromised. Television news shows have aired segments featuring high-end restaurants that are enough to make you sick!

Tattooing and body piercing are ancient art forms that have continued through the centuries. In developed countries these practices have long been popular with sailors and bikers. Young people in the 1990s and the new millennium have brought the trend into the mainstream. Tattoo and body piercing parlors are found in many countries, including the United States; for a price you can get just about any part of your body tattooed or pierced (**FIGURE 1.19**). The risk of infection with a variety of microbes, particularly staphylococci, is a real possibility, and patrons are often at risk because of nonsterile instruments and poorly trained personnel. The CDC reported forty-four cases of methicillin-resistant *Staphylococcus aureus* skin infections in Ohio, Kentucky, and Vermont in 2004–2005 as a result of thirteen unlicensed tattooists, presumably because of the use of nonsterile equipment in these three states. *Mycobacterium haemophilus* has also been identified as a cause of infection following tattooing. Even if the establishment is certified by a local health authority, let the buyer beware!

■ Overview

This chapter makes the case that despite the optimism of forty or fifty years ago, microbial diseases have not been eliminated (with the single exception of smallpox) but continue to flourish as a major cause of mortality and morbidity around the world. The reasons for this are based on world population growth, urbanization, ecological disturbances, technological advances, microbial evolution and adaptation, and human behavior. A quotation from Donald A. Henderson during his tenure as associate director of the U.S. Office of Science and Technology Policy serves as an excellent way to close this chapter:

"The recent emergence of AIDS and dengue hemorrhagic infections, among others, [is] serving usefully to disturb our ill-founded complacency about infectious diseases. Such complacency has prevailed in this country [USA] throughout much of my career. . . . It is evident now, as it should have been then that mutation and change are facts of nature, that the world is increasingly interdependent, and that human health and survival will be challenged, ad infinitum, by new mutant microbes, with unpredictable pathophysiological manifestations."

■ Self-Evaluation

PART I: Choose the single best answer.

1. In 2011, world population was approaching
   a. 8 billion **b.** 7 billion **c.** 10 billion **d.** more than 12 billion

2. Which one of the following people warned of unchecked population growth?
   a. Satcher **b.** Stewart **c.** Malthus **d.** Darwin

3. In the United States, by 2050 estimates are that people over the age of sixty-five will constitute about what percentage of the population?
   a. 20% **b.** 28% **c.** 39% **d.** not predictable

4. The construction of the Central Railroad in Brazil led to an increase in
   a. leishmaniasis **b.** malaria **c.** tuberculosis **d.** Chagas disease

5. Which of the following is not an emerging or reemerging disease?
   a. common cold **b.** cholera **c.** bird influenza **d.** toxic *E. coli*

6. Which disease has been shown most conclusively to be linked to climate change?
   a. leishmaniasis **b.** *E. coli* infection **c.** leptospirosis **d.** malaria

7. Which of the following is a foodborne illness caused by a microbe?
   a. Dengue fever **b.** coccidioidomycosis **c.** salmonellosis **d.** tuberculosis

8. Tattooing carries a risk of infection. Which organism (or disease) is most likely to be involved?
   a. staphylococci **b.** *E. coli* **c.** meningitis **d.** cryptosporidiosis

9. Which of the following practices is used to prevent malaria?
   a. use of bed nets **b.** antibiotics **c.** vaccination **d.** blood transfusion

10. What is the biological vector that carries the pathogen that causes West Nile encephalitis?
    a. ticks **b.** mosquitoes **c.** fleas **d.** sandflies

PART II: Fill in the blank.

1. Most microbes are beneficial and many are essential to the cycles of nature without which higher forms of life could not exist. True or false?

   _____

2. In the United States infectious diseases are the leading cause of death. True or false? _____.

3. Name an infectious disease in the "top ten" worldwide. _____

4. Numerous studies have concluded that city dwellers get sick more often than their rural counterparts and that people living in poverty get sick less. True or false? _____

5. According to the U.S. Census Bureau, the world's largest urban area is _____.

6. Give an example of a drag-resistant bacterial disease. _____

7. What does the acronym IDP stand for? _____

8. Deforestation is a major factor in the eruption of emerging and reemerging diseases. True or false? _____

9. Diseases transmitted from animals to humans are called _____ diseases.

10. Infections are a part of civilization and some actually predate civilization. True or false? _____

PART III: Answer the following.

1. List five reasons why infections are emerging and increasing.

2. Choose two of the reasons you gave in question no. 1 and discuss them.

3. Why was the Aswan Dam a "disaster"?

4. Cairns described cities as "graveyards of mankind." Explain.

5. A number of quotations are cited in this chapter. Develop your own quotation that targets the problem of new, emerging, and reemerging infections.

6. Do you believe that a grade of C– should be given to the world for its efforts in coping with microbial diseases. What grade would you award, and why?

7. In 1967 the surgeon general of the United States declared it was "time to close the book on infectious diseases," but events proved otherwise. What was the basis of the surgeon general's remark?

8. Complacency is listed as a major factor responsible for the continued threat of microbial diseases. Describe some specific examples, including examples for which you and family members may be "guilty."

9. Create a list containing at least five drug-resistant diseases and discuss reasons for their development.

10. Explain how technological advances may pose public health risks.

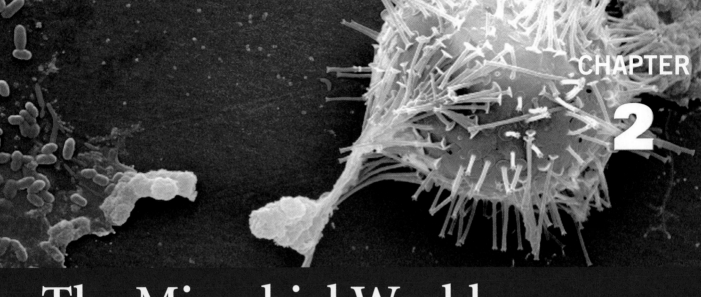

CHAPTER

2

The Microbial World

The Microbe is so very small
You cannot make him out at all,
But many sanguine people hope
To see him down a microscope.
Oh! Let us never, never doubt
What nobody is sure about!

—Hilaire Belloc, *More Beasts for Worse Children*

■ **Preview**

In the introductory section the term *microbe* has been used extensively because this text is about microbes, particularly those relatively few that are pathogens. The term *microbe* was not defined or even adequately described, but the six groups of microbes were named—prions, viruses, bacteria, protozoans, unicellular algae, and fungi. (Worms, biologically known as helminths, are frequently included in microbiology texts even though they are not microbes because a number of species cause infections resembling microbial infections.)

Some Basic Biological Principles

Cell Theory

To further understand microbes, whether pathogens or not, it is necessary to review a few very basic concepts of biology, because all microbes are biological packages with certain unique characteristics. **Cells** are considered the basic unit of life, based on the observations of Robert Hooke in 1665. Hooke used the word *cella* in his examination of cork, which revealed tiny compartments that reminded him of the cells in which monks lived. His studies ultimately gave rise to the cell theory, a fundamental concept in biology, as postulated by Matthias Schleiden and Theodor Schwann (1838) and Rudolf Virchow (1858). The major points of the cell theory are as follows:

1. All **organisms** are composed of fundamental units called cells.
2. All organisms are unicellular (single cells) or multicellular (more than one cell).
3. All cells are fundamentally alike with regard to their structure and their metabolism.
4. Cells arise only from previously existing cells ("life begets life").

"Life begets life" is a refutation of the doctrine of **spontaneous generation,** a concept that was disproved by the end of the nineteenth century. An understanding of the cell theory is the basis for an understanding of life, including microbial life. The cell theory does not apply to viruses and prions; they are described as *acellular, subcellular,* or as *biological agents,* terms that are used somewhat interchangeably. Nevertheless, as a matter of convenience license is sometimes taken, and they are described as microbes or microorganisms. Viruses are not considered by scientists as being "alive," but they come close; they are in that gray area between living and nonliving. Prions are even less biologically complex than viruses.

Metabolic Diversity

The term *life* is elusive and cannot be given an exact definition; at best, it can only be described. Nevertheless, several attributes are associated with living systems that, collectively, establish life. By one strategy or another all organisms exhibit these characteristics, summarized in TABLE 2.1. A major property of life is the ability to constantly satisfy the requirement for energy. It takes energy for every cell to stay alive, whether it is a single cell or a component of a multicellular organism; in the latter case each cell contributes to the total energy requirement of the organism. Your body constantly expends energy.

It takes energy to breathe even during sleep and for the heart to constantly push blood through an interconnected and tortuous maze of blood vessels. Because you don't fill up at the gas station, it's obvious that your energy is derived from the foods you eat. Through a complex series of biochemical reactions, the body metabolizes the **organic** compounds (proteins, fats, and carbohydrates) of your diet and releases the energy stored in their chemical bonds into a biologically available high-energy compound known as adenosine triphosphate (**ATP**); you live directly off of this and constantly replace it as you take in nutrients. Most organisms, including most microbes, are **heterotrophs**, meaning that they require organic compounds as an energy source;

TABLE 2.1 Characteristics of Life

Characteristic	Description
Cellular organization	The cell is the basic unit of life; organisms are unicellular or multicellular.
Energy production	Organisms require energy and a biochemical strategy to meet their energy requirement.
Reproduction	Organisms have the capacity to reproduce by asexual or sexual methods and in doing so pass on genetic material (DNA) to their progeny.
Irritability	Organisms respond to internal and external stimuli.
Growth and development	Organisms grow and develop in each new generation; specialization and differentiation occur in multicellular organisms.

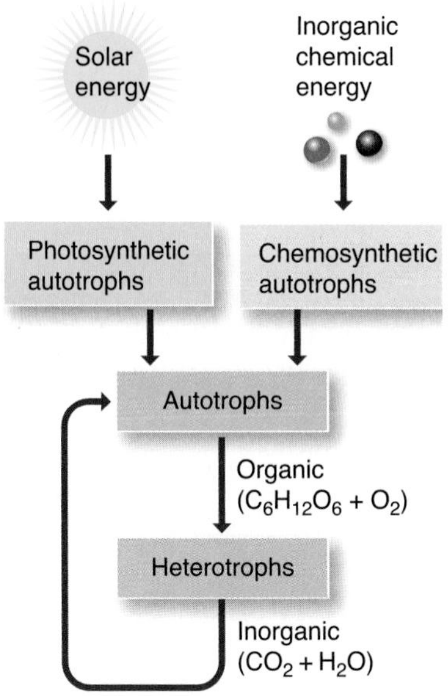

FIGURE 2.1 A pathway map showing heterotroph dependency on autotrophs and the autotrophs' energy sources.

humans are heterotrophs. Other microorganisms and plant life are **autotrophs** and do not require organic compounds, but they do require energy. Some are able to directly use the energy of the sun (**photosynthetic autotrophs**), and others derive energy from the metabolism of **inorganic** compounds (**chemosynthetic autotrophs**). In so doing autotrophs produce organic compounds and oxygen (O_2). Hence, heterotrophs are dependent on autotrophs for energy (FIGURE 2.1).

Requirement for Oxygen

In addition to metabolic diversity, organisms exhibit diversity in their O_2 requirement. The "higher" organisms that are more familiar to you are **aerobes**, meaning they require O_2 for their metabolic activities. Some bacteria are **anaerobes** and do not require oxygen; other anaerobes are actually killed by O_2. **Facultative anaerobes** are bacteria that grow better in the presence of O_2 but can shift their metabolism, allowing them to grow in the absence of O_2. Knowledge of the oxygen requirements of pathogens is important in clinical microbiology. For example, specimens from infections caused by bacteria suspected of being anaerobes must be transported and cultured under anaerobic conditions (FIGURE 2.2).

Genetic Information

The genetic information for the structure and functioning of all cells is stored in molecules of **deoxyribonucleic acid (DNA)**, a large and complex organic molecule. **Genes** are segments of the DNA molecule. Since the establishment of DNA as the hereditary material, the expression "life begets life" can be expanded to explain the mechanism by which a particular life form gives rise to the same life form; that is, tomatoes produce tomatoes, humans produce humans, and *Escherichia coli* produces *Escherichia coli*. Each of these groups has its characteristics embedded in DNA that confer its identity. The DNA is transferred, by a variety of reproductive strategies, from parent to offspring.

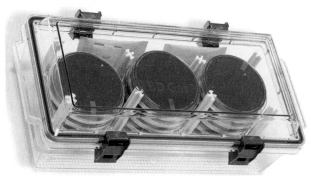

FIGURE 2.2 Culturing anaerobic bacteria. Some bacteria cannot grow in the presence of oxygen. The GasPak tray is a means of culturing anaerobes. Courtesy and © Becton, Dickinson, and Company.

■ What Makes a Microbe?

With these basic biological principles in mind, the term *microbe* (or *microorganism*) can now be better described. The question to be considered is what makes a microbe a microbe? As will become apparent, this question is not easily answered. Your first response may be "they are all too small to be seen without a microscope" or are **microscopic**. Wrong. At first thought this would appear to be true, but what about the algae and the fungi? Are fungi microscopic? No doubt you have seen molds (**FIGURE 2.3**) classified as fungi, growing on food left too long in the refrigerator or perhaps on a pair of old sneakers that you forgot about in the dank basement or hidden away in a dormitory closet. They are **macroscopic**; that is, they can be seen with the naked eye. Hence, "microscopic" is not a distinguishing microbial characteristic. To describe all microbes as being unicellular is also not correct because the fungi and many of the algal forms are macroscopic and clearly multicellular. (As pointed out later, some fungi, namely yeasts, are unicellular.) These organisms must be multicellular; if they were unicellular, that one cell would be enormous—a ridiculous idea!

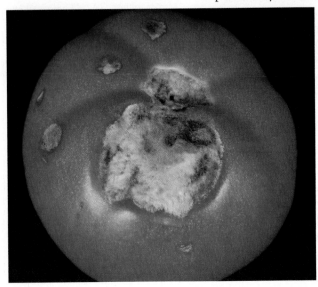

FIGURE 2.3 Mold growing on a tomato. © Jones & Bartlett Learning.

There are exceptions to the rule that all bacteria are unicellular and microscopic. This might seem like an amazing fish story, but in 1985 a large cigar-shaped microorganism was found in the guts of the Red Sea brown surgeonfish. This organism was subsequently identified as a bacterium, approximately a million times larger in volume than *E. coli*, and was christened *Epulopiscium fishelsoni*. Twelve years later, in 1997, an even more monstrous bacterium was discovered in sediment samples residing off the coast of Namibia (**BOX 2.1**); the organism has the tongue-twisting name *Thiomargarita namibiensis* and to date is one for the *Guinness Book of World Records*. They are visible to the naked eye.

To give you some idea of size relationships, if an ordinary bacterium were the size of a baby mouse, *E. fishelsoni* would be equivalent to a lion and *T. namibiensis* would be the size of a blue whale, the world's largest animal. The blue whale measures up to 90 feet (29 meters) and weighs about 120 tons. How many cells might make up such an enormous creature? The number would be in the trillions. Each of these cells exhibits the same fundamental life characteristics as the single microbe. Microbes are sometimes described as "simple" because many consist of only a single cell or are less than a cell (viruses and prions). Consider, however, that this single cell must fulfill all the functions of life. On the other hand, in a multicellular organism (like the whale), although each cell fulfills all the criteria for life, there is a "sharing" of function because of specialization into a variety of cell types (for example, muscle cells, nerve cells, and blood cells). Perhaps that makes life easier. Hence, single-celled organisms, and even those multicellular organisms consisting of only a small number of cells without evidence of true specialization, are simple only in the sense of numbers and not in a physiological (functional) sense.

So, if microbes are not necessarily microscopic and/or unicellular, then what is a microbe? There really is no unifying principle or precise definition. The term

BOX 2.1 — "Monster" Bacteria

If asked to describe bacteria, just about everyone would reply that they are too small to be seen without a microscope. However, in 1985 *Epulopiscium fishelsoni,* a giant bacterium that can be seen without a microscope, was discovered in the guts of surgeonfish in the warm waters of the Red Sea and off the coast of Australia. The organism can grow to about 500 micrometers, or about the size of the period at the end of this sentence. To give you some idea of size, one scientist projected that "if ordinary bacteria were mouse sized, *E. fishelsoni* would be equivalent to a lion." This organism is referred to as "epulos" and was originally thought to be protozoan-like. However, analysis of their DNA revealed that they are, in fact, bacteria.

In 1997 *Thiomargarita namibiensis* stole the prize for size from *Epulopiscium.* This "monster" bacterium, approximately the size of a fruit fly's head, was discovered in samples of sediment in the greenish ooze off the coast of Namibia in Africa. These spherical cells range from 100 to 750 micrometers in size. Dispersed throughout their cytoplasm are globules of sulfur. The bacteria tend to organize into strands of cells that glisten white from light reflected off their sulfur globules, which explains the name. *T. namibiensis* means Namibian sulfur pearl.

Both epulos and the sulfur pearl are anomalies in the bacterial world. The sizes of cells of all kinds, not only bacterial cells, are limited by the surface area of the membrane, because nutrients and waste are transported in and out of the cells by diffusion across the cell membrane. As cells increase in size, both volume and surface area increase, but surface area increases to a lesser degree than does volume. At some point the surface area becomes too limited to allow for sufficient diffusion between the cell and its environment.

So how did *E. fishelsoni* and *T. namibiensis* manage to become so big? What are the physiological adaptations? In the case of epulos, microscopic examination reveals that the cell membrane, rather than being stretched smoothly around the cell, is convoluted (wrinkled), resulting in "hills and valleys," a phenomenon that greatly increases cell surface. (This adaptation is not unique to bacterial cells; the surface of the human brain is highly convoluted, resulting in a greater surface area, a factor that correlates with species intelligence.) The large size of *T. namibiensis* is attributed to the presence of a large fluid-filled sac occupying over 90% of the cell's interior. The sac is packed with nitrate that the cell uses in its metabolism to produce energy, making it less dependent on constant diffusion across the membrane to transport nutrients and waste.

microbe, or microorganism, is a term of convenience used to describe biological agents, in a collective sense, that in general are too small to be seen without the aid of a microscope. The term is also used for microbes that are cultured and identified using similar techniques. Based on what has been presented here, it is clear that these descriptions are not always true. Some biologists consider microbes to be organisms that are at less than the tissue level of organization. This statement requires some explanation and is based on what is referred to as "biological hierarchy," or levels of biological organization.

Recall that a cell is the fundamental unit of biological organization and that groups of cells establish multicellularity. Consider the human, or any other multicellular animal or plant, and it is obvious that in addition to an increase in cell numbers, the process of differentiation and specialization has taken place. For example, over 200 cell types make up the human, including red blood cells, five categories of white blood cells, epithelial cells, connective tissue cells, nerve cells, and muscle cells. All these cells, as stated in the cell theory, share common fundamental characteristics, but superimposed on their basic structure and function is a specialization of structure and function. Cells of the same type constitute the **tissue** level of organization, as exemplified by nerve tissue, blood tissue, and connective tissue. Tissues in turn constitute **organs**, structures composed of more than one tissue

AUTHOR'S NOTE (RIK)

Some years ago I attended the annual meeting of the American Society for Microbiology in Miami Beach, Florida and overheard two airport baggage handlers commenting that about twelve thousand microbiologists were expected to attend. One asked the other, "What's a microbiologist, anyway?" to which the other replied, "Beats me! I suppose it's a small biologist." Several miles from the airport was a huge billboard with the words "Orkin Pest Control welcomes microbiologists." It was a memorable meeting.

FIGURE 2.4 Levels of biological organization. **(a)** Microbes. **(b)** Multicellular.

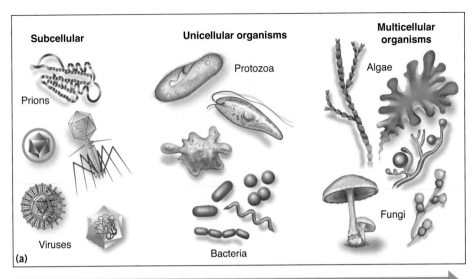

MICROBES

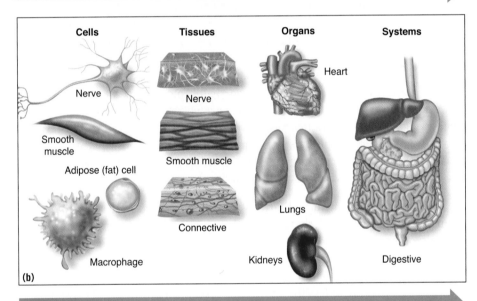

MULTICELLULAR

type; the heart, brain, stomach, and kidney are examples. Organs in turn constitute **organ systems**, a collection of organs that contribute to an overall function or functions. The digestive system, nervous system, respiratory system, excretory system, and reproductive system are examples familiar to you. This hierarchy is summarized as follows and is further illustrated in FIGURE 2.4: subcellular → cells → tissues → organs → organ systems.

All microbes are devoid of tissues. That is, they are all at the subcellular or cellular level of organization, although fungi and some algae hint at specialization and approach the tissue level of organization. Prions and viruses can be properly placed at the acellular or subcellular level, which, simply put, means that they are less than cells and are at the threshold of life.

Procaryotic and Eucaryotic Cells

Biologists recognize the existence of two very distinct types of cells, referred to as **procaryotic** and **eucaryotic** cells (Greek, *pro,* before, + *karyon,* nut or kernel, + *eu,* true). Procaryotic cells have a simpler morphology than eucaryotic cells and are primarily distinguished by the fact that there is no membrane around the nucleus. There is a nuclear area rich in DNA that serves as the carrier of genetic information, as in all cells, but that DNA is not enclosed within a nuclear membrane. This DNA-rich area is referred to as a **nucleoid** rather than as a true nucleus. Further, in procaryotic cells there are no membrane-bound cellular structures (organelles) in contrast to the cellular anatomy of the eucaryotic cells.

Procaryotic and eucaryotic cells are compared in TABLE 2.2 and in FIGURE 2.5. Bacteria are procaryotic microorganisms; protozoans, unicellular algae, fungi, and all other forms of life (except viruses and prions) are composed of eucaryotic cells.

Microbial Evolution and Diversity

Procaryotes date back 3.5 billion years and were the only life forms for 2.5 billion years; they are the ancestors of eucaryotes. Aristotle pondered the relationships among organisms, as do scientists today. In the eighteenth century the botanist Carolus Linnaeus classified all life forms as belonging to either the plant or the animal kingdom. (Students would be delighted if only this were the case today!) Microbes were largely ignored because little was known about them, but, because they had to be placed somewhere, they were considered plants, probably because those that had been observed possessed cell walls. Various schemes of classification have been proposed over the last few centuries, and taxonomy, the science of classification, became more and more complex. In 1866 Ernst Haeckel proposed a three-kingdom system—animals, plants, and a new kingdom, Protista, a collection to accommodate microbes. In the light of modern biology, it became apparent that

TABLE 2.2 Comparison of Procaryotic and Eucaryotic Cells

Characteristic	Procaryotes	Eucaryotes
Life form	Bacteria, Archaea	All microbial cells (with the exception of bacteria, viruses, and prions) and all other cells
Nucleus	DNA chromosome but not enveloped by membrane	Chromosome present and enveloped by membrane
Cell size	About 1–10 micrometers	Over 100 micrometers
Chromosomes	Single circular DNA (two chromosomes in a few)	Multiple paired chromosomes present in nucleus
Cell division	Asexual binary fission; no "true" sexual reproduction	Cell division by mitosis; sexual reproduction by meiosis
Internal compartmentalization	No membrane-bound internal compartments	Organelles bound by membrane
Ribosomes	Smaller than eucaryotic cells and not membrane bound	Membrane bound and free

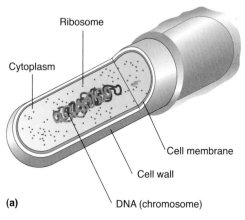

FIGURE 2.5 Schematic drawings of **(a)** a eucaryotic cell and **(b)** a procaryotic cell.

(a)

Ribosome

Cytoplasm

Cell membrane

Cell wall

DNA (chromosome)

(b)

Centrioles

Microtubules

Flagellum

Golgi apparatus

Nuclear pore

Basal body

Lysosome

Free ribosomes

Mitochondrion

Nuclear envelope

DNA (chromosomes)

Nucleolus

Ribosomes attached to endoplasmic reticulum

Plasma membrane

Actin filaments

Cilia

Rough endoplasmic reticulum   Smooth endoplasmic reticulum

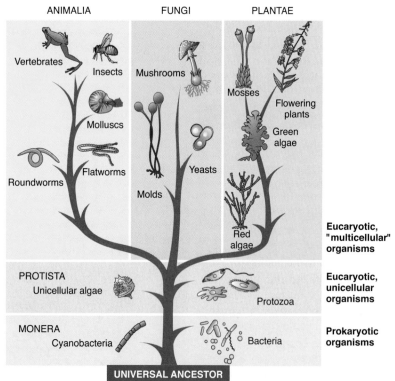

FIGURE 2.6 Whittaker's five-kingdom system.

ANIMALIA

Vertebrates

Insects

Molluscs

Flatworms

Roundworms

FUNGI

Mushrooms

Yeasts

Molds

PLANTAE

Mosses

Flowering plants

Green algae

Red algae

Eucaryotic, "multicellular" organisms

PROTISTA
Unicellular algae

Protozoa

Eucaryotic, unicellular organisms

MONERA
Cyanobacteria

Bacteria

Prokaryotic organisms

UNIVERSAL ANCESTOR

even three kingdoms were not enough. In 1969 a five-kingdom system was proposed by Robert Whittaker and initially accepted by most biologists. This classification describes organisms as belonging to the kingdoms Monera, Protista, Fungi, Animalia, and Plantae (FIGURE 2.6). Recall that microbes consist of six groups accommodated in one of Whittaker's five kingdoms as follows: Bacteria are classified as Monera, protozoans and unicellular algae are classified as Protista, and fungi are classified as Fungi. Note that viruses and prions are not considered in this scheme of classification, because they are neither procaryotic nor eucaryotic cells but are subcellular.

The 1950s ushered in the tide of molecular biology, and its wake introduced new techniques. Biologist Carl Woese and his colleagues at the University of Illinois focused in on ribosomal ribonucleic acid (rRNA) as a "fingerprint" to identify shared characteristics of microbes and thus gain insight into their relatedness, which in turn would point to their evolutionary history.

In 1990 Woese, along with Otto Kandler and Mark L. Wheelis, proposed a novel scheme of classification based on Woese' analysis. The Woese system assigns all organisms to one of three domains or "superkingdoms"—the **Bacteria**, **Archaea** (formerly *Archaebacteria*), and **Eucarya** (FIGURE 2.7), all of which arose from a single ancestral line. (All of Whittaker's five traditional kingdoms can be reassigned among the three domains.) The *Bacteria* and the *Eucarya* first diverged from an ancestral stock, followed by the divergence of the *Archaea* from the *Eucarya* line. The domains differ remarkably from one another in their chemical composition and in other characteristics, as summarized in TABLE 2.3. The term *bacteria* as commonly used includes both the bacteria and the archaea.

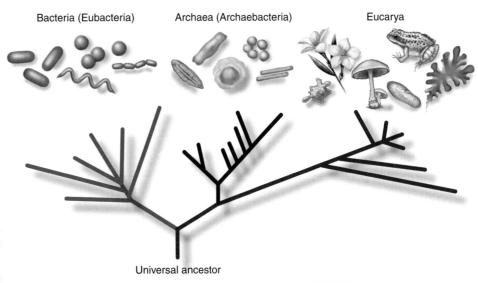

FIGURE 2.7 Woese's three-domain system.

It should be apparent that classification, particularly at the level of microorganisms, is not cast in concrete but is constantly under revision as new information becomes available. It is a credit to the scientific process that reevaluation is the name of the game. Admittedly, it is confusing, but to quote William Shakespeare (who probably never even took a course in biology), "What's in a name? That which we call a rose by any other name would smell as sweet." So you need not sweat it too much! No matter what the classification, bacteria were the first forms of life on Earth. Fossilized bacteria have been discovered in **stromatolites**, stratified rocks dating back 3.5 billion to 3.8 billion years, a long time ago in the history of the estimated 4.6 billion-year-old planet Earth. When life arose, the Earth's ancient atmosphere contained little or no free oxygen but consisted principally of carbon dioxide and nitrogen with smaller

TABLE 2.3 Comparisons of *Bacteria*, *Archaea*, and *Eucarya*[a]

Domain	Membrane-bound Nucleus	Cell Wall	Antibiotic Susceptibility	Characteristic
Bacteria	No	Present	Yes	Large number of bacterial species
Archaea	No	Present	No	"Extreme" bacteria growing in high-salt environment and at extreme temperatures
Eucarya	No	Variable	No (some exceptions in fungi)	Algae (most), fungi, protozoans, "higher" animals and plants

[a]Major differences are present in the biochemistry of cell walls, cell membranes, genetic material, and structures in the cytoplasm.

amounts of gases, including hydrogen (H$_2$), hydrogen sulfide (H$_2$S), and carbon monoxide (CO). This ancient atmosphere, devoid of O$_2$, would not have supported life as we know it. Only microbes that were able to meet their energy requirements with non-oxygen-requiring chemical reactions populated the primordial environment. The early microbes were photosynthetic and used water and carbon dioxide (CO$_2$) in photosynthetic reactions, resulting in the production of O$_2$ and **carbohydrates**. This process was responsible for the generation of O$_2$ in the Earth's atmosphere approximately two billion years ago.

Since their origin on Earth billions of years ago, bacteria have exhibited remarkable diversity and have filled every known ecological niche. Yet, according to some estimates, fewer than 2% of microbes have been identified and even fewer have been cultured and studied. Perhaps you can recall the old Mother Goose nursing rhyme "Peas Porridge Hot" which states "some like it hot and some like it cold." (Ask your instructor to sing it while you fall asleep in class!) It's like the diversity in bacteria. Bacteria belonging to the domain *Archaea* continue to be found in environments once considered too extreme or too harsh for life at any level. In most cases these organisms, **extremophiles**, cannot be grown by existing culture techniques; evidence of their presence has been obtained by molecular biology techniques that allow scientists to examine minute amounts of their deposited ribonucleic acid (RNA). Some like it hot and are called **hyperthermophiles** ("heat lovers"). Some hyperthermophiles have been identified in the hot springs in Yellowstone Park where the temperatures exceed 70°C. Some microbes do best at temperatures even higher, above 100°C. *Thermus aquaticus* was isolated from hot springs in the 1960s. It produced an enzyme (*Taq polymerase*), which is essential to a very important technique (polymerase chain reaction; PCR) in molecular biology allowing for rapid DNA synthesis and sequencing. *Pyrococcus furiosus* lives in boiling water bubbling from undersea hot vents and freezes to death in temperatures below 70°C (FIGURE 2.8a). Some extremophiles, the **psychrophiles**, like it cold. Psychrophiles have growth temperatures lower than −20°C and are happy in Arctic and Antarctic environments (Figure 2.8b). Finally, if you like tongue-twisters, try your tongue around *Psychromonas ingrahamii* and *Colwellia psychrerythraea*, both at home and living comfortably in frigid waters. Some bacteria are extreme **halophiles** ("salt lovers") (FIGURE 2.9), and some produce methane gas in their metabolism. These bizarre examples indicate that many of the archaea live at the extremes of life zones (BOX 2.2). Archaea have not been implicated as disease producers and are not further considered in this text.

In 2010, NASA scientists reported the discovery of bacteria that could substitute arsenic for phosphorous, a necessary component of DNA. If correct, the finding

FIGURE 2.8 **(a)** *Pyrocoecus furiosus,* a highly heat-resistant bacterium. © Eye of Science/Photo Researchers, Inc. **(b)** Psychrophilic *Methanococcoides burtonii* discovered in 1992 in Ace Lake, Antartica, can survive in temperatures as low as −2.5°C. © Dr. M. Rohde, GBF/Photo Researchers, Inc.

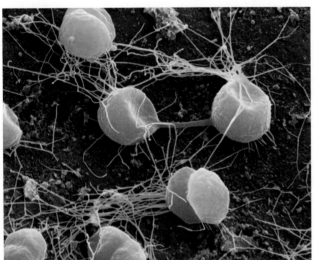

(a)

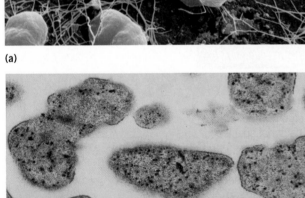

(b)

FIGURE 2.9 The Dead Sea. This sea has a salt concentration well above that found in the Great Salt Lake in Utah; it lies farther below sea level than any other terrestrial spot on Earth. You can lie on your back and float without any effort. Amazingly, this extreme environment is home for a variety of halophilic bacteria. Author's photo (RIK).

BOX 2.2 — Some Bizarre Bacteria

Some microbes exhibit an unusual lifestyle and remarkable characteristics that provide fascinating stories and illustrate the tremendous diversity of the microbial world.

Consider *Deinococcus radiodurans,* a bacterium further described in Box 2.3, that can survive a dose of radiation greater than 3,000 times the dose that can kill a human. The Dead Sea, characterized by its extreme salinity, is erroneously named; it is not dead at all but teems with salt-loving (halophilic) bacteria. You will be surprised to learn that microbes can grow in your car's battery acid or that some bacteria thrive on arsenic. How about magnetotactic bacteria? They manufacture minute, iron-containing magnetic particles used as compasses by which the organisms align themselves to the Earth's geomagnetic field. These curious microbes prefer life in the deeper parts of their aquatic environment where there is less oxygen. Their magnetic compass points the way.

And then there are the as yet unnamed bacteria living in symbiotic partnership with giant tube worms, as long as 2 meters, living in the hydrothermal vents of the ocean floor. As these worms mature, their entire digestive tract disappears, including their mouth and anal openings. Now that presents a problem, and it's bacteria to the rescue! The tissues of the worm are loaded with bacteria that obtain energy from the surrounding chemical environment sufficient for their own needs and for those of their worm hosts. In turn, the worms provide a safe harbor for the bacteria, ensure an adequate environment for energy production, and provide nitrogen-rich waste materials, allowing for synthesis of microbial cellular components—a great mutualistic arrangement.

Here is a strange story about *Serratia marcescens,* a bacterium whose colonies form a deep red pigment when grown in moist environments. In 1263 in the Italian town of Bolsena a priest was celebrating Mass. When he broke the communion wafers he found what he thought was blood on them and assumed it to be the blood of Christ. Given the lack of scientific knowledge during the Dark Ages, it is understandable the event was regarded as a miracle. It was not a miracle at all. The red-pigment-producing *S. marcescens* had contaminated the wafers during their storage in the dampness of the ancient church (**FIGURE B2.2**). Nevertheless, Raphael's painting *The Miracle of Bolsena,* depicting this event, hangs on a wall in the Vatican. Here's another story of an unusual bacterium, *Shewanella,* reported in June 2011. These microbes use metal ions in place of oxygen in their metabolism and in so doing, can minimize some toxic metals from migrating into soil and groundwaters—a nice example of bacteria used for clean-up.

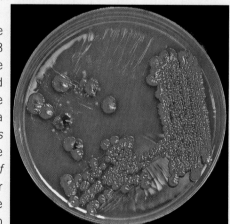

FIGURE B2.2 A culture of *Serratia marcescens.* Courtesy of Jeffrey Pommerville.

BOX 2.3   Conan the Bacterium

Conan the Barbarian, a 1982 movie starring Arnold Schwarzenegger, was the first Conan movie. In this fantasy story, from the mythical age of sword and sorcery, Arnie portrays Conan as only Arnie can do!

Deinococcus radiodurans has been nicknamed "Conan the bacterium"; it is one of nature's "toughest cookies." It can survive the rigors of being completely dried out, have its chromosomes disrupted, and be exposed to 1.5 million rads of radiation, a dose 3,000 times greater than that which would kill a human. Further, it can transform toxic mercury into a less toxic form, a feature especially useful at nuclear waste sites. According to Owen White of the Institute for Genomic Research in Rockville, Maryland, "The Department of Energy is very jazzed about *D. radiodurans,* because the

agency has a pretty big toxic cleanup problem at its waste development sites." Genes from bacteria that can digest toxic waste but cannot survive radiation have been genetically engineered into *D. radiodurans,* resulting in bacteria that can transform toxic mercury into a nontoxic form and unstable uranium into a stable form. These genetically engineered bacteria are powerful tools in cleaning up the three thousand waste sites containing millions of cubic yards of contaminated soil and contaminated groundwater estimated to be in the trillions of gallons. The ability of *D. radiodurans* to repair its own DNA is of interest to biologists because the process provides an insight into the mechanisms of aging and into the biology of cancer.

would open up a new form of life on earth. Other scientists were skeptical about the study, so the role of arsenic-loving bacteria remains open. Here is another strange story. *Deinococcus radiodurans* is a bacterium that can withstand 3,000 times more gamma radiation than that which would kill a human because of its unique ability to repair its damaged DNA. The scientists who study *D. radiodurans* have dubbed it "Conan the bacterium" (BOX 2.3).

You may not like to hear this, but the human mouth is considered one of the most diverse ecosystems and rivals the biological diversity of tropical rainforests. Within the past few years, scientists at Stanford University have discovered 37 new organisms in the mouth, pushing the total to more than 500. These new microbes were found in the scum (plaque) in the deep gum pockets between teeth. (Your dentist would love this tidbit!) Their presence remained unknown simply because traditional culture methods do not allow their growth. Enterprising microbiologists (sometimes known as plaque pickers) extracted DNA from plaque and mapped out DNA sequences, revealing bacteria that had not been previously known to inhabit the mouth. In fact, some new bacterial species were identified, supporting the statement that less than 2% of the microbial population has been identified.

A comprehensive global microbial survey to identify microbes that make up the biosphere is underway thanks to the cooperative effort of the National Science Foundation and the American Society for Microbiology. These two organizations are establishing a network of biodiversity research sites or "microbial observatories." Other international efforts are in the works to develop a worldwide microbial inventory of genetic sequences.

The origin of life on Earth continues to be a fascinating and mind-boggling question to which the explanation is purely speculative. The general consensus among scientists is that the "primordial soup" hypothesis is the most likely explanation. In this hypothesis, organic compounds formed from a specific combination of atmospheric gases collected in water and sparked by an energy source. How exactly this happened is still debated.

Another intriguing possibility is the hypothesis that Earth was seeded by life forms from Mars, the Red Planet. Photographs taken from the orbiting Mars *Global Surveyor* spacecraft indicated the possibility of water just below the surface of the planet. If, in fact, Mars has water, it is possible that the planet entertains, or entertained, life. According to the Laboratory for Atmospheric and Space Physics at the University of Colorado at Boulder, "Mars meets all the requirements for life." The possibility that life originated on Mars and was subsequently carried to Earth is plausible. Meteors and meteorites are constantly bombarding the Earth and some, originating from Mars' surface, could have transported ancestral procaryotic cells. Bacteria have been cultured out of Siberian and Antarctic permafrosts that have been in the deep freeze for millions of years. The National Aeronautic and Space Administration is now planning the Mars Sample Return Mission, which will bring Martian rocks back to Earth, and this will help to resolve the question of the beginnings of life. A famous and historic press conference was held at the National Aeronautic and Space Administration in Washington, D.C. on August 7, 1996, announcing that scientists had found evidence of ancient microbial life in a Mars meteorite known as ALH84001. Bear in mind, however, that the evidence was viewed by some authorities as weak and remains highly refuted.

■ Introducing the Microbes

Although there is no clear definition of microbes, it is time to introduce those biological agents that fall under the microbial umbrella (FIGURE 2.10 and TABLE 2.4). Algae are not discussed in detail in this text beyond this section, although they are highly significant in terms of food chains and other beneficial aspects. Further, some fungi are human pathogens, and many contribute to the death toll of patients with AIDS. With the exception of viruses and prions, all microbes have both DNA and RNA, as do all cells.

Microbes are measured in very small units of the metric system called **micrometers** (equal to one millionth of a meter), abbreviated as μm, and **nanometers** (equal to 1 billionth of a meter), abbreviated as nm. A meter is equivalent to about 39 inches, so a micrometer is equal to one millionth of 39 inches. These numbers are probably not very meaningful to you, because they do not allow you to appreciate the size of microbes relative to more familiar

FIGURE 2.10 The microbial umbrella.

TABLE 2.4 Comparison of Microbial Groups[a]

Characteristic	Archaea	Bacteria	Protozoans	Fungi	Unicellular Algae
Cell type	Procaryotic	Procaryotic	Eucaryotic	Eucaryotic	Eucaryotic
Size	Microscopic	Microscopic[b]	Microscopic	Macroscopic	Microscopic
Cell wall	Present	Present	Absent	Present	Present
Reproduction	Mostly asexual (binary fission)	Mostly asexual (binary fission)	Sexual and asexual	Sexual and asexual	Asexual
Energy process	Variable	Mostly heterotrophic	Heterotrophic	Heterotrophic	Autotrophic

[a]Viruses and prions are not cells and, therefore, are not included.
[b]There are a few exceptions.

objects, but there are some points that may help you think in this scale. A spoonful of fertile soil contains trillions of microbes, and the number of microbes that can be accommodated on the period at the end of this sentence is in the millions. FIGURE 2.11 indicates the relative size of microbes. The monstrous bacteria described earlier are exceptions. Bacteria are many times smaller than eucaryotic cells but are about fifty times (or more) larger than viruses.

FIGURE 2.11 Comparison of sizes of different kinds of microorganisms (not drawn to scale).

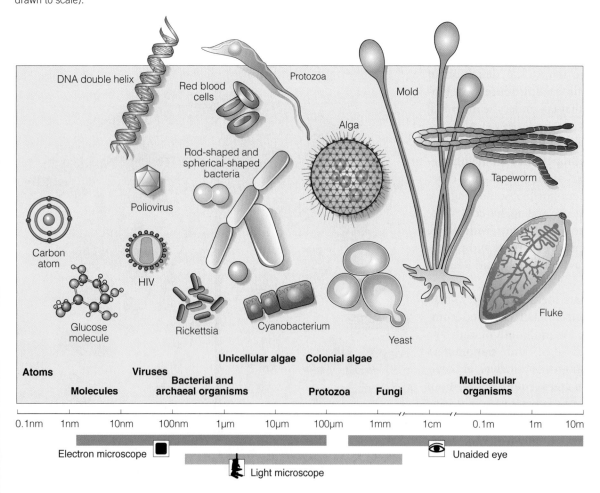

Smallness has its advantages. Smallness provides a large surface area per unit volume, allowing for rapid uptake of nutrients from the environment. *E. coli,* for example, has a surface-to-volume ratio about twenty times greater than that of human cells. And now for the introduction—from least to the most complicated.

Prions

Prions are the most recent addition to the microbial list; some texts continue to place them with viruses for lack of a better place. But the awarding of the 1997 Nobel Prize to Stanley Prusiner, who discovered these agents, legitimized them as separate entities. The word *prion* is an abbreviation for pro-teinaceous infectious particles. Prions are protein mole-cules and are devoid of both DNA and RNA; their lack of nucleic acid is their major (and most puzzling) biological property. Prions exist normally, primarily in the brain, as harmless proteins. Abnormal prions convert normal pro-teins into infectious, disease-producing proteins responsi-ble for mad cow disease and dementia type diseases in humans and in other animals. Questions remain unan-swered regarding their biology.

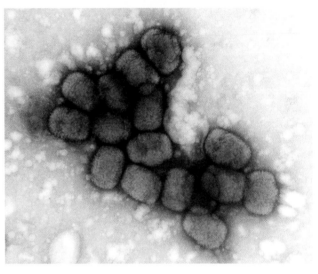

FIGURE 2.12 A transmission electron micrograph of smallpox (Variola) viruses. Courtesy of Dr. Fred Murphy/CDC.

Viruses

As frequently noted, viruses are not organisms; Figure 2.4 indicates their subcellular position. Two major distinguish-ing characteristics of viruses are that, in contrast to cells, they contain either RNA or DNA (never both) and, further, they are submicroscopic particles and can be seen only with an electron microscope (**FIGURE 2.12**). Some have an addi-tional coat or envelope encompassing them. Viruses are de-scribed as **obligate intracellular parasites**, meaning they must be (obligate) inside living cells (intracellular) to repli-cate; they are not capable of autonomous replication. They take over the metabolic machinery and reap the benefits of energy production, without any expenditure of energy, by the host cell. Perhaps this is the ultimate in parasitism.

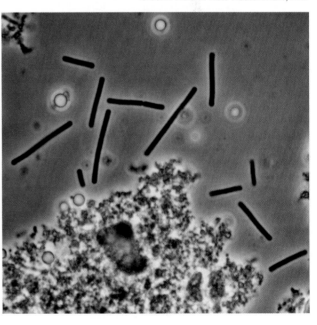

FIGURE 2.13 *Lactobacillus* bacteria. © John Walsh/Photo Researchers, Inc.

Bacteria

Bacteria are the best known of the microbial. They are mi-croscopic, unicellular, procaryotic, and have cell walls (with the exception of a single subgroup, the *Mycoplasma*). They reproduce asexually by binary fission. In terms of size, they can be seen with a regular (light) microscope (**FIGURE 2.13**). Many bacteria are heterotrophs and use organic com-pounds as a source of energy. Others are autotrophs and use the energy of the sun, whereas some derive energy from the use of inorganic substances. Although a number of

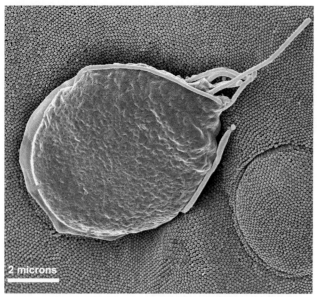

FIGURE 2.14 A colorized scanning electron micrograph of a flagellated *Giardia* protozoan adhering to an intestinal epithelial cell. Courtesy of Dr. Stan Erlandsen/CDC.

FIGURE 2.15 Freshwater diatoms. Courtesy of Robert W. Pillsbury.

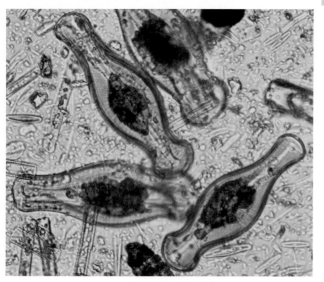

bacteria are pathogens and are the major subject of this text, the vast majority of bacteria are nonpathogenic and play essential roles in the environment without which life would not be possible.

Protozoans

Protozoans are unicellular and eucaryotic and are classified according to their means of locomotion (FIGURE 2.14). Their energy generation requires the utilization of organic compounds. Many diseases, including malaria, sleeping sickness, and amebic dysentery, are caused by protozoans.

Algae

Algae are photosynthetic eucaryotes and in the photosynthetic process produce oxygen and carbohydrates used by forms requiring organic compounds. Hence, they are highly significant in the balance of nature. **Dinoflagellates** and **diatoms** are examples of unicellular algae and fall under the umbrella of microbes (FIGURE 2.15). Dinoflagellates (plankton) are the primary source of food in the oceans of the world. Some algae are pathogenic for humans indirectly. For example, the toxin produced by the dinoflagellates that causes red tide, *Gymnodinium breve,* causes neurological disturbances and death in humans as a result of our consumption of fish and shellfish that had fed on the dinoflagellates. Another species of dinoflagellate, *Pfiesteria piscicida,* also referred to as the "cell from hell," threatened the fishing industry in the eastern United States in 1997. A bloom of these algae resulted in the release of large amounts of neurotoxin, causing neurological symptoms in fishermen and fear among consumers.

Fungi

Fungi are eucaryotes. Morphologically, they can be divided into two groups, the yeasts and the molds. The yeasts are unicellular and are larger than bacteria; many reproduce by budding. Molds are the most typical fungi and are multicellular, consisting of long, branched, and intertwined filaments called **hyphae**. In early schemes of classification fungi were considered plants, primarily because they have cell walls. However, their cell wall composition is quite different from that of plants and from the cell walls of bacteria. Fungi are highly significant in terms of food chains and have certain beneficial aspects. Some are pathogenic and cause diseases that are difficult to treat; others play a highly significant role as **opportunistic pathogens** (organisms that are not usually considered to be pathogens), because, as in AIDS, when the immune system is depressed they cause disease. Molds played a major role in the aftermath of Hurricanes Katrina and Rita, rendering houses uninhabitable. Patches of mold threaten the prehistoric paintings of animals in the Lascaux Cave in Dordogne region of southwest France and museum curators constantly need to guard against mold intrusions.

Mushrooms are a well-known group of fungi. Their diversity is unusual as demonstrated by a wide range of size, colors, and patterns on their cap (FIGURE 2.16). Many species are edible. The terms *bugs* and *germs* are part of our popular speech but have no scientific meaning. It should be clear from the above descriptions that each group of microbes is distinct from the others. When your physician diagnoses you as having a "bug," you might ask what kind.

FIGURE 2.16   Two mushrooms growing on a tree stump in a field. Author's photo (TS).

Overview

The microbial world is remarkable for its extreme diversity, as is evident in the distinct characteristics of the six microbial groups—prions, viruses, bacteria, protozoans, unicellular algae, and fungi. Further, within each group there is considerable diversity. Not all microbes are unicellular and microscopic; some are multicellular and macroscopic and others are subcellular and microscopic. In recent times "monster" bacteria have been found that are unique in being unicellular and macroscopic, a rare combination. Viruses and prions are subcellular and are not considered life forms. There is no clear definition of what makes a microbe a microbe, but it is clear that they are all at less than the tissue level of biological organization.

All bacteria are procaryotic, and all other microbes are eucaryotic. (Viruses and prions are not cells and are neither procaryotic nor eucaryotic.) Taxonomy evolved from a two-kingdom system (in which bacteria were considered plants) to a five-kingdom system, with various other schemes along the way; the trend has been toward recognizing the uniqueness of microbes. Woese proposes a classification system based on rRNA analysis and assigns bacteria to one of three domains and reflects their evolutionary history.

Since their origin on Earth microbes have adapted to extreme ecological diversity and can be isolated from all environments. Some live at the extremes— from hot springs to permafrost. All organisms must meet a basic requirement for energy, and microbial evolution has fostered a diversity of strategies. Some microbes obtain energy from organic compounds, whereas others use the energy of the sun or derive their energy from the metabolism of inorganic compounds.

The major characteristics of each of the six microbial groups show that each category is distinctive. The popular terms *bugs* and *germs* are used in a collective sense, but there is no basis for lumping these diverse microbial agents together. Further, these terms have a negative connotation, because they are usually used to describe microbial diseases, but it is important to remember that only a handful of microbes are disease producers.

PART I: Choose the *single* best answer.

1. A major distinction between procaryotic and eucaryotic cells is based on the presence of

   a. a cell wall **b.** DNA **c.** a nuclear membrane **d.** a cell membrane

2. Most bacteria are considered to be

   a. harmful **b.** anaerobes **c.** autotrophs **d.** heterotrophs

3. The smallest of these units of measurement is

   a. millimeter **b.** nanometer **c.** micrometer **d.** centimeter

4. Which of the following are obligate intracellular parasites?

   a. bacteria **b.** viruses **c.** unicellular algae **d.** diatoms

5. Which one of the following does not have nucleic acid in its structure?

   a. viruses **b.** diatoms **c.** bread mold **d.** prions

6. The five-kingdom system of taxonomy is credited to

   a. Haeckel **b.** Woese **c.** Whittaker **d.** Darwin

7. According to Woese,

   a. *Eucarya* arose from *Archaea*.

   b. *Archaea* arose from *Eucarya*.

   c. *Bacteria, Archaea,* and *Eucarya* all arose independently.

   d. None of the above is correct.

8. This bacterium produces an enzyme that catalyzes DNA synthesis.

   a. *E. coli* **b.** *Thermus aquaticus* **c.** *Serratia marcescens* **d.** *Dienococcus radiodurans*

9. Which of the following microbes are not disease producers?

   a. fungi **b.** bacteria **c.** archaea **d.** protozoa

10. This bacterium produces a red pigment and contaminated communion wafers during a mass in 1263.

    a. *Staphylococcus aureus* **b.** *Serratia marcescens* **c.** *Yersinia pestis*

    d. *Bacillus cereus*

PART II: Fill in the blank.

1. Bacteria, viruses, fungi, and protozoans are microbes. Name another group that falls under the microbial umbrella. _____

2. The cell theory is credited to _____.

3. Compounds of carbon are called _____ compounds.

4. Organisms that do not require organic compounds are called

   _____.

5. The "energy compound" is called _____.

6. Strict anaerobes are killed by _____.

7. The term _____ is used to describe organisms too small to be seen without a microscope.

8. _____ are the ancestors of eucaryotes.

9. Most bacteria reproduce by _____.

10. _____ are infectious protein molecules.

PART III: Answer the following.

1. Criticize the terms *bugs* and *germs* as used in a collective sense to describe microbes. List the categories of microbes, and write a one-sentence description of each.

2. What makes a microbe a microbe?

3. What is the relevance to microbiology of Shakespeare's "What's in a name? That which we call a rose by any other name would smell as sweet."?

4. Heterotrophs are dependent on autotrophs. Why is this the case?

5. The archaea can survive extreme environments. Why?

6. Fungi play a highly significant role as opportunistic pathogens. Define opportunistic pathogens and who is most susceptible to these types of infections.

7. Explain why viruses and prions are not cells and are neither prokaryotic nor eukaryotic.

8. Explain why some microbes can survive without oxygen present.

9. Compare and contrast eukaryotic and prokaryotic cells.

10. Why do some bacteria form symbiotic partnerships with other life forms?

3

Beneficial Aspects of Microbes: The Other Side of the Coin

Topics in This Chapter

There is no field of human endeavor, whether it be in industry or in agriculture, whether it be in the preparation of foodstuff or in connection with problems of shelter and clothing, whether it be in the conservation of human and animal health and the combating of disease, where the microbe does not play an important and often a dominant part.

—Selman A. Waksman, 1943

■ Preview

It would be understandable if you are biased against microbes; you may have more reason to hate and fear them than to love them. In this chapter, however, the take-home message is that only a few microbes are disease producers and many are beneficial in our daily lives. The goal of microbiologists is not to annihilate all microbes but to eradicate pathogens or at least to minimize their impact and the burden of microbial disease by avoiding circumstances leading to a collision course. After all, microbes were the first inhabitants of our planet; they are the

Photo © foto.fritz/ShutterStock, Inc.

senior citizens from which evolution to eucaryotic cells and multicellularity proceeded. Microbes are the largest component of Earth's biomass and are present in the most extreme habitats of life, as described in this text.

Microbes make the planet's ecosystems go around. They are the foundation of the **biosphere**, and many act as **decomposers** or **scavengers**. Bacteria are the underpinnings of the **biogeochemical cycles**. Their role in these cycles is unseen and taken for granted, but without microbes the cycles couldn't be completed and life, ultimately, would cease. The cycles occur without our initiative or our intervention. In fact, there is a danger that our increasing technology could inadvertently interfere with and shut down certain cycles of nature as a result of nonbiodegradable products and pollution of the environment.

Since antiquity, societies the world over learned to harness microbes for their beneficial aspects long before there was any awareness of a microbial world. Societies were content with the empirical evidence that certain practices simply "worked." The production of distilled spirits (alcoholic beverages) and a variety of food products, including breads, yogurt, and cheeses, are examples. Yogurt, which contains live bacterial cultures, was prescribed centuries ago for "stomach ailments" and, in some cases, seemed to do the trick. As knowledge of microbes and the enzymes they produce in their metabolism increases, the manufacture of alcoholic beverages and foodstuffs becomes increasingly sophisticated, resulting in an increasing array of fermented food products.

Microbes are powerful biological research tools because of the relative ease of culturing and obtaining them in large populations in a short period of time. Evidence establishing DNA as the genetic material is the result of experiments using bacterial viruses (bacteriophage) and bacterial cells. Industry, particularly the pharmaceutical industry, has learned to harness microbes for the production of many products, including antibiotics, vaccines, genetically engineered therapeutics, pesticides, and a large variety of other compounds to control or eliminate microbes.

Bioremediation, the use of microorganisms to clean up polluted environments, is on the increase; microbes played a role in reducing the impact of the *Exxon Valdez* oil spill that occurred off the coast of Alaska on March 24, 1989; they were also employed as remedial agents in the Gulf Coast oil spill in April 2010. After reading this chapter, perhaps you will rethink your love–hate relationship with the microbial world.

Microbes in the Environment

Microbes as Decomposers

Perhaps at some point in your life you had an aquarium with goldfish or tropical fish. If so, you recall the necessity of properly maintaining the aquarium. You may not have realized it at the time, but the aquarium was a simulated **ecosystem**—a population of organisms in a particular physical and chemical environment (FIGURE 3.1). The fish and the plants are the added **biotic** components, whereas the chemical and physical environment constitutes the **abiotic** component. You will recall paying attention to the light source, temperature, acidity, and cleanliness of the water to maintain a healthy and balanced ecosystem. You can assume the

(a)

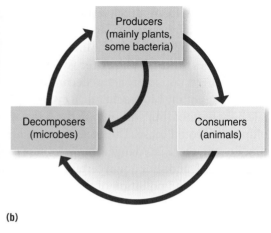
(b)

FIGURE 3.1 **(a)** A fish tank is an artificial ecosystem. © Matt Jones/ ShutterStock, Inc. **(b)** The cycle of life in an ecosystem: producers, consumers, and decomposers.

presence of microbes as additional biotic components. In the fish tank, the green plants are considered the **primary producers** because of their photosynthetic capabilities that result in the production of organic compounds and oxygen; the fish are the **consumers** and take oxygen from the water and exhale carbon dioxide and use organic compounds as their nutrient source. The bacteria and fungi are the decomposers and are the link between the producers and the consumers. The microbial population decomposes waste materials of the fish and dead leaves of the plant and, in so doing, functions as recyclers. And so it is in nature—witness plant debris, animal wastes, and the bodies of dead animals.

The greatest recyclers of all time are microbes, without which life would be a dead end and, ultimately, would cease. Imagine a huge garbage dump into which materials are deposited daily and continue to accumulate year after year and generation after generation. Under these circumstances, and without recycling, the resources of the planet would soon run out. If you walk through swampy areas, you may detect the unmistakable odor of methane—marsh gas—resulting from the bacterial action on decomposing materials. The microbes involved as scavengers are nonpathogenic and free-living. The resources on Earth are limited and are recycled through food webs with microbes as the decomposers (FIGURE 3.2). Nature has always practiced this, but it has only been in the past thirty or forty years that society has realized the inextricable link between populations, soil, water, air, and energy, all of which are interdependent and dependent on microbes.

FIGURE 3.2 Microbes are the ultimate decomposers. © Michael J. Thompson/ ShutterStock, Inc.

Microbes and the Biogeochemical Cycles

As previously stated, bacteria are the basis for the biogeochemical cycles; the processes involved in the recycling of carbon, nitrogen, sulfur, iron, and phosphorus, resulting in the return of these elements to nature for reuse. These cycles are discussed separately but are linked. The carbon and nitrogen cycles are presented as examples.

Carbon Cycle

Carbon atoms are key elements in living systems and are found in proteins, carbohydrates, fats, and DNA. Most of the carbon used by organisms is present in association with carbon dioxide, a simple compound consisting of one carbon atom attached to two oxygen atoms. Photosynthetic organisms capture the sun's energy and use it for the conversion of atmospheric carbon dioxide and, along with water, produce glucose and other energy-rich carbohydrates. Hydrogen and water are necessary reactants in photosynthesis. An important spin-off of photosynthesis is the release of oxygen from the carbon dioxide back into the atmosphere.

Plants are associated with photosynthesis, but some microbes are also photosynthetic and are the primary producers in the ocean. **Chlorella** is a photosynthetic alga that is found on the surface of ocean water, and **cyanobacteria** are photosynthetic bacteria.

Carbon, captured as carbon dioxide, is ultimately recycled back to its elemental form through food chains. Cellulose, a polymer (chain) of glucose (sugar) molecules, is an energy-rich carbohydrate product of photosynthesis. Bacteria produce enzymes that are able to break down cellulose into single molecules of glucose. **Herbivores** (grazers) feed on plants but lack the necessary digestive enzymes to break down the cellulose. Bacteria come to the rescue! They are a part of the normal flora residing in the intestinal tract of grazers, and their enzymes digest cellulose, allowing these animals to use cellulose as an energy source. Bacteria in the intestinal tract of termites, as another example, allow termites to lunch on your house. Ultimately, the grazers are fed on by predators, including humans, but one way or another they and their waste material enter the food web. From there, microbial decomposition takes over, and the cycle is completed (FIGURE 3.3).

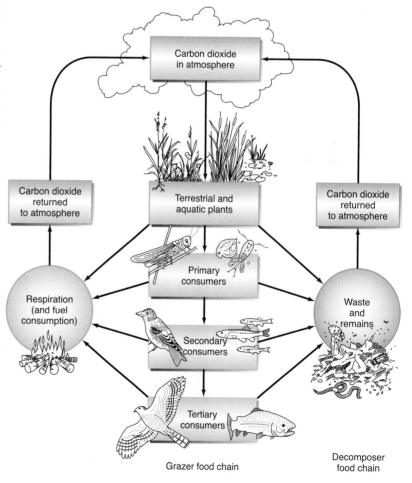

FIGURE 3.3 The carbon cycle. Microbes are essential in the conversion of atmospheric carbon dioxide to organic compounds and back to the atmosphere.

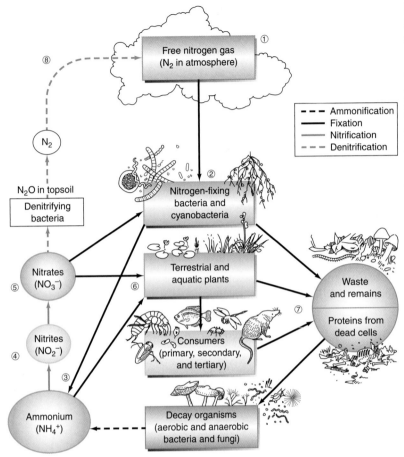

FIGURE 3.4 The nitrogen cycle. Microbes are essential in the conversion of atmospheric nitrogen to organic compounds and back to the atmosphere.

① Free nitrogen gas (N₂ in atmosphere)

- - - Ammonification
—— Fixation
——— Nitrification
- - - Denitrification

② Nitrogen-fixing bacteria and cyanobacteria

Terrestrial and aquatic plants

Consumers (primary, secondary, and tertiary)

Decay organisms (aerobic and anaerobic bacteria and fungi)

Ammonium (NH₄⁺) ③

Nitrites (NO₂⁻) ④

Nitrates (NO₃⁻) ⑤ ⑥

⑦ Waste and remains / Proteins from dead cells

N₂O in topsoil — Denitrifying bacteria

N₂ ⑧

Nitrogen Cycle

Nitrogen is a constituent of amino acids, the building blocks of proteins and of the nucleic acids of microbes, plants, and animals. It is the most common gas in the atmosphere (about 80%), but atmospheric nitrogen cannot be tapped by animals or by most plants. Here again microbes come to the rescue; only bacteria can convert, or fix, nitrogen into a usable form and, ultimately, recycle it back to the atmosphere. The **nitrogen cycle** is illustrated in FIGURE 3.4.

The process begins with the fixation of atmospheric nitrogen and its conversion to ammonia by **leguminous** plants. These are plants that have swellings or nodules along their root systems containing *Rhizobium* and other nitrogen-fixing bacteria (FIGURE 3.5). Peas, soybeans, alfalfa sprouts, peanuts, and beans are examples of leguminous plants. The association of nitrogen-fixing bacteria and leguminous plants is an example of **symbiosis**.

The next phase of the nitrogen cycle is called **nitrification**; in this process ammonia is converted into nitrates, the nitrogen form most used by plants. Members of the bacterial genera *Nitrobacter* and *Nitrosomonas* carry out these processes. Other bacteria, as well as fungi, decompose plants and animals and their waste products and in the process convert nitrogen into ammonium, giving meaning to the expression "death yields life."

AUTHOR'S NOTE (RIK)
You might be interested in seeing these resident bacteria. All you need to do is dig up peas, a patch of clover, or some other leguminous plant, being sure to take some of the root system. Wash away the soil and crush a nodule onto a clean slide. Add a drop of water and a dye, such as methylene blue, and spread the preparation with a toothpick or a matchstick onto the slide to establish a thin film. Allow the preparation to dry and examine it under a microscope. You will observe bacilli, probably members of the genus *Rhizobium*. If you perform a Gram stain, the bacterial population will be dominated by gram-negative (pink) bacilli.

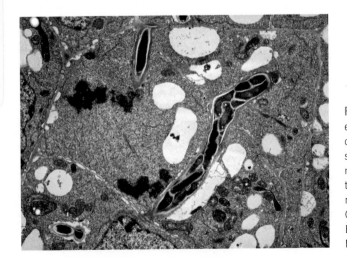

FIGURE 3.5 Transmission electron micrograph of a cross section through a soybean (*Glycine max*) root module. The bacteria infects the roots and establishes a nitrogen-fixing symbiosis. Courtesy of Louisa Howard, Dartmouth College, Electron Microscope Facility.

Urine is a waste product particularly rich in nitrogen. Finally, **denitrifying** bacteria are responsible for the return of nitrogen to the atmosphere as nitrogen gas. Horticulturists and agriculturists have long realized the importance of nitrogen in growing flowers and food crops and use a variety of fertilizers containing nitrogenous compounds.

Other Cycles

In addition to the carbon and nitrogen cycles, the movement of other elements, including sulfur, phosphorus, and iron, through ecosystems in a cyclical manner depends on microbial communities.

It is really microbes that "make the world go around." Their role in these cycles, which take place in all imaginable ecosystems and at all extremes of temperature, demonstrates the necessity of microorganisms for sustaining life on Earth.

Microbes in Food Production

Foods

Mushrooms have long been recognized as tasteful and nutritionally rich foods whether raised in mushroom farms or picked "in the wild." But let the mushroom pickers beware: some species of mushrooms are extremely toxic to the point of death. There are a number of dangerous myths about the distinction between edible and toxic (nonedible) mushroom species including:

- Poisonous mushrooms cause a silver spoon to turn black.
- Poisonous mushrooms have a pointed cap.
- Poisonous mushrooms taste bad.

The medical benefits of mushrooms are numerous. Studies have shown them to be beneficial as antimicrobial agents, anticholesterol agents, cognitive stimulants, a vitamin source, and in lowering blood sugar levels. Interestingly, before synthetic dyes, mushrooms were the source of numerous textile dyes.

Algae

These microbes are food sources in some societies. China consumes more than 70 species of algae including fat choy, a vegetable that when dried resembles long, black human hair and has a soft texture. (Sounds good! Order it next time!)

Food Production

The next time you shop at the market, look around at the shelves of foods for those that depend on microorganisms for their production or products that contain live bacterial cultures. Examples can be found in just about all categories of foodstuffs (TABLE 3.1). FIGURE 3.6 presents items dependent upon microbes for their production. Use your imagination and come up with your own microbial banquet. Now it is time to delight in the fanciful images that come to mind as you think about the wonderful foods and beverages whose stimulating tastes and aromas are the result of microbial activities.

AUTHOR'S NOTE (RIK)

Look at a variety of fertilizers, and you will see three numbers, for example, 22-3-12. The first number pertains to the nitrogen content, the second pertains to the phosphorus content, and the third pertains to the potassium content. Some farmers might not add fertilizer to their soil but instead may include leguminous crops, which are plowed under during the off-season as a way of enriching the nitrogen content of the soil.

TABLE 3.1 Foods Produced by Using Microbes

Milk Products	Meats	Breads	Miscellaneous Products	Alcoholic Beverages
Cheese	Bologna	Sourdough bread	Sauerkraut	Beer
Yogurt	Salami	Numerous other breads and rolls	Pickles	Wine Saki
Buttermilk	Country-cured ham		Olives	Distilled spirits (e.g., brandy, whiskey, rum, vodka, gin)
Kefir	Sausage		Vinegar	
Acidophilus milk			Tofu	
Sour cream			Soy sauce	
			Kimchi	

As stated in the introduction of this chapter, people have used microbes for the production of fermented foods, knowingly or unknowingly, for thousands of years. **Fermentation** is a series of chemical reactions mediated by enzymes of a variety of strains of bacteria and yeasts that break down sugars to small molecules (most commonly lactic acid or ethanol and carbon dioxide). A characteristic of all organisms is the ability to meet their energy requirement to stay alive. Most procaryotic microbes (and most organisms) use energy-rich organic foodstuffs and oxygen and through a complex cyclical series of biochemical reactions "extract" the energy inherent in the bonds of that food and convert it into adenosine triphosphate (ATP)—a readily available form of energy. Fermentation is a metabolic path by which some microbes, primarily yeasts, are able to shift their metabolism in the absence of oxygen to produce small but sufficient amounts of ATP. Although it is true that fermentation, an anaerobic process, is a far less efficient manner for the conservation of energy, it is an evolutionary advantage in that it allows for survival under anoxic conditions.

FIGURE 3.6   Thanks to the microbes! Bon appetit! Author's photo (RIK).

The end products of fermentation, such as lactic acid, carbon dioxide, and ethanol, are of considerable value in commercial food and alcoholic beverage industries as further described in this chapter. In particular, lactic acid and ethanol may act to inhibit the growth of unwanted microbes, thus acting as food preservatives (extending the useful life of these foods). The choice of microorganism to carry out fermentation determines the taste and aroma of the product. The specific microbial strain referred to as the "starter" culture and the process used for many products are carefully guarded secrets; starter cultures are handed down in families from generation to generation. In the survey of the microbial world, fungi were briefly presented. Yeasts are unicellular fungi and are efficient fermenters of sugar into alcohol and

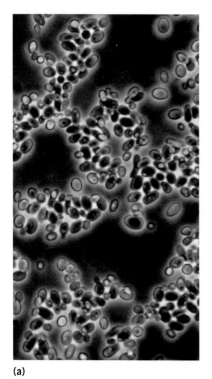

(a)

(b)

FIGURE 3.7   Yeast. **(a)** Yeast is used in baking to get the dough to "rise" as a result of carbon dioxide production. Courtesy of Peter Doe, Edinburgh, Scotland, www.flickr.com/photos/peteredin. **(b)** Baker's yeast (*S. cerevisiae*) is available in various forms at supermarkets. Courtesy of Fleischmann's Yeast. Used with permission.

carbon dioxide, a property exploited in the production of breads and alcoholic beverages. *Saccharomyces cerevisiae* is one commonly used yeast.

Bread Products

Fermentation by yeasts produces the gas carbon dioxide (and ethanol that evaporates during baking), which causes bread dough to rise and increase in size or leaven before baking. If you want to make your own bread, you can buy packages of live yeasts inexpensively at the supermarket (FIGURE 3.7). Bread has long been a staple in primitive societies. Bread samples dating back to 2100 B.C. are on display at the British Museum.

The importance of yeast as the leavening agent was already known. Unleavened bread is central to the 3,000-year old story of Passover, a Jewish celebration commemorating the time when the pharaoh of Egypt freed the Israelites from bondage. The people left in haste, without time to bake bread for their journey. They took the raw dough and baked it on rocks under the hot sun. With no yeast to raise the dough, it produced flat crackers called matzo, an unleavened bread (FIGURE 3.8).

FIGURE 3.8   Most bread uses yeast, which causes the bread to rise. Matzo uses no yeast, so the bread stays flat. © Roman Jigaev/ShutterStock, Inc.

Dairy Products

Cheese production, too, dates back thousands of years, as do fermented milk beverages that have been promulgated for centuries to treat a variety of intestinal

Yogurt is a centuries-old food from Eastern Europe and a particularly popular food in the United States. A look in the dairy section of a supermarket attests to the popularity of yogurt. The shelves are stacked with no-fat, low-fat, "fruit on the bottom," and flavored yogurts, including mocha latte, peach, and apricot. "Do-it-yourself" yogurt-making kits are readily available. Why the yogurt craze? The answer is simple: It is good for you. Many people attempting to lose weight consume a container of yogurt as a meal. Yogurt is an excellent source of calcium, some vitamins, and protein as indicated on the label. It can be low in fat, low in calories, or low in both fat and calories. The choice is yours!

What is in yogurt that, according to some, makes it a beneficial food? Most yogurts, and many other fermented milk products, contain live bacteria. Look at the label; it will state "live, active cultures," or words to that effect, depending on the brand. *Lactobacillus acidophilus* and other lactobacilli are the predominant live cultures; other bacteria are present as listed on the yogurt container. A gram of yogurt contains about one million lactobacilli; a 6-ounce container, intended as a single serving, weighs 170 grams and, therefore, contains about 170 million live bacteria. Generally speaking, about one billion live *L. acidophilus* cells are necessary for effectiveness.

Elie Metchnikoff, best known for his work in immunity, developed the theory that toxic bacteria in the intestinal tract were a component in the aging process. He proposed that lactic acid was a key factor in longevity and drank sour milk as a source of lactic acid on a daily basis. Ultimately this led to the production of a variety of fermented foods, including yogurt, and probiotics.

Foods supplemented with live microbes are called probiotics, defined by the U.S. Food and Drug Administration as "live microorganisms which, when administered in sufficient quantities, may improve health." They are primarily in dairy products, but probiotics are also available as tablets or capsules that can be purchased in pharmacies, health food stores, markets, and other retail outlets. We tend to think of the presence of microbes in the intestinal tract, other than the normal flora, as detrimental to our health. The proposed beneficial effects of probiotics are based on the assumption that consumption of live lactobacilli and certain other bacteria complements, or in some cases partially replaces, the normal microbial flora. Advocates of probiotics claim that benefits include reduction in blood pressure, regression of tumors, reduction in allergy, decreased duration of diarrhea, and decreased gas production. Individuals on antibiotic therapy sometimes suffer from yeast infections, most commonly manifested in the mouth and in the vagina. There is evidence suggesting that consumption of yogurt and other probiotics during antibiotic therapy may be of value in preventing and treating yeast infections. Many probiotic consumers are convinced that regular consumption improves their health.

There are skeptics, however, who maintain that the claimed benefits associated with probiotics are exaggerated and based on weak science, primarily a poor understanding of the intestinal flora. Some skeptics say "they [probiotics] go in at one end of the digestive tract and come out the other, and hopefully something good happens along the way." Take a look at the bags of dog food when you are in the market and note that many contain probiotics. Probiotic supplements are available for cats as well. Those who are not convinced of the value of yogurt and similar products call for research conducted in a scientific manner. Meanwhile, if you enjoy your yogurt, continue to eat it!

tract disorders from constipation to flatulence (gas) (BOX 3.1). A variety of fermented milk products exist, many with centuries-old origins in the Middle East (Table 3.1 and FIGURE 3.9). They vary in their texture, taste, and aroma, depending on the type of milk, incubation period, and, most significantly, the microbial culture used to carry out the fermentation process. Different species of lactobacillus and a few species of streptococci are commonly used. Cultured buttermilk is generally made by adding *Streptococcus cremoris* and *Leuconostoc citrovorum* to pasteurized milk. Other microbes produce buttermilk with different flavors. Certain species of lactobacilli or streptococci added to cream results in sour cream. *Lactobacillus bulgaricus* and *Streptococcus thermophilus* are commonly used to produce yogurt and yogurt drinks. Kefir is a cultured milk product made from the milk of cows, sheep, goats, or buffalo and kefir grains (gelatinous clumps of bacteria and yeasts). Kefir dates back many centuries to the shepherds of the Caucasus Mountains, who discovered that fresh milk carried in goatskin bags is sometimes fermented into an effervescent beverage. (The Caucasus Mountains are between the Black and Caspian Seas and range through Georgia, Armenia, Azerbaijan, and the southwest region of Russia.)

It is not mice but bubbles of carbon dioxide produced by the fermentative activity of bacteria that put the holes in Swiss cheese. Cheeses are classified as soft, semisoft, hard, and very hard (FIGURES 3.10 and 3.11). Over a thousand varieties of cheeses exist in countries around the world, and cheeses from The Netherlands (FIGURE 3.12), Switzerland, Italy, and France are particularly popular.

The texture, aroma, and taste of cheese depend primarily on the production process and the microorganisms used. Some varieties of cheese have a wonderful aroma, whereas others really stink! Cheese is made by adding lactic acid–producing bacteria and the enzyme rennin (or bacterial enzymes). The lactic acid sours the milk, and the enzymes coagulate casein, a protein in milk, to the solid curd portion and a watery portion known as whey. The terms "curd"

FIGURE 3.9 Fermented food products. Yogurt and a variety of other products are produced by microbes, which carry out fermentation. Most yogurts contain "live" cultures.
© Jones & Bartlett Learning.

FIGURE 3.10 A variety of cheeses. Their texture, aroma, and taste are the results of the strain of microbe and the fermentation process used.
© Agita/ShutterStock, Inc.

FIGURE 3.11 Cheeses of the world. An array of cheeses from many countries is displayed in a supermarket. Author's photo (RIK).

(a)

(b)

FIGURE 3.12 (**a** and **b**) A cheese factory in a village near Amsterdam, The Netherlands. Author's photos (RIK)

AUTHOR'S NOTE (RIK)

Believe it or not there is now a new standard allowing grade-A Swiss cheese to have smaller holes, or "eyes," to keep the cheese from getting tangled in high-speed slicing machines. The older standard required that the eyes had to be 11/16 to 13/16 inch in diameter, but new regulations reduce minimum eye size to 3/8 (6/16) inch.

and "whey" are used in the old nursery rhyme about Little Miss Muffett: "Eating her curds and whey." The curd is pressed to remove the whey. Cottage cheese and cream cheese are packaged and sold without further ripening. Other cheeses can be ripened without the addition of other microorganisms, whereas for some cheeses additional microbes are added during the ripening process. Spores of the mold *Penicillium roqueforti* are added during the production of blue cheese and Roquefort cheese. In producing Swiss cheese, bacteria known as propionibacteria are added for the desired taste. The length of time allowed for ripening and the microbes involved in the ripening process determine the consistency of the cheese.

Wine, Beer, and Other Alcoholic Beverages

The next time you drink a cold and refreshing beer, enjoy the fragrance and flavor of a fine wine on your taste buds, celebrate an important event (like getting a high grade in this course) by sipping on champagne, or lie on a beach drinking a frozen daiquiri, remember that none of these alcoholic beverages would have been possible without the fermentation of a variety of sugars and grains carried out by strains of *S. cerevisiae* and other yeasts. Wines and other alcoholic beverages have been imbibed as far back as 6000 B.C. and have been used in religious ceremonies dating back many centuries.

Enology is the science of wine making. Most wines are derived from the sugary juice extracted from grapes, but other fruits can be used; even dandelions are used to make wine. The extracted juice is usually treated with sulfur dioxide to kill naturally occurring yeasts that would produce uncontrolled and undesirable fermentation products. The yeast strain is added, and fermentation is allowed to proceed for a few days at a temperature between 20° and 25°C, followed by the aging process, which is carried out in wooden casks and takes weeks, months, or even years. During the aging process the flavor, aroma, and bouquet of the wine develop due to the production of a variety of compounds resulting from the metabolism of the yeast. A number of factors are involved in the quality of the wine, including characteristics of the grapes, the strain of yeast, the casks in which the wine is aged, and the duration of aging. Wine connoisseurs pride themselves on knowing a particular year for a fine vintage wine and are prepared to pay hundreds of dollars for this treasure (BOX 3.2).

BOX 3.2 — Wine Tasting

Some people are true connoisseurs of wines (or put on a good act). Next time you have occasion to dine in a fancy restaurant, watch the antics of patrons who appear to be sophisticated in choosing, tasting, and approving a wine once it is brought to the table. Their expressions, as they ceremoniously sniff the cork of the opened bottle and roll the first taste of the wine around in their mouths, are almost comical. Note, also, the manner of presentation of the bottle of wine to the diner; it is held in a way to prominently display the label on which is clearly stated the year. Although the actions of the tasters may seem to the uninitiated somewhat snobbish and frivolous, the rating of wines is a very serious business. A fine bottle of wine in a restaurant could cost several hundred dollars. Wine-tasting clubs for the amateur and for the professional are popular.

When tasting wine, connoisseurs consider sweetness; wines are characterized as "sweet" when their taste is dominated by sugars and "dry" when other flavors mask the sugar. Acidity is another attribute; the words "tart," "crisp," and "fresh" are part of the jargon. "Astringency" refers to what is known as the "bitterness" of the wine. Some wine tasters break the process down to the five basic components of color, swirl, nose, taste, and finish:

Color: Color is a reflection of the type of grape used as the source, the age of the wine, and the aging process. White wines increase in color with aging, whereas the color of red wine decreases.

Swirl: Gently swirl the glass of wine to oxygenate the wine. This releases those beautiful aromas characteristic of a good wine and complements the taste. By swirling, the wine is allowed to "breathe"; this can also be accomplished by uncorking the bottle and letting it sit open for a while before drinking.

Nose: Swirling the wine releases the aroma, or "bouquet." Here some subjective and quite imaginative terms such as *bountiful, cherry, heady,* and *nutty* are used. Some people sniff the cork, but the smell can be just as well detected from the wine in the glass. The main point is that wine may have some unpleasant odors.

Taste: Take a small sip of wine from your glass, taking care not to swallow it. Let it bathe the taste buds on your tongue. Your taste buds are sending signals to your brain. Are the sensations evoked pleasant ones?

Finish: The finish is the summation of the previous steps. Is the wine satisfying, mellow, and laced with a pleasant taste and no unpleasant aftertaste? If so, give the waiter a pleasant nod and be prepared to pay the price, and be generous in your tip.

Now you can talk like a wine connoisseur and (if you are of legal drinking age) on the next occasion impress your companions as you delicately inspect the color, gently swirl, fashionably "nose" the wine, and then settle back a few seconds for the "finish." You may then say to your waiter, "Yes, the color is perfect, the bouquet is superb, and it has a perfect crispness."

A variety of wines to please every taste is available (FIGURE 3.13). All grapes have white juices; red wines are made from red grapes, and the color is due to the pigments of the grape skins. Sweet wines are those in which fermentation is stopped while a significant amount of sugar is still present; in dry wines little sugar remains. Sparkling wines result from continued fermentation that takes place in bottles.

In the 1860s the French wine industry was in a state of chaos and collapse due to poor-quality wines. Emperor Napoleon III called on Louis Pasteur to seek a solution, and within only about three years he determined that the problem was due to contamination of the wines. His solution was simple—heat the wine to 50° to 60°C, a process later applied to milk and other food products that is now referred to as pasteurization. Pasteur's manuscript, *Études sur le Vin* (*Studies on Wine*),

FIGURE 3.13 A drink for all occasions. Alcoholic beverages, including wines, beers, and distilled spirits, are available for all tastes and occasions. Their production depends on the fermentation process carried out by yeasts. © iofoto/ShutterStock, Inc.

was published in 1866, and his experience with "sick wines" played a role in his shift to studying disease in humans.

Whereas wines are produced primarily from grape juices, beers are products of the fermentation of cereal grains, including barley (the most common), wheat, and rice. The grains are "malted" by being moistened and kept warm until partially germinated, which begins the enzymatic breaking down of the starch to simpler carbohydrates. The malt is then oven dried; the length of the drying process contributes to the flavor and determines the final color of the beer. The dried, cracked grains are steeped in hot water (mashed) to extract the sugars, starches, and other flavor compounds. The resulting liquid, called wort, is boiled to sterilize it, stop enzyme activity, and establish flavor. Hops (dried flowers of the female *Humulus lupulus* vine or their extract) are added at various times for flavor, aroma, preservative qualities, and retention of the head (the foam at the top of a glass of beer). After filtration and cooling, yeast (usually a strain of *S. cerevisiae*) is added to ferment the wort, resulting in production of ethyl alcohol, carbon dioxide, and distinctive aroma and flavor compounds.

Brandy, whiskey, rum, vodka, and gin are referred to as **distilled spirits**. Their production resembles that of wine fermentation. A raw product is used as the starting point; it is fermented by yeast species and then aged in casks. After fermentation, distillation is carried out, yielding a product with a higher alcohol content than beer or wine. The alcoholic content of beer is usually 4% to 6%, that of wine is about 12% to 13%, and that of spirits ranges from 40% to 50%. Scotch whiskey results from the fermentation of barley and rye, brandy results from the fermentation of wine or fruit juice, vodka results from the fermentation of potatoes or grains such as rye or barley, and rum results from the fermentation of molasses.

■ Harnessing Microbes as Research Tools

Microbes (except viruses and prions) offer to biologists packets of life complete with enzymes, energy-generating mechanisms, nucleic acids, structure, and reproductive ability. Even viruses, although not cellular, have some of these properties and are equally important as biological tools. Biologists have capitalized on the fact that microorganisms are easy and inexpensive to grow and reproduce rapidly. As knowledge of the microbial world and techniques to manipulate microbes became available over the past century, experimentation with microbes increased, resulting in many of the major advances in biology. Virtually all fields in biology, and many aspects of physics and chemistry, have been enhanced by exploration with microbes. Genetics and molecular biology, in particular, are beneficiaries of the use of microbes in the laboratory. **Genetic engineering**, also known as **recombinant DNA technology**, is a product of these studies.

One of the greatest achievements of the twentieth century was the success of the **Human Genome Project**—the mapping of the approximately 25 thousand genes in the 23 pairs of human chromosomes. The announcement of the completion of this genetic human blueprint amazed the world and captured the headlines. Earlier efforts at mapping microbes played a major role in this triumph.

The Human Genome Project was initiated in 1990 with the mission of mapping and sequencing the entire human genome—a genetic human blueprint with enormous potential impact on humankind in the coming years. Without microbes, none of this would have happened. The Microbial Genome Program was initiated in 1994 with the goal of sequencing the genomes of medically, environmentally, and industrially significant microbes; this program will lead to further success in harnessing these microbes for the benefit of humans. Genetics has come a long way since Mendel's mid-nineteenth century observation on the inheritance of color and other characteristics in plants.

A concept that has emerged in recombinant DNA technology is **gene therapy**. Gene therapy is the insertion of modified DNA into a patient's cells to treat disease. A malfunctioning gene may be replaced with a correctly functioning one. A major setback to gene therapy, however, occurred in 1999 when an eighteen-year-old man died four days after the initiation of gene therapy as a result of organ failure, presumably due to a severe immune response to the viral vector. Nevertheless, with certain restrictions imposed by government agencies, gene therapy has continued with some very positive results. The National Cancer Institute, a component of the National Institutes of Health, in 2006 successfully engineered lymphocytes, a category of white blood cells, to target and attack melanoma cancer cells; in that same year an international group using gene therapy succeeded in treating two patients with a disorder affecting a particular type of white blood cells. Experimentally, in 2005, researchers at the University of Michigan cured laboratory-induced deafness in guinea pigs by injecting them with a genetically engineered virus carrying a gene to stimulate growth of hair cells in the cochlea; possibly, the procedure will be effective in humans (BOX 3.3).

BOX 3.3 Extreme Gene Therapy

An interesting and bizarre idea, using the techniques of genetic engineering, was reported in the *Washington Post* in March 2001. According to the article, researchers interested in developing a cure for cystic fibrosis, an inherited lung disease, hoped to deliver new genes, a strategy referred to as gene therapy, by using viruses for "delivering therapeutic payloads." Viruses gain access into cells and, perhaps, can act as shuttle vehicles, an idea that has also been proposed for other applications. Researchers planned to combine pieces of two of the world's deadliest viruses, Ebola virus, which causes hemorrhagic fever (bleeding disease), and HIV, the cause of AIDS. The choice of these two viruses as partners is based on Ebola virus' predilection to attach to lung cells and on HIV's reputation for persisting in the body. The hybrid virus would be used to deliver new genes into the cells of the lungs in patients with cystic fibrosis. The idea of this dynamic duo was frightening. Robert Gallo, co-discoverer of HIV, stated, "I wouldn't want this thing put into me." Gallo warned of harmful immune reactions and the possibility that the hybrid virus could combine with HIV to create a new monster.

On the other hand, W. French Anderson, a pioneer and leader in gene therapy, and other biologists considered the proposed hybrid virus to be safe. "It's not even HIV anymore, it's just pieces. Ebola sounds horrible, but this has nothing to do with the Ebola virus that knocks out all your defense mechanisms and kills you," according to Anderson. Whether or not the researchers ever attempted the therapy is unclear.

The *New England Journal of Medicine* reported in April 2008 that the world's first gene therapy for a type of inherited blindness was safe and improved eye sight after a clinical trial in the United Kingdom. The potential for gene therapy is enormous; the best is yet to come.

Harnessing Microbes in Industry

In industry, including the pharmaceutical industry, the challenge is to harness microbes as factories and extract their metabolic products. The growth medium depends on the particular microbe and the desired products. The list of products is impressive (TABLE 3.2). Antibiotics and other medicinals, food additives, a variety of chemicals, cleaning products, enzymes, proteins, carbohydrates, nucleic acids, and yeasts are examples. Certain characteristics of microbes promote their use as microbial factories:

- The high ratio of surface area to volume leads to rapid replication; the product yield depends on the number of microbes maintained under optimum conditions.
- Microbes are versatile and can be grown in vats on a large scale and under a variety of growth conditions.
- Some products can be produced only by microbes.
- Microbes produce a large variety of enzymes that can be harvested to obtain desired products.
- Microbes can be genetically engineered to produce biological products that are used in the prevention and treatment of cardiac disease and other medical problems.
- Microbes can be genetically engineered to increase their productivity.
- Microbial factories are cost-effective.

In the early years of antibiotic production, only about 5 milligrams of penicillin could be recovered per liter of culture, whereas new strains of *Penicillium* have increased the yield to over 50,000 milligrams per liter. Industrial microbiologists are always on the hunt for microorganisms that synthesize new products or synthesize known products at a greater yield, as for example, microbial mutants that are not able to control synthesis of a particular product and thereby produce "overruns." The search for antibiotic-producing microorganisms in soil continues, particularly in light of the antibiotic resistance problem.

A field trip to an industrial plant is a worthwhile experience. You cannot help being amazed at the sheer magnitude—the fermenter vats are two or more stories high—and complexity of industrial microbiology and the skills of the bioengineers. Many companies, including breweries, offer tours of their facilities (and free samples!).

Composting takes advantage of the natural decomposition process to turn organic matter into rich soil (FIGURE 3.14). Aerobic bacteria are the prime decomposers aided by a variety of insects and worms. Industrial composting is a big

FIGURE 3.14 A large compost heap in a garden. © jeff gynane/ ShutterStock, Inc.

TABLE 3.2 Products of Genetic Engineering

Product and/or Microbe	Function
Products used in human medicine	
Alpha interferon (*E. coli*)	Treatment for some viruses
Insulin (*E. coli*)	Treatment for diabetes
Human growth hormone (*E. coli*)	Treatment for pituitary dwarfism
Interleukin-2 (*E. coli*)	Stimulation of immune system
Tumor necrosis factor (*E. coli*)	Treatment of certain cancers
Epidermal growth factor (*E. coli*)	Treatment of skin wounds and burns
Hepatitis B vaccine (*S. cerevisiae*)	Vaccine used in prevention of hepatitis B
Products used in animal husbandry	
Bovine growth hormone (*E. coli*)	Increases weight gain and milk production
Porcine (swine) growth hormone (*E. coli*)	Increases weight gain
Genetically engineered microbes	
Pseudomonas fluorescens	Carries genes from *Bacillus thurigiensis* that produce an insect poison
Pseudomonas syringae (ice-minus bacterium)	Engineered to remove protein that initiates ice formation on plants and affords protection from frost

(and wormy) business and is a valuable strategy in reducing the volume of wastes dumped in overcrowded landfills. Home composting has been on the rise for over the past two decades, and a variety of composting drums and bins are on the market.

The genetic engineering of microbes has enabled the growth of biotechnology and pharmaceutical companies that produce medicinals, including vaccines, antibiotics, hormones, and immune regulatory factors. **Human insulin** and **human growth hormone** are remarkable examples of genetic engineering products (FIGURE 3.15), as are some of the newer recombinant DNA vaccines. Human insulin is now produced in *Escherichia coli* by cloning the human "insulin gene" into its genetic material. Prior to genetic engineering, this insulin was produced from the pancreas of slaughtered cows and pigs; it was not as effective as human insulin and more expensive to produce. Before genetic engineering, human growth hormone for the treatment of dwarfism was obtained from the pituitary glands in human cadavers. Each patient received hormone from two or three batches per year. Each batch was derived from a pool of approximately sixteen thousand cadaver pituitary glands and posed a severe limitation of supply. Gene therapy using viral vectors (**virotherapy**) is the wave of the future and has already taken place on a trial basis. Viruses are obligate intracellular parasites, meaning their ecological niche is within a host cell accounting for their usefulness as vectors to deliver replacement genes into defective cells.

The agricultural industry uses bioinsecticides—a preparation containing microbes or toxins produced by them. Strains of *Bacillus thuringiensis* are used in a bioinsecticide, commonly called Bt, to control caterpillars and other leaf-eating

AUTHOR'S NOTE (RIK)

Over a long teaching career I have enjoyed a variety of awards and honors for which I am appreciative. But the one I would cherish the most, yet to come my way, is being named (or at least nominated) as "American Composter of the Year"!

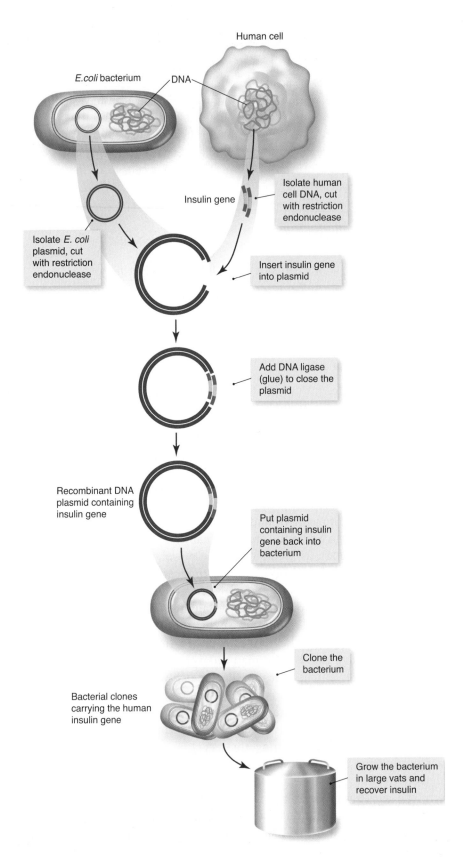

FIGURE 3.15 Production of insulin by genetic engineering. The human insulin gene is cloned and expressed in *E. coli*.

insects. Each strain produces a particular protein that is toxic for one or a few related species of insect pests; the protein binds to the larval gut, causing the insect to starve. Bioinsecticides are available that will kill moth, mosquito, fly, and other larvae. Silver leaf and greenhouse white flies are a problem for horticulturists, but it is a pesticide from the fungus *Paecilomomyces fumosoroseus* (another tongue-twister) that comes to the rescue! Some bioinsecticides are virus derived as, for example, baculovirus, which acts exclusively against tomato fruitworm. Bioinsecticides are available to control mosquitoes, Japanese beetles, crickets, and grasshoppers. These products, unlike many chemicals, are highly selective, of low toxicity for humans, and relatively safe for the environment.

◼ Harnessing Microbes for Bioremediation

Visit Prince William Sound off the coast of Alaska after a storm and you will see beaches tarnished with an oil slick. On March 24, 1989 the oil tanker *Exxon Valdez* ran aground (BOX 3.4). Immediately, massive efforts took place to clean up the oil and rescue the thousands of seabirds, sea otters, seals, bald eagles, killer whales, and salmon and herring eggs affected by the spill. The cleanup was mostly effective in removing the oil on the surface of the beaches, but it had little effect on the oil that had sunk beneath the sand and rocks. It is impossible to estimate the damage caused to microbial life and to food chains. One of the techniques used to clean up the mess was bioremediation (FIGURE 3.16).

The Environmental Protection Agency defines **bioremediation** as the act of adding materials to the environment, such as fertilizers or microorganisms, to increase the rate at which natural biodegradation occurs. It is important to realize that natural biodegradation would occur for many products but at a much slower rate. Biodegradation can also be enhanced by the spraying of nutrients on beaches or on other problem sites to foster the growth of the indigenous microbes to accelerate degradation of the pollutant; this process is called **bioaugmentation**.

Environmental disaster struck again in the form of an oil spill in the Gulf of Mexico on April 20, 2010; this event was about twenty times more devastating than the *Exxon Valdez* incident. The spill was the result of an oil gusher stemming from the explosion of the *Deep Water Horizon* oil drilling rig in the Gulf about forty miles from Louisiana. Eleven workers were killed and sixteen injured. The oil continued to flow for 3 months releasing almost 5 million barrels. On July 15, 2010, the gushing wellhead was finally capped. The method involved the use of chemical dispersants at the wellhead to facilitate microbial digestion of the oil in order to keep the oil beneath the surface. The logic had its critics including the National Oceanic and Atmospheric Administration.

Microbes exhibit diversity in many ways, as previously noted, underscoring their seemingly unlimited potential in terms of bioremediation. This has spawned the growth of companies in the private sector specializing in remedial

FIGURE 3.16 Rocks cleaned by bioremediation (right) compared with uncleaned rocks (left) covered with oil from the *Exxon Valdez* spill. © Science VU/Visuals Unlimited, Inc.

BOX 3.4 _An Account of the *Exxon Valdez* Oil Spill_

The *Exxon Valdez* departed from the Trans-Alaska Pipeline terminal at 9:12 p.m. on March 23, 1989. William Murphy, an expert ship's pilot hired to maneuver the 986-foot (300-m) vessel through the Valdez Narrows, was in control of the wheelhouse. At his side was the captain of the vessel, Joe Hazelwood. Helmsman Harry Claar was steering. After passing through Valdez Narrows, Pilot Murphy left the vessel and Captain Hazelwood took over the wheelhouse. The *Exxon Valdez* encountered icebergs in the shipping lanes, and Captain Hazelwood ordered Claar to take the *Exxon Valdez* out of the shipping lanes to go around the icebergs. He then handed over control of the wheelhouse to Third Mate Gregory Cousins with precise instructions to turn back into the shipping lanes when the tanker reached a certain point. At that time, Claar was replaced by Helmsman Robert Kagan. For reasons that remain unclear, Cousins and Kagan failed to make the turn back to the shipping lanes, and the ship ran aground on Bligh Reef at 12:04 a.m. on March 24, 1989. Captain Hazelwood was in his quarters at the time.

The National Transportation Safety Board investigated the accident and determined that the top three probable causes of the grounding were

1. Failure of the third mate to maneuver the vessel properly, possibly due to fatigue and excessive workload.
2. Failure of the captain to provide a proper navigation watch, possibly due to impairment from alcohol. The captain was seen in a local bar and admitted to having some alcoholic drinks; a blood test showed alcohol in his blood even several hours after the accident. The captain has always insisted that he was not impaired by alcohol. The state charged him with operating a vessel while under the influence of alcohol. A jury in Alaska, however, found him not guilty of that charge.
3. Failure of the Exxon Shipping Company to supervise the captain and provide a rested and sufficient crew for the *Exxon Valdez.*

The amount of oil spilled was 10.8 million gallons (257,000 barrels or 38,800 metric tons), roughly equivalent to 125 Olympic-sized swimming pools. The ship was carrying 53,094,510 gallons, or 1,264,155 barrels (8,015,825 metric tons). The *Exxon Valdez* spill is the largest ever in the United States but ranks thirty-fourth largest worldwide. It is widely considered the number-one spill worldwide in terms of damage to the environment, however. The timing of the spill, the remote and spectacular location, the thousands of miles of rugged and wild shoreline, and the abundance of wildlife in the region combined to make it an environmental disaster well beyond the scope of other spills.

The spill stretched from Bligh Reef to the tiny village of Chignik, 460 miles (740 km) away on the Alaska Peninsula. The oil affected approximately 1,300 miles (2,100 km) of shoreline. Two hundred miles was heavily or moderately oiled (meaning the impact was obvious); 1,100 miles (1,760 km) was lightly or very lightly oiled (meaning light sheen or occasional tar balls). By comparison, there is more than 9,000 miles (17,200 km) of shoreline in the spill region.

The cleanup efforts spanned more than four summers before it was called off. Some beaches remain oiled today. At its peak, the cleanup effort included 10,000 workers, about 1,000 boats, and roughly 100 airplanes and helicopters, known as Exxon's army, navy, and air force.

To clean the rocky beach shoreline, dozens of people used fire hoses to spray high-pressure cold water and hot water onto the rocks and sand (**FIGURE B3.4ab**). The water, with floating oil, would trickle down to the shore. The oil would be trapped within several layers of boom and either be scooped up, sucked up, or absorbed using special oil-absorbent materials. The hot water treatment was popular until it was determined that the treatment could be causing more damage than the oil. Small organisms were being cooked by the hot water.

Mechanical cleanup was attempted on some beaches. Backhoes and other heavy equipment tilled the beaches to expose oil underneath so that it could be washed out. Many beaches were fertilized to promote growth of microscopic bacteria that eat the hydrocarbons. Known as bioremediation, this method was successful on several beaches where the oil was not too thick. Exxon says it spent about $2.1 billion on the cleanup effort. It is widely believed,

(a)

(b)

FIGURE B3.4   (a) Using high-pressure water jets to clean oil from seashore rocks. Photo courtesy of the *Exxon Valdez* Oil Spill Trustee Council. (b) Clean up after the Alaskan oil spill. Photo courtesy of the *Exxon Valdez* Oil Spill Trustee Council/NOAA.

however, that wave action from winter storms did more to clean the beaches than all the human effort involved.

No one knows how many animals died outright from the oil spill. The carcasses of more than thirty-five thousand birds and one thousand sea otters were found after the spill, but because most carcasses sink this is considered to be a small fraction of the actual death toll. The best estimates are 250,000 seabirds, 2,800 sea otters, 300 harbor seals, 250 bald eagles, up to 22 killer whales, and billions of salmon and herring eggs.

There are three primary ways that oil kills wildlife:

1. The oil gets on the fur and feathers and destroys the insulation value. Birds and mammals then die of hypothermia (they get too cold).
2. They eat the oil, either while trying to clean the oil off their fur and feathers or while scavenging on dead animals. The oil is a poison that causes death.
3. The oil affects them in ways that do not lead to a quick death, such as damaging the liver or causing blindness. An impaired animal cannot compete for food and avoid predators.

A professional team and dozens of volunteers, including veterinarians, set up a cleaning facility and recovery facility (FIGURE B3.4c). Dawn dishwashing detergent was the cleaning agent of choice. Of twenty-three species listed as injured by the spill, only two have been declared "recovered." Lingering injuries continue to plague most species.

And what became of the *Exxon Valdez?* The ship was repaired and renamed the *Sea River Mediterranean;* it is used to haul oil across the Atlantic Ocean. The tanker is banned by law from ever returning to Prince William Sound.

Adapted from *Frequently Asked Questions about the Oil Spill, Exxon Valdez* Oil Spill Trustee Council, 2001.

FIGURE B3.4c   Many seabirds were killed by spilled oil. Photo courtesy of the *Exxon Valdez* Oil Spill Trustee Council.

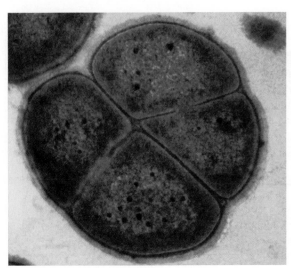

FIGURE 3.17 A transmission electron micrograph of *Deinococcus radiodurans.* Courtesy of Dr. Michael Daly, Department of Pathology at Uniformed Services, University of the Health Sciences.

services; for example, a company in Florida uses a soil cleaner containing emulsifiers to cause oil, gas, diesel, and other contaminants to be more easily broken down by soil microorganisms into carbon dioxide and water. Consider the use of a bioremediation system using oil and soap-eating bacteria in a car wash in California, allowing recycling of water instead of discharging it into a septic system. Genetic engineering allows for the creation of microbes specifically designed for bioremediation. *Deinococcus radiodurans* is used in highly radioactive nuclear wastes to digest toluene and toxic mercury (FIGURE 3.17). Bioremediation offers an efficient and relatively low-cost alternative approach as compared with excavation followed by a treatment strategy.

It is difficult to believe that microbes can bring about the decomposition of so many materials, but this is another illustration of microbial diversity. There are several advantages to the use of microbes as recyclers to clean up the environment, including cost-effectiveness, self-destruction once the conditions are improved, and minimal disruption of the environment. Bioremediation will be further advanced and play an important role in coping with ecological disturbances.

The possibilities of bioremediation are exciting and seemingly unlimited as microbes from diverse habitats are identified and as new microbes are genetically engineered with the enzymatic capability of breaking down environmental contaminants, including toxic products, petroleum products, landfill wastes, soils, polychlorinated biphenyls, plastics, paper, concrete and even disposable diapers.

There is some evidence that global warming, an ecological disturbance with worrisome consequences, may be at least partially slowed down by *Synechococcus*, a cyanobacterium that decreases carbon dioxide released in industrial processes, counteracting the greenhouse effect. *Methylosinus trichosporium* is another naturally occurring microbe that conceivably could be harnessed to reduce the process of global warming; the organism breaks down the chlorofluorocarbon gases that are products of refrigerants, air conditioners, foam packaging, and spray can propellants. Chlorofluorocarbons are "greenhouse" gases that contribute to global warming and deplete the earth's protective ozone layer.

Sewage and Wastewater Treatment

Until the 1900s and the realization that microbes were the causative agents of a variety of (waterborne) diseases, communities discharged their raw, or untreated, sewage into nearby river, streams, and marine waters. (The source of cholera epidemics in London in the 1800s was untreated wastewater in the Thames.) However, microbes also play an essential role in the treatment of raw sewage and wastewater, the goal of which is to prevent fecal pathogens from contaminating clean water. Marine waters, too, need to be protected from fecal contamination. It is not uncommon, particularly in the warmer months of the year, for beaches to have high bacteria counts in the water. This may be primarily the result of flooding and overflow from catch basins and tidal barriers and from careless hygienic habits of persons using the waters for recreation. Because so many human microbial diseases are waterborne, municipalities are required to establish and maintain facilities to treat their sewage for the protection of the community.

Microbes play an essential role in treating the sewage before it is finally discharged into receiving waters. The process can be divided into three stages: primary, secondary, and tertiary (FIGURE 3.18). Initially, raw sewage is filtered through screens to remove sticks, plastics, and other large pieces of debris; the wastewater then flows into a primary treatment settling tank in which heavy particulate matter, referred to as the primary sludge, settles to the bottom of the tank. The primary stage is a physical one to accomplish separation of the liquid portion from the solid and particulate matter (which constitutes less that 1% of the wastewater). The microbes are put to use in the secondary treatment to carry out digestion. In this process the liquid portion from the primary tank is passed into the secondary tank; the wastewater is then aerated by a trickling filter. The wastewater passes over and trickles through a bed of fist-sized rocks during which time biofilms composed of bacteria, protozoa, algae, and viruses carry out aerobic and anaerobic fermentative degradation. An alternative trickling filter system is the activated sludge process; the effluent from the primary treatment tank is aerated by bubbling air through it in the secondary tank. Aerobic microbes, primarily the bacterial species *Zoogloea ramigera,* grow in aggregates called flocs and degrade the matter. As in the trickling process, the fluid wastewater from the activated sludge unit passes into the secondary clarifier unit and is chlorinated to kill many of the remaining pathogens as a further safety feature.

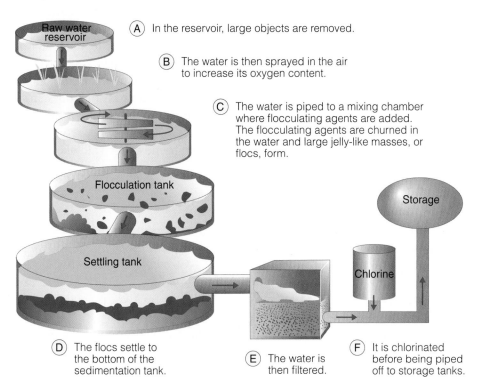

(A) In the reservoir, large objects are removed.

(B) The water is then sprayed in the air to increase its oxygen content.

(C) The water is piped to a mixing chamber where flocculating agents are added. The flocculating agents are churned in the water and large jelly-like masses, or flocs, form.

(D) The flocs settle to the bottom of the sedimentation tank.

(E) The water is then filtered.

(F) It is chlorinated before being piped off to storage tanks.

FIGURE 3.18 Steps in sewage treatment.

Secondary treatment is successful in removing organic matter from wastes but can be a problem because it does not remove the inorganic byproducts of the microbial degradation, which may act as nutrients and stimulate algal blooms. Tertiary wastewater processes can effectively reduce these inorganic materials, but this treatment is expensive and not always used.

A mall of factory outlets in Wrentham, Massachusetts, is a pioneer in "gray" water recycling. It recycles 90% of its 35 thousand gallons of water a day into clean water that can be reused for toilet flushing. Gray water is nonindustrial wastewater from domestic sources such as sinks, tubs, showers, and bathtubs. It contains lower levels of organic matter and nutrients than does wastewater because urine, fecal material, and toilet paper are not present. About 50% to 80% of wastewater is gray water that can be safely recycled to flush toilets, irrigate landscape, and in certain industrial processes as a cost-effective alternative to "clean water."

Overview

This chapter presents positive aspects of human association with the microbial world, a world whose presence is unseen but is manifest in many ways every day. Microorganisms constitute a major component of Earth's biomass, and their role in the cycles of nature underpins life itself.

Earlier civilizations learned by experience to harness the metabolic activities of microbes in the production of foods, including fermented milk products and alcoholic beverages. In more recent years, as the science of microbiology developed, the diversity of microbes has been further exploited. Microbes are used as research tools, and have spawned many disciplines in biology. They are factories for the production of many useful products, including products of genetic engineering, and as agents of cleaning by means of bioremediation. *The Microbes' Contribution to Biology* (1956) described the importance of microorganisms to the study of biology. If that text were updated today, many more pages would be necessary.

Because of the barrage of microbial threats, perhaps there is too much emphasis on strategies to wipe out, destroy, sanitize, or scrub away microbes. In so doing, useful, life-sustaining microbes are killed, thereby "throwing out the baby with the bath water." The following quotation serves as an appropriate conclusion to this chapter: "If you take care of your microbial friends, they will take care of your future."

Self-Evaluation

PART I: Choose the single best answer.

1. The greatest recyclers of all time are
    a. plants b. microbes c. animals d. humans
2. Nitrification is characterized by
    a. conversion of nitrates into ammonia b. fixation of atmospheric oxygen
    c. conversion of ammonia into nitrates d. return of nitrogen to atmosphere

3. Red wines (unlike white) are the result of
   a. red juices of red grapes b. aging process c. skins of red grapes
   d. white grapes to which red skins have been added
4. A primary producer
   a. takes in oxygen and releases carbon dioxide b. uses photosynthesis to produce organic compounds and oxygen c. decomposes waste material and produces methane d. is a free-living scavenger microbe
5. A sweet wine is characterized by
   a. distillation b. little remaining sugar c. fermentation in a bottle
   d. a relatively high sugar concentration
6. Which series of chemical reactions is mediated by enzymes of a variety of strains of bacteria and yeasts that break down sugars to small molecules?
   a. symbiosis b. fermentation c. photosynthesis d. respiration
7. A gram of yogurt contains about one million
   a. staphylococci b. clostridia c. lactobacilli d. mycobacteria
8. This bacterium plays a major role in sewage treatment.
   a. *Streptococcus* b. *Clostridium* c. *Zoogloea* d. *Bacillus*
9. Which of the following foods is not produced by using microbes?
   a. sour cream b. soy sauce c. beer d. mayonaise
10. This may be of value in preventing yeast infections during antibiotic therapy.
    a. wine consumption b. probiotics c. gene therapy d. starvation

PART II: Fill in the blank.
1. Plants with nodules along their root systems are called _____ plants.
2. Foods supplemented with _____ are called probiotics.
3. _____ is the genus and species of the yeast frequently used in making bread and alcoholic beverages.
4. _____ was genetically engineered to remove protein that initiates ice formation on plants and affords protection from frost.
5. Wines are produced from fruit juices, whereas beers are produced from fermentation of _____.
6. _____ is the science of wine-making.
7. Gene therapy uses genetically engineered _____ to deliver genes into a patient's cells.
8. Microbes are essential in the conversion of atmospheric _____ to organic compounds and back to the atmosphere.
9. _____ bacteria convert free nitrogen gas to ammonium.
10. _____ bacteria convert ammonium to nitrites.

PART III: Answer the following.

1. Distinguish between bioremediation and bioaugmentation.
2. Using specific examples, explain the expression "Microbes make the world go around."
3. Cite and briefly explain the ways microbes can be used in research and in industry.
4. List at least five medical benefits of mushrooms.
5. Cite examples of fermentation end products and whether any of them have commercial value.
6. Name the five basic components of wine used by wine tasters.
7. Speculate as to why it might be advantageous for a wild mushroom to be toxic for human consumption.
8. What is the significance of *Rhizobium?*
9. What causes bread to "rise"?
10. Discuss the importance of genetically engineering microbes.

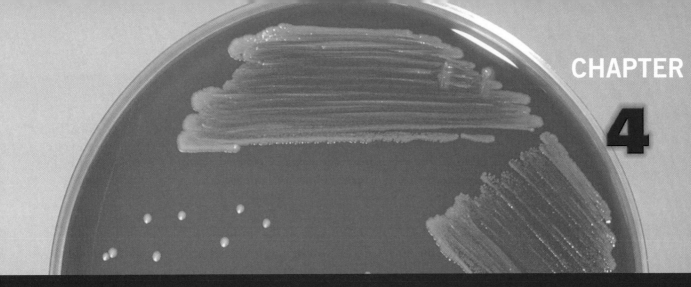

Bacteria

What marvels there are in so small a creature.

—Antonie Van Leeuwenhoek

■ Preview

The above quotation is the words of Leeuwenhoek, a Dutch merchant by vocation and a lens grinder by avocation, in a report to the Royal Society of London in 1675. Leeuwenhoek's development of a simple microscope allowed him to see and describe "**small animalcules.**" Microbes date back to the antiquity of life. A scientist at California Polytechnic Institute reported the isolation of ancient bacteria from a bee in a specimen of amber, a hardened resin from ancient pine trees, dating back 25 million to 40 million years. This account is reminiscent of the novel and movie *Jurassic Park,* in which a scientist extracted dinosaur DNA from entombed mosquitoes that had fed on dinosaurs.

The notion that bacteria are simple because they are unicellular is far from the truth; they may be small, but they are not simple. They are complex biological entities. The fact is, within a matter of only several hours after the invasion of a human body by some pathogenic bacteria, severe illness and even death may result. This is equally true of infected elephants and whales, whose mass is measurable in tons and yet they fall victim to microscopic microbes. It would take a fantastic number of bacteria to equal one ton.

The anatomy of the bacterial cell as an organized and functional structure meeting the characteristics associated with life is described in this chapter. Under appropriate conditions some bacteria can undergo **binary fission** in as little as

Photo courtesy of Dr. W.A. Clark/CDC.

20 minutes, resulting in huge populations in 24 hours. An appreciation of the bacterial growth curve is necessary to understand the dynamics of growth when only a few pathogens are introduced into the body.

Many known bacteria are easily grown in the laboratory, facilitating the diagnosis of bacterial infection. Mycoplasmas, chlamydiae, and rickettsiae are atypical, "oddball" bacteria, and their unique characteristics are described later in the chapter. They cause a variety of diseases in humans.

■ Cell Shapes and Patterns

Examination of bacterial cells under a microscope reveals the presence of a variety of shapes and patterns of arrangement (FIGURE 4.1). The majority are either rod shaped, known as **bacilli** (singular, bacillus); spherical, known as **cocci** (singular, coccus); or spiral shaped, known as **spirilla** (singular, spirillum). For the most part bacilli tend to occur as single cells, but some species form chains. Characteristic groupings of cocci are more common and are useful in identification. **Streptococci** are chains of cocci resembling a string of pearls; **staphylococci** look like a bunch of grapes; **diplococci** occur in pairs; and **tetrads** are groupings of four. Undoubtedly, the terms *strep* and *staph* are familiar to you. Spiral or curved rods are categorized as spirilla (rigid helix), **spirochetes** (flexible helix), and **vibrios** (comma-shaped curved rods). Other bacteria have been described as star shaped, triangular, flat, or square. Microscopic determination of shape and pattern is often the first step in the identification of bacteria.

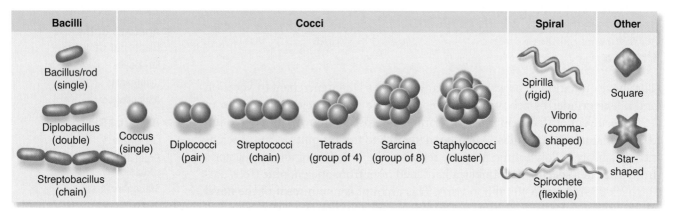

FIGURE 4.1 Bacterial cell shapes and patterns.

■ Naming Bacteria

Expectant couples spend hours searching through baby-naming books, conferring with family and friends, and otherwise agonizing over their choice of the perfect name for their expected child, particularly the firstborn. Biologists have established names for microbes and have also put considerable thought into doing so, because the names frequently identify important characteristics. As you read through this book, you may feel overwhelmed by the long and difficult-to-pronounce names, but at least you can impress your family and friends.

Microbes are named in accordance with the **binomial** system of nomenclature established by Carolus Linnaeus in 1735. This system is not restricted to microbes but applies to all organisms. The names are Latinized, and each organism carries two names (binomial), the first designating the **genus** and the second designating the **species**. For example, humans are *Homo* (genus) *sapiens* (species), and the fruit flies most commonly used in genetic studies are *Drosophila melanogaster*. Note that the first letter of the genus name is always capitalized, but the species name is not. Further, both the genus and species names are italicized.

Unfortunately for the student, microbiologists have not been consistent in assigning names. Some microbes are named in honor of the scientist responsible for first describing them, whereas others indicate the microbe's habitat, shape, associated disease, or combinations of these factors. A few examples make this clear. *Rickettsia prowazekii* (the cause of epidemic typhus) is named in honor of Howard T. Ricketts and Stanislaus von Prowazekii, both of whom died of the disease they were studying. *Legionella pneumophila* is named after an outbreak of pneumonia at an American Legion convention in 1976. *Escherichia coli* is named in commemoration of Theodor Escherich and identifies the *coli* (large intestine) as the organism's habitat.

AUTHOR'S NOTE (RIK)
I would love to have a microbe or disease named after me!

The use of descriptive morphological terms (e.g., *bacillus*) can lead to confusion. *Bacillus anthracis* belongs to the genus *Bacillus* (with a capital "B") and morphologically is a bacillus (with a lowercase "b"). But *E. coli* is also a bacillus ("b") belonging to the genus *Escherichia*. There are many other bacilli that do not belong to the genus *Bacillus*. So all members of the genus *Bacillus* are bacilli but not all bacilli are classified under *Bacillus*. The same case can be made for streptococci and diplococci. *Diplococcus pneumoniae* strains are all diplococci, but so is *Neisseria gonorrhoeae*. (The genus *Diplococcus* has been renamed *Streptococcus*.)

Anatomy of the Bacterial Cell

To appreciate the challenge posed by bacteria, it is necessary to be familiar with their structure and properties. The fact that they are procaryotic cells sets them apart from the other microbes. The structures that compose these cells are outlined in TABLE 4.1 and illustrated in FIGURE 4.2; note that some anatomical features are not common to all bacteria.

Envelope

The bacterial **envelope** consists of the **capsule, cell wall,** and **cell** (outer) **membrane** (in gram-negative bacteria). A capsule is not present in all species, but the cell wall (with the exception of the mycoplasmas, to be described later in the chapter) and the cell (plasma) membrane are structures present in all bacteria.

Capsule

When present, the capsule is not integral to the life of the cell. In fact, the capsule can be easily removed by treating a culture with appropriate enzymes or by manipulating the presence of nutrients in the culture; in either case the cells grow just as well in the laboratory but lack capsules. The progeny of these noncapsulated cells

TABLE 4.1 Anatomical Features of the Bacterial Cell

Structure	Function
Cell envelope	
Capsule[a] (glycocalyx, slime layer)	Promotes virulence in some cases
Cell wall (with the exception of mycoplasmas)	Constrained structure that confers shape and provides tensile strength
Cell membrane	Controls movement of molecules into and out of cell; referred to as gatekeeper
Cytoplasm[b]	
Nucleoid	DNA-rich area not enclosed by a membrane
Plasmids[a]	Nonchromosomal DNA that confers properties, including antibiotic resistance
Spores[a]	Confers extreme resistance to environmental factors
Ribosome	Protein synthesis
Chromosome	Determinant of genetic traits
Inclusion bodies	Storage and reserve supply of nutrient materials
Appendages	
Flagella[a]	Motility
Pili[a]	Adhesion to surfaces, possible bridge for transfer of DNA

[a]Structure not found in all bacterial cells.
[b]Area within membrane; contains numerous organelles.

FIGURE 4.2 A "composite" bacterial cell. The asterisks indicate structures not present in all bacteria.

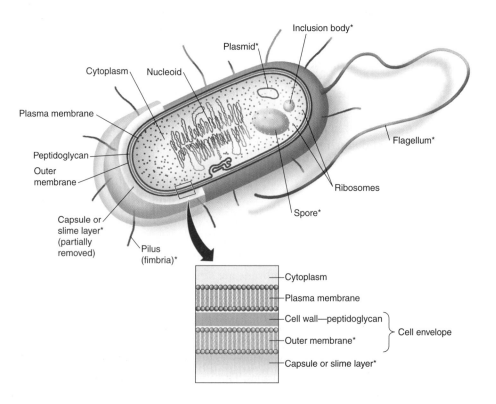

will have capsules. In some species the presence of a capsule promotes **virulence** (the capacity to produce disease), as, for example, in the bacillus that causes anthrax (*B. anthracis*). The streptococci responsible for strep throat and necrotizing fasciitis (flesh-eating streptococcal disease) are another example of capsulated organisms. The presence of the capsule interferes with **phagocytosis**. Phagocytosis is an important body defense mechanism by which bacteria are ingested and killed by a variety of phagocytic cells. The capsular material may become so thick that it resembles a **slime layer**; it is like trying to grab on to a slippery fish. Occasionally, hunks of this mucus-like layer can be found in the broth used to grow the culture. A condition known as "**ropy milk**" is a nuisance to the milk industry because of the growth of the nonpathogen *Alcaligenes viscolactis* and the shedding of its slime layer into milk. The fact that this organism is not a disease producer is of little comfort as your tongue and palate unexpectedly encounter this ropy, mucus-like material.

Cell Wall

Cell walls are characteristic of all bacteria (again, with the exception of the mycoplasmas) and are a structure shared with plant, algae, and fungal cells. The cell wall is a rigid, corset-like structure responsible for the characteristic shape of the cell. Further, it confers resistance to the cell from the inward **diffusion** of water. Without a strong cell wall, the membrane would swell, and lysis (cell bursting) would occur. In the 1940s penicillin came into widespread use despite the fact that its mechanism of action was not yet understood. Several years later scientists discovered that penicillin interferes with the ability of cells to synthesize normal cell walls, making these cells subject to lysis. This is the mechanism by which the "wonder drug" penicillin and related antibiotics indirectly deliver a death blow to certain bacteria. A number of other antibiotics target bacterial cell walls, leading to the death of these cells. (Because human cells do not have cell walls, these antibiotics are, generally, of less toxicity.)

Bacteria are divided into two groups based on differences in the chemistry of their cell walls, namely, gram positive (g+) and gram negative (g–) as demonstrated by the Gram stain, a procedure introduced by Hans Christian Gram in 1884. The staining procedure is an important first step in identifying the specific bacterial organism as the cause of infection. A specimen from a culture source or from a swab is taken from an individual (e.g., skin, mouth) and spread onto the surface of a slide and then flooded with crystal violet, a purple-bluish dye. Both g+ and g– cells stain a purple-bluish color after application (about 30 seconds) of the crystal violet. Iodine is the next reagent (about a minute), followed by addition of alcohol (30 to 60 seconds)—the differential stage; the alcohol dehydrates a backbone layer of the g+ cell wall preventing the loss of the crystal violet iodine complex. In g– cells the alcohol dissolves the relatively abundant outer layer of the lipopolysaccharide bilayer and makes small pores in the inner layer component characteristic of g– cells through which the crystal violet iodine complex diffuses outward. At this stage these g– cells appear colorless. The addition of safranin, a pink-reddish counterstain in contrast to the initial purplish-blue crystal violet, results in these cells appearing as a pink-reddish color. In contrast, in g+ bacteria the initial crystal violet-iodine complex is retained preventing the entrance of safranin, and, hence, these cells are a purplish-blue color (FIGURE 4.3).

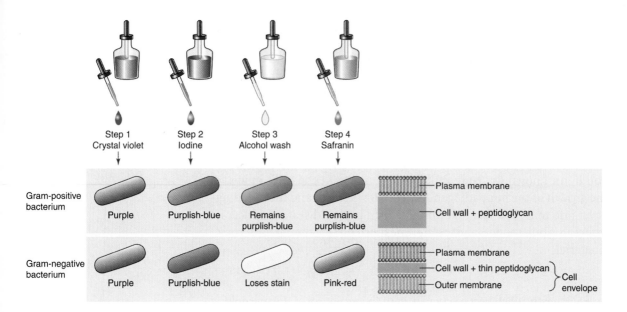

FIGURE 4.3   Gram stain.

Perhaps you, or someone you know, have been diagnosed with a gram-positive or a gram-negative infection. The distinction is important, because the choice of antibiotics is strongly influenced by the Gram stain reaction of the causative bacteria (TABLE 4.2). For example, penicillin and a variety of related antibiotics are effective against g– but not g+ cells; on the other hand, the antibiotic ciprofloxacin is

TABLE 4.2   Diseases Produced by Gram-Positive and Gram-Negative Bacteria

Bacterium	Disease
Gram Positive	
Bacillus anthracis	Anthrax
Clostridium botulinum	Botulism
Clostridium tetani	Tetanus
Corynebacterium diphtheriae	Diphtheria
Listeria monocytogenes	Listeriosis
Staphylococcus aureus	Skin infections, toxic shock
Streptococcus mutans	Tooth decay
Streptococcus pyogenes	Streptococcal sore throat; scarlet fever
Gram Negative	
Escherichia coli O157:H7	Bloody diarrhea
Campylobacter jejuni	Gastroenteritis
Legionella pneumophila	Legionnaires' disease
Salmonella typhi	Typhoid fever
Salmonella enterica	Salmonellosis
Neisseria gonorrhoeae	Gonorrhea
Yersinia pestis	Plague
Vibrio cholerae	Cholera

effective against gram-negative bacteria. Examples of broad-spectrum antibiotics are the tetracyclines and the chloramphenicols. The differentiation between g+ and g– cells is based on the chemistry of their cell walls.

Cell walls of g+ and g– bacteria are composed of **peptidoglycan**, a backbone that gives the cell wall its characteristic rigidity. Some of the molecules associated with the bacterial cell wall are not found elsewhere in nature. Cell walls of g+ bacteria have a thick peptidoglycan layer, whereas this layer is relatively thin in g– cells. Additionally, in g– cells there is an **outer membrane** external to the peptidoglycan layer that is associated with virulence. When these cells undergo lysis, an endotoxin that causes fever and damage to the host is released from the outer membrane.

Cell Membrane

The passage of molecules between the bacterial cell and its external environment is controlled by the cell membrane. To understand this point, think of chains of streptococci residing on the surface of your pharynx (FIGURE 4.4). (They are probably quite content, and why shouldn't they be?) The surface is a warm, moist, and nutrient-rich environment that is optimal for bacterial multiplication. Oxygen and nutrients diffuse into the "strep" cells through their cell membranes to allow for metabolic processes consistent with life and reproduction. During their growth and reproduction, as a natural part of their metabolism they secrete a number of products, some of which are toxic. It is these toxic products that produce the sore throat and fever associated with a "strep throat."

The cell membrane is **selectively permeable**, meaning that not all molecules are able to pass freely into or out of the cell. The structure of the membrane,

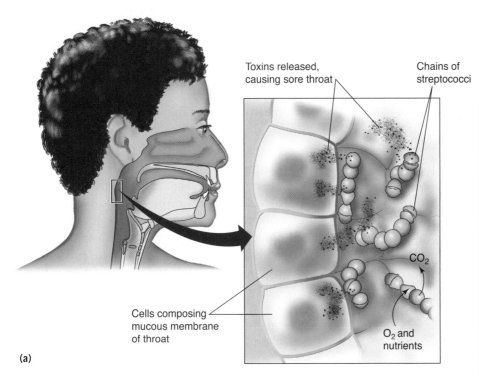

Toxins released, causing sore throat

Chains of streptococci

CO$_2$

Cells composing mucous membrane of throat

O$_2$ and nutrients

(a)

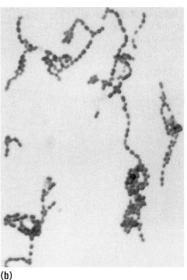

FIGURE 4.4 **(a)** Streptococci growing on the surface of the pharynx. **(b)** Streptococci chains. Courtesy of the CDC.

(b)

inherent in its chemistry and physical arrangement, accounts for its "gatekeeper" function; it is double-layered and has pores and transport molecules instrumental in the movement of materials between the cells and their environment. The physical processes of diffusion and osmosis, plus the size and the charge of the molecules, govern the selective permeability of the membrane. The bacterial cell membrane is the site for energy-generating reactions of the cell and also contains enzymes that are instrumental in cell wall assembly and membrane synthesis. A cell membrane is not unique to bacterial cells but is common to all cells, procaryotic and eucaryotic, and has a remarkably similar structure in all cells.

Cytoplasm

The **cytoplasm** is the part of the cell enclosed within the cell membrane, and within it are numerous cellular constituents that function in cell growth and multiplication. Figure 4.2 illustrates the variety of structures found within this region.

Nucleoid

The bacterial cell lacks a nuclear membrane, and, hence, it is a procaryotic cell. Within the cytoplasm there is a DNA-rich area, demonstrated by staining procedures and electron microscopy, which is referred to as a **nucleoid**; there is no membrane defining this area. It is relatively easy to extract bacterial DNA from *E. coli* and some other bacteria. Most bacterial cells contain a single **chromosome** present as a circular, double-stranded stretch of DNA. (The organism *Vibrio cholerae* has two circular chromosomes. A few bacteria have been found to contain a linear chromosome, and some have both a linear and a circular chromosome.) This is in contrast to eucaryotic cells, in which the DNA is organized into "multiple" discrete linear bodies, the chromosomes. (Humans, for example, have 23 pairs or 46 chromosomes.) The mechanism of cell division in bacteria is binary fission, which, simply put, means "splitting in two" (FIGURE 4.5).

Plasmids

Plasmids are small circular molecules of nonchromosomal DNA that are found in some bacteria; they are located in the cytoplasm (Figure 4.2) and are independent of the chromosomal DNA. Under laboratory conditions plasmids are not essential to the cell; it is possible to "cure" (eliminate) them by laboratory manipulation without killing the cell. Plasmids may consist of only a few to a large number of genes, and they code for the production of significant properties. Some plasmids carry genes that confer virulence. Plasmids may be transferred from one bacterial

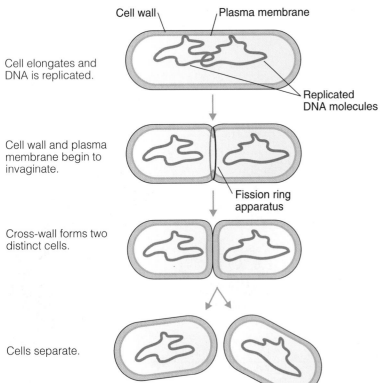

FIGURE 4.5 Binary fission.

Cell wall — Plasma membrane

Cell elongates and DNA is replicated.

Replicated DNA molecules

Cell wall and plasma membrane begin to invaginate.

Fission ring apparatus

Cross-wall forms two distinct cells.

Cells separate.

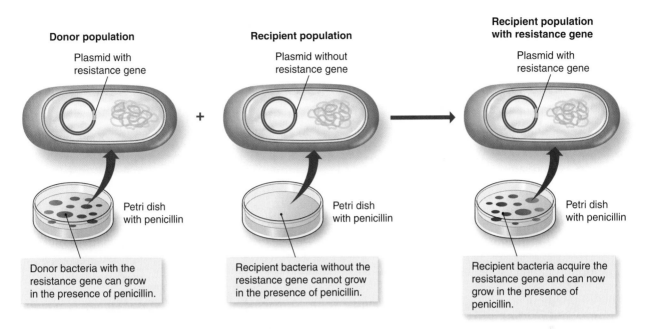

Donor population
Plasmid with resistance gene

Recipient population
Plasmid without resistance gene

Recipient population with resistance gene
Plasmid with resistance gene

Petri dish with penicillin

Donor bacteria with the resistance gene can grow in the presence of penicillin.

Petri dish with penicillin

Recipient bacteria without the resistance gene cannot grow in the presence of penicillin.

Petri dish with penicillin

Recipient bacteria acquire the resistance gene and can now grow in the presence of penicillin.

FIGURE 4.6 The infectious nature of plasmids.

cell to another, and because of this they are referred to as **infectious agents** (see text on bacterial genetics).

Other plasmids carry genes that confer antibiotic resistance; these plasmids are referred to as **R (resistance) factors**. Imagine, therefore, a population of bacteria in your intestinal tract in which, initially, only a few cells have plasmids conferring resistance to three different antibiotics. It is possible that within only several hours plasmids may infect other cells, resulting in most of the microbial population becoming multiple drug resistant (FIGURE 4.6). Some plasmids allow for the exchange of DNA between bacterial cells. Because plasmids are self-replicating DNA segments, the donor cells retain their plasmids while transferring copies to the recipient cell.

Spores

The genera *Bacillus* and *Clostridium*, both of which contain disease-producing species are spore producers. (The term *spore* refers to the spore that forms within the bacterial cell.) **Spores** are extremely hardy structures that are highly resistant to heat, drying, radiation, and a variety of chemical compounds, including alcohol (FIGURE 4.7). Most bacteria are killed in boiling water within only a few minutes, but this is not true of bacteria that make spores; they remain viable even after boiling for a few hours. Spores have been recovered from soil samples dating back several thousand years. They are in a state of dormancy with little or no metabolism. In fact, a handful of scientists maintain that the first life forms on Earth were bacterial spores that had drifted from some distant planet. The spore is actually a condensed form of the cell, including the cell's DNA, encased within a several-layered, chemically complex coat. A plant seed is somewhat analogous. If you plant a tomato seed, you get a tomato plant because the seed contains the genetic information

FIGURE 4.7 A transmission electron micrograph of *Bacillus anthracis* showing cell division **(A)**, and a spore **(B)**. Note the hard-shell capsule that provides protection for the spore. Courtesy of Dr. Sherif Zaki and Elizabeth White/CDC.

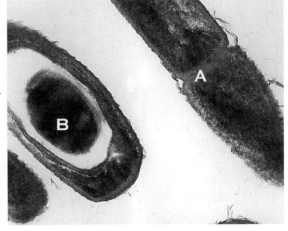

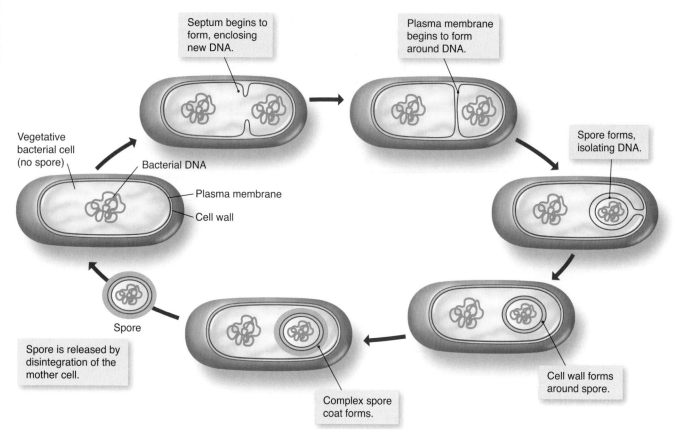

Septum begins to form, enclosing new DNA.

Plasma membrane begins to form around DNA.

Vegetative bacterial cell (no spore)

Bacterial DNA

Plasma membrane

Cell wall

Spore forms, isolating DNA.

Spore

Spore is released by disintegration of the mother cell.

Complex spore coat forms.

Cell wall forms around spore.

FIGURE 4.8 The spore cycle.

coding for that particular species of tomato. Spore-forming bacteria without spores are termed ***vegetative cells***. In the laboratory, "hard times" (for example, old cultures or deficient growth media) induce the vegetative cells to produce spores. Providing these sporulating cells with more optimal growth conditions results in their germinating as vegetative cells (FIGURE 4.8). Note that bacterial spores, unlike spores of fungi, are not involved in multiplication. Also, in contrast to fungal spores that can be killed at 60°C, *Bacillus* and *Clostridium* spores can withstand 100°C. The rule is "one cell, one spore, one cell." The organism *B. anthracis* is a spore former. The spore's hardiness and infectivity make it a top candidate for biological warfare and terrorism. You may have seen news accounts of threatened terrorism using the anthrax bacillus; some proved to be hoaxes, but others, unfortunately, have been the real thing.

The spore-forming anaerobic genus *Clostridium* includes three species that cause serious and potentially life-threatening diseases of humans: tetanus (lockjaw), botulism (a type of food poisoning), and gas gangrene. The stage for tetanus is set when you step on a rusty nail carrying spores and sustain a puncture wound, providing anaerobic conditions for germination of the spores. (It does not have to be a nail, nor does the object need to be rusty. Have you had your tetanus booster shot?) Botulism is a serious and potentially lethal disease caused by *Clostridium botulinum*. Those who are thinking of doing their own home canning with vegetables from their garden need to be aware that spores are resistant to boiling. The food must be cooked in a pressure cooker after being packed into jars and sealed carefully. The point is that "home canners" need to be fully aware of what they are doing and cannot take shortcuts. Occasionally, commercially canned foods can be contaminated with *botulinum*; a "bulging" can is a tell-tale sign and should be immediately brought to the attention of the store manager.

Appendages

Flagella

Flagella confer motility. They are composed of a protein called **flagellin** and are found in some species of bacilli and cocci. One or more flagella are arranged in clumps or are distributed all over the cell; the particular arrangement of the flagella is consistent for those species that have flagella (FIGURE 4.9). Each flagellum is like a long helix and measures over 10 micrometers, many times longer than the length of the cell. They are extremely thin and difficult to stain and to observe under a light microscope. The anatomy of flagella reveals that they are anchored by their **basal bodies** to the cell wall and membrane from which they extrude through the cell wall. The motility of live and unstained bacteria can be observed by suspending a drop of culture on a slide and examining this **hanging drop** under a microscope.

The flagellum rotates like a propeller and is driven by rings in the basal body. Descriptive and colorful terms, including *runs, tumbles, twiddles,* and *tweedles,* are used to describe bacterial motility. Receptor sites are present within the cell resulting in directed movements—a process called **chemotaxis**—toward or away from a chemical stimulus. Some bacteria "swim" toward an attractant (glucose) and away from repellents (acids). It's amazing to think they can detect minute changes in their environment. This is remarkable behavior. Motility appears to play a role in the ability of some disease-producing bacteria to spread through the tissues.

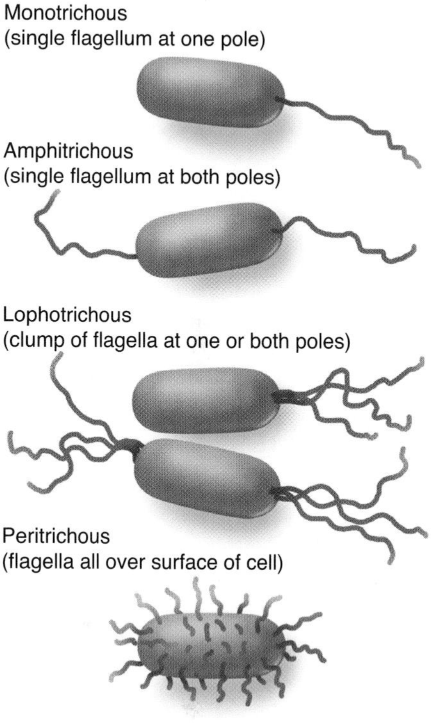

Pili

Pili are composed of the protein **pilin** and, like flagella, extrude through the cell wall and are present in many g– species. They are shorter, straighter, and thinner than flagella. There may be only a single pilus or up to several hundred pili per cell; in some bacteria, including those that cause the sexually transmitted disease gonorrhea, pili serve as adhesins and anchor the bacteria to the mucous membrane of the vagina or the penis. This is an early step in the **colonization** and subsequent establishment of disease. Other pili, referred to as **sex pili**, function in genetic exchange by forming a bridge between cells through which DNA passes from donor to recipient (review bacterial genetics); this is a major factor in the spread of antibiotic resistance in bacteria among different species.

Bacterial Growth

For unicellular organisms, both procaryotic and eucaryotic, the terms *growth* and *multiplication* are used synonymously. In contrast, for multicellular organisms (consider the human), growth refers to an increase in the size of an individual as

Monotrichous
(single flagellum at one pole)

Amphitrichous
(single flagellum at both poles)

Lophotrichous
(clump of flagella at one or both poles)

Peritrichous
(flagella all over surface of cell)

FIGURE 4.9 Structure and arrangement of flagella.

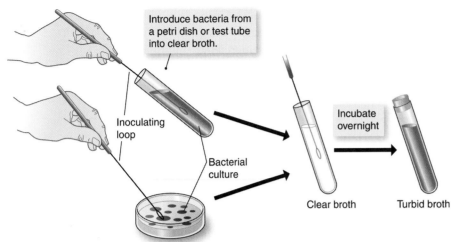

FIGURE 4.10 Overnight growth in a broth tube—clear to turbid.

a unit and is the result of multiplication at the cellular level (mitosis) leading to an increase in the total number of cells in an individual, but not in the number of individuals. A population, microbial or otherwise, if allowed to grow unchecked, would take over the Earth and, ultimately, exterminate all other forms of life. Can you imagine a universe populated entirely by *E. coli*?

A system of checks and balances limits population size. This system includes abiotic factors, such as availability of oxygen (O_2), temperature, and water, and biotic factors, including suitable habitat, predator–prey relationships, and competition. A course in ecology usually considers these factors as they apply to plants and animals, but these same concepts apply to microbes as well. Interaction between bacterial species is just as important as it is between plants and animals. (Microbial ecology is a specialized course that might be of interest to a microbiology major.) For example, as Fleming observed in 1928, the product secreted by the mold *Penicillium chrysogenum* (later termed *penicillin*) is inhibitory to various bacteria. Penicillin is a defense mechanism for the mold in the same sense that poisonous secretions of some jellyfish or the venom of certain snakes serve as defense mechanisms in their predator–prey relationships.

Bacterial growth provides insight into the dynamics of population growth as it occurs in nature, including in the body (FIGURE 4.10). Growth curves illustrate what occurs when *E. coli* incubating in the mayonnaise in the tuna salad at the beach is ingested and multiplication of the microbe occurs in your intestinal tract. The worst is yet to come when four or five hours later on the way home, family members, usually children, might develop nausea, vomiting, and diarrhea; to add to this hypothetical scenario, consider that you might be in the middle lane of a congested highway with no opportunity to pull over.

To prepare a growth curve, samples are taken at frequent intervals and population counts are performed by allowing the cells to develop into (macroscopic) **colonies** that can be counted (FIGURES 4.11 and 4.12). (One-hour intervals are

FIGURE 4.11 Growth curve of *E. coli*.

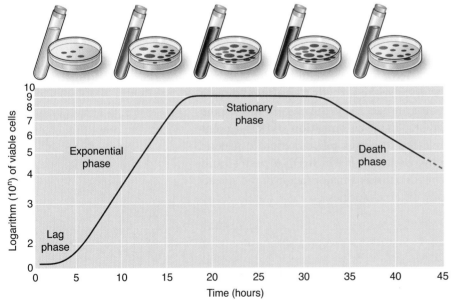

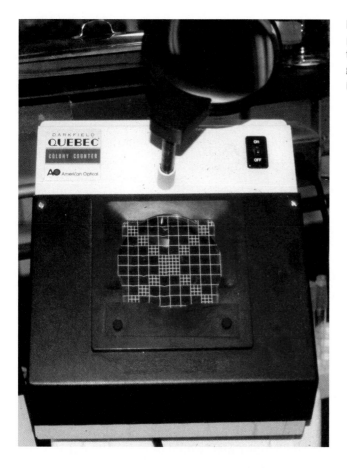

FIGURE 4.12 A colony counter. Agar plates with colonies are placed on the counting grid. A magnifying glass facilitates counting. Author's photo (RIK).

convenient—except for the student who needs to culture-sit throughout the night and into the next day.) Based on these counts, a growth curve is constructed illustrating the four major phases.

Lag Phase

The lag phase is a "get ready for growth" stage and represents time of adaptation to the new environment; adaptation also occurs in actual infection. Under both circumstances there may be an initial drop in the count because of adjustment and survival of only a part of the microbial population. In infection, immune mechanisms of the body more readily kill those cells that are "less fit" for survival so only the fittest survive—a nice example of Darwinism at the microbial level.

Exponential (Logarithmic) Phase

During the "log" phase there is an explosion of growth governed by the **generation time**—the time it takes a cell to undergo binary fission, producing two "new cells." *E. coli,* for example, has a generation time of only twenty minutes, assuming optimal growth conditions. At the conclusion of each generation time, the population is doubled.

TABLE 4.3 Microbial Generation Times

Microbe	Generation Time[a]
Bacteria	
Escherichia coli	20 minutes
Staphylococcus aureus	30 minutes
Clostridium botulinum	35 minutes
Mycobacterium tuberculosis	18 hours
Treponema pallidum	33 hours
Protozoans	
Leishmania donoviani	10 hours
Giardia lamblia	18 hours

[a]Times are approximate.

As an example, consider the dynamics of growth starting with a single *E. coli* bacillus. How many would you have in a 24-hour period during which 72 generations would have occurred? After the first 20 minutes there would be 2 cells, 20 minutes later (elapsed time of 40 minutes) there would be 4 cells, and after another 20 minutes (elapsed time of 60 minutes) there would be 8 cells. Although this may not be particularly impressive, consider the solitary cell has now reached a population of 4,096 cells in 4 hours, or just 12 generations. What about after 10 hours, or after 24 hours, and so on? Mathematically, this can be expressed as $2^1, 2^2$, and 2^3, respectively, with the exponents 1, 2, and 3 representing the number of generations. Therefore, at the end of 24 hours the total population would be 2^{72}, and at 48 hours (2^{144}) the population's mass would be greater than that of the Earth! Calculate the actual population and you will be amazed. Most of the common bacteria involved in human disease have short generation times. At the other end of the spectrum is the spirochete *Treponema pallidum,* the causative agent of syphilis, which has a generation time of 33 hours. Examples of generation times are given in TABLE 4.3.

There is a limit set by the conditions within the test tube, as there is in any functioning ecosystem, including exhaustion of nutrients, space, and accumulation of toxic materials. Sometime after about 10 or 12 hours, depending on the number of cells started with, adverse conditions set in and binary fission slows down, leading into the stationary phase.

Stationary Phase

What are the adverse conditions that arrest the growth of a culture? Having started with a single species there is no species competition, but there is a competition for the depleting supply of nutrients and O_2. Toxic materials, including carbon dioxide (CO_2), accumulate. Binary fission continues to occur but at a rate about equal to the death rate, and thus the culture has now reached a plateau, so there is no net increase in numbers. The culture has just about had it at this time, but as far as the individual bacterial cells are concerned, the worst is yet to come—the dreaded death phase!

Death Phase

The adverse conditions in the test tube become more and more pronounced. The number of cells dying exceeds the rate of binary fission. It still might be possible to save the culture by simply transferring it into a larger tube with fresh medium or, alternatively, by taking a small sample of the culture that should contain some viable cells and inoculating a fresh tube of medium. If the organism is a spore former (*Bacillus* or *Clostridium*), spore formation may have resulted, and the cells may remain viable for years in the spore stage.

Significance of Bacterial Growth

The dynamics of growth is of more than academic interest. What is the significance of all of this in terms of the natural state of infection? Think of the intestinal tract, the ear, an area of skin, or the pharynx as a test tube representing a particular ecological niche, each with its own distinctive environment. Certainly, the "ecosystem" of the intestinal tract differs markedly from that of the middle ear. In most cases symptoms of infection do not appear until there is a relatively large bacterial population. How do you apply this to the growth curve? Under most circumstances the initial introduction of bacteria into the body escapes notice, and this represents the lag phase. Within only several hours, depending on the particular organism and its generation time, a dramatic increase in the bacterial population takes place. It is during this exponential phase that you experience the symptoms associated with that infection, such as lethargy, abdominal pain, nausea, vomiting, diarrhea, headaches, and fever.

Intervention, by taking an antibiotic or other medicines as a supplement to the immune system, is an attempt to interrupt the logarithmic phase, enter the stationary phase, and, as quickly as possible, hasten the microbial death phase. In the absence of intervention, the body's immune system responds to the presence of the microbial pathogen by calling into play defense mechanisms, which attempt to destroy the invading microbes, thereby halting the continuation of the logarithmic phase. Should the body's defenses fail, the log phase continues, and the person will die of overwhelming infection.

Culturing Bacteria: Diagnostics

What happens to the throat swab that your physician uses to swab your pharynx, possibly when "strep throat" is suspected, or to the urine, fecal, or blood specimens that might be taken when bacterial infection is considered? (You cannot forget the throat swab! It is quite normal to gag, because, in fact, the swab does hit the spot that is responsible for the gag reflex. It is important to obtain a sample from your pharynx and not from the roof of your mouth.) The purpose of obtaining a specimen is to grow the suspect organism so that it can be identified. The swab or a sample of the specimen is placed into a tube of liquid nutrient broth and/or streaked across the surface of a **Petri dish** containing nutrients and agar (derived from seaweed). Agar is a semisolid material with a consistency slightly more firm than that of Jell-O™. Frau Hesse is credited with the use of agar

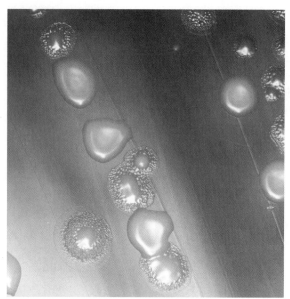

FIGURE 4.13 Isolation. The specimen is streaked onto the surface of an agar plate, which is then incubated to allow development of colonies. Courtesy of Dr. Richard Facklam/CDC.

as a solidifying agent of bacteriological media in the early 1800s; it remains an isolation technique used in microbiology research and diagnostic laboratories throughout the world.

Growth on **agar** (FIGURES 4.13 and 4.14) allows the production of colonies with recognizable and identifiable properties. Pure cultures are obtained by the streaking of a loopful of bacterial culture across the surface of an agar-containing Petri dish. Each colony is a "pure" culture and is the result of the binary fission of a single cell (at least theoretically). Colony characteristics, including size and pigmentation (Figure 4.14), are valuable clues in identification. For example, the streptococcus that causes strep throat is easily identifiable on primary cultures on sheep blood agar (agar to which sheep blood has been added). The streptococci secrete a product (hemolysin) that destroys the sheep red blood cells surrounding the colonies. This results in clear zones because of the lysis of the red blood cells and confirms the diagnosis of a sore throat caused by streptococci (FIGURE 4.15).

In most cases further studies are necessary to identify the organism. A wide variety of broth and agar media is available for growth and identification, and the particular selection depends on the source of the specimen. After all, the bacteria that are commonly associated with intestinal tract infections are different from those that cause skin, urinary tract, or respiratory tract infections. After incubation, the cultures are examined for characteristic properties (FIGURE 4.16). Gram stains are performed and are important tools to aid in identification. Further, the isolate is usually spread onto the surface of an agar plate, antibiotic disks

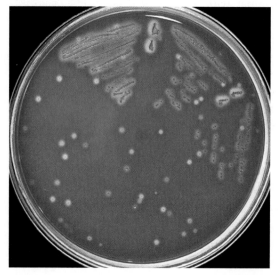

FIGURE 4.14 Colony characteristics. Species of bacteria display characteristic colony morphology, including pigmentation, texture, and size. Courtesy of Richard R. Facklam, Ph.D./CDC.

FIGURE 4.15 Diagnosis of strep throat. The streptococci responsible for strep throat are easily identified. A swab of the throat is streaked onto the surface of an agar plate enriched with sheep blood. The presence of clear zones around the areas of growth establishes the diagnosis. Author's photo (RIK).

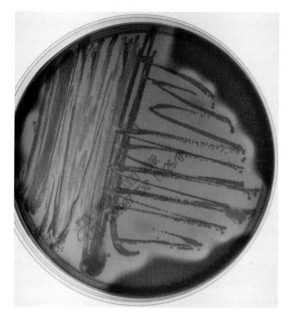

are added (FIGURE 4.17), and the plate is incubated and examined after growth has occurred. Zones of inhibition (no growth) occur around those disks to which the bacteria are sensitive, because the antibiotic leaches from the disk into the moist agar surface. This is an important tool for the physician in prescribing an appropriate antibiotic; other factors, including toxicity, cost, and duration of therapy, are taken into account.

The process of identification may take only 24 hours, as in the case of streptococci; microbes with longer generation times may require extended incubation to achieve sufficient growth before moving on to the next stage. In the case of tuberculosis, bacterial identification takes several weeks because of the long (18-hour) generation time characteristic of *M. tuberculosis* (Table 4.3).

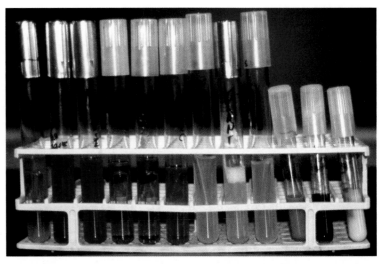

FIGURE 4.16 A variety of media in test tubes are used to determine metabolic properties of bacteria that are useful in identification. Author's photo (RIK).

Diagnostic microbiology has come a long way since the days of Frau Hesse. "Short-cut" techniques are available for diagnosing a variety of diseases caused by microbes or viruses including urinary tract infections (UTIs), diarrhea (e.g., salmonellosis and shigellosis), strep throat, syphilis, malaria, schistosomiasis (worm), rabies, influenza, and hepatitis.

Certain bacterial infections caused by g– bacteria are identified using rapid and cost-saving multitest procedures. Instead of inoculating a number of separate test tubes to determine the biochemical test results of a pure culture (FIGURE 4.18), a plastic strip known as an API-20E test strip is inoculated with a bacterial suspension. It contains twenty mini test tubes (cupules) with dehydrated substrates (media). During inoculation, the media is reconstituted and the strips are allowed to incubate for 18 to 24 hours. The bacterial metabolism causes color reactions that are immediate or revealed with the addition of reagents. The reactions are read by sophisticated automated instrumentation that assigns a specific code, the Analytical Profile Index, from which the initials API are derived. The API identifies the microbe to its genus and species more quickly leading to a rapid diagnosis and appropriate treatment (Figure 4.18). Persons suffering from Montezuma's Revenge are grateful!

It is not always possible to culture the suspect organism, particularly if the infected individual is already receiving antibiotics. Further, as too frequently happens, the urine, throat swab, or blood specimen may have been left at room temperature too long, resulting in its drying out, or the time necessary for the organism to grow out may impose too long a delay. Fortunately, there is another avenue of diagnosis available. Microbes leave telltale "footprints" of their presence by eliciting the production of specific antibodies by the host. Antibodies are produced in response to molecules called antigens that are part of the microbe or that are secreted (toxins). Because antibodies are specific and react only with the antigen

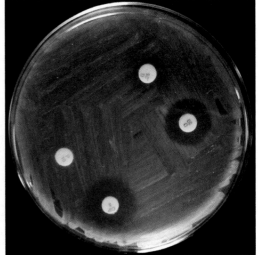

FIGURE 4.17 Determination of antibiotic sensitivity. The specimen is spread onto the surface of an agar plate, antibiotic disks are added, and the plate is incubated. Zones of inhibition (no growth) around a disk indicate the bacteria are sensitive. The disk is coded with the name and concentration of the antibiotic. Note that the spectrum of activity varies for each culture. Author's photo (TS).

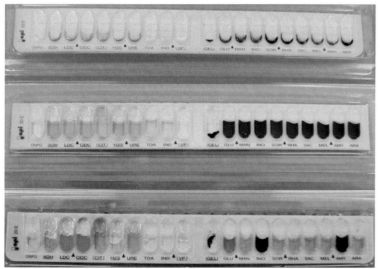

FIGURE 4.18 API gram-negative bacterial identification. Upper strip: uninoculated; middle strip: inoculated; lower strip: results after 24 hours of incubation at 37°C. Note the color changes in the cupules after incubation/growth of the microbe in the test strip. Similar testing is available to identify gram-positive bacteria. Author's photo (TS).

that caused their production, the presence of antibodies helps to establish the identity of the bacteria involved.

The tools of molecular biology have opened research and diagnostics in microbiology and, particularly, in virology. The application of molecular biology has added precision and rapidity in a cost-effective manner to diagnostic procedures. Essentially molecular biology as applied to diagnostics is focused on getting to the "heart" of the microbe (i.e., its genetic content). Diagnostic bacteriology has lagged somewhat in using this "new" approach because traditional culture methods and the more recent development of rapid tests have served well over the years.

Sore throats are common among kids. About 10% of all sore throats are caused by *Streptococcus pyogenes*. **Strep throat** is highly contagious and if untreated can increase the individual's risk for developing complications such as rheumatic fever, meningitis, and diseases affecting the heart and kidneys. **Rapid strep antigen tests** are used to quickly determine if *Streptococcus pyogenes* is the culprit, allowing for immediate antibiotic therapy if the test is positive. The throat and tonsils are swabbed to collect bacteria from the infected area for testing. The swab is placed in a plastic tube and drops of reagents are added to extract streptococcal antigens, followed by drops of antibodies against *Streptococcus pyogenes* that capture the antigens, resulting in a colored band or line in the reaction zone of a test strip (FIGURE 4.19). The test

FIGURE 4.19 Photograph of rapid strep antigen test results. The tube with the test strip on the left side is a positive control [positive result: *Streptococcus pyogenes* antigens are present in the sample. It contains a colored band in the upper test (or T) region and colored band in the lower control (or C) region]. A band in the test and control regions is a positive result. The tube in the middle represents a negative control; a colored band forms in the control region but not in the test region. The patient strip (M) on the right displays a band in the control but not the test region. It is a valid negative result. Author's photo (TS).

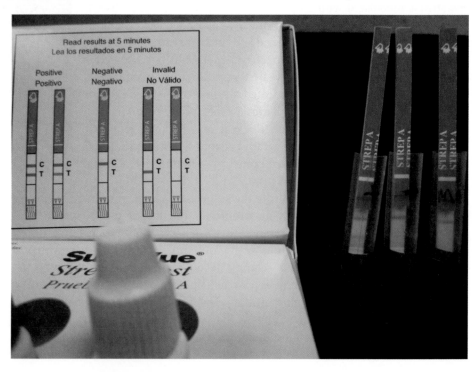

results are available within 10 to 15 minutes. A normal or negative test means that *Streptococcus pyogenes* may not be present. Following a negative result, an additional throat swab is taken and used to inoculate a blood agar plate (Figure 4.15). This is a safeguard against false negative reactions (meaning someone actually has a strep throat infection even though the rapid strep antigen test result was negative). A positive result indicates an infection. Culture follow-up is not required and the patient is treated with antibiotics. The next time a doctor or nurse swabs the back of your throat—an unpleasant experience—try hard to resist gagging as you anticipate your rapid strep test results!

Medical laboratory microbiologists need to be highly trained to identify suspected disease-producing bacteria properly. The techniques are simple, but interpretation of the laboratory findings can be complicated. The microbiologist must be thoroughly familiar with the normal bacterial inhabitants of the body to distinguish suspected disease producers. Many directors of today's clinical microbiology laboratories have a Ph.D. or M.D. degree, or both.

Oddball (Atypical) Bacteria

The foregoing description applies to "typical" bacteria, but there are three important "oddball" groups of bacteria, each of which possesses distinctive properties distinguishing it from the typical bacteria and from each other. These groups are significant disease producers in humans. They are **mycoplasmas**, **chlamydiae**, and **rickettsiae**; their properties are summarized in TABLE 4.4.

Mycoplasmas

Mycoplasmas are described as bacteria without cell walls; they are the "smallest of the small." Perhaps you have heard the term *walking pneumonia*; the causative agent is *Mycoplasma pneumoniae*.

Chlamydiae

The chlamydiae are slightly larger than the mycoplasmas and can be seen with a light microscope. These coccoid bacteria are obligate intracellular parasites meaning that they can only grow within a cell, a property shared with rickettsiae and viruses. Transmission is from person to person. Chlamydiae are responsible for a variety of diseases, including urethritis (inflammation of the urethra), trachoma (a common cause of blindness), and lymphogranuloma venereum (a sexually transmitted disease).

TABLE 4.4 Comparison of Mycoplasmas, Chlamydiae, and Rickettsiae

Bacterial Group	Size (micrometers)	Cell Wall	Obligate Intracellular Parasite
Mycoplasmas	0.3–0.8	Absent	No
Chlamydiae	0.2–1.5	Present	Yes
Rickettsiae	0.8–2.0	Present	Yes

Rickettsiae

Rickettsiae are rod-shaped bacteria, the largest of the three groups of atypical bacteria and can be seen with a light microscope. With the exception of Q fever, diseases caused by these organisms are acquired through the bites of mosquitoes, ticks, fleas, lice and other arthropods.

Like the chlamydiae they are obligate intracellular parasites. They are commonly grown in the yolk sac of (live) chicken embryos for laboratory study. According to some historians, typhus fever (as distinct from typhoid) defeated Napoleon's army in the Russian invasion of 1812. Perhaps you have read *The Diary of Anne Frank*, the story of a teenage girl who hid from the Nazis with her family in a secret attic between 1942 and 1944. The family was betrayed and sent to Bergen-Belsen, a German concentration camp. Anne died there of typhus fever at the age of 15, along with hundreds of others shortly before the camp was liberated by Allied forces (FIGURE 4.20).

FIGURE 4.20 A memorial statue of Anne Frank in Amsterdam. Anne Frank died of typhus fever at the age of fifteen in Bergen-Belsen, a German concentration camp. Author's photo (RIK).

Overview

Bacteria are a distinct group of unicellular microbes dating back to antiquity and were, possibly, the beginnings of life. They are small, but they are not simple. Bacteria display a diversity of shapes and patterns of arrangement. Most are bacilli, cocci, and spirilla. Cocci are subdivided into streptococci, staphylococci, diplococci, and tetrads.

Bacteria have a complex anatomy consisting of an envelope, cytoplasm, and appendages. The envelope includes the capsule, the cell wall, and the cell membrane. The cytoplasm is the region of the cell within the cell membrane and includes a nucleoid, plasmids, and spores. Appendages are flagella and pili, both of which extrude through the cell wall. The capsule, plasmids, spores, flagella, and pili are not present in all bacteria.

Bacteria exhibit considerable diversity in generation time—the time it takes a cell to undergo binary fission resulting in two new cells. Under optimal laboratory conditions, the generation time for *E. coli* is 20 minutes, whereas the generation time for *Treponema pallidum* is 33 hours. The four major phases of the curve are the lag phase, exponential growth phase, stationary phase, and death phase. Bacterial growth curves provide insight into the dynamics of bacterial growth that occur in an infected individual. The fact that most bacteria can be readily cultured in the laboratory facilitates identification of the causative organism, allowing appropriate treatment. Mycoplasmas, rickettsiae, and chlamydiae are three groups of atypical bacteria, each of which includes pathogens (disease-producing microbes).

Self-Evaluation

PART I: Choose the single best answer. Questions 1 to 4 are based on the following key:

    a. capsule **b.** cell wall **c.** cell membrane **d.** pili

1. Accounts for Gram stain reaction _____
2. Responsible for gatekeeper function _____

3. A deterrent to phagocytosis _____

4. Important in adhesion to host cells _____

5. Assume that the organism *Robertus krasnerii* has a generation time of twenty minutes and that you are starting with one cell. How many generations will it take to achieve a population of 2,048 cells?

  a. cannot determine **b.** 10 **c.** 11 **d.** 12

6. In examining a Petri dish culture with antibiotic sensitivity disks, the greatest sensitivity would be demonstrated by the disk with the

  a. most growth in the vicinity of the disk **b.** smallest zone of no growth

  c. largest zone of no growth **d.** no zone of inhibition

7. The property of "hardiness" is associated with

  a. flagella **b.** cell wall **c.** capsule **d.** spores

8. Which group of bacteria is unique in lacking a cell wall?

  a. procaryotes **b.** chlamydiae **c.** rickettsiae **d.** mycoplasmas

9. G+ bacteria possess many layers of this chemical component

  a. starch **b.** peptidoglycan **c.** protein **d.** sialic acid

10. Which small circular nonchromosomal molecules replicate independently inside of certain bacteria.

  a. inclusion bodies **b.** flagella **c.** plasmids **d.** fimbria

PART II: Fill in the blank.

  1. Cocci that tend to form chains are called _____.

  2. Corkscrew-shaped spiral bacteria are called _____.

  3. A defense mechanism of the body is the ability of certain cells to engulf and destroy bacteria. This process is known as _____.

  4. _____ and _____ are examples of spore-forming genera.

  5. _____ is a rapid test used in the doctors office to detect *S. pyogenes* antigens.

  6. _____ is credited with the use of agar in bacteriological media.

  7. A bacterium that has flagella all over its surface is described as _____.

  8. _____ is a form of cell division used by all bacteria.

  9. It is best to treat a bacterial infection with antibiotics while it is in _____ phase (hint: rapidly dividing cells).

  10. _____ are responsible for a variety of diseases including urethritis, trachoma, and lymphogranuloma.

PART III: Answer the following.

  1. Draw and label a "composite" bacterial cell. Indicate what structures are not found in all bacterial cells.

  2. The term *bugs* is conventionally used to describe all microbes. If you are ill and diagnosed with a "bug," what are the limitations of this diagnosis?

3. The chances are that *Clostridium tetani* will not cause tetanus when an individual sustains a cut but might when a puncture wound occurs. Why is this the case?

4. Why is it necessary to use pure cultures of bacteria in the diagnostic laboratory?

5. Spores are extremely hardy structures. Create a list of at least four environmental conditions or processes aimed to kill vegetative cells but will have no or little effect on bacterial spores.

6. List the atypical or oddball bacteria. What physical characteristics make them different from typical bacteria?

7. Draw and label six different bacterial shapes. Include the two shapes of the majority of bacteria and put an asterisk next to them.

8. As a general rule, any raw food may contain *Clostridium perfringens*. Explain.

9. Draw a typical bacterial growth curve and label the four major phases.

10. Explain why it is not always possible to culture a suspect microbe from an infected person?

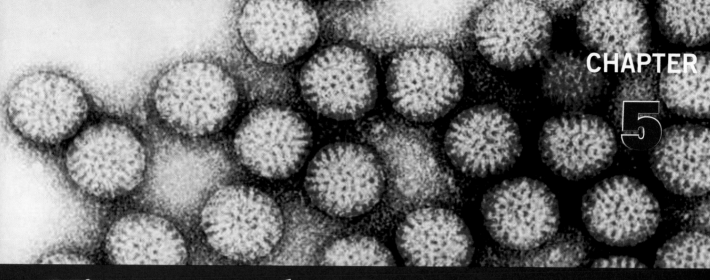

Viruses and Prions

Viruses are interesting because they are neither inanimate nor living; a virus is a roadway between brute matter and living organism. Nothing living lies so close to the middle of life—and to its solution—as viruses.

—Wolfhard Weidel

I'm stopped. I'll never eat another burger. (referring to prions)

—Oprah Winfrey

■ Preview

Viruses are subcellular infectious agents responsible for a variety of diseases in humans. Familiar diseases such as smallpox, poliomyelitis, and influenza are but a few examples. It was not until the 1930s that viruses were described in terms of their biological and chemical properties. Viruses are obligate intracellular parasites, meaning they must gain access into cells to replicate. Viruses are particularly unique in that they have either DNA or RNA, whereas cells contain both DNA and RNA. The nucleic acid in viruses is wrapped in a protein coat, and some have an envelope around the coat; others have protruding spikelike structures. A generalized cycle of viral replication follows the sequence of adsorption, penetration, replication, assembly, and release, but the strategies differ depending on the specific virus. Because of their obligate intracellular parasitism, cell culture and fertile chicken eggs are used to grow viruses. The diagnosis of a specific viral disease is

Photo courtesy of Dr. Erskine Palmer/CDC.

usually based on the patient's symptoms rather than on isolation of the virus. Bacteriophages are viruses that infect bacterial cells; these viruses have served as models toward understanding the dynamics of host–parasite relationships that occur in human viral diseases. The use of bacteriophages to treat bacterial infections, particularly since the increase in antibiotic-resistant bacteria, has reemerged as a possible therapy. Gene therapy and virotherapy as a cancer-fighting treatment is moving forward more quickly into medical practice.

Prions are infectious proteins capable of causing fatal neurodegenerative diseases in animals and humans. They are subcellular agents unique in that they lack nucleic acids.

■ Viruses as Infectious Agents

What biological agents are smaller than a cell, not considered to be alive, able to be transmitted from person to person, potentially fatal, and have caused the largest recorded outbreak of disease on a worldwide scale (pandemic) in the history of humankind? A key word in the title of the chapter is a giveaway to the answer to this riddle: viruses. The **pandemic** referred to is the influenza (Spanish flu) catastrophe of 1918, which killed fifty million people around the world in only one year. AIDS, too, is a pandemic viral disease that began in the latter part of the twentieth century and is devastating worldwide, particularly in Africa. According to the director of the United Nations Children's Fund, "by any measure the HIV-AIDS pandemic is the most terrible undeclared war in the world with the whole of sub-Saharan Africa a killing field."

Viruses are subcellular (simpler than a cell) and, therefore, can replicate (their sole claim to life) only when circumstances result in their gaining access to a cell. They have no life outside the cell and exist only as "freeloaders." The host cell is taken prisoner and turned into a virus-producing machine. This is the basis for describing viruses as obligate intracellular parasites. They literally need to "get a life." On the positive side, the fact that viruses take up intracellular existence is being exploited in genetic engineering and gene therapy by using "tamed" viruses as vehicles for the delivery of "good" genes into cells to replace defective or nonexistent genes and **virotherapy** to treat cancers.

Everyone has experienced numerous bouts of common viral infections, including colds, chickenpox, influenza, and gastroenteritis. Human immunodeficiency virus (HIV), Ebola virus, hantavirus, and West Nile virus cause diseases that have been widely publicized (FIGURE 5.1). In 1997 a potential outbreak of avian influenza in Hong Kong, dubbed by the media as the bird-flu virus, caused a scare around the world. Subsequently, between 2003 and 2011 the World Health Organization documented 566 cases with 332 deaths—a staggering mortality rate of 59%. Millions of chickens and other birds have been slaughtered in an attempt to halt the spread. Fortunately, that scare proved to be unwarranted, but the potential threat has reemerged. In October 1999 an outbreak of 30 cases with at least 5 deaths caused by a mosquito-borne virus occurred in New York and was initially presumed to be due to the St. Louis encephalitis virus. The real culprit, however, turned out to

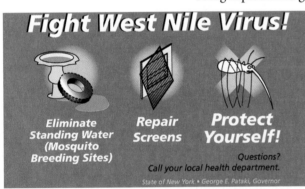

FIGURE 5.1 Public health departments are responsible for making the public aware of many diseases, including the illness caused by the West Nile virus. Reproduced with permission of the New York State Department of Health.

be the West Nile virus, which had never before been seen in the Americas. How the virus came to New York remains an open question.

Ebola hemorrhagic fever garnered front page headlines when it broke out in the Democratic Republic of the Congo in 1996 and reemerged in September 2007, causing 187 deaths out of 264 cases. It returned again in 2008 and 2011 in Uganda. In 2011 cases of mumps and measles reemerged in Canada, U.S., United Kingdom, Europe, Australia, and New Zealand, spreading by air travel. In most of the aforementioned locations these were the first mumps or measles cases in twenty or more years. These examples attest to the occurrence of new and emerging infections. Smallpox is a terrible viral disease that decimated large segments of the population. The disease was described as early as about 10,000 B.C. Examination of the body of Pharaoh Ramses V, who died in 1157 B.C., revealed pustule-like scars on the face, neck, and shoulders, which appeared to have been caused by *Variola* (the virus that causes smallpox). Shamefully, *Variola* was used in biological warfare long before its identity was known. Some historians claim that in the 1760s English troops, under the guise of appeasement, gave the Native Americans blankets from a smallpox hospital in hopes the Native Americans would contract the disease. Smallpox was the first disease for which immunization was available, thanks to Lady Mary Wortley Montagu's promotion of a procedure to prevent smallpox known as variolation during the 1720s and the pioneering work of vaccination by Edward Jenner in 1796. Smallpox is the first and, to date, the only disease to be eradicated from the face of the earth. The word *eradicated* is to be emphasized as compared with the word *controlled*. The irony is that this triumph renders the world susceptible to the threat of smallpox, because the virus still exists in a couple of government laboratories and could find its way into the hands of terrorists.

The term *virus* can be found in the mid-1800s writing of Pasteur and others who preceded him long before the germ theory of disease was established and the biology of viruses was known. *Virus* means "poison" or "slime" and has a negative connotation. By the 1940s viruses had been well described, and the term now has a biological basis. In more recent years *virus* has become part of computer jargon and indicates something that is easily spread (infectious) and destroys data.

Based on the work of the Russian scientist Dimitri Ivanowsky in 1892, viruses were described as "filterable." Ivanowsky filtered extracts of tobacco leaves with symptoms of tobacco mosaic disease through filters that were known to retain bacteria. He then injected these extracts into healthy tobacco plants and observed symptoms of tobacco mosaic disease (FIGURE 5.2). It was clear that the infectious agent was, indeed, smaller than bacteria. Several

FIGURE 5.2 **(a)** Discovery of "filterable" viruses. **(b)** A transmission electron micrograph of tobacco mosaic virus. © Phototake/Alamy Images.

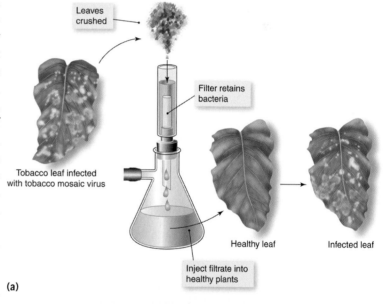

Leaves crushed

Filter retains bacteria

Tobacco leaf infected with tobacco mosaic virus

Healthy leaf

Infected leaf

Inject filtrate into healthy plants

(a)

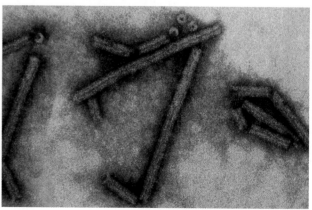

(b)

years later the Dutch microbiologist Martinus Beijerinck also demonstrated that tobacco mosaic disease was caused by a filterable agent. The term *filterable* was eventually dropped. A breakthrough occurred in 1935 when Wendell M. Stanley, an American biochemist, purified and crystallized the tobacco mosaic virus, opening the door for more sophisticated examination of a purified virus in an effort to further elucidate its structure. The invention of the electron microscope, at about the same time, allowed viruses to be viewed.

All cells are potential hosts for viruses. Even *Escherichia coli* and other bacteria serve as hosts. Bacterial viruses are known as bacteriophages (usually shortened to "phages"). Phages, because of the ease of growing the bacteria that serve as their hosts, have served as models for studying animal and plant viruses. Our current knowledge of viruses, in the molecular biology of cells has been largely gained by studying bacterial viruses.

Virus Structure

Among microbes, viruses are the second smallest; prions are the smallest. Bacteria are described as microscopic, meaning they can be seen only with a light microscope. Viruses, however, are in a different realm, the realm of the electron microscope, and they range in size from about 20 to 350 nm. Well over one thousand phage particles can fit into an *E. coli* cell. The 2003 discovery of mimiviruses living inside of amoebas in a cooling tower changed the definition of virus size and genetic material. Mimivirus was not only larger than previously seen viruses, but its genetic material was similar to atypical bacteria such as mycoplasmas, chlamydiae, and rickettsiae. In 2008, a nearly identical but larger virus named Mamavirus was described followed by the 2011 discovery of the largest relative named Megavirus (BOX 5.1). If there is a Mamavirus, will there be a Papavirus? What will the next ginormous virus be called— Gigavirus? (Not Gingivitis?) These larger viruses are found in aquatic environments and represent the exception to the traditional size criteria. This text will focus on the characteristics and diseases caused by the vast majority of traditional viruses.

As viewed through an electron microscope, viruses are simple in structure and may consist only of a nucleic acid wrapped in a protein coat (FIGURE 5.3), whereas others may have an envelope around the coat and, additionally, "spikes." The larger viruses (e.g., Variola (smallpox) virus) overlap the size of the smaller bacteria, and although they contain a few enzymes, they are obligate intracellular parasites. A complete viral particle is referred to as a **virion**.

Nucleic Acids

Viruses are unique in that they have either RNA or DNA genomes but never both. Cells, whether procaryotic or eucaryotic, contain both DNA and RNA. In the RNA viruses the RNA serves as the repository for genetic information. For example, the influenza virus is an RNA virus that replicates to produce more influenza viruses. Hence, the genetic information must be encoded in the RNA. Further, some viruses are unique in that their genome is single-stranded DNA (ssDNA); others have double-stranded RNA (dsRNA). Other than in these viruses, DNA is double stranded and RNA is single stranded. Hence, in terms of nucleic

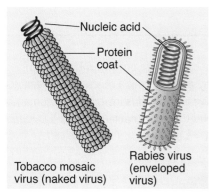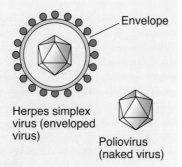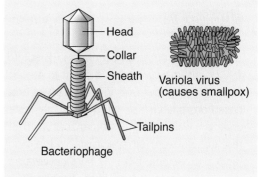

Nucleic acid	Envelope	Head
Protein coat		Collar
		Sheath
		Variola virus (causes smallpox)

Tobacco mosaic virus (naked virus)

Rabies virus (enveloped virus)

Herpes simplex virus (enveloped virus)

Poliovirus (naked virus)

Tailpins

Bacteriophage

Helical viruses

Icosahedral viruses

Complex viruses

(a)

FIGURE 5.3 **(a)** Basic virus structure. **(b)** A negative stain transmission electron micrograph (TEM) showing a smallpox virus particle (310,000×). Courtesy of J. Nakano/CDC. **(c)** A colorized, negative-stained TEM depicts the ultrastructural details of the virus that caused an outbreak of swine flu (H1N1) in Mexico and other countries in April, 2009. Courtesy C. S. Goldsmith and A. Balish/CDC. **(d)** Colorized TEM of adenovirus. Courtesy of Dr. G. William Gary, Jr./CDC. **(e)** A negative-stained TEM of a number of filamentous Marburg virions (100,000×). Courtesy of Dr. Erskine Palmer, Russell Regnery, Ph.D./CDC.

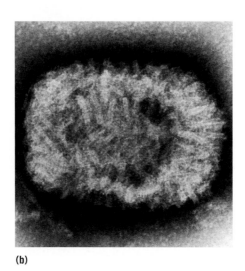

(b)

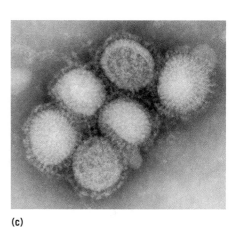

(c)

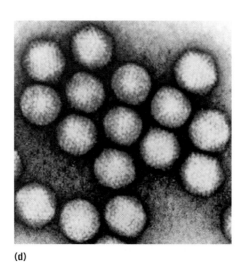

(d)

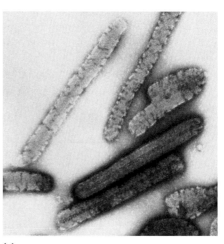

(e)

BOX 5.1 Big and Bizarre Viruses

A pneumonia outbreak occurred in Bradford, England in 1992 caused by what was thought to be bacteria isolated from a cooling tower. Its identification as a bacterium was based on its large size and the fact that it stained g+. The organism was named "Bradford coccus" (FIGURE B5.1). The researchers were all specialists in the study of oddball (atypical) bacteria, but they were not aware that they had stumbled upon a microbe that changed the classic definition of a virus.

As a part of their studies they were using water samples from a cooling tower in an effort to co-culture amoebas with Bradford cocci. They are symbiotic organisms that failed to culture separately. The investigators revealed that the so-called Bradford cocci lived within amoebas. Attempts to find and sequence a particular genetic marker (16S rRNA) used as a tool to help in distinguishing bacteria and viruses were fruitless. However, under the electron microscope the coccus looked like a hairy, large insect virus known as an "iridovirus." The Bradford coccus today is known as a Mimivirus—short for "mimicking microbe."

Similar hairy viruses, called Mamavirus from a Paris water cooling tower and Megavirus isolated from seawater at a marine station at Las Cruces (Chile) have been

identified. Here's another bizarre one: in 2010, the contact lens solution used by a 17-year-old French woman suffering from an aggressive eye **keratitis** infection was contaminated with two different types of bacteria, and an amoeba carrying a new giant virus called Lentille (TABLE B5.1).

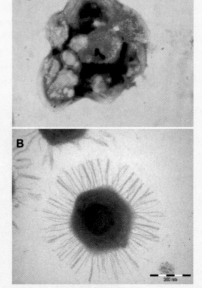

FIGURE B5.1 (A) Gram stain of virus in amoeba. (B) Electron micrograph of Mimivirus. Photos reprinted with permission from the American Society for Microbiology (*ASM News*, July 2005, p. 279.) Photos courtesy of Didier Raoult.

TABLE B5.1 Size Ranges of Big and Bizarre Viruses

Name	Source	Isolated	Capsid Size (nm) ("hairs" not included)
Megavirus	Seawater	Chile	520
Lentille	Contact lens solution of patient with keratitis	France	500
Mamavirus	Cooling tower	France	450
Longchamps	Decorative fountain	France	450
Terra1	Soil	France	420
Mimivirus	Cooling tower	England	400
Bus	Cooling tower	France	400
Pointerouge1	Seawater	France	390

acid content, there are four categories of viruses: dsDNA, ssDNA, dsRNA, and ssRNA. This fact is important for the classification of viruses (TABLE 5.1), along with other properties. Small viruses have only a few genes, whereas larger viruses have a few hundred genes. *E. coli,* by comparison, has approximately 4,000 genes, and human cells are estimated to have about 25,000 genes, possibly more.

TABLE 5.1 Virus Classification Based on Nucleic Acid Composition

Nucleic Acid	Virus(es) (Disease)
dsDNA	Human papillomavirus, Epstein-Barr virus, adenovirus, herpes simplex virus, varicella-zoster virus (chickenpox and shingles), *Variola* virus (smallpox)
ssDNA	Parvovirus B19 (possibly slapped-cheek disease)
ssRNA	Hepatitis A virus, poliovirus, norovirus, rubella virus, Ebola virus, influenza virus, West Nile encephalitis virus
dsRNA	Rotavirus, Colorado tick fever virus, reovirus

Protein Coat

The protein coat is called the **capsid**, and the term ***nucleocapsid*** refers to the nucleic acid genome plus the protein coat. The coat, in turn, consists of protein units called **capsomeres**. Three arrangements of these capsomeres have been described: **helical**, **icosahedral**, and **complex** (Figure 5.3).

Helical viruses consist of a series of rod-shaped capsomeres that, during assembly, form a continuous helical tube containing the nucleic acid characteristic of the particular virus. The tobacco mosaic virus is an example of a helical virus. Imagine eating a turkey wrap; the shell is like the protein coat, and the turkey is the nucleic acid.

In most polyhedral viruses the capsomeres form **icosahedrons**—three-dimensional, twenty-sided, triangular structures that confer a geodesic-appearing shape to the virus. These viruses, based on their appearance under the electron microscope, are also described as spherical viruses. Triangles, structurally stable geometric configurations in which the stress is evenly distributed, are used by engineers to create geodesic structures. On a trip to Epcot Center at Disney World in Florida, there is no escaping viewing Spaceship Earth (FIGURE 5.4), a giant geodesic dome at the entrance. At the Antarctic research station at the South Pole, a geodesic dome houses one of the facilities. It follows that the arrangement of the capsomeres in polyhedral viruses confers a structurally sound design.

Some of the commonly studied bacteriophages are complex viruses; they consist of a polyhedral head, a helical tail, and tail fibers that serve for attachment to the host cell. These viruses have an intricate anatomy and resemble a spacecraft. There are still other viruses that are not easily categorized as either polyhedral or helical and are considered atypical or complex viruses (e.g., *Variola* virus [causes smallpox]).

FIGURE 5.4 The Spaceship Earth at Disney World's Epcot Center is an example of an icosahedral geodesic dome. © Songquan Deng/123RF.

Viral Envelopes

Some viruses, during the process of emerging from the host cell, acquire a piece of the host cell's plasma membrane, which constitutes an additional layer covering the capsid. These viruses are called **enveloped viruses**. The envelopes are modifications of the host cell membrane in which some of the membrane proteins are replaced with viral proteins. Some of these appear as **spikes** because they protrude from the membrane and are important for the attachment of viruses to host cells. The AIDS virus (HIV)

TABLE 5.2 Clinical Classification of Viruses

Category	Tropisms (Tissue Affinities)	Diseases
Dermotropic	Skin and subcutaneous tissues	Chickenpox, shingles, measles, mumps, smallpox, rubella, herpes simplex, warts and cervical cancer (caused by papillomaviruses)
Neurotropic	Brain and central nervous system tissues	Rabies, arboviral encephalitides (eastern equine encephalitis, St. Louis encephalitis, La Crosse encephalitis, West Nile encephalitis), polio, Nipah encephalitis
Viscerotropic	Internal organs	Yellow fever, AIDS, hepatitis A and B, infectious mononucleosis, dengue fever, gastroenteritis (e.g., norovirus)
Pneumotropic	Lungs and other respiratory structures	Influenza, common cold, respiratory syncytial disease, severe acute respiratory syndrome (SARS)

has spikes that attach to strategic white blood cells (T lymphocytes) of the immune system, leading to serious damage to the individual's ability to fight infection. Viruses lacking an envelope are referred to as **naked** or **nonenveloped** viruses (Figure 5.3).

Viral Classification

Superficially, viruses are classified as plant, animal, or bacterial viruses in accordance with the host organism they infect. Clinically, they may also be classified according to the organ or organ system involved (**TABLE 5.2**), for example, **dermotropic** (skin), **viscerotropic** (internal organs), or **pneumotropic** (respiratory system). This latter classification, however, is limited by the fact that some viruses can infect more than a single organ or organ system. Current classification schemes are based on the biology of the virus, starting with their nucleic acid content. Despite the fact that viral taxonomy (classification) is complex, a workable scheme is based on the type of nucleic acid (ssDNA or dsDNA or ssRNA or dsRNA), capsid structure (helical or polyhedral), number of capsomeres, and other chemical and physical properties. Over five thousand viruses have been described by the **International Committee on Taxonomy of Viruses,** and each has been assigned to one of 87 recognized families. Who knows how many viruses are out there?

Viral Replication

The fact that viruses are obligate intracellular parasites makes them harder to grow and study. Although it is easy to grow bacteria in lifeless medium in test tubes or on Petri plates, the culture (growth and replication) of viruses requires the presence of living cells; these techniques are described later in this chapter. This is particularly complicated in the case of animal and plant viruses but, nevertheless, is routinely done in laboratories specialized in these techniques. Bacteriophages, on the other hand, are much easier to study because their host cells are bacteria, which can be readily grown. The system that has been most extensively investigated and that has served as a model

TABLE 5.3 Generalized Viral Replication Cycle[a]

Stage	Description
Adsorption (attachment)	Viruses attach to cell surface receptor molecules by spikes, capsids, or envelope.
Penetration	Entire viral particle or only nucleic acid enters via endocytosis or by fusion with cell membrane.
Replication	Process is complex, and details depend on particular viruses and their nucleic acid structure. Replication may occur in the nucleus or cytoplasm, viral nucleic acid replicates, and genes are expressed, leading to production of viral components.
Assembly	Components are assembled into mature viruses.
Release (exit)	Viruses are extruded from host cell by budding (HIV) or lysis of host cell membrane.

[a]Specific strategies vary with particular viruses.

for animal viruses is the T4 bacteriophage (a complex dsDNA phage) and *E. coli*. The T4 viruses enter the *E. coli* cell and use the cell's metabolic machinery to form a large number of new T4 viruses, which are released by causing the lysis of their bacterial hosts. These new virions are now capable of infecting other *E. coli* cells.

The replication of T4 in *E. coli* is a model for the replication of viruses in general, although there is considerable variation in the strategies demonstrated by specific viruses. The replication cycle is divided into five stages (TABLE 5.3 and FIGURE 5.5): **adsorption**, **penetration**, **replication**, **assembly**, and **release**. Depending on the particular virus, there is variation in the time required to complete the cycle. For example, the length of the cycle for poliovirus is only 6 hours, whereas it is 36 hours for a herpesvirus. The replication cycle is of more than academic interest. This process occurs when you have a viral infection.

FIGURE 5.5 Replication cycle of a bacterial virus.

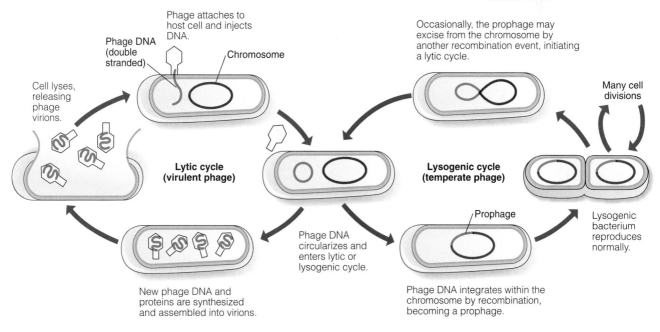

Phage attaches to host cell and injects DNA.

Phage DNA (double stranded)

Chromosome

Occasionally, the prophage may excise from the chromosome by another recombination event, initiating a lytic cycle.

Cell lyses, releasing phage virions.

Many cell divisions

Lytic cycle (virulent phage)

Lysogenic cycle (temperate phage)

Prophage

Lysogenic bacterium reproduces normally.

New phage DNA and proteins are synthesized and assembled into virions.

Phage DNA circularizes and enters lytic or lysogenic cycle.

Phage DNA integrates within the chromosome by recombination, becoming a prophage.

Adsorption

As an obvious prelude to replication, the virus must penetrate the cell, a process that first requires adsorption onto the host cell surface. Bacterial cells have receptors on their cell walls that serve as sites of attachment. Animal cells lack cell walls, and receptor molecules are embedded within their cell membranes. Viruses do not seek out a particular host cell; it is a matter of chance encounter. To view it another way, the virus must "dock" with a receptor molecule onto which it fits (FIGURE 5.6). Consider a piece from a jigsaw puzzle; it can only fit into an appropriate complementary shape. Specificity is thereby required and establishes host range. For example, *E. coli* strain B serves as the host for a particular phage, called T4, whereas *E. coli* strain C serves as the host for phage ΦX174. T4 does not infect C, nor does ΦX174 infect B, despite the fact that these two *E. coli* strains are virtually identical. Yet, there is a difference between these strains in the receptor molecules, which are specific for the tail fibers of either T4 or ΦX174. Hence, the host for T4 is *E. coli* B, and the host for ΦX174 is *E. coli* C. Specificity is always the name of the game to some degree, but it may not always be as fussy as in the *E. coli*–phage system.

In animal and in human viruses, too, specificity establishes host range. HIV has spikes on its surface that dock with human receptors (CD4 and co-receptors) on the surface of a particular type of white blood cell (T lymphocyte), allowing, ultimately, the replication cycle of the virus to be completed. During this process the cells are destroyed, rendering the immune system severely impaired. The hemagglutinin spikes on the surface of influenza viruses bind to sialic acid receptor sites present on cells of the respiratory tract, including the lungs.

Penetration

At this point the virus is positioned on the surface of the host cell. This, in itself, is of little consequence because viruses are obligate intracellular parasites. Therefore, one way or another, penetration of the host cell is necessary for translation of viral proteins and replication of viral genetic material. Keeping in mind that it is the viral nucleic acid, whether ssDNA, dsDNA, ssRNA, or dsRNA, that carries the

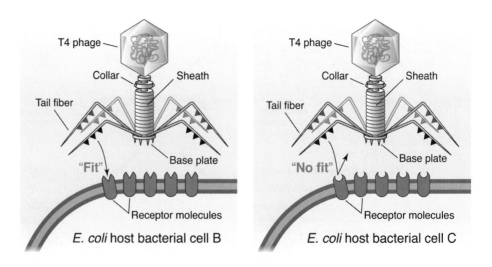

FIGURE 5.6 Bacteriophage "docking" with receptor molecules.

(a) Eucaryotic Cell

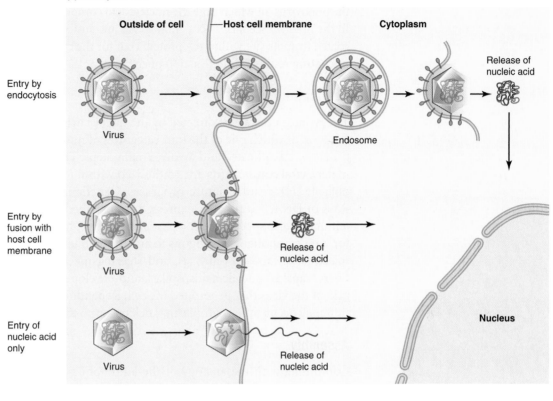

Outside of cell — Host cell membrane — Cytoplasm

Entry by endocytosis

Virus

Endosome

Release of nucleic acid

Entry by fusion with host cell membrane

Virus

Release of nucleic acid

Entry of nucleic acid only

Virus

Release of nucleic acid

Nucleus

(b) Procaryotic (Bacterial) Cell

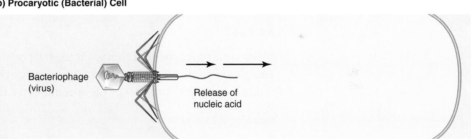

Entry of nucleic acid after lysozyme degrades small portion of cell wall

Bacteriophage (virus)

Release of nucleic acid

genetic message, the critical factor is that the nucleic acid must enter the cell (FIGURE 5.7). Perhaps the most intriguing mechanism of penetration is observed in some of the phages, including the T4 phage, of *E. coli* B. This phage initially attaches to the bacterial cell wall by phage tail fibers and docks its base plate on the cell wall of its bacterial host. The base plate contains the enzyme **lysozyme**, which degrades a portion of the peptidoglycan of the cell wall. The phage DNA is transferred across the cell wall into the bacterium by an unknown mechanism. The capsid remains on the outside of the cell and is of no further consequence.

In the case of animal or human viruses, the nucleic acid is not injected into the host cell. Rather, some enveloped viruses enter the cell by a process called **endocytosis** in which the complete virion is engulfed by the host cell and subsequently contained within a **vesicle**. This is the case for *Variola* (smallpox) virus and other poxviruses. In other enveloped viruses, **fusion** of the envelope with the cell membrane occurs and the nucleocapsid enters as, for example, in the mumps and

FIGURE 5.7 Strategies of viral penetration of cells. **(a)** Animal or human viruses penetrating a eucaryotic cell. **(b)** Bacteriophage nucleic acid entering bacterial host cell.

measles viruses. Still a third strategy exists for some of the naked viruses, including the poliovirus, in which only the nucleic acid component enters the cell's cytoplasm. In the first two scenarios (i.e., endocytosis and fusion), the nucleic acid must be uncoated from its site within the protein coat for the cycle to continue. This process of **uncoating** results from the action of enzymes of the host cell or of the virus.

Replication

The replication stage is directed by the nucleic acid of the virus. The details are a function of which one of the four categories of nucleic acid the virus contains; the process is complicated and involves many steps. The important point is that, ultimately, viral components are synthesized within the host cell, again reflecting the obligate intracellular nature of viruses. Genetic information, whether stored in RNA or DNA, is ultimately expressed (transcribed and translated) into viral protein molecules specific for the particular virus. Think of the host cell as a factory for viral components analogous to an automobile factory. Instead of making fuel injectors, dashboards, airbags, and steering mechanisms, the products are tail fibers, capsids, nucleic acids, spikes, and envelopes. The blueprint is the nucleic acid of the virus that penetrated the cell. Depending on the specific virus, production may occur in the cytoplasm, nucleus, or possibly in both production sites.

Assembly

Continuing with the analogy of the automobile plant, now that the parts have been made, the virus is assembled into a functional structure in a manner similar to that of a production line (FIGURE 5.8).

Release

Mission accomplished! At this point the host cell is teeming with newly formed and complete virions, all potentially infective for host cells. The mechanism of release of virions varies as a function of the specific virus. In some cases release results in the

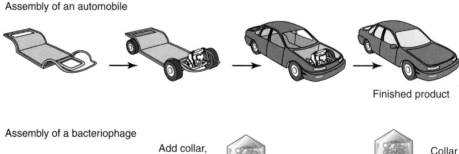

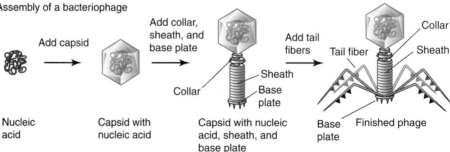

FIGURE 5.8   Viral assembly.

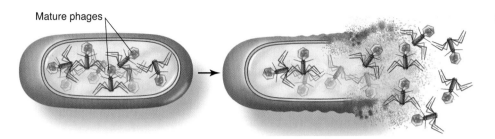

Mature phages

death of the host cell. For example, with some of the phages (e.g., T4) the release of new phage particles, referred to as the burst, is the result of the enzymatic splitting open (lysis) of the cell membrane (**FIGURE 5.9**). As many as a few hundred new phage virions may be released. The entire process from adsorption to burst averages about twenty to forty minutes in phages. In other cases death is not the outcome.

How is release accomplished in animal or in human viruses? Nonenveloped animal viruses are released from the cell by a process similar to the release of some phages involving lysis and death of the host cell.

Release is somewhat more complicated in the case of enveloped viruses (**FIGURE 5.10**). Generally, before release the expression of viral genes brings about the production of viral envelope components in some cases. **Budding** or **extrusion** is, essentially, the reverse of endocytosis. The virus pushes its way through the membrane and pinches off a piece of the membrane–spike complex, which now envelopes the nucleocapsid. Hence, the envelope consists of elements encoded by viral genes and by host cell membrane genes. Unlike lysis, budding is a gradual and continuous process and does not necessarily kill the cell. Ultimately, cell death is due to the takeover of the cell's machinery and accumulation of viral components. The number of mature virions released is enormous, ranging from a few thousand for poxviruses to over a hundred thousand for poliovirus. Each of these virions is an infective particle capable of invading a healthy cell and causing damage symptomatic of the particular virus. All it takes is a few viral particles, which, by chance encounter with host cells bearing appropriate receptor molecules, initiate infection and within a short time cause symptoms, possibly death.

As stated earlier all viruses, including phages, follow a similar sequence of events in their replication cycle (adsorption, penetration, replication, assembly, and release). The strategies differ; this is particularly true in the replication stage and depends on the nucleic acid content. It is remarkable that the minute amount of viral nucleic acid that gains access to the host directs the events leading to new viruses. Whatever the uniqueness of the strategy, the significant point is that viral replication proceeds at a rapid

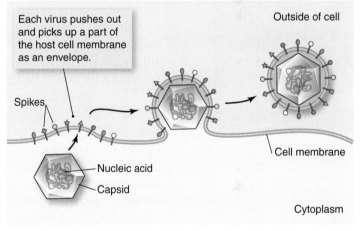

Each virus pushes out and picks up a part of the host cell membrane as an envelope.

Spikes

Nucleic acid

Capsid

Outside of cell

Cell membrane

Cytoplasm

(a)

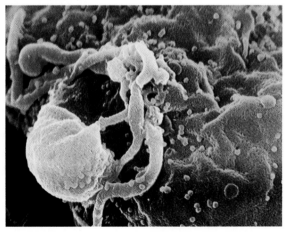

(b)

rate; the production of new virions numbers in the thousands. Considering that each virus is infective, allowing for ongoing cycles involving more and more host cells, it is not difficult to understand the overwhelming effects that occur when you acquire a viral infection. Antibacterial drugs (antibiotics) are ineffective against viruses, because, unlike bacteria, viruses lack a cell wall and other structures unique to bacteria that are targeted by antibiotics. There are some antiviral drugs, but they are not as effective against viruses as antibiotics are against bacteria.

■ Host Cell Damage

Imagine what it must be like in a host cell that is invaded by a virus? It would probably be an awesome experience to witness the devastation. In many cases the outcome is death or at least cell damage. The virus leaves its specific imprint, known as the **cytopathic effect** (**CPE**), evidenced by cell deterioration. The particular CPE, viewed under a microscope, is helpful in the identification of many viral infections in cell cultures. **Cell culture** is a method of growing viruses to be described in the next section. The microscopic observation of virus-infected cells may show a characteristic CPE such as the rounding of the cells, cell shrinkage, detachment, cell lysis and fusion, or **syncytia** formation. Syncytia are giant multinuclear cells that are believed to allow viruses to spread from cell to cell more readily. A number of different viruses cause syncytial formation including HIV and measles virus (FIGURE 5.11). In some cases CPE is applicable to tissue specimens taken directly from an infected individual or at autopsy.

Rabies is a viral disease that can be transmitted from certain wild animals, including skunks, raccoons, bats, mongooses, and foxes, directly to humans or dogs or from dogs to humans. Consider the situation when a raccoon whose bizarre behavior is suggestive of rabies bites a dog or a human. After the raccoon is shot or captured and killed, brain tissue is examined for the presence of **Negri bodies**. Negri bodies are a CPE that is seen in about 50% of brain tissues examined from rabid animals. (An animal suspected to be rabid should not be shot in the head.) The gold standard to diagnose rabies in animals is the **direct flourescent antibody** (dFA) test used to detect rabies proteins (antigens) in brain tissue. Other CPEs are diagnostic for particular viruses that cause smallpox, herpes, and the common cold.

■ Cultivation (Growth) of Viruses

Cultivation of bacteriophages is simple, based on the fact that easily grown bacterial cells serve as convenient host cells and meet the requirements for obligate intracellular parasitism. However, the growth of animal viruses is hindered by the necessity to use animal cells as host cells. In earlier years this presented a serious problem for virologists. Viruses were studied by injecting them into laboratory animals and observing the resulting symptoms; animals were killed so that the effects of infection could be observed at the organ and tissue levels, particularly the development of characteristic CPEs. The use of live animals was expensive and posed ethical issues, and these problems remain today. Animals are still used in certain circumstances, but **embryonated (fertile) chicken eggs** (FIGURE 5.12) and cell culture (FIGURE 5.13) are the primary methods that have largely replaced animal use.

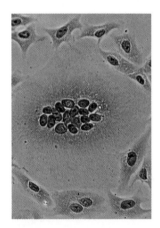

FIGURE 5.11 A giant or "multinuclear" cell (also referred to as syncytia), a mechanism of viral spread. Courtesy of Shmuel Rozenblatt, Tel Aviv University, Israel.

Embryonated (Fertile) Chicken Eggs

Some viruses can be grown in embryonated chicken (or other) eggs, a discovery made in the 1930s. The fertile chicken egg is a convenient and relatively inexpensive alternative to inoculation of live animals. The embryo is enclosed within the protective eggshell and, further, is covered by the shell membrane, providing a sterile and nutrient-rich environment for the developing embryo. There are several different sites and membranes within the egg, each representing a unique ecological niche that supports the growth of particular viruses. With the use of sterile techniques, a hole is drilled through the shell, and the viral preparation is injected into the selected cavity, onto the appropriate membrane, or into the embryo itself, depending on the particular virus or suspected virus under study. Viral multiplication may be detected by the death of the embryo, pocks or lesions on membranes, or abnormal growth of the embryo. The viruses are recovered by harvesting fluids from the egg.

For the manufacture of flu vaccine, influenza virus is grown in embryonated chicken eggs. As a result the vaccine may contain egg proteins, prompting the question as to whether you are allergic to eggs to avoid an allergic reaction to the protein present in the flu vaccine. Manufacturing an avian flu (H5N1) vaccine is more problematic because the virus kills chicken embryos. For this reason, companies have developed cell culture lines that are permissive to avian flu virus infection.

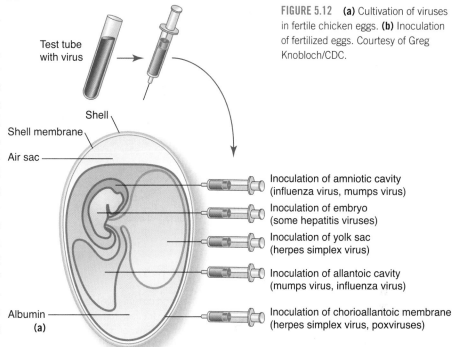

Test tube with virus

Shell
Shell membrane
Air sac
Albumin
(a)

Fertile chicken egg

Inoculation of amniotic cavity (influenza virus, mumps virus)
Inoculation of embryo (some hepatitis viruses)
Inoculation of yolk sac (herpes simplex virus)
Inoculation of allantoic cavity (mumps virus, influenza virus)
Inoculation of chorioallantoic membrane (herpes simplex virus, poxviruses)

FIGURE 5.12 **(a)** Cultivation of viruses in fertile chicken eggs. **(b)** Inoculation of fertilized eggs. Courtesy of Greg Knobloch/CDC.

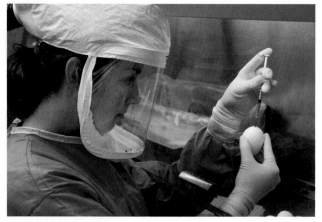

(b)

Cell Culture

The application of cell culture techniques to the growth of viruses was widespread by the early 1950s. These techniques opened the door to the advancement of animal virus studies because the viruses could now be grown at lower costs, rapidly, and on a large scale. Cells are cultured in a variety of sterile laboratory containers, including flasks, tubes, and Petri dishes, containing a complex mixture of nutrients appropriate to the particular cells under culture. The viral suspension is then introduced into the culture flask; viral growth is detected by

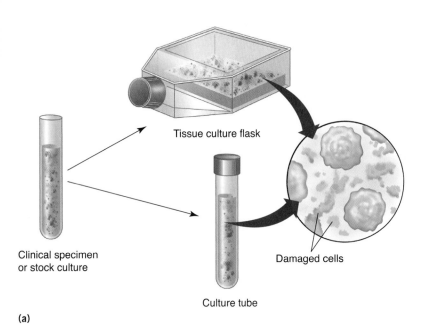

Tissue culture flask

Culture tube

Clinical specimen
or stock culture

Damaged cells

(a)

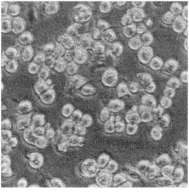

(b)

(c)

FIGURE 5.13   **(a)** Cultivation of viruses in cell culture. The flask and culture tube contain layers of live host cells in nutrient broth to support viral growth. **(b)** Uninfected monkey kidney cells viewed by low-power light microscopy. The confluent monolayer of cells remains attached and intact to the bottom of a tissue culture dish. Author's photo (TS). **(c)** Monkey kidney cells infected with vaccinia virus. CPEs include rounding and detachment of the cells from the tissue culture dish. Author's photo (TS).

(a)

(b)

AUTHOR'S NOTE (RIK)

The Longwood Avenue section of Boston is world famous for its hospitals and research centers, including the John F. Enders Pediatric Research Laboratories (FIGURE 5.14). While at the Harvard School of Public Health in 1999, I passed the building daily on my way to classes and, on several occasions, went into the lobby to view the magnificent and inspiring portraits of John Enders and his colleagues. Each time I experienced a sense of awe, admiration, and inspiration for their work, having grown up in the days when the fear of their children developing polio was uppermost in the minds of parents. The eradication of this ravaging disease is close at hand, delayed largely because of internal strife in Africa.

FIGURE 5.14   **(a)** Portrait of John Enders in his laboratory. Courtesy of the Children's Hospital Boston Archives, Boston, MA. **(b)** The John F. Enders Pediatric Research Laboratories. Author's photo (RIK).

microscopic observation to detect CPEs and lysis of cells as evidence of viral infection and replication (Figure 5.13).

John Enders, Thomas Weller, and Frederick Robbins discovered in 1948 that poliovirus could be induced to grow in various types of cell culture, making possible the large-scale growth of the poliovirus and leading to the development of the Salk and Sabin polio vaccines (FIGURE 5.14). Previously, the virus could be cultured only in live monkeys, a fact that posed severe limits on polio research. Enders, Weller, and Robbins shared the 1954 Nobel Prize in Physiology or Medicine for their pioneering work. This is an excellent example of the potentially significant payoff of basic research for medicine and health (BOX 5.2).

Diagnosis of Viral Infection

Although cell and embryonic egg techniques are available and routine in the hands of trained laboratory personnel, there is usually no attempt to culture the viruses from those with suspected viral infections, as is routinely done for bacterial infections. There are several reasons for this, including cost and limited viral laboratory facilities, but a prime reason is that results of cell culture take too long to be useful for clinical diagnosis. Nevertheless, definitive tests are available when indicated. The presence of **antibody** molecules in the patient's blood may be of diagnostic value; these molecules are produced in response to (foreign) **antigens** (e.g., viral capsid proteins), and their presence is an indication of current or past infection. Accurate identification of a virus may be essential for the protection of the public health, as in the 1999 encephalitis outbreak in New York, when tissue specimens taken at autopsy identified the virus not as the St. Louis encephalitis virus but rather as the West Nile encephalitis virus.

The majority of viral infections are based on the patient's symptoms, including fever, general aches and pains, weakness, nausea, and muscle fatigue. If there is a tell-tale rash, as in measles, or other characteristic signs, such as the skin lesions associated with chickenpox, the diagnosis is more apparent and more accurate. Usually, the patient has to settle for vague and meaningless terms and expressions like "it's a bug, 24-hour grippe, flu, or intestinal virus." Recovery from the more common viral infections usually occurs without complications, thanks to the immune system.

In view of the fact that most acute viral infections are not identified using routine commercial tests, there remain reasons why definitive laboratory tests are needed and developed in terms of public health. Commercial test kits are used to screen blood donations for HIV, hepatitis B, hepatitis C, and West Nile viruses, reducing the spread of these viral diseases through blood transfusions. Viral testing is used to monitor the effectiveness of vaccination programs. Rapid tests for viral diseases such as West Nile encephalitis enables authorities to initiate anti-mosquito control measures. Most hospital clinical laboratories perform very few viral diagnostic tests.

Molecular technology is now routine in the form of commercialized kits in specialized virology diagnostics laboratories. Some approaches used to detect and identify bacterial pathogens in the laboratory are modified to diagnose infections caused by viruses. Techniques used to detect viruses from clinical specimens involve microscopy, detection of patient antibodies against a specific virus, viral antigen detection, viral nucleic acid detection and isolation, or detection of the virus suspect in cell cultures. Light microscopy is used to visualize CPEs of virus-infected cells whereas electron microscopy is used to observe individual virus particles. Electron microscopy is especially useful to detect viruses that can't be cultured in the laboratory. **Enzyme-linked immunosorbent assays (ELISAs)** are designed to detect viral antigens or antibodies against viruses present in patient

There are many examples in biology in which basic research has led to major advances in biology and medicine. The research of John Enders, Thomas Weller, and Frederick Robbins is one such case. Their discovery that poliovirus could be grown in test tube cultures of a variety of nonneural tissue opened the door for research on virus vaccines in addition to polio. Previously, poliovirus could be grown only in culture of human embryonic brain tissue. Nerve tissue is difficult to obtain and to culture, imposing severe limitations. The discoveries of Enders, Weller, and Robbins made possible the development of the research for the Salk vaccine in 1955 and the Sabin vaccine in 1959. Polio is about to be eradicated from the face of the earth. Little more than a generation ago in the United States and across the world, summer was likely to be a season of dread because of the fear that polio might strike. Thank you, John Enders, Thomas Weller, and Frederick Robbins.

The following is excerpted from the text of the presentation speech by S. Gard, member of the Staff of Professors of the Royal Caroline Institute.

"Your Majesties, Your Royal Highnesses, Ladies and Gentlemen.

The principles of cultivation of bacteria were laid down in the late 1870s by Robert Koch. Since that time the bacteriologists could study systematically the diseases caused by bacteria, isolate the causative agents in pure culture, and make themselves familiar with their nature. With the aid of the culture technique they . . . could produce therapeutic sera and prophylactic vaccines. . . .

Turning to the virus diseases we meet an entirely different picture . . . many virus diseases are on the increase, a tendency particularly evident in poliomyelitis. . . . Poliomyelitis in this country is now responsible for almost one fifth of all deaths from acute infections. . . .

It is not difficult to find the reason why the virologists have failed where the bacteriologists were so successful. . . . Unlike bacteria and other microorganisms, virus is incapable of multiplying in artificial lifeless culture media.

Then, in 1949 there appeared from a Boston research team a paper, modest in size and wording but with a sensational content. John Enders . . . and his associates Thomas Weller and Frederick Robbins reported the successful cultivation of the poliomyelitis virus in test-tube cultures of human tissues. A new epoch in the history of virus research had started. . . .

. . . Other scientists had previously attacked the problem with very moderate success. It was generally

serum. ELISAs are based on antibodies binding to antigens and subsequently detected by a resulting color change. The enzyme reacts with its substrate to produce a color change. Like the rapid strep test, rapid influenza tests distinguish influenza virus infection from a number of other viruses in about an hour (FIGURE 5.15).

Molecular diagnostics is the new gold standard for the diagnosis of viral encephalitis and viral meningitis and is performed on patient cerebrospinal fluid collected by a "spinal tap." **Polymerase chain reaction (PCR)** is used to amplify

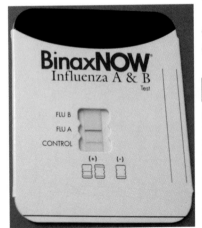

FIGURE 5.15 Rapid influenza A & B test. This patient is suffering from an influenza A infection. Author's photo (TS).

TABLE 5.4 Most Useful Diagnostic Procedures Used for Viral Infections

Diagnostic Approach	Virus
Microscopy	Varicella zoster, arenaviruses, poxviruses
ELISA (to detect viral antigens or antibodies)	Influenza A and B, HIV, hepatitis B and C viruses, papillomaviruses
PCR-based method	Rabies virus, rotaviruses
Cell culture	Rhinoviruses, mumps virus, rubella virus

held that the final word had already been said by Sabin and Olitsky who in 1936 tried to grow the virus in Maitland cultures of various tissues from chick embryos, mice, monkeys, and human embryos. Their results remained completely negative except in cultures of human embryonic brain tissue in which the virus at least seemed to maintain its activity. These findings were taken as a definitive confirmation of the accepted concept of the virus as a strictly neurotropic agent, i.e., capable of multiplying in nerve cells exclusively. Accordingly, the hopes of a practicable method for the cultivation of the poliomyelitis virus were temporarily shelved. Of all tissues, nerve tissue is the most specialized, the most exacting, and consequently the most difficult to cultivate. As, at that, there seemed to be no alternative to the use of human brain tissue, the general resignation is easily understood.

. . . Enders, Weller, and Robbins decided to repeat Sabin and Olitsky's experiment with an improved technique. In their first experiments they used human embryonic tissue. To the great surprise of everybody . . . they registered a hit in their first attempt. The virus grew not only in brain tissue but equally well in cells derived from skin, muscle, and intestines. Furthermore, in connection with the multiplication of the virus, typical changes appeared in the cellular structure, finally leading to complete destruction, easily recognizable under the microscope. This observation furnished a convenient method of reading the results. . . . Enders, Weller, and Robbins found that . . . all tissues except bone and cartilage seemed to be equally suitable. Finally they tried to isolate the virus from various specimens directly in tissue cultures. This was likewise achieved. In the latter observation probably the greatest practical importance of their discoveries is to be found. The virologists finally had a tool in the same class as the culture technique of the bacteriologists.

Dr. John Enders, Dr. Frederick Robbins, Dr. Thomas Weller. Karolinska Institute has decided to award you jointly the Nobel Prize for your discovery of the capacity of the poliomyelitis virus to grow in test-tube cultures of various tissues. Your observations have found immediate practical application on vitally important medical problems, and it has made accessible new fields in the realm of theoretical virus research."

viral nucleic acids and has replaced brain biopsies. Variations of PCR are also used to monitor viral loads of AIDS and hepatitis C patients in the management of their antiviral therapy. TABLE 5.4 lists examples of viral pathogens and the most useful diagnostic approach.

As an anonymous alternative to testing in a clinic, home kits purchased online (via the Internet) are available to diagnose sexually transmitted diseases (**STDs**) such as HIV and hepatitis C. A 2011 investigation by the Medicines and Healthcare Products Regulatory Agency (MHRA) revealed the home kits yield inaccurate results (e.g., false negative results). The MHRA planned to shut down at least six websites based in the UK selling home/self kits and warned against home/self testing.

■ Phage Therapy

You are now familiar with the devastating effect of T4 phage on *E. coli* (FIGURE 5.16). The phage can wipe out an entire *E. coli* culture. Hence, the possibility of using bacterial viruses to control bacterial infections may not seem so far-fetched and

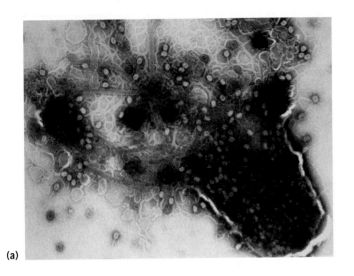

(a)

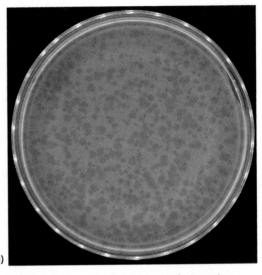

(b)

FIGURE 5.16 The lytic activity of phage. Bacteriophage invade and kill *E. coli*. **(a)** Transmission electron micrograph of lysis of a bacterium by T4 bacteriophages. The T4 (light ovals with stalks) infect the bacterium, using the cell's genetic machinery to code for its own multiple replication. Crowded with T4 progeny, the cell's plasma membrane bursts, destroying the cell and releasing phage to infect other cells. 40,000×. © Biozentrum, University of Basel/Science Photo Library/Photo Researchers, Inc. **(b)** A lawn of *E. coli* with plaques (lighter-colored areas). Courtesy of Giles Scientific, Inc., www.biomic.com.

has been realized since their discovery. *Arrowsmith*, a 1926 Pulitzer Prize winning novel by Sinclair Lewis, includes in the plot the discovery of a bacteria-destroying phage. Hundreds of bacterial species, including many pathogens, are subject to viral infections, frequently resulting in bacterial cell damage or death. Obviously, then, the bacterial growth cycle is interrupted.

The use of phage has been a model for understanding the dynamics of interactions between animal and plant viruses and their hosts. However, the earlier interest in bacterial viruses was based on their use as therapeutic agents to fight bacterial infections. To put it another way, investigators realized the possible potential of phages to kill pathogenic bacteria (BOX 5.3). Phage therapy, although controversial, was increasing as a mode of treatment in the early part of the twentieth century but was essentially abandoned in the Western world when antibiotics became available in the 1940s. However, research and implementation of phage therapy continue primarily in Eastern Europe. Georgia, a part of Russia, remains as a stronghold of human phage therapy. In the past several years there has been renewed worldwide interest, sparked by the highly significant problem of antibiotic resistance. In the United States and in other countries in addition to Eastern Europe, several biotechnology companies are actively pursuing phage therapy. To date, the U.S. Food and Drug Administration has not approved human phage therapy. The logic is there, and, perhaps, phage therapy has a future. Recent clinical trials indicate renewed interest in phage therapy, including a study that demonstrated the effectiveness of phage therapy in reducing the mortality of infection (*Pseudomonas aeruginosa*) in mice. Further, clinical trials using phage at the Royal National Throat, Nose and Ear Hospital in London for the treatment of ear infections (otitis) as well as clinical trials at the Southwest

BOX 5.3 __A Return to Phage Therapy_

Phage therapy might be the answer to the crisis resulting from the misuse of antibiotics in human medicine and on the farm to promote the growth of farm animals. Antibiotic resistance has been cited as a significant factor in emergence and reemergence of microbial diseases. Meanwhile, biologists are constantly on the search for new antibiotics and new therapies, including a renewed interest in phage therapy.

Phage therapy has an almost 100-year history. Frederick Twort (1915) and Felix d'Herelle (1917) independently isolated "filterable" agents that could "eat" bacteria. They observed small clear areas, called plaques (Figure 5.16b), that appeared on the surface of bacterial lawns grown on agar plates when exposed to phage. D'Herelle is credited with coining the term *bacteriophage*, frequently shortened to *phage*. He realized the potential therapeutic use of phage as a natural process to control bacterial infections. In the 1930s, the government ordered wells in India to be treated with phages to prevent cholera.

Perhaps, had the antibiotic era not occurred in the 1940s, the early promise of phage therapy might have further developed. However, in Eastern Europe, including Russia, phage research continued over the years, particularly in research institutes in Poland and in Tbilisi, Georgia. Phage therapy was used to successfully treat Russian soldiers suffering from dysentery both during and after WWII. In fact, an Intestiphage® cocktail is routinely used today in Georgia to prevent gastrointestinal infections in pediatric hospitals.

Successes in the early years of phage therapy were reported, but so were disappointments and failures. In retrospect, these negative results may well have been due to flawed procedures resulting from the relatively meager knowledge available at the time regarding phage–bacteria interactions. There are several inherent advantages in the use of phages that warrant continued research in their potential to combat bacterial infections:

- Phages can be targeted to specific bacteria, minimizing disturbance to the normal bacterial flora of the body. (For example, patients on antibiotics may develop oral or vaginal thrush, a painful condition caused by yeasts that are part of the normal flora because of their overgrowth resulting from an antibiotic-induced suppression of the normal bacterial flora.)
- Phages are easily grown and purified and are relatively inexpensive.
- Small doses can be used because their numbers increase exponentially as they spread from bacterium to bacterium.
- There is no toxicity because phages invade bacterial cells and not human cells.
- The replication and activity of phages introduced into the body are self-limited. Once their bacterial targets are destroyed, they gradually disappear from the body.

In recent years researchers have reported on promising new phage-based approaches to combat bacterial infections as indicated in the text. A group in Canada is studying phage therapy as a means to cure mastitis, an inflammation of the udder in dairy cattle. Researchers at Texas A&M University and at the Rockefeller University have harvested the enzymes that phages use to disintegrate cell walls of their bacterial hosts, allowing for phage release. Their logic is that these same enzymes can be used to attack the bacteria from the outside by cell wall destruction. The Rockefeller University's Vincent Fischetti, who studies the streptococci that cause streptococcal sore throat and other streptococcal infections, states, "It kills the target bacteria instantly. It's amazing—instantly. It does this by punching holes in the cell walls. We can take ten million organisms in a test tube, add a very small bit of (phage) enzyme, and five seconds later, they are all dead. Nothing other than chemical agents can kill bacteria this quickly." Other researchers are focusing on phages to control bacterial infections in food crops. Scientists at the University of Florida at Gainesville have found that tomato plants treated with phage before their inoculation with a microbial pathogen are afforded dramatic protection. A 2010 study conducted at the University of Minho (Braga, Portugal) introduced phage cocktails into chicken feed of *Campylobacter*-infected chickens. Results suggest that phages can be used to reduce the colonization of *Campylobacter* species in poultry, thereby reducing the incidence of campylobacteriosis associated with the consumption of contaminated chicken meat for commercial sale. Pharmaceutical and biotechnology companies are now expanding research funds for the development of phage therapy.

So, who knows? The time may not be too distant when you will be using live active phage particles as a throat gargle, a nasal or throat spray, a liniment or cream to be rubbed on your skin, eye drops, or a liquid poured directly onto or into a wound.

Regional Wound Care Center in Texas dealing with wound infections caused by a variety of bacteria and other studies serve as indicators of research interest in phage therapy.

The use of phage in the treatment of methicillin-resistant staphylococci is being explored. One researcher, referring to the potential applications of phage biology in general, states, "The best is yet to come." The potential extends into several areas, including agricultural practices. Over seventy years have passed since the early trials of phage therapy during which time the knowledge of phage biology garnered from their role in the development and advancement of molecular biology greatly increase their potential in human therapy. The future of modern medicine in the U.S. could include using topical phage creams, drinking phage cocktails, or being injected with a culture of bacteriophages!

▓ Virotherapy

Gene therapy is a beneficial application of microbes. One of the biggest challenges of gene therapy is delivering the functional or "good" gene(s) to the correct cells or tissues. For example, if a gene is required to function in the liver, the genes must be targeted to the liver cells and not harm other cells. How is it known that the gene is targeted to the liver and not the big toe? The answer is that there are definitive, noninvasive reporter genes that track patient gene therapy.

Viruses are obligate intracellular parasites making them good gene delivery "vehicles" or vectors. They can be engineered to attach and enter specific types of cells and not harm normal cells. Either the engineered virus is directly injected into the patient or the patient's cells are removed, cultured in the laboratory, incubated with the engineered virus, and transplanted back into the patient. If the virus can insert the correct gene into the nuclear DNA of the cell, it functions indefinitely in the patient. TABLE 5.5 contains a list of popular virus vectors used in gene therapy trials.

Cancer treatment is now at a crossroads. The traditional options to treat cancer are surgery, radiation, and chemotherapy. Even today, with most

TABLE 5.5 Features of Popular Gene Therapy Viral Vectors		
Virus Vector	Cell Target	Integration into Cell's Nuclear DNA?
Retrovirus	Infects dividing cells	Yes, random integration
Adenovirus	Infects dividing and nondividing cells	No
Adeno-associated virus	Infects dividing and nondividing cells. Requires a "helper" virus to replicate inside of cells	Yes, 95% of the time integrates into chromosome 19
Herpes simplex virus	Can infect cells of the nervous system	No, but stays in the nucleus for a long time
Vaccinia or related poxvirus	Can infect dividing cells	No

FIGURE 5.17   Viral oncolysis.

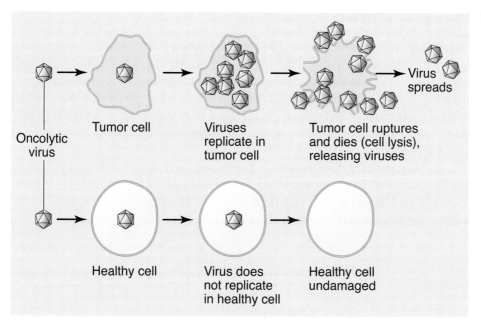

advanced cancers, radiation and chemotherapy are toxic and remission is brief because of the emergence of chemo-resistant cells. **Virotherapy** is emerging as a new treatment of incurable cancers. Virotherapy is the use of **oncolytic (cancer-killing) viruses** to destroy cancer cells. Oncolytic viruses replicate inside of cancer cells and kill them while sparing nearby healthy cells (FIGURE 5.17). Even before viruses were visualized with the electron microscope, physicians reported observations of malignant tumors regressing during a viral infection. One of the most cited examples is a 1904 report by Dr. George Dock describing a 42-year-old woman with leukemia who went into remission following a bout of influenza. In a 1953 report, a four-year-old boy suffering from leukemia experienced a similar remission after the onset of chickenpox. Reports in the 1970s described remissions coinciding with measles infection. Over the next few decades the scientific community pursued the idea of using viruses to treat cancer. Most of the experiments involved curing mouse tumors, but, unfortunately lacked quality control.

Advances in genetic engineering of viruses, increased understanding of the immune system, virus–host cell interactions, and the isolation of many animal viruses that are harmless in humans but can infect and kill human cancer cells have pushed virotherapy forward. Studies revealed that virotherapy could be used positively in combination with radiation, chemotherapy, and targeted antibodies. The first oncolytic virus was approved in 2005 for the treatment of head and neck cancer. The number of research publications reflects the pace of virotherapy. During the 1990s, there were a handful of publications. From 2008 to 2010, a PubMed search reveals an average of eighteen papers per month. There is no doubt that virotherapy will be a part of future cancer therapies. A day may come when the viruses can be used to treat skin cancer, brain gliomas, and other cancers. Oncolytic viruses will be used to defeat one of modern man's most evasive and pernicious threats—cancer.

■ Biology of Prions

Name a particle smaller than a virus that lacks both DNA and RNA and poses a potential threat to the world's beef supply. The answer is **prions**, a shorthand term for "proteinaceous infectious particles." They cause bovine spongiform encephalopathy, or mad cow disease (MCD) in animals, and the human brain disease, variant Creutzfeldt-Jakob disease (vCJD), and related diseases in a variety of animals. Prions as infectious agents and prion diseases will be covered in more detail in the text on viral diseases.

The history of prion diseases is fascinating and dates back to 1957 when Dr. D. Carleton Gajdusek of the U.S. National Institutes of Health, and Dr. Vincent Zigas and Dr. Michael Alpers of the Australian Public Health Service described "kuru," a prion disease seen only among the Fore Highlanders of Papua, New Guinea (BOX 5.4).

BOX 5.4 _Diary Entries of D. Carleton Gajdusek_

D. Carleton Gajdusek was instrumental in leading the research that uncovered "slow (unconventional) viruses" in the field of virology. His journey began in 1957 during studies on a strange neurological disease known as "kuru" or "laughing death" among the Fore tribe of Papua, New Guinea. The Fore tribe was a group of 35,000 people living in 160 remote villages in a remote isolated area in the Eastern Highlands of Papua, New Guinea. They were a Stone Age culture and practiced ritual cannibalism as part of a funeral rite. The first part of Gajdusek's investigation involved mapping the geographical boundaries in which kuru was present while documenting the clinical signs of disease followed by transmission experiments. The fieldwork was harrowing and fraught with great danger, particularly considering that Gajdusek had been raised in a privileged and well-educated family. Few could endure the hardships he suffered. Several excerpts from his diary emphasize his devotion and passion for this work.

An excerpt about the ritual of cannibalism:

"When a body was considered for human consumption, none of it was discarded except the bitter gall bladder. In the deceased's old sugarcane garden, maternal kin dismembered the corpse with a bamboo knife and stone axe. They first removed the hands and feet, then cut open the arms and legs to strip out the muscles. Opening the chest and belly, they avoided rupturing the gallbladder, whose bitter contents would ruin the meat. After severing the head, they fractured the skull to remove the brain. Meat, viscera, and brain

were all eaten. Marrow was sucked from the cracked bones, and sometimes the pulverized bones themselves were cooked and eaten with green vegetables. In the North Fore, but not in the South, the corpse was buried for several days, then exhumed and eaten when the flesh "ripened," and the maggots could be cooked as a separate delicacy.

Thus little was wasted, but not all bodies were eaten. Fore did not eat people who died of dysentery or leprosy, or who had yaws. Kuru victims, however, were viewed favorably—the layer of fat on those who died rapidly heightening the resemblance of human flesh to pork, the most favored protein. Nor were all body parts eaten by everyone. For instance, the buttocks of Fore men were reserved for their wives, while female maternal cousins received the arms and legs."

October 24, 1957 Diary Entry about Bush Camp en route to So'o from Weme, Yar Pawaiian:

"As I type, I am febrile, with markedly tender left inguinal adenopathy, secondary to the numerous small waxing and waning tropical ulcerations on my leg and foot. I have peripheral vasoconstriction starting with pale nail-beds and the beginning I believe, of a chill. I have mild diarrhea and moderate malaise and generalized hyperesthesia. In addition, the sprain of my right foot (two lateral metatarsals) reveals, on removal of the pressure bandage, far more residual edema than I thought would be present. Thus, with innumerable complaints,

---

Prions are proteins found in mammal brains. Like all proteins they are composed of specific sequences of amino acids, and their presence is the result of gene expression in the same way that eye color is a manifestation of protein pigments conferred by eye color genes. In their misfolded state, prions are highly stable and resist freezing, drying, and heating at usual cooking temperatures; they are resistant to pasteurization and conventional sterilizing temperatures. TABLE 5.6 describes the biological distinctions between bacteria, viruses, and prions. The *PrP* gene, located on the human chromosome 20, encodes the prion protein (PrP, also called PrP^c; PrP^{sc} refers to the abnormal misfolded form responsible for neurodegenerative changes).

Stanley Prusiner received the 1997 Nobel Prize in Physiology or Medicine for his discovery of prions. According to the prion theory, **bovine spongiform encephalopathy (BSE)**, CJD, vCJD, mad cow disease, and other **transmissible spongiform encephalopathies (TSEs)** are the result of abnormally folded prions that latch onto normal prions and convert them into altered and misfolded forms,

hypochondriacal concentration thereupon, and a bit of real illness, I have started myself on 3.0 gms sulfadimidine and shall shortly take another 1.0 gms.

We asked the village boys [here at So'o] about crocodiles in the river, and they say there are many but they do not come up on the bank into the village. However, the shore is less than 100 yards from our rest home and the village, and we are not over 20 feet above the stream level. The boys tell us that very recently a village youth was taken by a crocodile just here at the canoe landing, as he was bent over the stream washing. The croc got his foot, then crushed his thigh and dragged him in and under, his body was not recovered.

Kuru is obviously more important than all expectations, intrigue, and imagination to the disease is certain. That it offers astonishment, intrigue, and imagination to the neurophysiologist (Prof. Eccles) and neurosurgeon (Simpson) beyond anything they have seen in their careers, we had expected; but this is interesting confirmation of the suspicions we had: that is something which modern medicine cannot afford to pass by as a simple medical oddity—there is too much chance here for real advances in our understanding of human physiology and disease if we can "crack" kuru, genetic or otherwise." (Courtesy of the National Library of Medicine.)

The transmission experiments conducted by Gajdusek and his colleagues were key in determining that kuru was caused by an infectious agent. A variety of animals were inoculated with brain tissue from diseased kuru victims. The animals were observed for six months and nothing happened without incident. The experiments were repeated with chimpanzees (the closest relative to humans). A chimp named Georgette was inoculated with brain tissue from a boy named Eiro who died from kuru. Another chimp, Daisy, was also inoculated with brain tissue from an 11-year-old girl, Kigea, that died from kuru. Two years after inoculation, the chimps came down with kuru-like symptoms.

Gajdusek proposed that the transmissible agent was a "slow virus" or "unconventional virus." In 1976, he won the Nobel Prize in Physiology or Medicine for his kuru-related work on "Slow and Temperate Viruses." By 1958, Gajdusek settled down and joined the Institute of Neurology and Communicative Disorders and Stroke. Kuru was put on the back burner for about twenty years and the causative agent was never identified. The mystery was solved. In 1997, Stanley Prusiner won the Nobel Prize in Physiology or Medicine for the discovery of prions, the first new kind of pathogen to be discovered in 100 years.

For further information see:
20th Century Microbe Hunters by Robert I. Krasner. Jones and Bartlett Publishers, Sudbury, MA. 2008.
Kuru, The Science and the Sorcery: A Medical Detective Story that Links Animal Diseases to Terrifying Fatal Human Diseases. Siamese Production Co., Australia. 2010. Available online at http://www.kuru-doco.com/.

TABLE 5.6 Comparison of Bacteria, Viruses, and Prions

Organism	Characteristics	Infectious	Genetic Material	Immunogenic	Resistance
Bacteria	Procaryotic cell	Yes	DNA and RNA	Yes	Moderate[a]
Viruses	Protein coat and nucleic acid, noncellular	Yes	DNA or RNA	Yes	Moderate[a]
Prions	Abnormally folded protein, noncellular	Yes	No DNA, No RNA	No	No effective treatment

[a]Depends on the particular test condition and the specific bacterial or viral species.

FIGURE 5.18 (a) Tertiary structure of a normal prion (PrP^c). (b) A misfolded prion (PrP^{SC}). The misfolding allows the proteins to clump together and contribute to disease in ways that are not fully understood. (c) Conversion of normal prions to infectious prions.

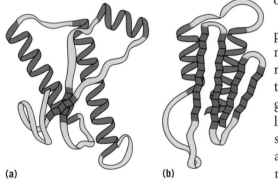

(a)

(b)

(c)

A prion protein (PrP^{SC}) interacts with a normal prion precursor (PrP^c).

The PrP^c is converted to PrP^{SC}.

Additional PrP^{SC} convert more PrP^c into PrP^{SC}.

PrP^c throughout the neuron are converted into PrP^{SC}. As the neurons die, vacuolar areas form in the grey matter.

a property that reveals their infectious nature (FIGURE 5.18). The expression "one bad apple spoils the barrel" is applicable. Robert Louis Stevenson's novel, *Dr. Jekyll and Mr. Hyde*, can be used as another analogy. Dr. Jekyll is the "good guy," and Mr. Hyde is the "evil beast" living within Jekyll. The consequence of these abnormally folded prions in the brain is the loss of motor coordination, dementia, other neurological symptoms, and death. Autopsy reveals a Swiss cheese-like or sponge-like brain full of holes. There is no treatment—there is no cure.

The BSE–vCJD puzzle is not complete; not all the pieces of the puzzle are in place, nor have they been identified. The idea that abnormal prions are able to convert other molecules to an altered form is a revolutionary hypothesis. Nucleic acids are the building units of genes, the hereditary material found in all microbes that is passed on from generation to generation. Genes encode proteins, which are essential for life's activities. Prions lack genes but are, nevertheless, infectious, representing a new agent of infection. Some skeptics of the prion hypothesis argue that there is no definitive proof that a slow virus is not the trigger necessary for initiation of conversion of proteins to a mutant infectious agent. Further, they assert, there is no evidence that prions, free of viruses, synthesized in a test tube produce disease. But, in the 1960s it had been suggested that an infectious agent lacking genetic material might be responsible for disease. One landmark study indicated that brain tissue removed from sheep with scrapie, a neurological disease, remained infectious, even after radiation that would destroy DNA or RNA. The most significant missing piece of the prion puzzle has to do with the trigger that induces the normal protein to misfold. Many scientists believe that the prion protein alone is the infecting agent. Prusiner acknowledges the need for further experiments and the possibility of a "missing factor" that might chaperone PrP into an abnormal shape. Perhaps Prusiner's missing factor will turn out to be a virus yet to be discovered. The Nobel Committee was not bothered by the unanswered questions. The deputy chair stated, "The details have to be solved in the future. But no one can object to the essential role of the prion protein in those brain diseases." Further, prions may be linked to Alzheimer's and Parkinson's diseases as well as to other neurological conditions.

Overview

Viruses are subcellular infectious agents. They are obligate intracellular parasites that subvert the host cell's metabolic machinery for their own replication. A unique aspect of viruses is that they contain either DNA or RNA, whereas cells contain both nucleic acids. Further, viral nucleic acid can be ssDNA, dsDNA, ssRNA, or dsRNA.

Some viruses consist of only a protein coat—the capsid—containing nucleic acid, whereas others have an envelope. Viruses are helical, icosahedral, or complex. They typically range in size from 20 to 350 nm but there are exceptions (e.g., Mimivirus). Bacterial viruses—bacteriophages—are models for understanding viral replication, a process that consists of five stages: adsorption, penetration, replication, assembly, and release. Viruses that cause human diseases exhibit these stages with variations in the strategies used, depending on the specific virus.

Viruses, as obligate intracellular parasites, cannot be grown on nonliving medium and so are cultured in fertile chicken eggs or in cell culture, both offering a supply of living cells allowing for replication. Identification of the specific virus-causing infection is impractical and routinely not done. Diagnosis, in most cases, is based on the symptoms of disease.

Treatment of bacterial diseases with bacteriophages sounds bizarre but is based on sound logic. The idea is attractive because of the growing problem of antibiotic-resistant bacteria. Viruses are used as delivery vehicles for gene therapy. Virotherapy is a new field advancing toward the treatment of incurable cancers.

Prions are infectious proteins that are very stable in the environment. They cause neurodegenerative diseases with long incubation periods known as **transmissible spongiform encephalopathies,** such as kuru, mad cow disease, and variant CJD.

Self-Evaluation

PART I: Choose the single best answer.

1. Tobacco mosaic virus was crystallized by
    a. Enders **b.** Pasteur **c.** Ivanowsky **d.** Stanley
2. The "turkey wrap" analogy applies to what shaped virus?
    a. complex **b.** polyhedral **c.** icosahedral **d.** helical
3. The units of the viral protein coat are called
    a. capsids **b.** helical **c.** capsomeres **d.** envelopes
4. Viruses are known to infect
    a. plants **b.** bacteria **c.** humans **d.** all organisms
5. The envelope of an animal virus is derived from the _____ of its host cell.
    a. cell membrane **b.** cell wall **c.** cytoplasm **d.** nucleus
6. The general steps in a viral multiplication cycle include (1) release, (2) penetration, (3) adsorption, (4) replication, (5) assembly. The correct sequence of these steps is
    a. 2, 3, 4, 5, 1 **b.** 2, 1, 3, 4, 5 **c.** 3, 2, 4, 5, 1 **d.** 1, 2, 3, 4, 5

7. Virotherapy converts viruses into _____-fighting agents.
   a. cancer b. diarrhea c. cholesterol d. fat
8. What are the infectious proteins that cause transmissible spongiform encephalopathies?
   a. Viroids b. Virophages c. Prions d. Prunes
9. Cellular changes observed in cell cultures infected with viruses are called
   a. clumping effects b. cytopathic effects c. necrotic effects
   d. pathogenic effects
10. What is the new gold standard for diagnosis of viral encephalitis?
   a. cell culture isolation b. rapid testing c. molecular diagnostics
   d. animal testing

PART II: Fill in the blank.

1. Based on nucleic acid categories, how many viral groups are there? _____.
2. A complete viral particle is called a _____.
3. Viruses can be grown in tissue culture or in _____.
4. The acronym PCR stands for _____.
5. _____ viruses replicate inside of cancer cells and kill them while sparing nearby healthy cells
6. _____ are ineffective in treating viral infections.
7. The acronym ELISA stands for _____.
8. _____ are viruses that infect bacteria.
9. All viruses require a living _____.
10. A _____ virus lacks an envelope and is relatively stable in the environment.

PART III: Answer the following.

1. Describe the five stages of viral replication.
2. The genetic material of viruses is unique. Discuss this statement.
3. Why did the discovery and understanding of viruses lag about forty years behind knowledge about bacteria?
4. How does the possibility of being treated for a bacterial infection with live bacteriophage strike you? Describe the logic of this strategy.
5. What do you think about the possibility of being treated for advanced cancer with an oncolytic virus? Describe the logic of this strategy.
6. List four categories of viruses along with their tissue affinities and examples of viral diseases (hint: Table 5.2).
7. Explain how it was ruled out that Mimivirus was not an atypical bacterium.
8. Compare and contrast: bacteria, viruses, and prions.
9. Explain why the development of cell cultures to propagate viruses is such a boon to virology research.
10. Explain why viruses can be found everywhere on Earth.

The telegram reads:

WESTERN UNION TELEGRAM

CLASS OF SERVICE
This is a fast message unless its deferred character is indicated by the proper symbol.

W. P. MARSHALL, PRESIDENT

SYMBOLS
DL=Day Letter
NL=Night Letter
LT=International Letter Telegram

1201

The filing time shown in the date line on domestic telegrams is STANDARD TIME at point of origin. Time of receipt is STANDARD TIME at point of destination

MB012 :(40).

M CDU058 72 PD INTL FR=CD 1/50=CD STOCKHOLM VIA RCA 30
PROFESSOR JOSHUA LEDERBERG DEPARTMENT OF GENETICS 1540=.
UNIVERSITY OF WISCONSIN MADISON (WIS)=
1958 OCT 30 AM 9 46
THE CAROLINE INSTITUTE HAS DECIDED TO AWARD THIS YEARS
NOBEL PRIZE IN PHYSIOLOGY OR MEDICINE WITH ONE HALF TO
GEORGE WELLS BEADLE AND EDWARD LAWRIE TATUM JOINTLY FOR
THEIR DISCOVERY THAT GENES ACT BY REGULATING DEFINITE
CHEMICAL EVENTS AND THE OTHER HALF TO YOU FOR YOUR
DISCOVERIES CONCERNING GENETIC RECOMBINATION AND THE
ORGANIZATION OF THE GENETIC MATERIAL OF BACTERIA=
STEN FRIBERG RECTOR=

THE COMPANY WILL APPRECIATE SUGGESTIONS FROM ITS PATRONS CONCERNING ITS SERVICE

Bacterial Genetics

We wish to suggest a structure for the salt of deoxyribose nucleic acid (DNA). This structure has novel features which are of considerable biological interest.

—James D. Watson and Francis H.C. Crick

Topics in This Chapter

- DNA Structure
- DNA Replication
- Proof of DNA as the Genetic Material
- Transcription: DNA to mRNA
- Translation: mRNA to Protein
- Gene Expression
  Chromosomes
- Bacterial Genetics
  Mutations
  Recombination
  Transformation
  Transduction
  Conjugation

Preview

Microbial diversity, particularly as related to bacteria, is the result of the expression of gene mutations and gene recombinations. The gene is the basic unit of heredity and is a segment of a chromosome consisting of tightly coiled DNA; most bacteria have a single chromosome. The processes of **transcription** and **translation** of DNA culminate in protein expression. Sexual reproduction and mutation account for variation in eucaryotic life forms. Bacteria, however, are procaryotes and reproduce asexually; their marked diversion is the result of mutations and the recombinational processes of **transformation**, **transduction**, and **conjugation**. In transformation foreign ("naked") DNA is integrated into competent cells; transduction is characterized by bacteriophage-mediated transfer of DNA, and in conjugation DNA is transferred from donor to recipient during physical contact. These processes are some of the tools for genetic engineering and biotechnology. Synthetic biology is an advancement that allows for the manipulation of genes to create new and unnatural biological products or to manipulate existing biological products into systems that function in a natural way.

Telegram photo reproduced with permission of the Nobel Assembly for Physiology or Medicine.

DNA Structure

This section and the sections leading up to Bacterial Genetics are intended as a review of the basic genetics typically covered in an earlier biology course. The general information is provided as a lead-in to the more specialized area of bacterial genetics.

Deoxyribose nucleic acid (DNA) is, as the name implies, a nucleic acid (as is RNA) and consists of two chains of nucleotides. Each nucleotide is composed of three building blocks, namely, a phosphate, a sugar, and a nitrogen-containing base (FIGURE 6.1). The sugar is deoxyribose, a five-carbon sugar. The phosphate

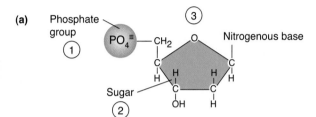

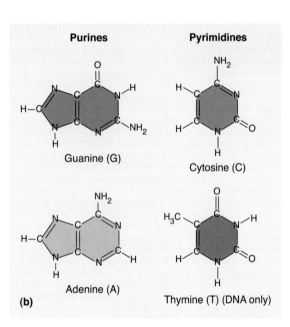

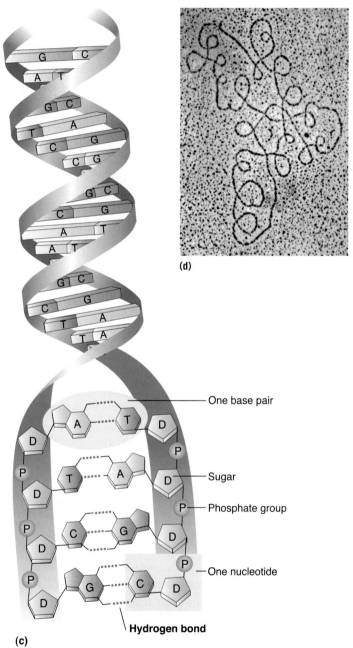

FIGURE 6.1   The structure of DNA.
(a) The nucleotide. Each nucleotide is composed of (1) one to three phosphate groups, (2) a sugar, and (3) a nitrogenous base. (b) The four nitrogenous bases in DNA.
(c) The Watson and Crick model (D = deoxyribose sugar).
(d) Supercoiled DNA—strands from lysed bacterial cell. Courtesy of Nirupam Roy Choudhury, International Centre for Genetic Engineering and Biotechnology (ICGEB).

124   **PART 1**   The Challenge

and the deoxyribose molecules are the same in all nucleotides; the variable part of each nucleotide is the nitrogenous base of which there are four—adenine (A), guanine (G), cytosine (C), and thymine (T). Hence, nucleotides are referred to as A, G, C, and T. A and G are similar in structure and are classified as purines; T and C are closely related and are called pyrimidines. The sugar and phosphate molecules are the backbone of a sequence of chemically joined nucleotides; the nitrogenous bases poke inward. The two strands twist to form a super-coiled double helix and are held together by hydrogen bonds between the bases. Think of the structure as a ladder in which the sides—the vertical aspect—are the sugar phosphate backbone and the rungs—the horizontal aspect—are the nitrogenous bases. Imagine now that the sides are twisted to establish a double helix. The two strands of DNA are anti-parallel, meaning they "run" in opposite directions. As stated, the strands are held together by the bases, but in a complementary way, meaning that A and T are a "fit" as are G and C, resulting in an equal number of A and T nucleotides and, similarly, an equal number of G and C nucleotides. The rule of complementarity is credited to the research of Erwin Chargaff in 1950.

DNA Replication

Before cell division, DNA replicates, resulting in the progeny receiving their "fair and equal share" of DNA (FIGURE 6.2). The starting point of replication is the enzymatic breakdown of the hydrogen bonds holding the two strands together (Figure 6.1). The two strands begin to unzip at specific sites, leaving exposed bases on each strand to which free nucleotides in the "soup" of the nucleus attach, following Chargaff's rule of complementary base pairing (A–T and G–C). As unzipping proceeds, each original strand serves as a template for a new strand (the enzyme DNA polymerase synthesizes this new DNA strand). The result is two double helices, each consisting of an old and a new strand. The replication process is called semiconservative because each helix conserves one old (original) strand and one new strand. In the event of an error, enzymatic-mediated correction usually occurs; if not, a mutation has arisen. The process is complex and the analogy of unzipping a jacket may help. As unzipping proceeds, the double-stranded zipper separates into its two (single) components exposing the "teeth" upon which another single zipper (strand) is built (FIGURE 6.3).

FIGURE 6.2   Binary fission.

The foregoing description of DNA replication is typical of eucaryotic cells in which there are multiple linear pairs of chromosomes. In most bacteria, however, there is a single circularized loop-like chromosome, resulting in some unique characteristics of DNA replication as compared with eucaryotic cells. The chromosome is enzymatically "nicked" and unzipped into a V-shaped, two-pronged

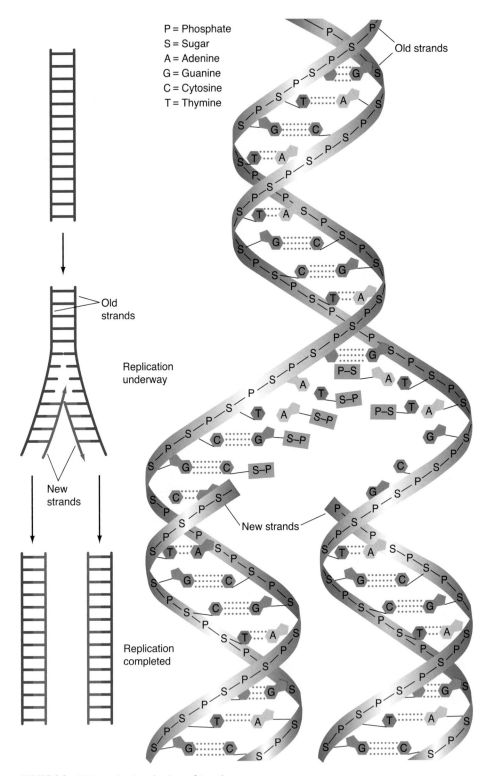

FIGURE 6.3 DNA replication. Analogy of two zippers.

replication fork along which new nucleotides are enzymatically assembled on each prong in a complementary fashion (FIGURE 6.4). The result is two double helixes, each of which ends up in a daughter cell during binary fission.

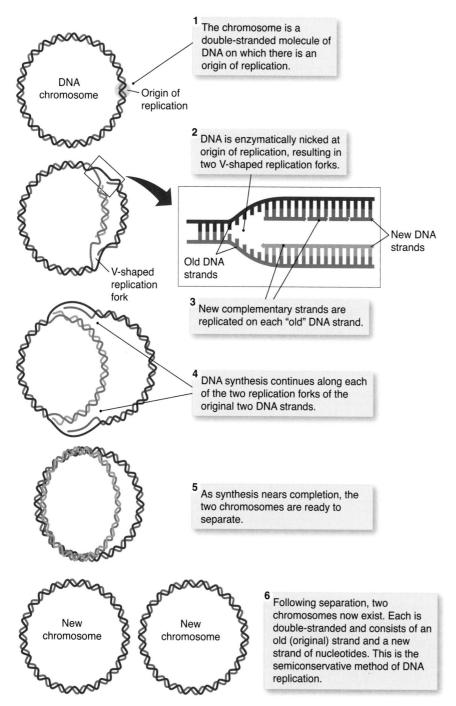

1 The chromosome is a double-stranded molecule of DNA on which there is an origin of replication.

DNA chromosome

Origin of replication

2 DNA is enzymatically nicked at origin of replication, resulting in two V-shaped replication forks.

V-shaped replication fork

Old DNA strands

New DNA strands

3 New complementary strands are replicated on each "old" DNA strand.

4 DNA synthesis continues along each of the two replication forks of the original two DNA strands.

5 As synthesis nears completion, the two chromosomes are ready to separate.

New chromosome

New chromosome

6 Following separation, two chromosomes now exist. Each is double-stranded and consists of an old (original) strand and a new strand of nucleotides. This is the semiconservative method of DNA replication.

FIGURE 6.4 Semiconservative replication. One chromosome replicates into two double-stranded chromosomes.

■ Proof of DNA as the Genetic Material

The proof that DNA is the genetic material was based on research using bacteria and bacterial viruses. Ask students who discovered DNA and its significance and the likely answer will be James Watson and Francis Crick. In 1953 James Watson and Francis Crick described this double helix structure in an article that appeared in the science journal *Nature* (BOX 6.1). Their brilliant work, announced in 1953, was not the *discovery* of DNA but rather the determination of the structure of the DNA molecule. DNA, however, was first isolated in 1869 by Friedrich Meischer from fish sperm and from pus in wounds. At the time there was no hint that **nuclein** (DNA) would prove to be the repository of genetic information; biologists had begun to speculate on the nature of the genetic material, but all bets were on protein.

Eighty-one years after Meischer, three classical experiments, all using microbes, established DNA as the genetic material, the "blueprint" of life. In 1928 Frederick Griffith, an English bacteriologist, was investigating the role of streptococci in pneumonia. As Griffith observed, there are two strains of streptococci, each with distinctive characteristics. Cells of one strain (the S strain) are encapsulated, produce "smooth" (S) colonies, and are virulent for mice. In contrast, cells of the other strain (the R strain) exhibit "rough" (R) colonies, are not encapsulated, and are not virulent (avirulent) for mice (FIGURE 6.5).

These characteristics serve as markers of differentiation. Griffith observed as expected, that heat-killed S bacteria (incapable of growth) did not cause pneumonia and death in mice. Surprisingly, however, when a mixture of heat-killed S bacteria and viable avirulent R streptococci was injected into mice, the mice

---

BOX 6.1 — Watson and Crick: *A Structure for Deoxyribose Nucleic Acid*

In April 1953 the science journal *Nature* published James Watson's and Francis Crick's article, *A Structure for Deoxyribose Nucleic Acid*. The article took up slightly more than one page of the journal, yet it became the benchmark of molecular biology. What follows are excerpts from their article.

"We wish to suggest a structure for the salt of deoxyribose nucleic acid (DNA). . . . The structure has two helical chains each coiled round the same axis. . . . Both chains follow right-handed helices, but the two chains run in opposite directions. . . . The bases are on the inside of the helix and the phosphates on the outside. . . . The novel feature of the structure is the manner in which the two chains are held together by the purine and pyrimidine bases. The planes of the bases are perpendicular to the fiber axis. They are joined

together in pairs, a single base from one chain being hydrogen-bonded to a single base from the other chain, so that the two lie side by side. One of the pair must be a purine and the other a pyrimidine for bonding to occur. . . . Only specific pairs of bases can bond together. These pairs are adenine (purine) with thymine (pyrimidine), and guanine (purine) with cytosine (pyrimidine). In other words, if an adenine forms one member of a pair, on either chain, then on these assumptions the other member must be thymine; similarly for guanine and cytosine. . . . It has not escaped our notice that the specific pairing we have postulated immediately suggests a plausible copying mechanism for the genetic material. . . ."

From J. D. Watson and F. H. C. Crick. A structure for deoxyribose nucleic acid. *Nature* 171 (1953): 737–738.

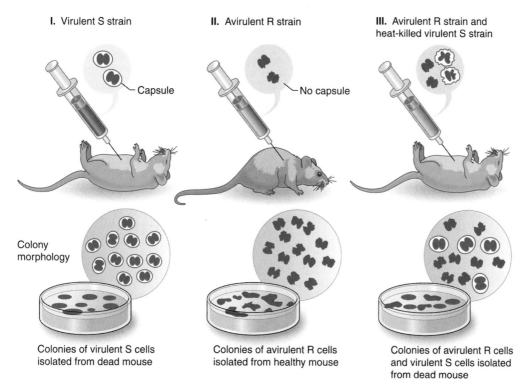

I. Virulent S strain

Capsule

II. Avirulent R strain

No capsule

III. Avirulent R strain and heat-killed virulent S strain

Colony morphology

Colonies of virulent S cells isolated from dead mouse

Colonies of avirulent R cells isolated from healthy mouse

Colonies of avirulent R cells and virulent S cells isolated from dead mouse

FIGURE 6.5 Characteristics of S and R strains of streptococci used in the Griffith experiment (1928).

developed pneumonia and died. By some mechanism the R cells were transformed into virulent S-like cells. Although Griffith was not able to explain the process by which the R cells picked up an unidentified factor from the S cells (this factor was named the transforming principle), it was the first demonstration of the transfer of genetic information.

Sixteen years later, in 1944, Oswald Avery, Colin MacLeod, and Maclyn McCarty identified DNA as the "active" factor in Griffith's transforming principle (**FIGURE 6.6**). They separated the carbohydrate, protein, DNA, and other components of the transforming principle from the S cells into separate fractions. To each fraction, R (avirulent) cells were added and injected into mice. The result was that only the DNA fraction mixed with the avirulent R cells established virulence. Confirmation experiments using the enzyme DNAase to degrade DNA negated the virulence of the R cell–DNA mixture, establishing DNA as the genetic material. There were still those in the scientific world that held to the idea that protein, not DNA, was the genetic material. Their logic was that proteins (discussed later in this chapter) consist of sequences of twenty different amino acids allowing for greater diversity than that possible in a four-nucleotide alphabet.

Finally, in 1952 Alfred Hershey and Martha Chase confirmed DNA as the genetic material (**FIGURE 6.7**) in an experiment using bacterial viruses (bacteriophage) and *Escherichia coli*. In phage, the nucleic acid is enclosed within the protein coat. The phage replicates within *E. coli*, resulting in new phage units identical to the original infecting phage. In the Hershey–Chase experiment the phage DNA was labeled with radioactive phosphorous and the protein coat with radioactive sulfur to determine whether it was the protein coat or the nucleic acid

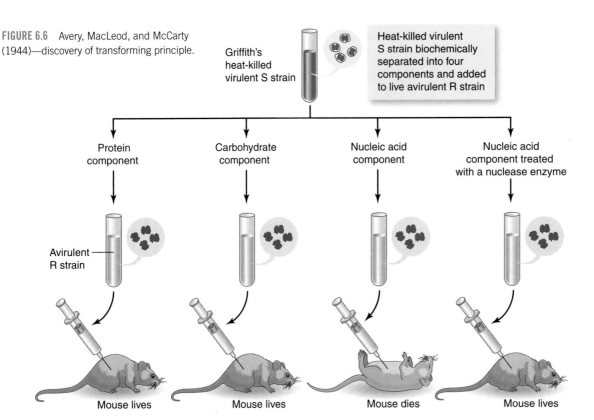

FIGURE 6.6 Avery, MacLeod, and McCarty (1944)—discovery of transforming principle.

Griffith's heat-killed virulent S strain

Heat-killed virulent S strain biochemically separated into four components and added to live avirulent R strain

Protein component

Carbohydrate component

Nucleic acid component

Nucleic acid component treated with a nuclease enzyme

Avirulent R strain

Mouse lives

Mouse lives

Mouse dies

Mouse lives

within the coat that directed the phage replication. Because radioactive DNA and not radioactive protein was found in the daughter phage, Hershey and Chase conclusively demonstrated that the DNA carried the genetic information, resulting in the synthesis of new and identical copies of the phage. Their investigations established DNA as the genetic material. Hence, by the time Watson and Crick worked out the structure of the DNA molecule in 1953, the significance of DNA as the genetic material had been established (TABLE 6.1). The elucidation of the structure of DNA was rapidly followed by cracking of the genetic code and determination of the mechanisms by which genes function.

TABLE 6.1 A Glimpse of DNA History: The Early Years

1869	Friedrich Miescher isolates nucleic acid from fish sperm and from pus cells obtained from discarded bandages.
1928	Griffith demonstrates first evidence of transfer of genetic material.
1943	Avery, MacLeod, and McCarty provide first evidence that DNA is the bearer of genetic information.
1952	Hershey and Chase demonstrate that phage carries DNA as genetic information.
1953	Double helix structure of DNA revealed by Watson, Crick, Franklin, and Wilkins.
1962	Nobel Prize awarded to Watson, Crick, and Wilkins for their discovery of the structure of DNA.

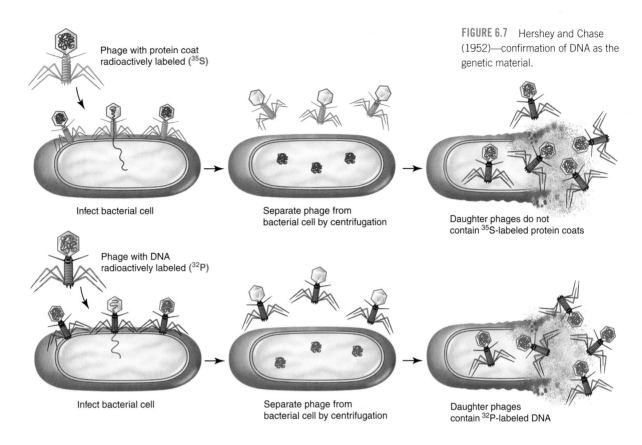

Phage with protein coat radioactively labeled (^{35}S)

Infect bacterial cell

Separate phage from bacterial cell by centrifugation

Daughter phages do not contain ^{35}S-labeled protein coats

Phage with DNA radioactively labeled (^{32}P)

Infect bacterial cell

Separate phage from bacterial cell by centrifugation

Daughter phages contain ^{32}P-labeled DNA

■ Transcription: DNA to mRNA

Having established the structure of DNA and its mechanism of replication, the mechanism by which DNA, a series of nucleotides, is expressed into series of **amino acids**, the building blocks of proteins, needs to be elucidated. The process of "sequence to sequence" is divided into the two steps of **transcription** and **translation**, both of which require RNA. The structure of RNA is similar to that of DNA, with the exceptions that RNA is single stranded, the nucleotide uracil (U) is substituted for the nucleotide thymine (T) and base pairs with adenine (A), and the sugar, ribose, a five-carbon sugar as in DNA, has one more oxygen in its structure than does deoxyribose. Based on its function, the RNA is called messenger RNA (mRNA).

Genetic information is transferred from DNA to mRNA during transcription (FIGURE 6.8). The process of transcription is similar to the mechanism by which DNA replicates, in the sense that one strand of DNA serves as a template for the mRNA strand. A major distinction is that in DNA replication the whole molecule is copied, whereas in making an mRNA transcript, only a short segment of single-stranded DNA is transcribed at a time. Transcription begins at a **promoter site** in a single strand of the DNA and ends at a **terminator** sequence. Within the segment between the promoter and the terminator the double-stranded DNA is unzipped (as in DNA replication), and one strand acts as a template for an (RNA polymerase) enzyme to make a complementary strand of mRNA. At the terminator sequence the enzyme and the new strand of mRNA detach from the DNA; the remaining

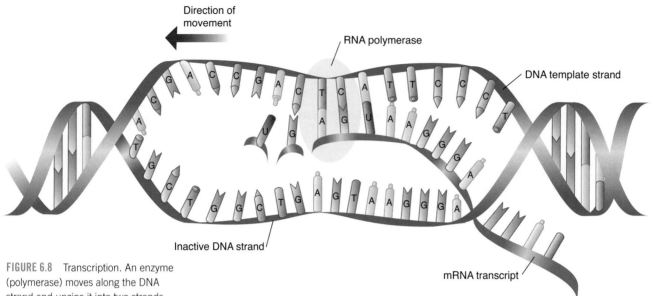

Direction of movement

RNA polymerase

DNA template strand

Inactive DNA strand

mRNA transcript

FIGURE 6.8 Transcription. An enzyme (polymerase) moves along the DNA strand and unzips it into two strands. Complementary RNA nucleotides are added to one strand, the DNA template, as unzipping proceeds.

single-stranded segment of DNA in the cell then converts into a duplex. The result is a strand of mRNA complementary to the template DNA strand—the blueprint—is formed.

Transcription takes place in the nucleus, but the final stage of protein synthesis, translation, takes place on the ribosomes; the mRNA molecules move from the nucleus to the cytoplasm. Thus the mRNA carries the message of DNA and is appropriately called "messenger."

■ Translation: mRNA to Protein

Translation (FIGURE 6.9) is the next and final stage in the expression of a gene. It is characterized by converting the genetic information encoded in mRNA (written in the four-letter alphabet of nucleic acids) into a strand of protein (with its 20-letter alphabet of amino acids). Transcription, therefore, refers to accurately copying the genetic information from DNA to mRNA (both in the language of nucleic acids), while translation refers to converting the genetic information encoded in mRNA into a "foreign" language (i.e., that of protein)—much as one would translate a menu from a French restaurant into English. The point of the central dogma is that the encoded information of the gene (DNA) flows into mRNA and then from the mRNA into the sequence of a protein.

Proteins are macromolecules and constitute about 50% of the dry weight of organisms. They are the most abundant compounds in protoplasm, function in a wide range of activities, and are extremely versatile molecules. Enzymes, biological catalysts, and antibodies are all proteins, as are certain structural compounds of cell membranes and other cell structures. Proteins function in gene regulation and in numerous complex reactions in the cell. Amino acids are the building blocks of protein. Each amino acid consists of a central core (carbon) to which is attached an amino (NH_2) and a carboxy group (COOH), as well as a side chain,

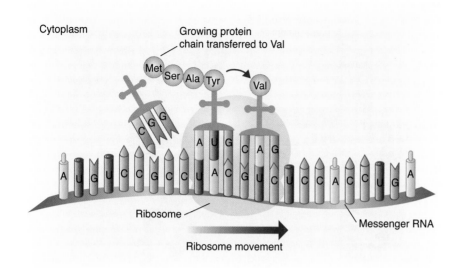

Cytoplasm

Growing protein chain transferred to Val

Met

Ser Ala Tyr

Val

Ribosome

Ribosome movement

Messenger RNA

(a)

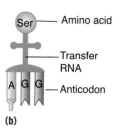

Ser — Amino acid

Transfer RNA

Anticodon

(b)

FIGURE 6.9 Translation. **(a)** The process of mRNA to protein. **(b)** tRNA with an attached amino acid.

represented by an R group (FIGURE 6.10). The R group may be as simple as a single hydrogen atom, as in the amino acid glycine, or complex as in phenylalanine. The distinctive part of each of the twenty naturally occurring amino acids is the chemistry of the R group. Amino acids are chemically joined together by peptide bonds to establish long chains of amino acids known as polypeptides (FIGURE 6.11).

Completion of the process of translation requires a second type of RNA similar to mRNA, called transfer RNA (tRNA) (FIGURE 6.12). This cloverleaf-shaped molecule transfers a specific amino acid molecule, one by one, to the ribosome to build a sequence of amino acids—a polypeptide—chemically linked together through peptide bonds. The question arises as to how the tRNA molecules and their attached amino acids are directed in a particular sequence corresponding to the mRNA sequence. The answer to this question goes back to mRNA. At a designated "start" point,

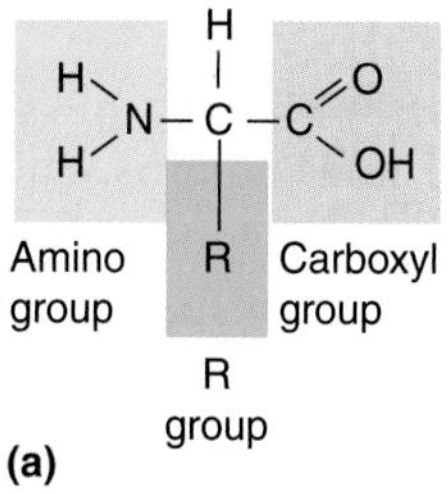

Amino group   R   Carboxyl group

R group

(a)

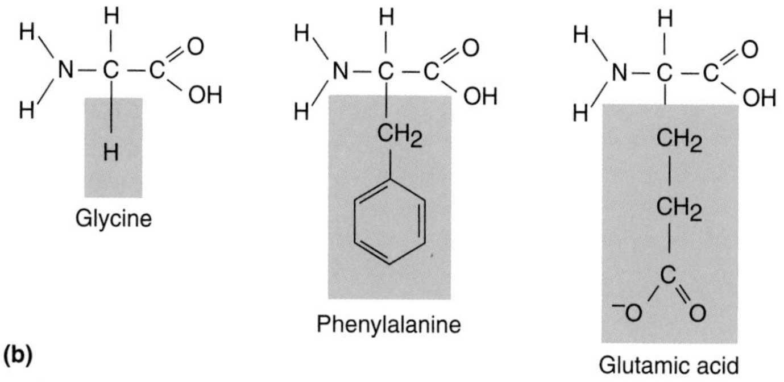

Glycine

Phenylalanine

Glutamic acid

(b)

FIGURE 6.10 Amino acids. **(a)** A generalized structure for an amino acid. **(b)** Three examples of amino acids.

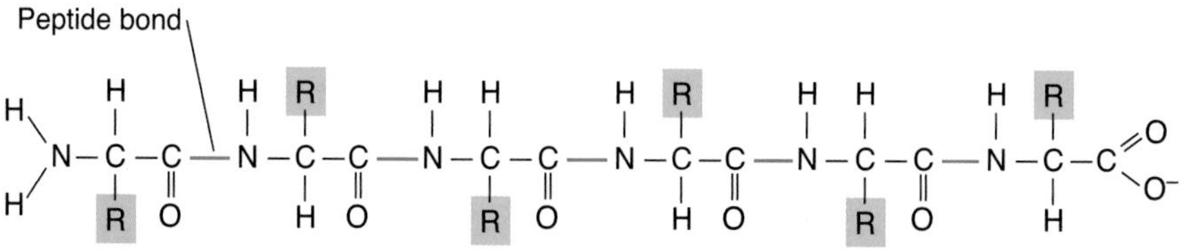

Peptide bond

FIGURE 6.11 Polypeptide bonds. Amino acids are bonded together by peptide bonds, establishing protein primary structure.

CHAPTER 6 Bacterial Genetics

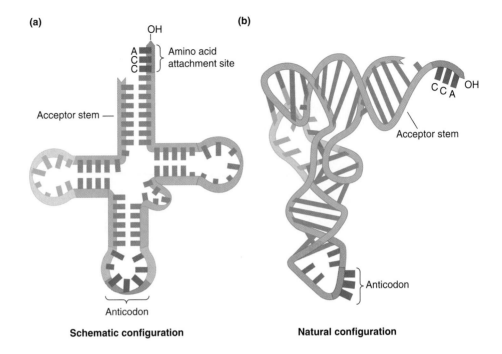

FIGURE 6.12 Transfer RNA (tRNA). **(a)** A "clover leaf" structure. **(b)** A more correct structure illustrating the folding of tRNA. In both **(a)** and **(b)** there are two business sites—the anticodon and the amino acid attachment sites.

(a)

OH

Amino acid attachment site

Acceptor stem

Anticodon

Schematic configuration

(b)

CCA OH

Acceptor stem

Anticodon

Natural configuration

the messenger RNA nucleotides are read on the ribosome in groups of three, referred to as **codons**. Each codon, in turn, relates to an **anticodon**, a series of three nucleotides on tRNA, in a complementary manner (A–U, G–C). Each tRNA molecules has a specific anticodon site to which is chemically bonded a specific amino acid. Hence, there are two "business sites" on each tRNA molecule. One is the anticodon region, consisting of three nucleotides, that can attach to an mRNA codon in a base-complementary fashion (the language of nucleic acids). The second is the binding site for a particular amino acid (the language of proteins) (Figure 6.12). Thus, the tRNAs carry the language and translate nucleic acid (mRNA) into protein, one amino acid at a time.

The codon dictionary is illustrated in **TABLE 6.2**. Note that the relationship is not always one codon–one amino acid. Consider the mathematics involved: If there are four nucleotide bases to choose from in building a codon of three bases, there are 64 possible combinations ($4^3 = 64$). That seems to be a problem because there are only twenty amino acids. Examination of Table 6.2 indicates that most amino acids are coded for by one or more codons, and, hence, the genetic code is referred to as redundant. The amino acids serine and glycine have four codons and arginine—the record breaker—has six codons. Also, one codon (AUG–methionine) serves as "start" (initiator) and three are "stop" (terminator) codons (UAG, UAA, and UGA), as seen in the codon dictionary (Table 6.2), and do not code for any amino acids (nor is there a tRNA that binds to the sequences).

The sequence of amino acids in a polypeptide determines the particular protein that results out of the thousands of proteins that exist. Even very small mutations in the genetic material can have a profound impact on the final protein that forms. Consider sickle cell anemia, a life-threatening disease in humans,

TABLE 6.2 The Genetic Code Decoder

The genetic code embedded in mRNA is decoded by knowing which codon specifies which amino acid. On the far left column, find the first letter of the codon; then find the second letter from the top row; finally read up or down from the right-most column to find the third letter. In *Bacteria,* AUG codes for formylmethionine.

Second base

First base	U	C	A	G	Third Base
U	UUU ⎤ Phe UUC ⎦ UUA ⎤ Leu UUG ⎦	UCU ⎤ UCC ⎥ Ser UCA ⎥ UCG ⎦	UAU ⎤ Tyr UAC ⎦ **UAA** ⎤ STOP **UAG** ⎦	UGU ⎤ Cys UGC ⎦ **UGA** STOP UGG Trp	U C A G
C	CUU ⎤ CUC ⎥ Leu CUA ⎥ CUG ⎦	CCU ⎤ CCC ⎥ Pro CCA ⎥ CCG ⎦	CAU ⎤ His CAC ⎦ CAA ⎤ Gln CAG ⎦	CGU ⎤ CGC ⎥ Arg CGA ⎥ CGG ⎦	U C A G
A	AUU ⎤ AUC ⎥ Ile AUA ⎦ **AUG** Met (START)	ACU ⎤ ACC ⎥ Thr ACA ⎥ ACG ⎦	AAU ⎤ Asn AAC ⎦ AAA ⎤ Lys AAG ⎦	AGU ⎤ Ser AGC ⎦ AGA ⎤ Arg AGG ⎦	U C A G
G	GUU ⎤ GUC ⎥ Val GUA ⎥ GUG ⎦	GCU ⎤ GCC ⎥ Ala GCA ⎥ GCG ⎦	GAU ⎤ Asp GAC ⎦ GAA ⎤ Glu GAG ⎦	GGU ⎤ GGC ⎥ Gly GGA ⎥ GGG ⎦	U C A G

Ala = alanine; Arg = arginine; Asn = asparagine; Asp = aspartate; Cys = cysteine; Gln = glutamine; Glu = glutamic acid; Gly = glycine; His = histidine; Ile = isoleucine; Leu = leucine; Lys = lysine; Met = methionine; Phe = phenylalanine; Pro = proline; Ser = serine; Thr = threonine; Trp = tryptophan; Tyr = tyrosine; Val = valine.

characterized by abnormal hemoglobin. The hemoglobin molecule consists of four polypeptide chains and a total of 584 amino acids. A mistake in amino acid number six is the source of this devastating condition. The number six amino acid is normally glutamic acid, but in sickle cell anemia valine is substituted because of a mutation in the gene that codes for hemoglobin. This is an example of a **point mutation**, in which one nucleotide is replaced with another nucleotide. Because of the redundancy of the genetic code, there is some "wobble" room for amino acid flexibility, allowing for minor variation. Consider, as an example, the amino acid serine, and that if a mutation were to occur that changed the codon in the mRNA sequence from UCC to UCA, serine would still be the amino acid added to the polypeptide. However, if the mutation caused UCC to become UUC, then phenylalanine would replace serine in the protein (Table 6.2), leading to the production of an incorrect protein. Other types of mutations are addressed later in this chapter.

Gene Expression

Both procaryotic and eucaryotic cells have a biochemical switch-like mechanism that allow genes to be turned on and off; it would be a waste of energy for genes to be constantly "turned on" when their products are not required. Major products of protein synthesis, as already mentioned, are enzymes, which are biological catalysts. Some enzymes are **constitutive enzymes**, which are constantly produced because the gene switch is always in the "on" position. On the other hand there are enzymes called **inducible enzymes**, and their production is the result of genes that can be turned on or off depending on the circumstances. The classical example of gene expression is in *E. coli* in which a particular enzyme (ß-galactosidase) is produced to break down the energy-rich lactose molecule into its two constituent sugars. As a matter of energy conservation, the gene that codes for the enzyme is only turned on when lactose is present. Obviously, it would be a waste for the enzyme to be produced in the absence of lactose. In this case the lactose acts as the inducer of the enzyme synthesis.

Another example of gene regulation is repression of gene expression. The mechanism is similar to inducer mechanisms. Repression, however, is triggered not by the substrate of the enzyme but rather by the end product of the reaction. A good example of this mechanism is the synthesis of the amino acid tryptophan; the presence of this amino acid switches off the genes for the further production of tryptophan.

The on and off switches are under the control of a group of functionally related genes known as **operons**. The major point of this brief description of gene expression is to emphasize that genes are not always in the "off" or "on" position but, rather, are regulated.

Chromosomes

Confusion exists regarding the terms "chromosomes" and "genes." Procaryotic cells have a single, circular genome (with few exceptions) mostly ranging from 166,000 base pairs to 12,200,000 base pairs (TABLE 6.3). In most cases, there is only

TABLE 6.3 Bacterial Disease, Genomes, and Genes[a]

Bacterium	Disease	Genome (Base Pairs)	Number of Genes
Bordetella pertussis	Whooping cough (pertussis)	4,086,186	3,816
Borrelia burgdoferi	Lyme disease	910,724	850
Chlamydia trachomatis	Trachoma	1,044,459	911
Clostridium tetani	Tetanus	2,799,251	2,373
Escherichia coli	Enteritis	4,938,920	4,585
Helicobacter pylori	Ulcers	1,667,867	1,566
Listeria monocytogenes	Food poisoning	2,944,528	2,926
Mycobacterium tuberculosis	Tuberculosis	4,403,837	4,189
Yersinia pestis	Plague	4,702,289	4,167

[a]Variation exists within species depending upon strain.

one copy of that chromosome but there are exceptions; for example, *Epulopiscium fishelsoni* has as many as 100,000 copies of its chromosome.

Mention the word "chromosome" and Gregor Mendel, an Austrian friar noted for his pioneering work on genetics in the 1850s and 1860s, comes to mind. Mendel is best known for experiments on patterns of inheritance in pea plants earning him the title "father of modern genetics." His many and varied contributions were not fully recognized until the early years of the twentieth century— some 60 years later.

A "chromosome" consists of long strands of DNA that are found in the nucleus of eucaryotic cells, while a "gene" is a piece or a segment of that DNA. The DNA, as has been noted, is, in turn, a sequence of nucleotide bases that, ultimately, through transcription and translation, prescribe a series of amino acids that are translated into proteins. The scenario is reminiscent of a portion of Jonathan Swift's (1733) "On Poetry: A Rhapsody":

> . . . *So, naturalists observe, a flea*
> *Has smaller fleas that on him prey;*
> *And these have smaller still to bite 'em;*
> *And so proceed ad infinitum. . . .*

The chromosome number is constant for each species, but does not differentiate each species; there is no pattern or logic for chromosome numbers. Consider a few examples, humans have 23 pairs of chromosomes, but so do guppies; mosquitos have six, field horsetails have 216, and a particular type of fern is the record breaker having 1,440 chromosomes (TABLE 6.4).

TABLE 6.4 Chromosome Numbers[a]

Species	Chromosome Number
Typical prokaryote	1
Mosquito	6
Rye	14
Pigeon	18
Earthworm	36
Bread wheat	42
Human	46
Guppy	46
Gorilla	48
Plum tree	48
Potato	48
Cow	60
Chicken	78
Goldfish	94
Field horsetail	216
Adder's tongue fern	1,440

[a]Variation exists within species.

■ Bacterial Genetics

With this basic background, the remainder of the chapter focuses on bacterial (procaryotic) genetics. Both mutation and recombination by transformation, transduction, and conjugation are the mechanisms by which new genes and new combinations of genes arise in procaryotic cells accounting for their diversity. Mutation is random and haphazard and may produce genes that are adverse to survival, whereas recombination is far more efficient because it deals with existing genes.

■ Mutations

As previously described, a mutation is a change in the nucleotide sequence. Some mutations are of little consequence to the microbe, whereas others have an adverse affect. Rarely, the mutation may have a beneficial effect. Considering the complex processes of transcription and translation, both of which involve a change from one language to another (DNA to mRNA and mRNA to protein), it is not surprising that errors—mutations—occur. Although cell mechanisms correct some of these mistakes, others slip through, and the error is passed on from generation to generation. **Point mutations** (previously discussed) are the simplest mutations, whereas other mutations involve **insertion** or **deletion** of nucleotides. Mutations can occur spontaneously, but chemical and physical agents called **mutagens** can also induce mutations. It is possible for a mutation to revert back to its original state. Different types of mutations are summarized in TABLE 6.5.

   Transposons, also called "jumping genes," in the genome are also responsible for mutations. Transposons are able to "jump" from one site on a chromosome to

---

TABLE 6.5 Mutations in Bacteria

Point mutation—a substitution of one nucleotide for another.

Nonsense mutation—one nucleotide change in a codon leading to a STOP codon; e.g., UAU and UAC code for amino acid tyrosine, but UAA and UAG are STOP codons.

Missense mutation—one nucleotide changes in a codon leading to a different amino acid, e.g., GUU codes for amino acid valine and GCU codes for amino acid alanine.

Insertion—addition of one nucleotide leads to a change in the reading frame of codons; e.g., the nucleotide sequence AGU CCA UUU ACC codes for the amino acid sequence of serine, proline, phenylalanine, and threonine. The addition of G in the second codon in front of A establishes the sequence AGU CCG AUU UAC, which codes for the amino acids serine, proline, isoleucine, and tyrosine.

Deletion—deletion of one nucleotide leads to a change in the reading frame (a frameshift mutation); for example, the nucleotide sequence AGU CCA UUU ACG codes for the amino acids serine, proline, phenylalanine, and threonine. However, if the nucleotide A in the second codon is deleted, the sequence is now AGU CCU UUA CG, resulting in the amino acids serine, proline, and leucine (a frameshift mutation).

Mutagens—physical or chemical agents introduced to cause mutation.
   Physical—ionizing radiation (X-rays and gamma rays), nonionizing ultraviolet light.
   Chemical—nucleotide analogs (similar to nucleotides).

another or from a chromosome to a plasmid or from a plasmid to a chromosome. Plasmids are extrachromosomal, typically circularized (some are linear), pieces of double-stranded DNA in the cytoplasm that replicate independently of the chromosome. They carry a few to a large number of genes and are not integral to the cell. It is possible to "cure" (eliminate) them without interfering with the viability of the cell. There are several kinds of plasmids, including those that code for antibiotic resistance and those that confer virulence factors. The fact that transposons are able to move into a plasmid opens up the possibility that they can jump from one bacterial cell to another and possibly even to a few eucaryotic cells. Transposons carry the genetic information allowing them to be mobile.

Recombination

An important distinction needs to be made between the significance of the process of **recombination** in bacteria as procaryotic cells and in eucaryotes, including eucaryotic microbes. Recombination in sexually reproducing organisms (eucaryotes) is **vertical** and involves the fusion of male and female type gametes. Each gamete carries half the chromosome number (haploid) as the result of **meiosis**. Therefore the offspring produced by gamete fusion have the full complement (diploid) number of chromosomes characteristic of that species. Human reproduction depends on the fusion of a sperm and an egg cell, each carrying 23 chromosomes, resulting in restoration of 46 chromosomes—the diploid number. Which one of the two identical chromosomes is selected from the 23 pairs of each parent is unpredictable but results in new recombinants based on "the genetic gamble." Obviously, recombination and reproduction are inextricably linked in eucaryotes.

But what about procaryotic microbes? The strategy is very different; recombination and reproduction are separate and distinct events. Bacteria reproduce by binary fission, an asexual process in which one cell splits into two, each of which splits into two and so on, resulting in a population in which all of the cells are identical (barring mutation).

Recombination of bacterial genes is separate from reproduction. There are three distinct processes of recombination: transformation, transduction, and conjugation (TABLE 6.6). Although these processes are distinct from each other, there are several factors they have in common:

- All are unidirectional.
- Multiplication is not an outcome.
- All are examples of horizontal versus vertical transfer.
- All are based on **homologous** recombination.
- All occur in nature, although described as laboratory phenomena.
- All require integration of foreign donor DNA into host DNA.
- All result in new gene combinations.

Transformation

Griffith's experiments (Figure 6.5) contributed to establishing DNA as the genetic repository of information. Transformation is the uptake of "naked" fragments of DNA released by dead cells; the fragments are linear and in the order of only

TABLE 6.6 Summary and Comparison of Transformation, Transduction, and Conjugation

Type	Characteristics	Discovery
Transformation	Uptake of naked DNA	Griffith, 1928
	Recipient cells need to be competent	
Transduction	DNA carried by bacteriophage from donor to recipient	Avery, Macleod, McCarty, 1944
	Generalized transduction result of DNA fragments in phage following lysis of bacterial host cell	
	Specialized transduction result of faulty breakout of prophage in the lysogenized host cell	
Conjugation	Contact between donor and recipient via F pilus	Lederberg, Tatum, 1946
	F^+ to F^-	
	Hfr to F^-	

about 10 to 20 genes. The upshot is that the recipient (host) cells (those that have picked up and integrated the donor DNA fragments) are now recombinant cells. In a bacterial population only about 10% to 20% of the cells are actually transformed; the major limitation is that the recipient cells must be able to take up DNA from the environment, that is, they must be **competent** for foreign DNA to enter through the cell wall and cell membrane. Although some cells are naturally competent to take up DNA, most are not competent and can be induced to competency in the laboratory by techniques that create pores in their membrane. These include electroporation (electric shock), chemicals, and hot and cold treatment. A large variety of gram-positive and gram-negative cells (particularly gram-positive) are naturally competent, meaning they take up and integrate fragments of donor DNA into a complementary region of their chromosomes, occurring most frequently near the end of the logarithmic growth phase. Naturally competent cells take up DNA from a variety of sources, including bacteria in other genera.

Entry of the foreign DNA into the recipient cell is not the end of the story. Although the donor DNA is now within the recipient, it must be integrated into the recipient DNA, displacing a piece of bacterial chromosome. This is where **homology** comes into play; homologous sequences are the key to integration. Recall that A and T (or A and U) are complementary nucleotides, as are G and C. The donor DNA fragment is positioned next to a complementary DNA sequence of the recipient. After enzymatic excision of the recipient fragment, the donor DNA replaces the excised fragment by "breakage and reunion." The process is complex and mediated by specific enzymes. Only a single strand of DNA enters into the recipient cell and is integrated into one strand of the host DNA, but in the next round of cell division one daughter cell retains the parent genotype and the other is the recombinant genotype. In general, the closer the relationship between the donor and the recipient strains, the more likely it is that there will be a greater number of homologous sequences and, hence, a greater number of genetic recombinants. If there is no "fit" between the donor and recipient DNA, the donor DNA is disintegrated.

Transduction

Transduction is another form of genetic exchange in which genes are transferred from donor host cells to recipient cells. The unique aspect of transduction is that the transfer agent, the vector, is a bacteriophage. Lytic phages are capable of infecting bacterial cells of appropriate specificity, in which case they follow the usual pattern of adsorption, penetration, synthesis, assembly, and release. Temperate (as opposed to lytic) phages exhibit a **lysogenic** cycle that is the basis for two types of transduction: **generalized transduction** or **specialized transduction**. Both types of transduction are the result of the packaging of bacterial DNA into the phage protein coat, the capsid, resulting in "defective" phages. These phages are defective in that they contain only bacterial DNA or a combination of both bacterial and bacteriophage DNA. When the defective phage infects another bacterial cell, the bacterial DNA from the first host becomes part of the genome of the second bacterial cell; recombination by transduction has occurred.

In the case of generalized transducing lytic phages, enzymes break the bacterial chromosome into many fragments (FIGURE 6.13). During assembly, each fragment has an equal chance of being packaged along with phage DNA into the phage head. At the completion of the replication cycle, most of the newly released particles are normal, that is, they contain only phage DNA, but a small percent are defective and carry random bacterial genes in addition to their phage genes, or the entire package of phage DNA is replaced by bacterial DNA. These phages with their accidentally packaged foreign bacterial DNA may infect other bacteria and thereby transfer bacterial genes into a new host bacterial cell. The hallmark of generalized transduction is that any bacterial chromosomal fragment can be transferred.

Specialized transduction, in contrast, is characteristic of temperate (lysogenic) phages (FIGURE 6.14). In these phages the viral nucleic acid is incorporated into the bacterial chromosome by homologous recombination. The incorporated phage DNA, or a piece of it, called a **prophage**, represents a state of "peaceful coexistence" between the virus and the bacterial host, and the prophage is replicated with each round of cell division as though as it were an integral part of the host chromosome. At some point, either under natural circumstances or laboratory induction, the phage nucleic acid is enzymatically excised from the chromosome and enters into a lytic cycle, ultimately releasing large numbers of new phage particles. Most of the time excision is precise, and the resulting new phage particles that are released contain only their full complement of phage genes. Rarely, about one in a million, the breakout is not "clean" and attached to the phage nucleic acid are a few genes from either side of the attachment site adjacent to the prophage. When the defective phage infects a new bacterial recipient, these bacterial traits are transferred. Whether generalized or specialized transduction has taken place, the overlying significance is the creation of new combinations of genes contributing to bacterial diversity.

FIGURE 6.13 Generalized transduction.

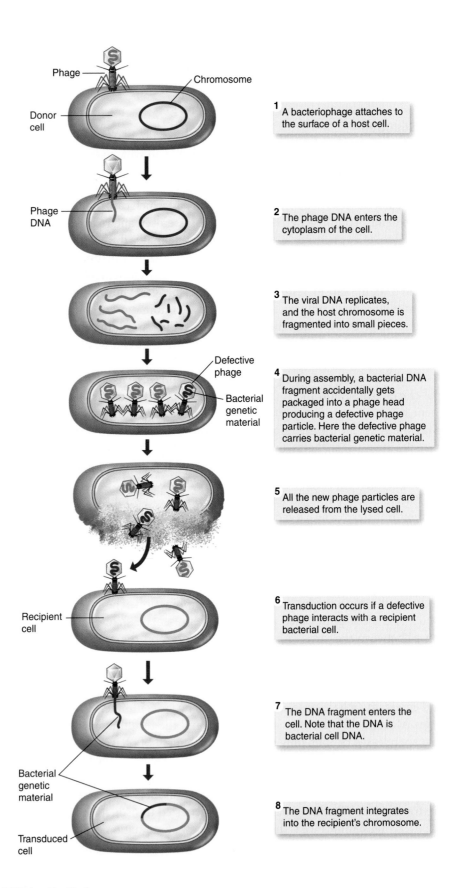

Phage

Chromosome

Donor cell

1 A bacteriophage attaches to the surface of a host cell.

Phage DNA

2 The phage DNA enters the cytoplasm of the cell.

3 The viral DNA replicates, and the host chromosome is fragmented into small pieces.

Defective phage

Bacterial genetic material

4 During assembly, a bacterial DNA fragment accidentally gets packaged into a phage head producing a defective phage particle. Here the defective phage carries bacterial genetic material.

5 All the new phage particles are released from the lysed cell.

Recipient cell

6 Transduction occurs if a defective phage interacts with a recipient bacterial cell.

7 The DNA fragment enters the cell. Note that the DNA is bacterial cell DNA.

Bacterial genetic material

Transduced cell

8 The DNA fragment integrates into the recipient's chromosome.

FIGURE 6.14   Specialized transduction.

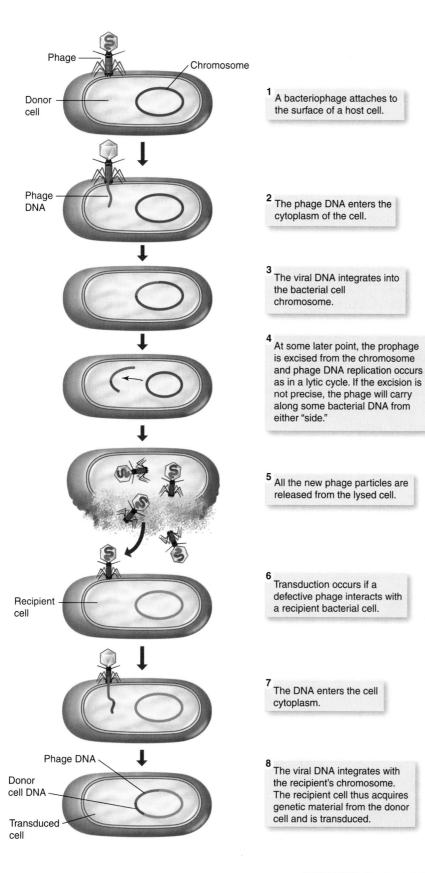

1 A bacteriophage attaches to the surface of a host cell.

2 The phage DNA enters the cytoplasm of the cell.

3 The viral DNA integrates into the bacterial cell chromosome.

4 At some later point, the prophage is excised from the chromosome and phage DNA replication occurs as in a lytic cycle. If the excision is not precise, the phage will carry along some bacterial DNA from either "side."

5 All the new phage particles are released from the lysed cell.

6 Transduction occurs if a defective phage interacts with a recipient bacterial cell.

7 The DNA enters the cell cytoplasm.

8 The viral DNA integrates with the recipient's chromosome. The recipient cell thus acquires genetic material from the donor cell and is transduced.

CHAPTER 6   Bacterial Genetics

143

Conjugation

Bacterial **conjugation** is another mechanism of genetic recombination characterized by direct cell-to-cell contact between a mating pair (FIGURE 6.15). Superficially, it appears to be a primitive form of sexual reproduction, but it lacks the characteristics to identify it as such. There is recombination of genetic material, but neither donor nor recipient produces gametes. Transmission is horizontal and one direction. Conjugation requires that the donor cell have a plasmid, called an F factor, meaning "fertility factor." Some F plasmids produce F pili (sex pili), which act as a bridge between donor and recipient cells. Those cells carrying the F plasmid are **F⁺ cells,** and those that lack the F factor are **F⁻** cells. The conjugation process is initiated when contact is made between the sex pilus of the F⁺ cell and receptor sites on the F⁻ cell; the pilus shortens, drawing the two mating cells closer together. Within only a matter of minutes, one strand of the donor DNA plasmid is enzymatically nicked and a single strand begins to cross over into the F⁻ recipient, a process that is completed in a few minutes (FIGURE 6.16). (Although it is convenient to think of the pilus as a channel for which the genetic material passes from donor to recipient, there is doubt about this, and the actual mechanism of transfer is not known.) The plasmids in both the donor and recipient are now single stranded. Based on complementarity, a new strand is synthesized around the templates in both cells, resulting in double-stranded DNA in donor and recipient. The F⁻ recipients have been converted to F⁺ status and are now able to function as F plasmid donors.

The transfer of genetic material is not limited to cells containing plasmids; chromosomal genes can also be transmitted from donor to recipient (FIGURE 6.17). If, as happens, the F plasmid in the donor is integrated into that same donor chromosome, a new cell type, called **HFr,** results. These Hfr cells exhibit a much higher frequency of combination with F⁻ (hence the abbreviation Hfr). The integration of the plasmid into the chromosome

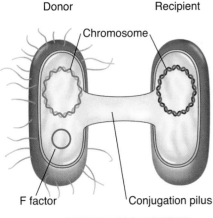

FIGURE 6.15 Bacterial conjugation: direct transfer of DNA.

Donor   Recipient

Chromosome

F factor   Conjugation pilus

Conjugation pilus connects donor and recipient

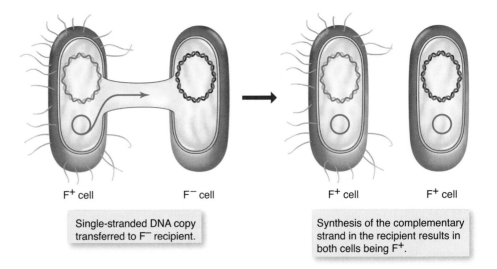

FIGURE 6.16 Bacterial conjugation F⁺ to F⁻. (The pilus may not act as a channel as shown.)

F⁺ cell   F⁻ cell   F⁺ cell   F⁺ cell

Single-stranded DNA copy transferred to F⁻ recipient.

Synthesis of the complementary strand in the recipient results in both cells being F⁺.

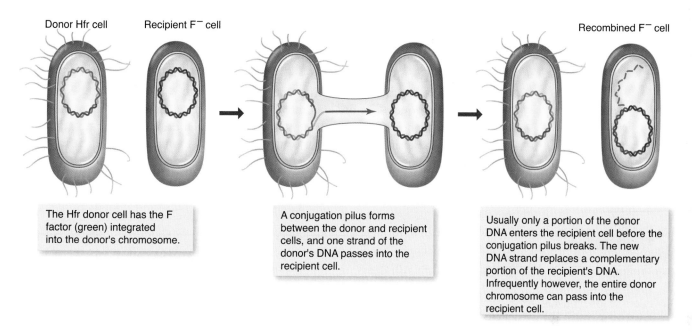

Donor Hfr cell   Recipient F⁻ cell   Recombined F⁻ cell

The Hfr donor cell has the F factor (green) integrated into the donor's chromosome.

A conjugation pilus forms between the donor and recipient cells, and one strand of the donor's DNA passes into the recipient cell.

Usually only a portion of the donor DNA enters the recipient cell before the conjugation pilus breaks. The new DNA strand replaces a complementary portion of the recipient's DNA. Infrequently however, the entire donor chromosome can pass into the recipient cell.

FIGURE 6.17 Bacterial conjugation: Hfr to F⁻. (The pilus may not act as a channel as shown.)

resulting in Hfr cells is reversible, leading to a mixed population of Hfr and F⁺ cells. When Hfr and F⁻ mating occurs, the first genes transferred are from the initiation site of the plasmid, followed by chromosomal genes in a linear sequence along the single-stranded DNA. The number of genes transferred is a function of the duration of conjugation and the fragility of the chromosome; only rarely is the entire chromosome transferred. The genes that dictate the cells donor state are the last to be transferred; therefore, the F⁻ cell will not usually become an F⁺ cell, but because it does carry some chromosomal genes, it is now referred to as a recombinant F⁻ cell. Although the details of conjugation may be complex, the significance is clear; conjugation is a major recombinant event leading to diversity.

Whatever the mechanism of genetic exchange in bacteria, transformation, transduction, or conjugation, along with mutations, account for bacterial diversity. Diversity, in turn, leads to increased ability to adapt to new environments. These new genes by mutation and new gene combinations are the raw material for Darwin's "survival of the fittest" at the genetic and molecular level. The "fittest" are those bacteria in a diverse population with genes that allow them to adapt, survive, and reproduce under hostile circumstances. As an example, consider that the emergence of methicillin-resistant *Staphylococcus aureus,* a current major public health problem, is an outcome of genetic exchange. In 1955 a number of species of *Shigella,* a cause of dysentery, suddenly became resistant to tetracycline, sulfanilamide, chloramphenicol, and streptomycin in Japan, causing great concern. Researchers found that strains of *E. coli,* also isolated from the intestinal tract, were resistant to the same four drugs. Investigations proved that the antibiotic-resistant genes were transferred from *Shigella* to *E. coli* via plasmids (conjugation). There are numerous other examples of transfer of antibiotic resistance as well as transfer of virulence factors by recombination events resulting from gene exchange.

A knowledge of the genetics of microbes is of prime importance in exploiting and harnessing them by genetic manipulation for their beneficial use. A number of useful products made possible by genetic engineering in human medicine, animal husbandry, and agriculture are on the market. Understanding the genetics of bacteria also increases the opportunity to understand and to cope with microbes that are disease producers. In order to more effectively forestall epidemics and pandemics, DNA sequencing, i.e., determining the exact order of the nitrogenous bases (thymine, guanine, cytosine, adenine), has contributed greatly to microbial genetics. To a large extent, biological warfare is based on the ability to engineer "designer bacteria" that

---

BOX 6.2   Does Life Really Beget Life?

Synthetic biology is exciting and has a "sky's the limit" potential, but conveys both promises and perils that need to be considered as advance takes place. The public has the right to ask: "Do we really need it?" or "How can it be contained?" or "What about the risk of creating a Frankenstein Monster?" Bioengineers are already developing an inventory of "biobricks"—biological parts containing genetic material. As an analogy, although not perfect, consider a group of children assembling a pile of logos of assorted shapes and colors into a variety of forms from the same pile of pieces in diverse ways.

An article published in 2009 by Michael Rodemeyer of the Woodrow Wilson International Center for Scholars entitled "New Life, Old Bottles: Regulating First-Generation Products of Synthetic Biology" refers to the "Golden Dilemma" faced by regulators charged with developing constraints. ("Goldilocks and the Three Bears" is an old nursery rhyme. Goldilocks finds the porridge in the bears' house "too hot" or "too cold" until, finally, she pronounces one "just right.") If regulators are too hesitant and precautionary, they interfere with the benefits of synthetic biology, but, on the other hand, if too lenient, their judgment could cause irrevocable damage. It's a juggling act, and needs to be "just right." President Obama expressed strong bioethical concerns and turned the matter over for review by the White House Bioethics Commission in a letter dated May 20, 2010 (FIGURE B6.2a). Subsequently, the Commission advised the President that appropriate guidelines and precautionary measures were in place (FIGURE B6.2b).

In 2002, it was announced that poliovirus (review viral diseases) had been created in a test tube by Eckard Wimmer and his colleagues at Stony Brook University School of Medicine (State University of New York). The public met this news with both "cheers and jeers." The development was hailed as a milestone and "the best is yet to come attitude," but also pronounced this event as the "Devil's Workshop" and as a challenge to Divine Power.

In 2010 Craig Venter, a "genomist" highly recognized for his leading role on the **Human Genome Project**, again entered center stage with an announcement that his laboratory had created the first life form, a bacterial cell that was able to produce proteins as an expression of its synthetic DNA and, thereby, opening a Pandora's Box to the development of designer bacteria for the production of biofuels, pharmaceuticals, and chemical and agricultural products. Not all sequences can be synthesized in a "willy-nilly" sense; some take months and other attempts have failed. Nevertheless, biologist George Church at Harvard believes that synthetic biology is on the march and continues to be more cost-effective at an accelerated rate.

But what about the other side of the coin—the perils. Consider that microbes:

- can be transmitted easily
- are capable of rapid reproduction
- (some) are pathogens
- can be accidentally released from containment
- can be intentionally released from containment
- can produce unintended environmental hazards
- can undergo mutation and evolution in the environment

produce large amounts of toxins that are better able to survive the hazards of the environment and that resist antibiotics. **Synthetic biology** is the "new kid" on the block, a new discipline resulting from the rapid development of techniques to manipulate genetic material. It aims to produce unnatural biological molecules from "scratch" and, possibly, create life, or to assemble naturally occurring biological molecules into systems that function unnaturally as the result of modifying existing systems of genetic material. Synthetic biology (BOX 6.2) uses an engineering approach to biology and has numerous definitions. Some scientists view this discipline as a "New Biological Revolution." "Designer genes" may become as popular as "designer jeans."

Synthetic biology and its implications on the cell theory threatens the benchmark of biology as proposed by Schleiden and Schwann (1838) and Virchow (1858), particularly that part that states: "Cells arise only from previously existing cells" or "life begets life."

THE WHITE HOUSE

WASHINGTON

May 20, 2010

Dr. Amy Gutmann
President and Christopher H. Browne
Distinguished Professor of Political Science
University of Pennsylvania
1 College Hall, Room 100
Philadelphia, Pennsylvania 19104-6380

Dear Dr. Gutmann,

As you know, scientists have announced a milestone in the emerging field of cellular and genetic research known as synthetic biology. While scientists have used DNA to develop genetically modified cells for many years, for the first time, all of the natural genetic material in a bacterial cell has been replaced with a synthetic set of genes. This development raises the prospect of important benefits, such as the ability to accelerate vaccine development. At the same time, it raises genuine concerns, and so we must consider carefully the implications of this research.

I therefore request that the Presidential Commission for the Study of Bioethical Issues undertake, as its first order of business, a study of the implications of this scientific milestone, as well as other advances that may lie ahead in this field of research. In its study, the Commission should consider the potential medical, environmental, security, and other benefits of this field of research, as well as any potential health, security or other risks. Further, the Commission should develop recommendations about any actions the Federal government should take to ensure that America reaps the benefits of this developing field of science while identifying appropriate ethical boundaries and minimizing identified risks. My Science and Technology Advisor, Dr. John P. Holdren, will be in communication with you about the scope and progress of your study.

I ask that the Commission complete its study within six months and provide me with a report with its findings, as well as any recommendations and suggestions for future study that the Commission deems appropriate. Given the importance of this issue, I request that the Commission consult with a range of constituencies, including scientific and medical communities, faith communities, and business and non-profit organizations.

It is vital that we as a society consider, in a thoughtful manner, the significance of this kind of scientific development. With the Commission's collective expertise in the areas of science, policy, and ethical and religious values, I am confident that it will carry out this responsibility with the care and attention it deserves.

Sincerely,

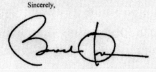

FIGURE B6.2a Letter from President Obama addressing concerns about synthetic biology.

PRESIDENTIAL COMMISSION FOR THE STUDY OF BIOETHICAL ISSUES

President Barack Obama
The White House
1600 Pennsylvania Avenue, NW
Washington, DC 20500

Dear Mr. President:

We are pleased to present to you this report, *New Directions: The Ethics of Synthetic Biology and Emerging Technologies*. In response to your request of May 20, 2010, this first report of the Presidential Commission for the Study of Bioethical Issues (PCSBI) examines the implications of the emerging science of synthetic biology, including the announcement in May of the successful creation of a self-replicating bacterial cell with a completely synthetically-replicated genome. It offers recommendations to ensure that America reaps the benefits of this developing field within appropriate ethical boundaries.

PCSBI approached this task through inclusive and deliberative engagement with ethicists, scientists, engineers, and individuals in faith, business, and non-profit communities. We held three public meetings, both in and outside of Washington, D.C., created an open forum for dialogue, and heard many diverse voices.

The Commission found that synthetic biology offers extraordinary promise to create new products for clean energy, pollution control, and medicine, to revolutionize chemical production and manufacturing, and to create new economic opportunities. With this promise comes a duty to attend carefully to potential risks, be responsible stewards, and consider thoughtfully the implications for humans, other species, nature, and the environment.

PCSBI concluded that synthetic biology is capable of significant but limited achievements posing limited risks. Future developments may raise further objections, but the Commission found no reason to endorse additional federal regulations or a moratorium on work in this field at this time. Instead, the Commission urges monitoring and dialogue between the private and public sectors to achieve open communication and cooperation.

The Commission recommends that the government, through a coordinated process or body within the Executive Office of the President, lead an ongoing review of developments, risks, opportunities, and oversight as this field grows. This review should be in consultation with relevant scientific, academic, international, and public communities, and whenever possible its results should be made public. We also recommend that reasonable risk assessment should precede any field release of synthetic organisms. We suggest support for public engagement, education, and dialogue to ensure public trust and avoid unnecessary limitations on science and social progress.

You gave the Commission a rare and exceptional opportunity to be proactive and forward looking in this first study. The Commission is grateful for the opportunity to serve you and the nation in this way. We would be happy to brief you if you have any questions about our recommendations.

Sincerely,

Amy Gutmann, Ph.D.
Chair

James Wagner, Ph.D.
Vice-Chair

1425 NEW YORK AVENUE, NW, SUITE C-100, WASHINGTON, DC 20005
PHONE 202-233-3960   FAX 202-233-3990 WWW.BIOETHICS.GOV

FIGURE B6.2b Chair of the United States Presidential Commission for the Study of Bioethical Issues, Dr. Amy Gutmann's response to President Obama's letter.

Overview

DNA is the repository of genetic information in procaryotic microbes. DNA is a double-stranded, supercoiled molecule. Genes are segments of DNA composed of the sugar deoxyribose, phosphate, and the nucleotides adenine (A), guanine (G), thymine (T), and cytosine (C). By 1953 three major proofs established DNA as the genetic material. Mutations and recombinational events of transformation, transduction, and conjugation, all of which provide for new genes and new recombinations of genes, account for diversity in bacteria. Transformation is characterized by uptake of naked DNA by competent cells, phages are the vectors for the transfer of DNA from a recipient to a donor cell in transduction, and conjugation is characterized by physical contact between a bacterial mating pair. Bacterial genetics is the basis for genetic engineering, the cornerstone upon which biotechnology is built. Synthetic biology, a relatively new and developing area, allows for the bioengineering of unnatural life to create biological molecules and the assembly of natural biological molecules into systems that function unnaturally.

Self-Evaluation

PART I: Choose the single best answer.

1. With reference to RNA, which is correct?
   a. A–T b. U–U c. U–T d. G–C
2. The "transforming principle" is attributed to
   a. Hershey and Chase b. Avery, MacLeod, and McCarty c. Griffith
   d. Griffin
3. Which expression describes transcription?
   a. mRNA to protein b. DNA to mRNA c. DNA to protein
   d. DNA to mRNA to protein
4. Anticodons refer to
   a. tRNA b. mDNA c. mRNA d. DNA
5. Operons serve to
   a. operate as on–off switches b. operate as initiator sites c. operate as termination sites d. screen for mutations
6. The transfer of agent of transduction is a
   a. pilus b. plasmid c. bacteriophage d. endospore
7. Incorporated phage DNA, or a piece of it, is called a
   a. prophage b. plasmid c. insertion mutation d. retrovirus
8. What provides room for amino acid flexibility during translation?
   a. wiggle b. wobble c. repair mechanisms d. rotation
9. Which of the following is not a characteristic of DNA?
   a. double-stranded b. supercoiled c. complementary d. contains uracil
10. What is/are the macromolecule(s) and constitute about 50% of the dry weight of organisms?
    a. DNA b. RNA c. Proteins d. Carbohydrates

PART II: Fill in the blank.

1. Genes that are continually expressed are called _____ genes.

2. Bacteria in which the F plasmid is integrated into the chromosome are called _____ cells.

3. _____ sequences are key to the integration into recipient DNA.

4. Bacterial cells must be _____ for foreign DNA to enter through the cell wall via transformation.

5. Jumping genes are also called _____.

6. _____ received the Nobel Prize for deciphering the structure of DNA.

7. In 1952, _____ performed experiments using radioactivity, bacteria, phage, and a blender to show that DNA is the hereditary or genetic material of a cell.

8. _____ is the cornerstone upon which biotechnology is built.

9. _____ allows bioengineering of unatural life creating biological molecules and the assembly of natural biological molecules.

10. _____ created poliovirus in a test tube.

PART III: Answer the following.

1. Discuss two of the three classical proofs establishing DNA as the genetic material.

2. What are the three major differences between RNA and DNA?

3. Compare the significance and meaning of multiplication and replication as applied to procaryotic and eucaryotic cells.

4. Why are "defective" phages so named?

5. A stretch of mRNA is GCUUACCGAUAC.
   a. What amino acid sequence results?
   b. The above mRNA sequence is derived from a sequence of DNA. What is that sequence?
   c. Assume a deletion of the first U occurs. List the resulting amino acids.

6. Name and draw a tRNA molecule, clearly indicating the two "business" ends.

7. List the different types of mutations. Which types are the most serious and why?

8. List the three main types of recombinational or genetic material transfer processes. Discuss how an avirulent bacterium could become pathogenic through one of these processes (Hint: Avery, MacLeod, and McCarty's elegant experiment, Figure 6.6).

9. Compare and contrast generalized and specialized transduction.

10. Define F$^+$, F$^-$, and HFr cell types and how they relate to bacterial mating.

Microbial Disease

Student contracts meningitis

Roter said the point of the e-mail was to provide awareness and prevention information.

Viral meningitis is an inflammation of the membranes that cover the brain...

MOVIES

You'll wash your hands

Man who handled rabid fox turns self in

INQUIRER STAFF WRITER

Facts About Meningitis

- Meningitis is an inflammation of the me branes on the brain and spinal cord
- Symptoms include: fever, severe head stiff neck, sensitivity to light, vomiting

Concepts of Microbial Disease

Author's photo (TS).

Given enough time a state of peaceful coexistence eventually becomes established between any host and parasite.

—Rene Dubos (1965)

Preview

"No man is an island" is a popular expression dealing with human interactions. Biologically, this expression can be extended to "no organism is an island." Wherever one looks in the biological world, associations between varieties of species are apparent. Even bacterial cells harbor viruses—bacteriophages. Humans are part of an invisible ecosystem of microbes and, like it or not, are a part of the food chain and serve for microbes as a microbial fast-food restaurant. A variety of microbes lives in peace on or in their human hosts, as relatively permanent inhabitants that constitute the normal flora or as transient visitors just passing through (BOX 7.1). Some almost always produce damage and disease to their host on contact, whereas others, including the normal flora, may produce disease only when the host immune defense mechanisms are compromised, as is the case in individuals with AIDS. Whatever the circumstance, the underlying factor is that there is a dynamic interplay between the number of microbes and their virulence mechanisms, as described in this chapter, and the immune mechanisms of the host. Microbial diseases are characterized by a sequential series of five stages.

The billions of bacteria, representing over five hundred species, that reside in the gut live in a mutualistic relationship with their host and contribute to their hosts' health in many ways including liver development, immune regulation, and capillary growth. Recent evidence indicates a link between gut flora and brain development and behavior. Microbes begin to colonize their hosts in early life, possibly during pregnancy. However, there is also increasing evidence that there may be a downside to the gut microbe—host dynamics. Obesity, inflammatory bowel disease, diabetes, and colorectal cancer are examples.

Some microbiologists see a therapeutic potential in the use of drugs that stimulate or change the "microbiota"

or flora of the intestinal tract. Probiotics and the eating of bacteria-enriched yogurt are attempts to influence the gut flora. According to Justin Sonnenburg at Stanford University School of Medicine and colleagues, more thought needs to be given to the genome of the flora as a huge array of drug targets to be exploited.

Ongoing research on mice led to the belief that gut microbes are linked to brain development. Experiments with mice that are germ free have been extremely useful. The mice appeared to be less anxiety prone and more exploratory as demonstrated by classic rodent behavioral tests: when they were administered normal gut microbiota, they exhibited greater anxiety and less exploratory behavior.

■ Biological Associations

The term *symbiosis* means "living together" and describes an association between two or more species. According to biologist Clark P. Reed:

"Symbiosis, particularly parasitism, is frequently regarded in distasteful terms by the hygiene-conscious citizen. We have a tendency to think of it as a peculiar and abnormal association of some lower organism with a higher one. There is an element of snobbishness in such a view, which must be quickly abandoned when a discerning look is taken of the living world. In nature there is probably no such thing as a symbiote-free organism. The phenomenon of symbiosis is quite as common as life itself."

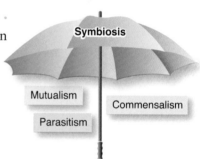

FIGURE 7.1 The symbiosis umbrella.

Symbiosis is like an umbrella with three possibilities: **mutualism**, **commensalism**, and **parasitism** (FIGURE 7.1). Mutualism may be the "best of all worlds" and is like a blissful marriage in which both members of the association enjoy benefits. A classic example of mutualism is the association between termites and the protozoans that constitute a part of the normal flora of the termite's intestine. Termites can cause severe damage to a house, a fact exploited by pest control companies with creative slogans such as "Bug off!" and other catchy phrases (FIGURE 7.2). The termites' ability to digest wood (cellulose, a carbohydrate component of plant cell walls) is based on enzymes secreted by its protozoan residents. In turn, the protozoans have it made; they live in a stable environment with plenty of food passing through. Consider *Escherichia coli*, a part of the human normal flora and an inhabitant of the human large intestine. These bacteria

FIGURE 7.2 An attention-getting landmark atop a building on Route 95 in Providence, Rhode Island. Author's photo (RIK).

FIGURE 7.3 Lichens are a symbiotic association of a fungus and an alga or cyanobacterium. Each organism provides benefits to the other. © Popovici Ioan/ShutterStock, Inc.

enjoy the comforts of home while producing vitamin K, a factor required for blood clotting. Lichens are commonly seen and represent a mutualistic relationship between two microbial forms, a fungus and an alga or cyanobacterium (FIGURE 7.3).

Commensalism is a relationship between two or more species in which one benefits and the other is indifferent, that is, neither benefited nor harmed. Consider, for example, most of the normal flora of the body, organisms that reside in or on the body and take their energy from the metabolism of host secretions and waste products but make no obvious contribution to the partnership with their host, nor do they damage their host. A humorous poem by W. H. Auden summarizes the commensal relationship enjoyed by the normal flora and the host (BOX 7.2). As pointed out in the poem, the body offers a wide variety of ecological niches, each of which has a population of particular bacterial species (FIGURE 7.4).

Microbes are not the only commensals to enjoy our bodies. Scrapings taken from the landscaped furrows of your forehead just below the hairline and examined microscopically might reveal mites crawling about. Mites are distantly related to spiders and are small enough to take up residence in sebaceous (sweat) glands and hair follicles. These examples illustrate that humans are part of the microbial food chain.

Whereas mutualism is at one end of the spectrum, parasitism—a relationship in which the parasite lives at the expense of its host—is at the other end. Parasitism is an association in which one species, the parasite, lives at the expense of the other, the host. There are at least as many definitions of the word "parasitism" as there are letters in the word itself, but they all point to harm or death of the host. The host (in this text) is the human, and microbes are the parasites. The term *parasite* is used to encompass all microbes, viruses, and worms that produce disease. On the other hand, some biologists restrict the term to protozoans (animal-like) and worms, and other biologists restrict it to worms.

The distinctions between mutualism, commensalism, and parasitism are not airtight. The borders between them are hazy and are frequently crossed. Think of

---

BOX 7.2 ___A Poem About Normal Flora___

For creatures your size I offer
  a free choice of habitat,
so settle yourselves in the zone
    that suits you best, in the pools
of my pores or the tropical
    forests of arm-pit and crotch,
in the deserts of my fore-arms,
    or the cool woods of my scalp
Build colonies: I will supply
    adequate warmth and moisture,

the sebum and lipids you need,
    on condition you never
do me annoy with your presence,
    but behave as good guests should
not rioting into acne
    or athlete's-foot or a boil.

From W. H. Auden, Excerpted from "A New Year Greeting," *Epistle to a Godson and Other Poems.* New York: Random House, 1972.

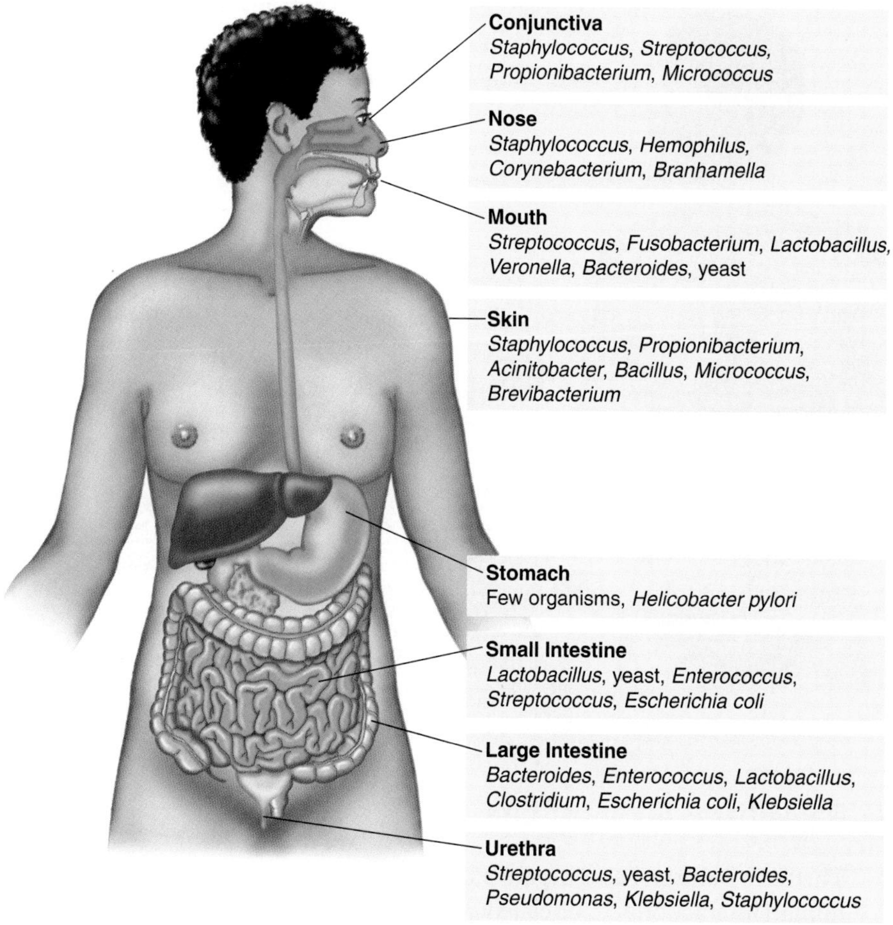

Conjunctiva
Staphylococcus, Streptococcus, Propionibacterium, Micrococcus

Nose
Staphylococcus, Hemophilus, Corynebacterium, Branhamella

Mouth
Streptococcus, Fusobacterium, Lactobacillus, Veronella, Bacteroides, yeast

Skin
Staphylococcus, Propionibacterium, Acinitobacter, Bacillus, Micrococcus, Brevibacterium

Stomach
Few organisms, *Helicobacter pylori*

Small Intestine
Lactobacillus, yeast, *Enterococcus, Streptococcus, Escherichia coli*

Large Intestine
Bacteroides, Enterococcus, Lactobacillus, Clostridium, Escherichia coli, Klebsiella

Urethra
Streptococcus, yeast, *Bacteroides, Pseudomonas, Klebsiella, Staphylococcus*

FIGURE 7.4 Examples of normal flora: mostly commensal organisms.

these relationships as being like a slippery seesaw (**FIGURE 7.5**). For example, in AIDS and other conditions resulting in a depressed immune system, nonpathogenic microbes, including the normal flora, are responsible for potentially fatal infections; these microbes are appropriately referred to as opportunistic pathogens. A major cause of death in AIDS is a fungal infection caused by the opportunist *Pneumocystis jiroveci*. Burn victims are at major risk of serious bacterial infection resulting from nonpathogenic microbes, including their own normal flora. Further, organisms that are part of the normal flora can take the slippery slide from commensalism to parasitism. For example, when *E. coli*, a normal resident of the intestine, gains access to another area of the body—as might happen if the intestinal wall is pierced during surgery—a potentially fatal condition called **peritonitis** may result (**FIGURE 7.6**).

Here's a bizarre relationship that truly escapes categorization as has been described; choose the one you think best! *Toxoplasma gondii* is a protozoan parasite in rats that blocks the neuronal circuitry that makes rats terrified of cats, but, even more strange, these rats become sexually attracted to the scent of cat urine. So the cats eat the rats and, in so doing, acquire the parasite.

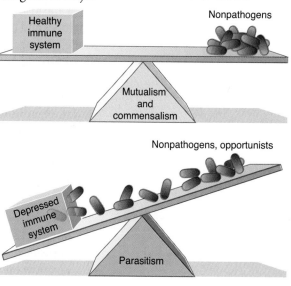

FIGURE 7.5 The seesaw of symbiosis—from mutualism to parasitism.

Healthy immune system

Nonpathogens

Mutualism and commensalism

Nonpathogens, opportunists

Depressed immune system

Parasitism

FIGURE 7.6 Peritonitis: microbes in the "wrong" place.

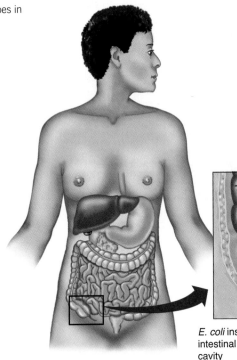
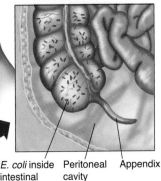
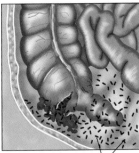

Normal appendix

Ruptured appendix

E. coli inside intestinal cavity   Peritoneal cavity   Appendix

E. coli in peritoneal cavity (peritonitis)

■ Parasitism: A Way of Life

The term *parasite* has a negative connotation and is sometimes used to describe human interactions. To be called a parasite is hardly a compliment. (You may have had the unpleasant experience of being part of a group project, perhaps a student presentation, in which an individual in the group "parasitizes" the others, meaning that he or she makes little contribution and takes advantage of everybody else's effort.)

Parasitism is the basis for microbial disease and for the challenge between microbes and humans. The parasite lives at the expense of the host and, in so doing, causes damage to the host. According to Lewis Thomas (1913–1993), an American physician, author, educator, and administrator known for his popular essays in biology and medicine, "Disease usually results from inconclusive negotiations for symbiosis, an overstepping of the line by one side or the other, a biological misinterpretation of borders." Thomas refers to the "slippery slide" previously described. Parasitism is a constant and delicate dance of biological adaptation in the Darwinian sense of survival of the fittest as both host and parasite coevolve.

Rats, Lice, and History, Hans Zinsser's intriguingly titled 1935 classic text, speaks of parasitism in a somewhat philosophical sense:

"Nature seems to have intended that her creatures feed upon one another. At any rate, she has so designed her cycles that the only forms of life that are parasitic directly upon Mother Earth herself are a proportion of the vegetable kingdom that dig their roots into the sod for its nitrogenous juices and spread

their broad chlorophyllic leaves to the sun and air. But these—unless too unpalatable or poisonous—are devoured by the beasts and by man; and the latter, in their turn, by other beasts and by bacteria. . . . The important point is that infectious disease is merely a disagreeable instance of a widely prevalent tendency of all living creatures to save themselves the bother of building, by their own efforts, the things they require. Whenever they find it possible to take advantage of the constructive labors of others, this is the direction of the least resistance, the plant does the work with its roots and green leaves. The cow eats the plant. Man eats both of them; and bacteria (or investment bankers) eat the man. . . . That form of parasitism, which we call infection, is as old as animal and vegetable life." (Reprinted from Hans Zinsser, *Rats, Lice and History*, courtesy of Little, Brown and Company.)

Zinsser's comments made over 70 years ago are remarkably true to this day.

Microbes as Agents of Disease

The bubonic plague, or Black Death, a bacterial disease, swept through Europe in the fourteenth century and killed millions of people. Smallpox, a viral infection, took a terrible toll for century after century but represents a triumph of public health; the World Health Organization declared it eradicated (the first and still only disease with that distinction) in 1980. Added to the misery produced by these microbial diseases was the fact that there were few preventive measures individuals and society could take. In those days the existence of bacteria was not even hinted at. Leeuwenhoek's observations of "animalcules" was reported in a letter in 1683 to the newly founded Royal Society of London. What, then, was the early thinking regarding the cause of the plague, smallpox, and many other infectious diseases? People considered disease and death a part of life and nature's way; the cause was often thought to be **miasmas**, a term meaning "bad air" or "swamp air." By the nineteenth century a link had been made between filth and the lack of sanitation and what is now referred to as infectious or microbial disease. This observation led to the recognition that nonhygienic conditions foster the prevalence of disease and disease vectors, including rodents, fleas, ticks, lice, and mosquitoes.

Toward the end of the nineteenth century Robert Koch, Louis Pasteur, and other early microbe hunters established the role of bacteria as causative agents of specific diseases. The French microbiologist, Pasteur, spent a good part of his life investigating diseases of wine and of silk moths, both of which were important to the economy of France. Later in his career Pasteur reasoned that if microbes could cause these maladies, perhaps they were responsible for certain diseases of higher animals and humans. In short order, he proved that microbes were responsible for fowl cholera and anthrax in sheep. His crowning achievement was immunization against the dreaded disease hydrophobia, now known as rabies. It was Koch, a German bacteriologist, who experimentally proved that specific microbes were the causative agents for specific diseases. Koch's work initially centered on tuberculosis, a disease that was rampant in his time. Koch established that this disease was caused by a bacterium, later named *Mycobacterium tuberculosis*; this was the first proven causal relationship between a microbe and a particular disease. Koch's

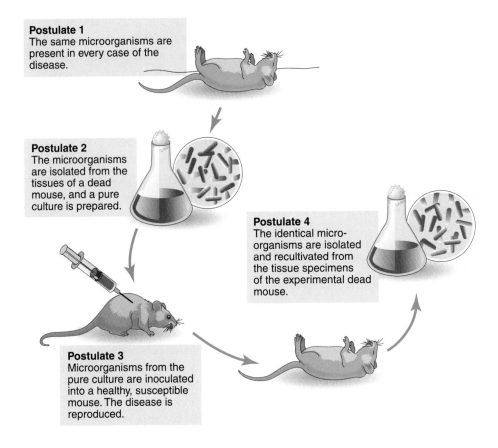

FIGURE 7.7 Koch's postulates.

Postulate 1
The same microorganisms are present in every case of the disease.

Postulate 2
The microorganisms are isolated from the tissues of a dead mouse, and a pure culture is prepared.

Postulate 3
Microorganisms from the pure culture are inoculated into a healthy, susceptible mouse. The disease is reproduced.

Postulate 4
The identical micro-organisms are isolated and recultivated from the tissue specimens of the experimental dead mouse.

work extended well beyond tuberculosis and led to the development of a series of four postulates, known as **Koch's postulates** that are still used to establish that a particular organism is the cause of a particular disease (FIGURE 7.7):

1. *Association:* The causative agent must be present in every case of specific disease.
2. *Isolation:* The causative agent must be isolated in every case of the disease and grown in pure culture.
3. *Causation:* The causative agent in the pure culture must cause the disease when inoculated into a healthy and susceptible laboratory animal.
4. *Reisolation:* Microbes identical to those identified in postulate 2 are isolated from the dead animal.

These postulates have been invaluable to medical microbiologists and physicians throughout the twentieth century and into the twenty-first century and continue to play an important role in the identification of new and reemerging infections. They are not perfect and sometimes have to be taken with a grain of salt. For example, the causative agent for the ancient bacterial disease leprosy was established long before an animal model was found. The unlikely model turned out to be the armadillo! Who would have thought of that creature as a candidate? Syphilis has long been known to be caused by a corkscrew-shaped (spirochete) bacterium, which still has not been routinely grown in culture medium. What about viruses? Viruses are obligate intracellular parasites that cannot be grown on lifeless culture media; nevertheless, there are countless examples of specific viral agents associated with specific diseases. (Remember that viruses were not described until the second and third decades of the twentieth century.) With the advent of the techniques of molecular biology, it is now possible to identify microbes through their DNA fingerprints without growth in culture, and, "classical" Koch's postulates may not always be fulfilled.

AUTHOR'S NOTE (RIK)
Frequently, to make a point, the authors have asked in their respective classes for a show of hands as to "Who feels tired and not so great?" Just about every hand goes up (including the author's). Is this the beginning of an infectious disease or conditions related to academic life, including preparing papers, studying (or lack of it), boredom, lack of sleep, and . . . (you can fill in the rest)?

Microbial Mechanisms of Disease

Attention now needs to be focused on the mechanisms by which microbes cause infections. There are many nonmicrobial, noninfectious diseases, including some cardiac diseases, nearly all cancers, metabolic diseases, mental disorders, nutritional deficiency disorders, and structural disorders, some of which are hereditary. A growing body of evidence indicates that some of these diseases may well be caused by microbes. A distinction needs to be made between the terms *infection* and *disease*. Some microbiologists use these two terms interchangeably, and that is the case in this text. Others use the term *infection* to describe the multiplication of microbes on or in the body without producing definitive symptoms; in this sense the normal flora cause infection. Disease is a possible outcome of microbial invasion, whether by normal flora or exogenous (from the outside) microbes, resulting in impairment of health to some degree. Again, the slippery slide principle exists, as it frequently does in biology, between the terms *infection* and *disease*. Early symptoms of many infectious diseases include lethargy and a sense of ill health. As experience indicates, infectious diseases run the gamut of severity from mild to moderate to severe to lethal. Some diseases are **subclinical** (asymptomatic), as evidenced by the presence of telltale antibodies against diseases that were never diagnosed.

The chance of acquiring a particular infection and the severity of the accompanying symptoms depend on three major factors: (1) **dose (n)**—the number of microorganisms to which the potential host has been exposed, (2) **virulence (V)**—microbial mechanisms or weapons, and (3) **resistance (R)**—the host immune system. There is a dynamic battle going on between disease-producing microbes and the resistance of the host that can be viewed as a tug-of-war or a seesaw. The equation $D = nV/R$ summarizes this struggle. D represents the severity of infectious disease, the numerator refers to the microbes (n = number of organisms, V = virulence factors), and the R in the denominator refers to resistance (immunity). Collectively, these factors determine whether disease occurs and, if so, its severity. Particularly significant, in most cases, are the early hours of challenge; if the microbes gain the upper hand over resistance, then they continue to multiply and damage the host, establishing disease.

Pathogenicity and Virulence

The number of organisms (n) gaining access to (in or on) the host plays a crucial role in determining whether or not disease will take place. For most bacteria there is a minimal dose, the **infective dose (ID)**, necessary to establish infection. Highly virulent organisms have smaller IDs, and in some cases fewer than a hundred bacteria

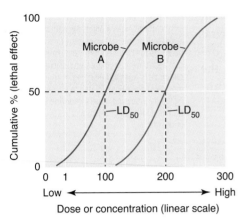

FIGURE 7.8 An LD$_{50}$ dose–response curve. The higher the LD$_{50}$ value, the more microbes it takes to kill, and, therefore, the less virulent the microbe is.

are needed. Those of low virulence can establish infection only when the number of organisms is large. All it takes is about 10 tubercle bacilli to establish infection, but as many as one million cholera bacteria and only as few as 10 to a thousand hepatitis A viruses are necessary for infection to result. The 50% lethal dose (**LD$_{50}$**) is a laboratory measurement of virulence determined by injecting animals with graded doses (varying concentrations) of microbes; the dose that kills 50% of the animals in an established time is the LD$_{50}$ (FIGURE 7.8).

The term *pathogen*, previously introduced, refers to the ability of microbes to produce disease. Some groups of microbes are considered to be highly pathogenic, including those that cause tetanus, botulinum, food poisoning, plague, AIDS, and Ebola fever. They almost always cause disease. (It should be kept in mind that the vast majority of bacteria are nonpathogens.)

Virulence is a measure of pathogenicity and encompasses those specific factors (for example, toxins) that enable pathogens to overcome host defense mechanisms and to multiply and cause damage. Even within the same species of microbes, there is variation in virulence as demonstrated by LD$_{50}$ determinations.

TABLE 7.1 Bacterial Mechanisms Of Virulence

Structure or Secretion	Function	Examples
Defensive strategies		
Adhesins	Fixation to cell surfaces and linings	*Neisseria gonorrhoeae* (attachment to urethral lining); *Streptococcus mutans* (causes tooth decay and sticks to surface of teeth)
Capsules	Interfere with uptake of bacteria (phagocytosis)	Anthrax, plague, streptococcal diseases (e.g., scarlet fever, "strep" throat, pneumococcal infection)
Miscellaneous	Interfere with uptake of bacteria (phagocytosis)	"Waxy coat" of tubercle bacilli; M protein in streptococcal cell walls
Offensive strategies		
Enzymes	Destroy integrity of tissue structure	Hyaluronidase (breaks down hyaluronic acid of connective tissue), hemolysins (bring about lysis of red blood cells), collagenases (break down collagen in connective tissue), leukocidins (destroy white blood cells)
Exotoxins	Specific activities that interfere with vital host functions	Botulinum toxin (one of the most potent toxins known; interferes with transmission of nerve impulse, resulting in flaccid paralysis); tetanus toxin (interferes with transmission of nerve impulses, resulting in irreversible muscle contraction)
Endotoxins	Produces shock-like symptoms, chills, fever, weakness	Structural components of gram-negative cells

Microbes gain entry into the body through portals of entry that include the skin and mucous membranes of the mouth, nose, genital tract, and blood. This subject is covered in the text on epidemiology and cycle of microbial disease as part of the cycle of disease.

Microbes that are able to establish disease have adaptive **defensive strategies** that allow them to escape destruction by the host immune system and **offensive strategies** (TABLE 7.1) that result in damage to the host. The distinction between these two strategies is not always clear, because some factors act both defensively and offensively. Think of it as being like a basketball team—to win the game (analogous to establishing infection), the team (microbes) must have effective defensive strategies to counter its opponent (the host immune system), but that is not enough. The team must also have effective offensive strategies to shoot baskets to score points. Each group of pathogens has evolved unique defensive and offensive virulence factors, many of which are present in more than one group. New molecular biology techniques have increased an understanding of the mechanisms of bacterial virulence.

Defensive Strategies

Bacterial Adhesins

Many pathogens possess cell surface molecules called **adhesins** (FIGURE 7.9) by which they adhere to receptor molecules at a portal of entry in a velcro-like manner. A clear example of this is represented by the causative agent of gonorrhea, *Neisseria gonorrhoeae;* it produces adhesins by which bacteria hold fast and colonize on the warm, moist membranes of the urogenital tract despite the periodic and forceful flow of urine. The bacteria that are primarily responsible for dental decay remain attached to the surface of teeth by adhesins despite the constant irrigation of the teeth by saliva. Bacteria lacking adhesins are subject to being washed or swept away and are not able to colonize the surface of the host cells lining the portal of entry.

Capsules and Other Structures

Some bacteria have capsules around their cell walls. In many cases the presence of the capsule material interferes with the process of phagocytosis, an important resistance mechanism of the host by which bacteria are engulfed by scavenger cells of the body. It can be experimentally demonstrated that removal of the capsule from encapsulated bacteria renders them more susceptible to phagocytosis (FIGURE 7.10). A number of bacteria, including those that cause plague, anthrax, pneumococcal pneumonia, and strep throat, have capsules. The presence of a capsule does not necessarily confer virulence, nor does the lack of a capsule exclude virulence. It is to be emphasized that bacterial virulence is due to a combination of virulence factors. The streptococci responsible for streptococcal sore throat and a variety of other diseases have a protein in their cell walls

(a) Pili

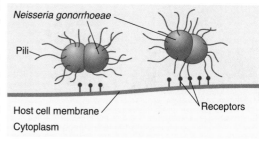

(b) Capsules

(c) Spikes

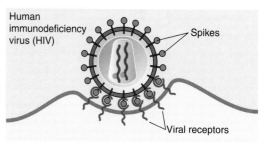

(d) Hooks or flagella

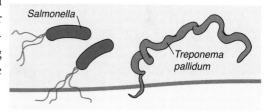

(e) Suction disks

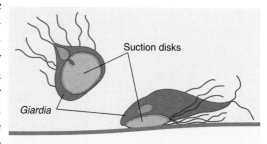

FIGURE 7.9 Examples of adhesins.

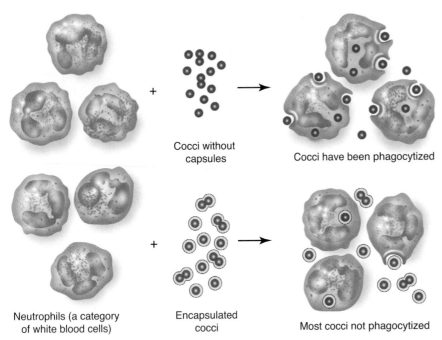

Cocci without
capsules

Cocci have been phagocytized

Neutrophils (a category
of white blood cells)

Encapsulated
cocci

Most cocci not phagocytized

FIGURE 7.10 Capsules and phagocytosis.

termed the **M protein**. Those streptococci containing the M protein display resistance to phagocytosis. Mutant strains not producing this factor and strains in which the M protein has been experimentally removed are readily engulfed by phagocytic cells. Note that the streptococci possess both capsules and M protein—a double whammy against phagocytosis. *Myobacterium tuberculosis,* the causative microbe of tuberculosis, is characterized by a distinctive "waxy" cell wall that allows the bacterium not only to resist digestion within phagocytic cells but, further, to multiply inside them. These are notable examples of structural components that act as virulence factors.

Antigenic Variation

Antigenic variation is a strategy exhibited by some microbes that allows them to evade the immune system of their host, another remarkable example of microbial evolution and adaptation. They are "masters of disguise." The influenza virus is a well-known example of a microbe that "changes its coat" by changing its surface antigens (FIGURE 7.11). A major function of the human immune system is to defend the body against microbial infection; certain cells of the immune system recognize components of microbes, called antigens, as foreign and produce antibodies specifically targeted against these antigens, marking the microbes for destruction. The upshot is that antibodies fail to recognize and to target these new coats. The

FIGURE 7.11 Changing one's coat. Trypanosomes can form new surface antigens not recognized by antibodies. This is an important evasion strategy.

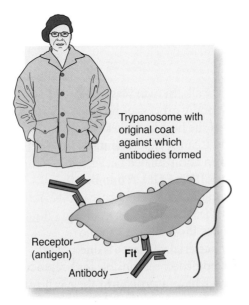

Trypanosome with
original coat
against which
antibodies formed

Receptor
(antigen)

Fit

Antibody

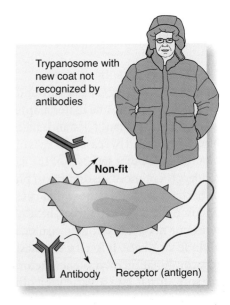

Trypanosome with
new coat not
recognized by
antibodies

Non-fit

Antibody

Receptor (antigen)

influenza virus' antigenic changes can occur within the course of a single year, explaining why vaccination (immunization) against the virus needs to be repeated each year. Trypanosome protozoans, responsible for African sleeping sickness, have genes for as many as a thousand distinct surface antigens and switch from one gene to another, resulting in disguise—quite a trick.

The causative agents of relapsing fever and gonorrhea, both bacterial diseases, are also masters of antigenic disguise. Antigenic variation is an effective strategy that allows microbes to evade the antibody defense system, an important component of the immune system.

Enzyme Secretion

One of the most remarkable examples of defensive strategies is exhibited by *Helicobacter pylori,* the causative agent of ulcers. An unusual adaptation allows this organism to grow in the extremely acid environment of the stomach. The hydrochloric acid in the stomach has a pH of 1–4; neutral pH is 7.0. This acid environment is intolerable for most microbes. How does *H. pylori* counteract the acidity? *H. pylori* manufactures the enzyme urease, which splits the normal component urea into ammonia and carbon dioxide. The ammonia helps to reduce the acidity, protecting the microbe in its own limited environment. *H. pylori* has been confirmed as the cause of most cases of peptic (gastric and duodenal) ulcers. *H. pylori* also secretes toxins (offensive strategy) that contribute to the formation of ulcers (FIGURE 7.12).

Offensive Strategies: Extracellular Products

A variety of products is produced and released by bacterial cells as they grow; these factors are referred to as extracellular substances. Some are classified as toxins, and some are classified as enzymes. Some exhibit both enzymatic and toxic activities. The distinction between toxins and enzymes is not always clear, so it is best to think of them collectively as extracellular products.

Exoenzymes

Some pathogens secrete **enzymes**, called spreading factors, which foster the spread of invading bacteria throughout the tissues by causing damage to host cells in their immediate vicinity and, in so doing, break down tissue barriers. Think of this action as being like that of a plow clearing a field. Seeds can be planted in the exposed earth and crops can grow. **Hyaluronidase** is an enzyme equivalent of a plow; it breaks down the "ground" substance, the hyaluronic acid content of connective tissue, reducing its viscosity, fostering the spread and penetration of microbes deeper into the tissues. **Collagenase** is another example of a spreading factor; it breaks down the structural framework of **collagen**, a vital part of connective tissue, resulting in gas gangrene and massive areas of necrotic (dead) tissue. Extensive tissue **debridement** (cutting away of necrotic tissue) or amputation may be necessary to halt infection and avoid death.

Hemolysins destroy red blood cells through destruction of cell membranes. **Kinases** break down clots, allowing entrapped bacteria to spread. (One kinase—**streptokinase**, a product of streptococci—is now an important therapeutic tool

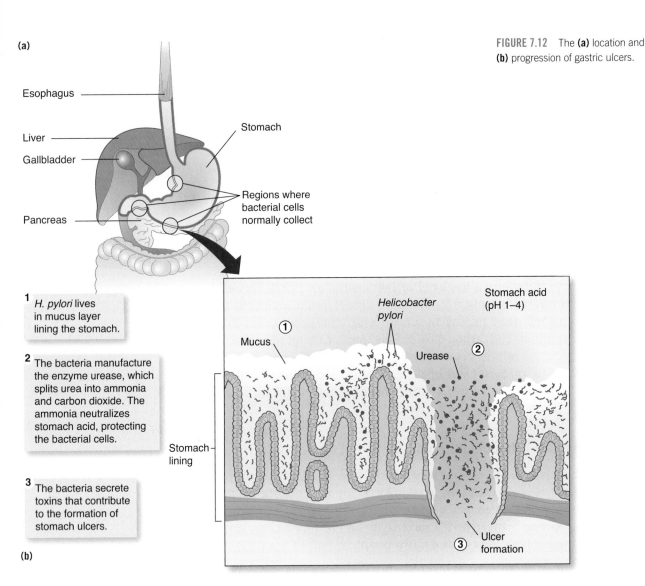

FIGURE 7.12 The **(a)** location and **(b)** progression of gastric ulcers.

(a)

Esophagus

Liver

Gallbladder

Pancreas

Stomach

Regions where bacterial cells normally collect

1 *H. pylori* lives in mucus layer lining the stomach.

2 The bacteria manufacture the enzyme urease, which splits urea into ammonia and carbon dioxide. The ammonia neutralizes stomach acid, protecting the bacterial cells.

3 The bacteria secrete toxins that contribute to the formation of stomach ulcers.

(b)

Stomach lining

Stomach acid (pH 1–4)

Helicobacter pylori

① Mucus

② Urease

③ Ulcer formation

for dissolving blood clots in patients suffering from heart attacks and has proved to be quite effective in preventing strokes and deaths.) **Coagulase** forms a network of threads around bacteria, affording protection against phagocytosis, and **leukocidins** destroy white blood cells.

Exotoxins

The word **toxin** conjures up thoughts of substances that are poisonous to the body. Toxins are major virulence factors for many pathogenic microbes; they are classified as either **exotoxins** or **endotoxins**. These products differ from each other in their chemical composition, modes of action, and nature of their release.

Exotoxins are protein molecules that are synthesized within the microorganism and released ("exo") into the host tissues during the growth and metabolism of the microbes. Exotoxin production is primarily associated with gram-positive organisms; the ability to produce toxins is called **toxigenicity**. Exotoxins are readily soluble in body fluids, are rapidly transported throughout the body, and may

TABLE 7.2 Disease Manifestations by Exotoxins

Disease	Exotoxin Activity and Result
Botulism	Neurotoxin: prevents transmission of nerve impulse, resulting in (flaccid) limp paralysis
Tetanus	Neurotoxin: prevents transmission of inhibitory nerve impulse, resulting in rigid contractions of skeletal muscles (lockjaw)
Clostridial food poisoning	Enterotoxin: diarrhea
Traveler's diarrhea *E. coli* (enterotoxigenic)	Enterotoxin: diarrhea
Scarlet fever	Cytotoxin: causes damage to blood capillaries, resulting in a red rash
Diphtheria	Cytotoxin: kills cells in throat, resulting in a buildup of dead cells; causes damage to heart
Whooping cough	Cytotoxin: kills ciliated epithelial cells in the respiratory tract
Gastroenteritis (*E. coli* O157:H7)	Cytotoxin: bloody diarrhea; kidney damage

act at sites quite distant from the site of colonization of the organism. They are highly toxic and specific in their activity. Exotoxins are grouped into three principal types:

1. **Cytotoxins**, which kill or damage host cells
2. **Neurotoxins**, which interfere with transmission of nerve impulses
3. **Enterotoxins**, which affect the cells lining the gastrointestinal tract, leading to diarrhea

TABLE 7.2 lists a number of bacterial toxins and their effects on the host. Exotoxins are among the deadliest poisons. It is estimated that 1 milligram of botulinum toxin (which causes botulism, a type of food poisoning) can kill half the population of a major city. This may be an exaggeration, so let's say it will kill only one-fourth of the population. What's the difference? Whatever the numbers, the potency of this toxin is remarkable! Toxoids are toxins that have been detoxified but retain their antigenicity. They are used in immunization against those diseases that are primarily a result of exotoxin activity. In addition to botulism, a number of potentially fatal bacterial diseases are characterized by the production of specific exotoxins, including diphtheria, tetanus, cholera, and pathogenic *E. coli* infections. These toxemic diseases are largely a manifestation of the toxin and not the presence of the bacteria; injection of toxin by itself into laboratory animals mimics the symptoms of the disease.

Although each toxin has a specific mechanism of activity (Table 7.2), the **AB model** (FIGURE 7.13) has been proposed to explain the general mode of activity of some toxins. According to this model, exotoxins are composed of two subunits referred to as the A (active) fragment and the B (binding) fragment. Isolated A fragments have been shown to be enzymatically active but lack the ability to bind and to enter cells; isolated B fragments are able to bind to target cells but lack toxicity. Hence, the AB complex is necessary for exotoxin activity; the activity of the A subunit follows the activity of the B subunit. The specific nature of the A fragment activity is a function of the particular exotoxin.

FIGURE 7.13 The AB model of exotoxin activity.

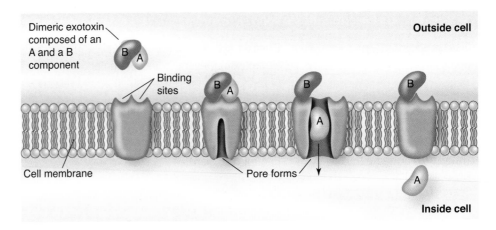

In the early years of microbiology, the effect of toxins on human patients and on laboratory animals was observed, but little was known about how they worked. With the advent of molecular biology, toxins have been studied in great detail to determine their mechanism of activity. FIGURE 7.14 illustrates the activity of the botulinum and tetanus toxins.

FIGURE 7.14 Activity of botulinum and tetanus toxins.

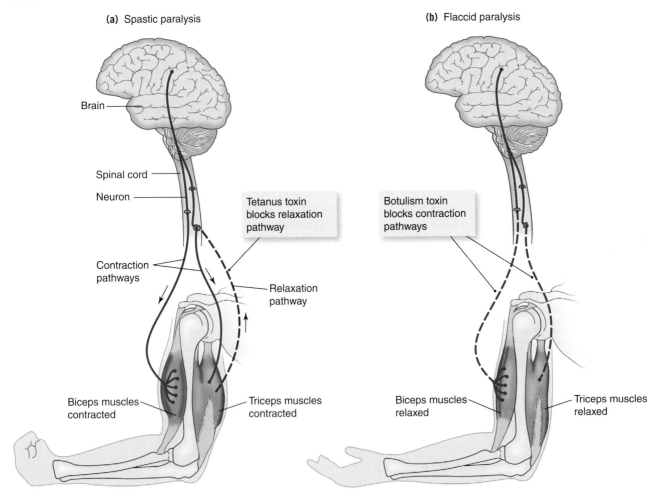

Bacteria may be infected by bacteriophages. In some cases the phage DNA becomes incorporated into the bacterial chromosome in a latent or dormant state and does not bring about lysis of the bacterium. The phage is termed a **prophage**, and the bacterium is said to be **lysogenized**. During cell division the prophage is replicated as a part of the chromosome, and the resulting daughter cells continue to harbor the phage DNA conferring new properties, possibly including toxin production, in the lysogenized bacteria. This phenomenon is referred to as **phage conversion** (FIGURE 7.15). For example, presence of the *tox* gene in lysogenized *Corynebacterium diphtheriae* results in production of an exotoxin that is responsible for the symptoms of diphtheria. When the bacteria are "cured" (the phage is removed), toxin production no longer takes place. Other exotoxins, including the erythrogenic, botulinum, and cholera toxins, are the result of prophage genes.

Endotoxins

Endotoxins are quite different from exotoxins, as summarized in TABLE 7.3. Unlike exotoxins, endotoxins are not usually released during microbial growth and metabolism; they are structural components of the outer membrane of gram-negative cells. They are released as these cells undergo disintegration, although some endotoxin material may be released during cell multiplication. Endotoxins are not proteins but are molecules known as **lipopolysaccharides**. As indicated in Table 7.2, each exotoxin has a specific action; in contrast, all endotoxins,

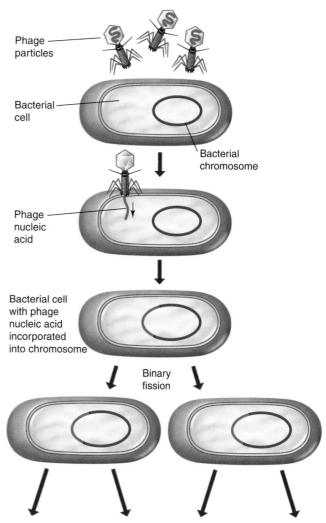

FIGURE 7.15 Mechanism of phage conversion.

TABLE 7.3 Characteristics of Bacterial Exotoxins and Endotoxins

Property	Exotoxins	Endotoxins
Site	Released from cell during growth and metabolism	Retained (for the most part) within outer membrane and released when cell disintegrates
Cell source	Primarily gram-positive cells	Gram-negative cells
Activity	Specific for each toxin	Essentially similar for all endotoxins
Chemical nature	Protein	Lipopolysaccharide
Toxicity	High toxicity	Minimal toxicity
Heat stability	Unstable; usually destroyed at about 60°C	Stable; can withstand temperatures of 100°C
Examples of diseases	Tetanus, scarlet fever, diphtheria, gangrene, botulism	Meningococcal meningitis, typhoid fever, salmonellosis

regardless of their source, produce the same host response, characterized by shock-like symptoms, chills, fever, weakness, formation of small blood clots, and possibly death. They are considerably less toxic than exotoxins but can cause severe symptoms during infections by gram-negative organisms. Ironically, individuals with gram-negative infections may experience exaggerated symptoms during antibiotic treatment because of the release of endotoxins from dead cells.

Virulence Mechanisms of Nonbacterial Pathogens

Viruses

Recall that viruses are subcellular obligate intracellular parasites. They are completely dependent on their host cells for survival and replication. On the defensive side, the fact that viruses are intracellular parasites allows them to hide from components of the immune system. The ability of some viruses, as for example, the influenza virus, to change their antigen coats affords a measure of defense against antibodies.

On the offensive side, death of the host cell may result from the lysis of the cell membrane caused by:

- Producing large numbers of replicating viruses
- Shutting down the host cell's protein synthesis
- Damaging the cell membrane of the host
- Inhibiting host cell metabolism

Some viruses have attachment molecules enabling them to dock with specific target cells, allowing for adsorption and penetration. For example, consider that HIV docks with strategic cells (T lymphocytes) of the immune system, causing the infected individual to become severely immunocompromised. Other viruses produce cytopathic effects that kill the host cell. Some viruses cause adjacent cells to fuse and form a syncytium (a network).

Some viruses leave telltale fingerprints called inclusion bodies. The bodies are "viral debris" consisting of viral parts (nucleic acids, coats, envelopes) in the process of assembly. These inclusion bodies are important in diagnosis, because they can be identified microscopically.

Eucaryotic Microbes (Including Helminths)

Many protozoans, fungi, and helminths are pathogens, but their mechanisms of virulence are not well defined. Many fungi secrete toxins and enzymes that cause damage to host cell tissues and aid in their invasion. *Giardia,* an important and widespread water-transmitted protozoan parasite that causes severe diarrhea, attaches to cells lining the small intestine by a virulence factor called an adhesive or sucking disk (Figure 7.9). The *Plasmodium* protozoan parasite responsible for malaria infects and reproduces within host red blood cells causing their rupture, and the trypanosome protozoan that causes sleeping sickness exhibits antigenic variation, a defensive strategy of virulence, as does influenza virus. Helminths are large extracellular parasites that can obstruct lymph circulation, leading to grotesque swellings in the legs and other body structures, a condition called elephantiasis. *Ascaris* worms can "ball up," obstruct the intestinal tract, and migrate into the liver. Waste products of helminths can cause toxic and immunologic reactions.

Stages of a Microbial Disease

As explained earlier the outcome of contact between a pathogen and a host (human) depends on the number of pathogens and their virulence factors versus the immune system of the host. Everyone has experienced the miseries of a common cold and of gastrointestinal disturbances manifested by vomiting and diarrhea or (perish the thought!) both. In most cases the disease runs its course, and recovery occurs. Treatment may help to alleviate the symptoms and shorten the duration of illness, but, nevertheless, five general stages (TABLE 7.4) take place, and their characteristics are significant in diagnosis and treatment. It should be noted that each stage is not necessarily distinct.

Incubation Stage

The **incubation stage** is the time between the pathogen's access to the body through a portal of entry and the display of signs and symptoms. During this time, although the infected individual is not necessarily aware of the presence of a pathogen, the microbes may be spread to others. The incubation time is quite consistent in some diseases but variable in others. The incubation times for several diseases are as follows: botulism, 12 to 36 hours; chickenpox, 2 to 3 weeks; staphylococcal food poisoning, 2 to 4 hours; (serum) hepatitis, 45 to 160 days; and AIDS, 6 months to about 12 years.

Prodromal Stage

The **prodromal stage** is relatively short and not always obvious. The symptoms are vague and mild and are frequently characterized by tiredness, headache, muscle aches, and "feeling lousy." (These symptoms may not be indicative of a disease at all, but the result of too much partying or the stress associated with exams, papers, and the routines of student life.) In some infections, the individual may be contagious during this stage.

Illness Stage

During the **illness stage** you might feel like you have been hit in the head with a hammer and wish you could die but aren't that lucky. In this period the disease

TABLE 7.4	Stages of Microbial Disease
Stage	**Description**
Incubation	Period between initial infection and appearance of symptoms; considerable variation among diseases
Prodromal	Period in which early symptoms appear; usually short and not always well characterized
Illness	Period during which the disease is most acute and is accompanied by characteristic symptoms
Decline	Period during which the symptoms gradually subside
Convalescence	Period during which symptoms disappear and recovery ensues

develops to the most severe stage, accompanied by typical signs and symptoms that may include fever, nausea, vomiting, chills, headache, muscle pain, fatigue, swollen lymph glands, and a rash. (What an impressive and depressing list!) This is the invasive time during which the tug-of-war between the pathogen's virulence factors and the host's immune system is taking place. It is a critical time in that, as in a tug-of-war, one side wins. Either recovery will be complete, or impairment or death of the host will result.

Stage of Decline

What a relief! You won the battle and are in the **stage of decline**. The signs and symptoms begin to disappear, the body returns to normal, and life is worth living again.

Convalescence Stage

During the **convalescence stage** recovery takes place, strength is regained, repair of damaged tissues takes place, and rashes disappear. In some cases healthy and chronic carrier states develop, as might happen with cholera and typhoid fever; the carrier state can exist for years.

Overview

This chapter considered microbial pathogens and their mechanisms of causing disease. "No organism is an island" is an expression implying that all organisms live in symbiotic association with other organisms. These associations can take the form of mutualism, commensalism, or parasitism and can slide from one to the other. Parasitism is the basis for microbial disease; the parasite lives at the expense of the host and causes damage to the host.

Throughout history microbes have caused anguish and death to individuals on a large scale. The bubonic plague, smallpox, and influenza were notorious for decimating large segments of the population, thereby influencing the course of civilization. Societies had no choice but to accept these plagues as part of life. The cause of these diseases was ascribed to vague notions of "miasmas, swamp air, and noxious odors." By the nineteenth century a link had been made between filth and a lack of sanitation to infectious disease, but it wasn't until the end of the nineteenth century that Koch, Pasteur, and other microbe hunters established the role of bacteria as agents of specific diseases.

The mechanisms of microbial disease have been described as a tug-of-war between (1) the number, or dose, of microorganisms to which a potential host has been exposed and their virulence factors and (2) the immune mechanisms of the host. The outcome of the host–parasite dynamic determines the severity of infection, varying from subclinical to death. Microbes vary considerably, even within a species, in their pathogenicity, as a function of their virulence factors. Pathogenic microbes have evolved defensive strategies by which they are able to withstand the immune system of the host and offensive strategies that cause damage to the host. Although there is considerable variation in the manifestation of illness, depending on the specific microbe and the immune system of the host, there is a general pattern of infectious diseases described as incubation, prodromal, illness, decline, and convalescence stages.

Self-Evaluation

PART I: Choose the single best answer.

1. Endotoxins are associated primarily with

    a. viruses **b.** gram-positive bacteria **c.** gram-negative bacteria **d.** none of the above

2. Capsules are

    a. adhesions **b.** exotoxins **c.** associated with viruses **d.** antiphagocytic

3. Which is *not* an example of a virulence factor?

    a. hyaluronidase **b.** collagenase **c.** hemolysins **d.** All the above are virulence factors.

4. Microbial diseases are the result of a biological association, which can best be described as

    a. commensalisms **b.** symbiosis **c.** mutualism **d.** parasitism

5. Which structure interferes with phagocytosis?

    a. M protein **b.** adhesions **c.** cell walls **d.** flagella

6. What potentially fatal condition that can happen if the intestinal wall is pierced during surgery, allowing bacteria of the intestines to gain access to another part of the body?

    a. sinusitis **b.** peritonitis **c.** tinnitus **d.** senioritis

7. Which strategies involve the production of extracellular products that foster the spread of bacteria throughout the tissues by causing damage to host cells?

    a. offensive **b.** defensive **c.** feeding **d.** evasion

8. Which strategies allow microbes to escape destruction by the host immune system?

    a. offensive **b.** defensive **c.** feeding **d.** evasion

9. *Corynebacterium diphtheriae* harbors a(n) _____ that contains a toxin gene responsible for the symptoms of diphtheria.

    a. prophage **b.** pigment **c.** enzyme **d.** plasmid

10. Which is the time between the pathogen's access to the body through a portal of entry and the display of signs and symptoms?

    a. illness stage **b.** prodromal stage **c.** incubation stage **d.** convalescence stage

PART II: Complete the following.

1. *E. coli,* a resident of the intestinal tract, produces vitamin K. What type of symbiotic association does this demonstrate?

2. What type of toxins interfere with transmission of nerve pulses?

3. What type of toxins affect the cells lining the gastrointestinal tract, leading to diarrhea?

4. What type of toxins kill or damage host cells?

5. What is the term used to describe a relationship between two or more species in which one benefits and the other is indifferent?

6. What is a process by which bacteriophage DNA is incorporated into the bacterial chromosome and confers new properties on the bacterial host?

7. Before microbes were known to cause infectious diseases, what did people think was the cause of disease and death?

8. What protein is present in the cell walls of streptococci that functions to resist phagocytosis?

9. What type of relationship is used to describe two microbes that benefit from each others' presence.

10. What kind of enzymes produced by bacteria break down clots, allowing entrapped bacteria to spread?

PART III: Answer the following.

1. Discuss Koch's postulates. Illustrate with a diagram.

2. The expression $D = nV/R$ expresses host–parasite relationships. Elaborate on this; identify each term (D, n, V, and R) in the equation.

3. Regarding symbiotic associations, a "slippery slide" exists. What does this mean? Give examples.

4. Explain the AB model that has been proposed to explain the general mode of activity of some toxins.

5. Compare and contrast exotoxins and endotoxins.

6. Antigenic variation is a significant microbial defense mechanism. Describe an example.

7. What does the term LD_{50} mean?

8. Name a "spreading factor" and describe its action.

9. List the five stages of microbial disease and describe what happens at each stage.

10. What does the acronym ID stand for, and what does it mean?

Epidemiology and Cycle of Microbial Disease

Dr. Erin Mears said, "The average person touches their face 3 to 5 times every waking minute. In between that we're touching door knobs, water fountains, and each other."

—From the movie *Contagion*, 2011

■ Preview

Microbes are a potential threat to the world population. This chapter describes the biology of microbes and the mechanisms by which they cause disease from a public health aspect and focuses on the factors responsible for infectious diseases in populations. Basic concepts of epidemiology are presented so that the occurrence and prevalence of disease in a particular environment at a particular time can be better understood. The existence of microbial disease requires a chain of linked factors that constitute the cycle of disease. These factors are reservoir, transmission, portal of entry, and portal of exit.

In hospitals and in long-term health care facilities, all factors involved in the cycle of infectious diseases are present in a concentrated way. These facilities are hotbeds for the transmission of microbes among hospital personnel, visitors, and patients.

Photo courtesy of the CDC.

Concepts of Epidemiology

Epidemiology is an investigative methodology designed to determine the source and the cause of diseases and disorders that produce illness, disability, and death in human populations. Epidemiologists have been dubbed "disease detectives"; they are among the first group to be dispatched by the Centers for Disease Control and Prevention (CDC) when the threat of an outbreak occurs anywhere in the world. Their sleuthing is directed at understanding why an outbreak of a disease is triggered at a particular time and in a particular place. Epidemiologists consider age distribution of the population, sex, race, personal habits, geographical location, seasonal changes, modes of transmission, and other factors. These parameters are used to design public health strategies for control and prevention of future outbreaks. Historically, epidemiology is based on an understanding of the causes and distribution of infectious diseases, but modern epidemiology has branched out to other public health problems, including alcohol and drug abuse, cancer, mental conditions, "road rage" and other acts of violence, and exposure to lead paint.

Epidemiology dates back to the time of Hippocrates (460–377 B.C.), who questioned the role of eating and drinking habits, the source of drinking water, lifestyle, and seasons of the year as factors related to causality of disease. He was astute enough to realize that the diseases now identified as yellow fever and malaria were associated with swampy environments where mosquitoes bred. The epidemiological studies of Edward Jenner, a country physician in England, proved that cowpox and smallpox were related and led to smallpox immunization in the late 1700s. In the mid-1800s another early epidemiologist-physician, Ignaz Semmelweis, showed that childbed fever resulted from physicians and other attendants who weren't washing their hands after dissecting corpses. They would proceed directly from the autopsy room to the delivery room (BOX 8.1).

John Snow was another early epidemiologist who laid the groundwork for modern methodologies of epidemiology. In 1849 a major epidemic of cholera occurred in the Soho district of London, causing about 500 deaths in the span of only 10 days. Cholera is a bacterial disease manifested by diarrhea so pronounced that life-threatening amounts of water are lost from the body in a short time, causing death from dehydration. Snow's epidemiological detective work showed that most of the cholera victims lived in the Broad Street area and drew their water from the Broad Street pump (FIGURE 8.1). Further investigation revealed that the pump was contaminated with raw sewage, and when the pump handle was removed, the cholera epidemic was halted.

A second outbreak of cholera occurred in London in 1854. Snow's sleuthing revealed that most of the cholera victims purchased their drinking water from the Southwark and Vauxhall Company; the company's source of water was the Thames River downstream from the site where raw sewage was discharged into the river. The Lambeth Company, another water supplier, obtained its water further upstream; the incidence of cholera in the population using Lambeth water

Semmelweis—"The Savior of Mothers." Author's photo (RIK).

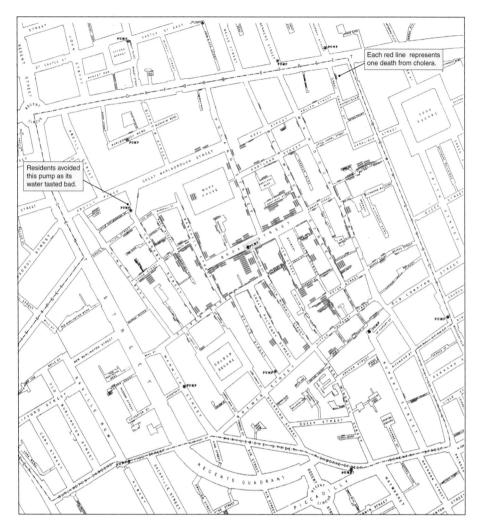

Each red line represents one death from cholera.

Residents avoided this pump as its water tasted bad.

(a)

FIGURE 8.1 **(a)** A portion of the map created by Dr. John Snow of the Broad Street area. The lines plot how many people died at each address. Courtesy of R. R. Frerichs' John Snow website: http://www.ph.ucla.edu/epi/snow.html, 2006. **(b)** The John Snow Pub and the "ghost of Broadwick Street" (formerly Broad Street). In the foreground is a replica of the "ghost," the Broad Street pump that was the source of contaminated water causing a major outbreak of cholera. © Wellcome Library, London.

AUTHOR'S NOTE (RIK)
In 1990 I visited the site of the old Broad Street pump. At the site is now a pub called the John Snow Pub in commemoration of the pump. I had a few beers and reminisced about history.

was much lower. The cause of the disease was not known and led to conjecture and imagination (FIGURE 8.2). The actual bacterial contaminant in the water in both outbreaks proved to be *Vibrio cholerae*.

Epidemiologists focus on the frequency and distribution of diseases in populations and classify diseases as **sporadic**, **endemic**, **epidemic**, and **pandemic**. Sporadic diseases are those that occur only occasionally and at irregular intervals in a random and unpredictable fashion. Typhoid fever, eastern equine encephalitis, and tetanus are examples. Diseases that are continually present at a steady level in a population and pose little threat to the public health are endemic diseases. The common cold, mumps, and whooping cough are endemic across the United States, whereas Lyme disease is endemic primarily in some New England states, but can be caught on a year-round basis. A disease is said to be epidemic when there is a sudden increase in the **morbidity** (illness rate) and in the **mortality** (death rate) above the usual,

(b)

BOX 8.1 ___Semmelweis and Childbed Fever_____

Ignaz Philipp Semmelweis (**FIGURE B8.1**) (1818–1865) graduated from the Vienna University Medical School in 1844 at the age of 26. He started his medical practice in Austria, specializing in obstetrics at a large maternity

FIGURE B.8.1 Ignaz Semmelweis, sometimes called the "savior of mothers." Before the establishment of microbes as causative agents of disease, Semmelweis realized that childbed fever was transferred from physicians to mothers during delivery. © National Library of Medicine.

hospital. He was not an easy doctor to work with. Dr. Semmelweis was cantankerous, arrogant, and abrasive. What he lacked in personality, he made up for in intelligence and his astute observation skills.

The death rate of healthy women from childbed fever was so high that women begged to deliver their babies at home rather than risk a hospital delivery. Semmelweis placed the blame, often with insults, on the doctors' and medical students' unsanitary procedures. He was ridiculed by his colleagues, his hospital privileges were limited, and his academic rank was reduced. "Bad blood" and mysterious forces were thought to be the cause of childbed fever and other diseases; these conditions certainly were not caused by the doctors. Upwards of 30% of women were dying of childbed fever after delivering babies in the hospital whereas women who delivered their babies at home with the aid of a midwife rarely died of childbed fever.

Childbed fever is sometimes called *puerperal fever*. Symptoms of the disease usually began on the second or third day after delivery. The new mothers experienced a violent shivering fit or fever followed by pain in the uterus radiating toward the abdomen that was tender to the touch. The pulse was rapid and the pain became excruciating. Patients complained that the pain was greater than that they suffered during labor. They stopped lactating. As the infection progressed, patients produced cloudy, putrid urine when they could urinate. They produced a foul smelling vaginal

FIGURE 8.2 "Monster Soup," commonly called Thames Water. A satirical cartoon created by William Heath, c. 1928. © Wellcome Library, London.

discharge. Some vomited and had diarrhea. Their tongues became white, and the patients became thirsty. As the disease progressed, the pain and agony were unbearable. As the suffering continued, they became confused and delirious. Some doctors ordered frequent bleeding or purging procedures and opium for the pain. This was a time before the role of bacteria in diseases was discovered. Today it is known that childbed fever is an infection of the endometrium following childbirth or an abortion caused by group A hemolytic streptococci.

After a woman died of childbed fever, their bodies were moved to nearby dissection rooms—the "death houses." Physicians and medical students would perform autopsies. They smelled of the death houses. They did not wear gloves. Their hands contained putrid matter from corpses. The bloodier and dirtier their laboratory coats, the prouder they became. The smell and filthy coats represented evidence of their superior skills as physicians and interns.

Dr. Semmelweis observed that the obstetricians and medical students performing autopsies on the wombs of postpartum women who died of childbed fever went directly to the delivery rooms to perform routine vaginal examinations. He believed they carried "death particles" with them to the birthing rooms. Because of his difficult personality, he used his authority to order all obstetricians and medical students to wash their hands with chloride of lime before entering the maternity ward. The results were dramatic! The doctors and interns no longer smelled of death. The morbidity rates plummeted to less than 2%!

Dr. Semmelweis was ignored and ridiculed by many in the medical community. He believed that cleanliness was critical in the hospitals; however, he could not handle criticism for his beliefs. He was difficult and dogmatic, retaliating by writing angry letters. His term of appointment at the hospital expired in 1849, and he was dismissed. Thirteen years later, he published his treatise, *The Etiology, Concept, and Prophylaxis of Childbed Fever,* which is dated 1861 but was actually published in 1860. The treatise of over 500 pages contains passages of great clarity interspersed with lengthy, muddled, repetitive, and bellicose passages in which he attacks his critics.

In 1865 Semmelweis was committed to an insane asylum. He had become an uncontrollable psychotic—possibly due to tertiary syphilis or Alzheimer's disease. He died at the age of 47, ironically, of a wound infection that may have been caused by the same bacterium that causes childbed fever. A tragic ending for a scientist ahead of his time. Irrespective of his difficult nature, he is credited with the practice of **hand washing**, a simple, standard aseptic technique. It is the single most important procedure in preventing hospital-acquired infections. In his honor, Semmelweis University, a medical school located in Budapest, Hungary, is named after him.

causing a potential public health problem. Throughout history, epidemics have resulted in more deaths than those caused by wars, and they have influenced the course of history. Plague has bedeviled humankind at least since the reign of Emperor Justinian in the sixth century; the fourteenth century epidemic was particularly devastating. Smallpox, cholera, and typhus fever are other examples of past epidemics. Epidemics may arise from an explosion of sporadic or endemic diseases or, it would seem, from out of nowhere. Pandemic diseases are those that spread across continents and may be worldwide; AIDS is a pandemic. Cholera has been responsible for pandemics on several occasions over time. In 1918 what was then referred to as swine influenza was perhaps the greatest pandemic of all times. The term *epidemic* has been borrowed to indicate a variety of conditions unrelated to infectious diseases that are present beyond the norm. For example, college officials talk of grade inflation as a problem of epidemic proportions, as are school violence, obesity, and cancer.

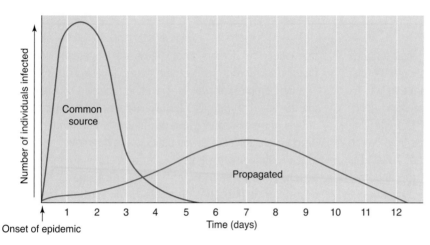

FIGURE 8.3 A comparison of the courses of common-source and propagated epidemics.

Epidemiologists have described the source and spread of epidemics as **common-source epidemics** or **propagated epidemics**. Common-source epidemics arise from contact with a single contaminated source and are usually associated with fecally contaminated foods and water. Typically, a large number of people become ill quite suddenly, and the disease peaks rapidly in the population. A propagated epidemic is the result of direct person-to-person contact; the microbe is spread from infected individuals to noninfected susceptible individuals. As compared with common-source epidemics, the number of infected individuals rises more slowly and decreases gradually. Chickenpox, measles, and mumps are examples of propagated epidemics. FIGURE 8.3 illustrates the courses of common-source and propagated epidemics.

The number of individuals in a population who are immune (nonsusceptible) to a particular disease as compared with those who are nonimmune (susceptible) is an important factor in the occurrence of epidemics. Immunity can be the result of having had a particular infection or of having been immunized. The term *herd immunity* (group immunity) refers to the proportion of immunized individuals in a population. Disease can only be spread to susceptible individuals; therefore, the smaller the number of susceptible individuals, the less opportunity for contact between them and infected individuals. Public health officials strive to maintain high levels of herd immunity against communicable diseases to minimize their progressing to epidemic status (FIGURE 8.4). Hence, immunization is required in the elementary grades against a variety of diseases; proof of an up-to-date immunization history is required for college admission.

The basic reproduction rate (denoted as R_0 or **R-nought**) is a measure of the potential for transmission, i.e., the mean number of secondary cases, occurring in a nonimmunized (susceptible) population in the wake of a particular infection. The population density and the duration of contagiousness and other factors need to be considered.

For an infection to spread, the R_0 value must be greater than 1; if less, the infection will die out. The greater the value, the greater is the chance of spread making it more difficult to establish measures of control. Pertussis (whooping cough), for example, has a high R_0 value between 12 and 17. TABLE 8.1 displays R_0 values for several infectious diseases. If you watched the movie *Contagion* (2011), recall that the infectious agent MEV-1 had an R_0 of 4 to 6 on the sixth day of the pandemic and cases increased exponentially over time.

As described, the epidemiologist's toolbox contains a number of strategies to deal with outbreaks from endemics, to epidemics, to pandemics, and to predict their course as components of epidemic theory.

A decrease in herd immunity can lead to reemergence of a disease. A case in point is the epidemic of diphtheria that occurred in the newly independent states

(a) Most of the cows are immunized. Spread of contagious disease is contained because the herd is immune.

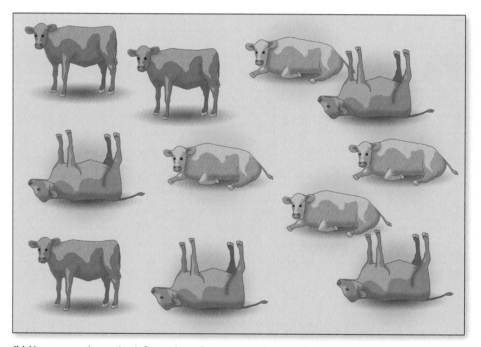

(b) No cows are immunized. Contagious disease spreads through the herd of cows.

FIGURE 8.4 Depiction of *herd immunity*. **(a)** Cows vaccinated against parainfluenza-virus 3 are protected from getting infected by the virus. **(b)** Cows not immunized against parainfluenza virus become infected and sick.

of the former Soviet Union in the early 1990s. A decline in the public health infrastructure resulted in fewer children receiving diphtheria vaccination and a decline in herd immunity. When the disease was introduced into the population, possibly by returning military personnel, diphtheria reached epidemic proportions.

TABLE 8.1 Basic Reproduction Rate (R_0) Within Human Populations

Disease	Type of Causative Agent	R_0
Measles	Virus	12–18
Pertussis	Bacterium	12–17
Diphtheria	Bacterium	6–7
Smallpox	Virus	5–7
Poliomyelitis	Virus	5–7
Rubella	Virus	5–7
Mumps	Virus	4–7
HIV/AIDS	Virus	2–5
Influenza A	Virus	2–3
Ebola hemorrhagic fever	Virus	1
Rabies	Virus	<1

Frequency and distribution with attention to age, gender, diet, lifestyle, and other factors describing a disease-afflicted population is a starting point in investigative epidemiology in order to establish the strategy necessary to halt an outbreak. Epidemiologists use a variety of graphs, charts, tables and maps to establish the parameters of an outbreak (FIGURE 8.5). (Note: these figures are intended only to display epidemiological tools, and not to cite specific diseases.)

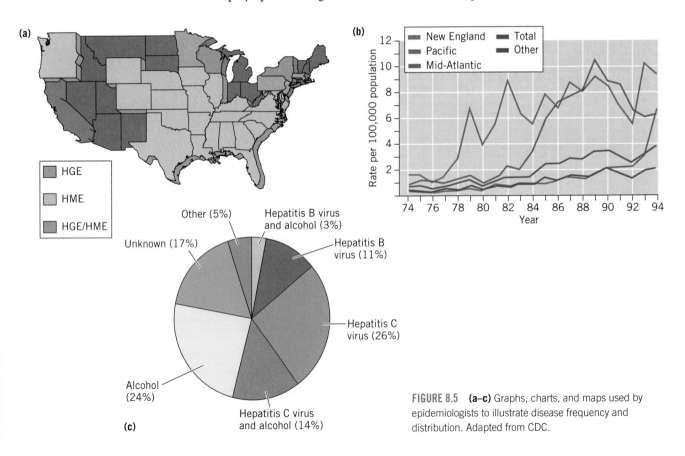

FIGURE 8.5 (a–c) Graphs, charts, and maps used by epidemiologists to illustrate disease frequency and distribution. Adapted from CDC.

Surveillance of disease outbreaks and of factors that could trigger outbreaks is an important mission of public health organizations throughout the world, including the WHO, the CDC, and agencies at the state and local levels. To keep track of these diseases in the United States, physicians are required to report cases of certain diseases, referred to as notifiable diseases, to their local health departments; these are then reported to the CDC. In 1994 49 diseases were listed as nationally notifiable; 69 diseases were reportable in 2011 (TABLE 8.2). The specific diseases are decided on at an annual meeting involving state departments of health and the CDC. An increase or a decrease in the number of notifiable diseases does not necessarily reflect a change in the health status but may be the result of reorganization each year. To further assist public health and medical personnel, the CDC publishes the journal *Emerging Infectious Diseases* as well as the *Morbidity and Mortality Weekly Report*, which contain data organized by states on morbidity and mortality of particular diseases in the United States and throughout the world.

Cycle of Microbial Disease

For infectious diseases to exist at the community level, a chain of linked factors needs to be present, somewhat reminiscent of a parade of circus elephants linked trunk to tail. These factors are reservoirs, modes of transmission, portals of entry, portals of exit, susceptible host, and infectious agent (FIGURE 8.6). An understanding of these factors is imperative to attempt to break the cycle somewhere along the path. For example, if insects are involved in transmission, then controlling their population is a target; for those microbes transmitted by drinking water, providing safe drinking water is a goal. Shrinking the reservoir (where the microbes exist in nature) is a potential target for other diseases. In some instances, a combination of targets is preferable.

For a particular microbial disease to exist there has to be a pathogen as the causative agent and a host in which the pathogen takes up residence. The potential for disease to occur and its outcome are a result of the complex interaction between the number of invading microbes and their virulence and the host immune system. Communicable diseases are infectious diseases in which the pathogen can be transmitted directly or indirectly from its reservoir to the host portal of entry.

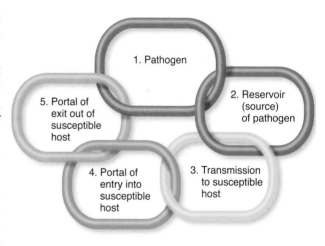

FIGURE 8.6 The cycle or chain of microbial infection.

Reservoirs of Infection

A **reservoir** is a site in nature in which microbes survive (and possibly multiply) and from which they may be transmitted. All pathogens have one or more reservoirs, without which they could not exist. Knowledge and identification of these reservoirs are important, because the reservoirs are prime targets for preventing, minimizing, and eliminating existing and potential epidemics. The facts that humans are the only reservoir of smallpox and that person-to-person transmission of smallpox takes place were key factors in the eradication of this disease.

TABLE 8.2 Major Infectious Diseases Designated as Notifiable at the National Level, United States, 2011

AIDS
Anaplasmosis
Anthrax
Arboviral neuroinvasive and
  nonneuroinvasive diseases
Babesiosis
Botulism
Brucellosis
California encephalitis
Chancroid
Chlamydia (eye, genital infections)
Cholera
Coccidioidomycosis
Cryptosporidiosis
Cyclosporiasis
Dengue fever
Dengue hemorrhagic fever
Dengue shock syndrome
Diphtheria
Eastern equine encephalitis
Ehrlichiosis
Giardiasis
Gonorrhea
Haemophilus influenzae, invasive
  disease
Hansen disease (leprosy)
Hantavirus pulmonary syndrome
Hemolytic uremic syndrome,
  postdiarrheal
Hepatitis A
Hepatitis B
Hepatitis C
HIV infection
Influenza-associated pediatric mortality
Jamestown Canyon virus
Legionellosis
Listeriosis
Lyme disease
Malaria
Measles
Meningococcal disease
  (*Neisseria meningitidis*)
Mumps
Neurosyphilis
Novel influenza A virus infections

Pertussis
Plague
Poliomyelitis, nonparalytic
Poliomyelitis, paralytic
Powassan encephalitis
Psittacosis
Q fever
Rabies
Rocky Mountain spotted fever
Rubella
Rubella, congenital syndrome
Salmonellosis
Severe acute respiratory
  syndrome-associated coronavirus
  (SARS-CoV disease)
Shiga toxin-producing *Escherichia coli*
  (STEC)
Shigellosis
Smallpox
Spotted fever: rickettsiosis
St. Louis encephalitis
Streptococcal disease invasive, group A
Streptococcal toxic shock syndrome
Streptococcus pneumoniae,
  drug-resistant invasive disease
Streptococcus pneumoniae, invasive
  disease nondrug resistant, in children
  less than five years of age
Syphilis
Tetanus
Toxic shock syndrome (staphylococcal)
Trichinellosis
Tuberculosis
Tularemia
Typhoid fever
Vancomycin-intermediate
  Staphylococcus aureus (VISA)
Vancomycin-resistant *Staphylococcus
  aureus* (VRSA)
Varicella (deaths only)
Varicella (morbidity)
Vibriosis
Viral hemorrhagic fevers
West Nile encephalitis
Western equine encephalitis
Yellow fever

Adapted from the CDC. *Summary of Notifiable Diseases*, 2011.

Additionally, humans are the only known reservoir for gonorrhea, measles, and polio. Animals, as well as plants and nonliving environments, also serve as reservoirs. In some cases the source of the pathogen is distinct from the reservoir and is the immediate location from which the pathogen is transmitted. For example, in typhoid fever the reservoir may be an individual with an active case of the disease who sheds typhoid bacilli in feces; the immediate source would be water or food contaminated with fecal material. On the other hand, in most sexually transmitted diseases the human body serves as both reservoir and source.

Active carriers are those individuals who have a microbial disease, whereas **healthy carriers** have no symptoms and unwittingly pass the disease on to others. **Typhoid Mary**, a cook and healthy lifetime carrier of typhoid fever, was responsible for about 10 outbreaks, 53 cases, and 3 deaths due to typhoid fever during her lifetime. **Chronic carriers** are those who harbor a pathogen for long periods after recovery, possibly throughout their lives, without ever again becoming ill with the disease. In the case of chronic (and healthy) carriers of typhoid fever, removal of the gallbladder may be effective in eliminating the carrier state; intensive therapy with antibiotics works in other cases. Tuberculosis is another disease in which carriers play a significant role. Depending on the particular infection, carriers discharge microbes via portals of exit, including respiratory secretions, feces, urine, and vaginal and penile discharges.

Domestic and wild animals serve as reservoirs for about 150 species of pathogenic microbes that can affect humans. These diseases are referred to as **zoonoses** (TABLE 8.3).

TABLE 8.3 Selected Zoonotic Diseases

Transmission by Arthropod Bites	Transmission Via Food, Water, or Animal Bites
Bacteria	***Bacteria***
Ehrlichiosis	Undulant fever
Relapsing fever	Leptospirosis
Lyme disease	Anthrax
Rocky Mountain spotted fever	Cat scratch fever
Plague	Tularemia
Typhus fever	***Viruses***
Viruses	Rabies
Yellow fever	Hantavirus disease
Eastern equine encephalitis	Viral gastroenteritis
West Nile virus disease	Severe acute respiratory syndrome (SARS)
La Crosse encephalitis	***Protozoans***
Rift Valley fever	Giardiasis
Dengue fever	*Cyclospora* infection
Protozoans	Toxoplasmosis
Babesiosis	
Sleeping sickness	
Malaria	
American trypanosomiasis	
Leishmaniasis	

Microbes of animals that are most closely related to humans have the greatest chance of making the "species leap" to humans or as having erased the species barrier. Consider, for example, the AIDS virus, which is thought to have a reservoir in chimpanzees and is now a human pathogen. Prions cause both mad cow disease and its human counterpart variant, Creutzfeldt-Jakob disease; prions jumped from cattle to humans. Perhaps all infectious diseases in humans originated in other species and jumped the species barrier. Monkeys are reservoirs for the microbes that cause malaria, yellow fever, and numerous other diseases in humans. The reservoirs for the spirochete bacteria that cause Lyme disease, a major problem in the northeastern United States, are deer and mice. Hantavirus pulmonary syndrome, a relatively new disease in the United States, uses a variety of rodent species, particularly the deer mouse, as reservoirs. Many mammals, including dogs, cats, raccoons, skunks, foxes, mongooses, and bats, serve as reservoirs for rabies.

Eradication of zoonotic diseases is particularly challenging and difficult because it is, ultimately, dependent on eradicating the reservoirs. Malaria is a protozoan disease transmitted by mosquitoes. Intensive spraying with the pesticide DDT (dichlorodiphenyltrichloroethane) in the 1940s markedly reduced the mosquito population and the number of malaria cases. The mosquitoes developed resistance to DDT eventually, leading to a reemergence of malaria.

FIGURE 8.7 Soil can be a reservoir for microbes and helminth eggs. This child is at increased risk for infection. © Bernardo Erti/Dreamstime.com.

Nonliving Reservoirs

Some organisms are able to survive and multiply in nonliving environments. Soil and water are the major nonliving reservoirs of infectious diseases. The tetanus bacillus and the botulinum bacillus, both members of the same bacterial group *Clostridium*, are spore formers and thus can survive for many years in soil. These organisms are part of the normal flora of horses and cattle and are deposited in their feces onto the soil. The use of animal fertilizers contributes to their distribution. Certain helminth (worm) parasites (for example, hookworms) deposit their eggs onto the soil, establishing a reservoir for human infection (FIGURE 8.7).

Contaminated drinking water and foods are major reservoirs for many microbes that cause gastrointestinal tract disease, ranging from mild to severe to fatal. The list includes bacteria, viruses, and protozoa (TABLE 8.4). Because of the potential for an outbreak of waterborne and foodborne illnesses, local departments of public health devote considerable attention to sanitary measures

TABLE 8.4 Water and Food as Reservoirs of Infection

Type of Microbe	Examples of Waterborne and Foodborne Infections
Bacteria	Salmonellosis, shigellosis, cholera, gastroenteritis
Viruses	Hepatitis A, poliomyelitis, viral gastroenteritis (e.g., norovirus)
Protozoa	Giardiasis, amebiasis, cryptosporidiosis
Worms	Ascariasis, trichinellosis, *Trichuris* infection

TABLE 8.5 Modes of Transmission	
Direct	**Indirect**
Contact (e.g., kissing, sneezing, coughing, singing, sexual contact)	Vehicles (fomites, e.g., doorknobs, eating utensils, toys, facial tissue)
Animal bites	Airborne (via aerosols created by, e.g., shaking bedsheets, sweeping, mopping)
Transplacental	Vectors (e.g., mosquitoes, ticks, flies)

designed to minimize risks; their activities include monitoring food service establishments, beaches, swimming pools, and certification of food handlers.

Transmission

The next link in the cycle of disease is **transmission**, the bridge between reservoir and portal of entry (Figure 8.6). Transmission is the mechanism by which an infectious agent is spread through the environment to another person. More simply put, transmission answers the question "How do you get the disease?" There are several modes of transmission, and they can be grouped into two major pathways, direct and indirect. Each of these, in turn, can be subgrouped into three categories (**TABLE 8.5**).

Direct Transmission

The most common type of **direct transmission** is person-to-person contact, in which the infectious agent is directly and immediately transferred from a portal of exit to a portal of entry. Sexual contact, kissing, and touching are the most common examples. Transmission is facilitated by contact of the warm, moist mucous membranes of one individual with the warm, moist mucous membranes of another, as occurs in sexually transmitted diseases. Sexual contact is an example of **horizontal transmission** (i.e., transmission from one person to another). Droplet transmission is also direct and horizontal and involves the projection of infected spray from coughing, sneezing, talking, and laughing onto the conjunctiva of the eyes or onto the mucous membranes of the nose or mouth. Influenza, whooping cough, and measles are spread by droplets. Droplets are about 10 micrometers or greater in diameter and travel less than 1 meter; as many as twenty thousand droplets may be produced in a sneeze (**FIGURE 8.8**). Think about that the next time you cough or sneeze directly into the crowded environment of the classroom and be certain to "spray" into the crook of your elbow and not into your hand.

A second type of direct and horizontal transmission involves animal bites, rabies being the most common example. The virus is directly transmitted from the saliva of the rabid animal onto the skin and underlying tissues. Finally, transplacental transmission is an example of **vertical transmission**, in which the pathogens are passed from mother to offspring across the placenta (AIDS, measles, and

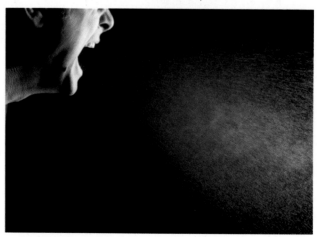

FIGURE 8.8 Droplet transmission. As many as 20 thousand droplets may be produced during a sneeze. It is important to carry a handkerchief or tissue and to cover your nose and mouth when sneezing. © James Klotz/ShutterStock, Inc.

chickenpox), in breast milk, or in the birth canal (syphilis and gonorrhea). Notice there are no intermediaries in all these categories of direct transmission. Microbes are transferred by the contact of portals of exit with portals of entry. These portals are described later in this chapter.

Indirect Transmission

Indirect transmission involves the passage of infectious material from a reservoir or source to an intermediate agent and then to a host. The intermediate agent can be living or nonliving. Vehicle-borne transmission is accomplished by food, water, biological products (organs, blood, blood products), and **fomites** (inanimate objects) as, for example, desk surfaces, doorknobs, or escalator rails.

Waterborne transmission is a serious problem throughout the world and is a major cause of death in many developing countries as a result of fecal–oral passage, in which pathogens are transmitted from the feces of one individual to another by hand-to-mouth transfer. Public, semiprivate, and private water supplies must all be carefully monitored for the presence of fecal pathogens.

Water and food can serve as both a reservoir and a transmitter of infectious agents. Because of the potential for an outbreak of waterborne and foodborne illnesses, local departments of public health devote considerable attention to sanitary measures designed to minimize risks; their activities include monitoring food packaging industries, food service establishments, beaches, swimming pools, and certification of food handlers.

Fomites play a significant role in the transmission of infectious agents. The list of fomites is seemingly endless and includes objects in common use, such as doorknobs, telephones, faucets, computer keyboards, and exercise equipment. Toys are fomites and contribute to illness in children wherever the toys are shared. Surgical instruments, medical equipment (for example, catheters, intravenous equipment, and syringes), bedding, and soiled clothing are also fomites. An interesting study involving soiled saris as fomites, conducted in 51 slum areas in Dhaka, Bangladesh, revealed a positive correlation between the number of misuses of dirty saris and episodes of childhood diarrhea.

The list of fomites and their role may cause you some concern, but there is at least a partial solution to the problem. The simple act of frequent hand washing has been shown to markedly reduce hand-to-mouth (and nose and eye) infection. Frequent wiping of tabletops and counters with disinfectants is effective and a sign of good hygiene in a sanitation-conscious restaurant. In health and exercise clubs it has become a widespread practice to wipe down exercise equipment after use; it is a sign of bad manners not to do so (FIGURE 8.9).

Airborne transmission by **aerosols** is the second type of indirect transmission. Aerosols are suspensions of tiny water particles and fine dust in the air; they are distinct from droplet nuclei, as they are smaller than 4 micrometers, travel more than 1 meter, and are small enough to remain airborne for extended periods. Aerosols cause outbreaks of Q fever, Legionnaires' disease, and psittacosis (from infected birds).

FIGURE 8.9 Exercise machines can be reservoirs for microbes. You may be exercising to maintain good health, but poor personal hygiene may place others at risk. Wipe down exercise machines after use. © Jones & Bartlett Learning.

The microbes in aerosols may not come directly from humans or animals but may be present in dust particles where they can survive for months. Most hantavirus pulmonary syndrome infections can be traced back to when the victim cleaned out mouse droppings from a dusty place such as a summer cabin. Bacteria and viruses can be disseminated by changing bed linens, sweeping, mopping, and other activities. Hospital personnel are keenly aware of this, as reflected in the practice of using wet mops and damp cloths to wipe surfaces.

The third type of indirect transmission is by vectors, living organisms that transmit microbes from one host to another. The term **vector** is sometimes more broadly used to cover any object that transfers microbes, but this is, strictly speaking, incorrect usage. Ticks, flies, mosquitoes, lice, and fleas are the most common vectors, and they belong to the same biological phylum, the **Arthropoda**, along with lobsters and crabs. (It may be difficult to understand what flies, fleas, ticks, and lobsters have in common—but the edibility of lobsters certainly sets them apart!) Arthropods are invertebrate animals with jointed appendages (*arthro* means joint, as in "arthritis" [inflammation of joints], and *pod* means foot, as in "podiatrist" [foot doctor]). Further, they all have segmented bodies and a hardened exoskeleton. The arthropods are members of the largest phylum and consist of many diverse species that are divided into four subphyla (TABLE 8.6). They are considered to be the most successful of all living animals in terms of the huge number of species and their distribution. FIGURE 8.10 is a diagram illustrating examples of direct versus indirect transmission.

Spiders, ticks, and mites hatch from eggs as six-legged larvae and undergo metamorphosis to eight-legged adults with two body segments and mouth parts adapted for the sucking of blood. Ticks transmit a variety of infectious diseases,

TABLE 8.6 The Phylum Arthropoda

Subphylum Chelicerata
 Scorpions
 Chiggers
 Spiders
 Daddy longlegs
 Mites[a]
 Horseshoe crabs
 Ticks[a]
Subphylum Crustacea
 Water fleas
 Isopods
 Fairy shrimp
 Crabs
 Copepods[a]
 Lobsters
 Barnacles
 Shrimp

Subphylum Hexapodia
 Insects (many subgroups)
  Flies[a]
  Fleas[a]
  Mosquitoes[a]
  Lice[a]
Subphylum Myriapodia
 Centipedes
 Millipedes

[a] Vectors of human disease.

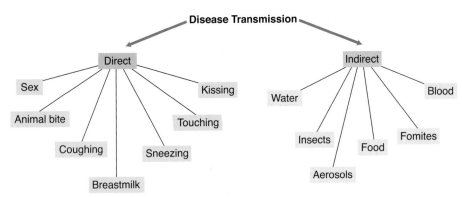

FIGURE 8.10 Direct versus indirect transmission of microbial diseases.

including Lyme disease, Rocky Mountain spotted fever, babesiosis, and ehrlichiosis. In addition to their role as vectors, some ticks are important reservoirs because they exhibit **transovarial transmission** (the passage of microbes into their eggs).

Insects are an extremely large and diverse group of arthropods with well over one million species. You may think of them as pests because (depending on the species) they bite, eat our crops, are bizarre-looking, some carry obnoxious odors, and are associated with uncleanliness. Insects have three body segments (the head, the thorax, and the abdomen) and six legs; some have one or two pairs of wings while others have two. Some have mouth parts adapted for puncturing the skin and sucking blood. "Kissing bugs" suck blood from their hosts and are vectors of Chagas disease, endemic in Central and South America.

Because many arthropods play a significant role in the cycle of infectious diseases, officials at public health departments know that arthropod control can lead to disease control. Mosquito abatement programs have been carried out on numerous occasions to control malaria. All the lower forty-eight states are threatened by West Nile virus, a mosquito-borne virus, resulting in insecticide spraying to control the mosquito population. Several other viruses that cause encephalitis (brain swelling and other neurological damage) belong to a group called arboviruses and are so named because they are arthropodborne (shortened to "arbo").

Arthropods can be either **mechanical vectors** or **biological vectors**. Mechanical vectors transmit microbes passively on their feet and other body parts; the microbes do not invade, multiply, or develop in the vector. Houseflies, for example, feed on exposed human and animal fecal material and then transfer microbes on their feet to food and eating utensils. Typhoid fever and other gastrointestinal diseases characterized by diarrhea or dysentery may be spread in this way. Covering of human and animal waste to avoid exposure to flies is an obvious answer, but this is not always possible in poverty-stricken areas, under wartime conditions, in refugee camps, and in other circumstances involving large groups of people when it is difficult to maintain good sanitation. Even in the best of circumstances, flies have access to dog feces and can mechanically transmit microbes to kitchen areas. It is disturbing to think that a fly that has just lunched on dog feces in your backyard or in a neighboring park may walk across the chicken salad that you prepare for a picnic. Cockroaches also serve as mechanical vectors; remember this when you see them marching across a kitchen counter. In December 2000 Chinese newspapers reported that Beijing was in the grip of a roach menace. Roaches were invading restaurants, hotels and motels, and even hospitals. Roaches carry more than 40 kinds of bacteria, some of which are pathogens. Cheap run-down motels are sometimes referred to as "roach motels."

Biological vectors, unlike mechanical vectors, are necessary components in the life cycles of many infectious disease agents and are required for the multiplication and development of the pathogen; transmission by biological vectors is an active process. As an example, when a mosquito picks up the malaria plasmodium parasite while taking a blood meal from an infected person, the parasites are not at an infective stage. Further development in the mosquito's body results in parasites that are now infective for hosts. Depending on the particular vector, parasites may be carried in the saliva and injected into the tissue while biting; other vectors have the nasty habit of regurgitating infectious secretions into and around the bite, and others defecate infectious material onto the bite area. Itching usually results, and scratching facilitates entry of the parasite. Mosquitoes, fleas, lice, and ticks are common biological vectors.

Mosquitoes can rightly be considered as "public health enemy number one" based on their transmission of bacterial, viral, protozoan, and worm diseases (TABLE 8.7). In a text about mosquitoes, the author wrote, "She doesn't aerate the soil, like ants and worms. She is not an important pollinator of plants, like the bee. She does not even serve as an important food item for some other animal. She has no 'purpose' other than to perpetuate her species. That the mosquito plagues humans is really, to her, incidental. She is simply surviving and reproducing."

TABLE 8.7 Diseases Transmitted by Arthropod Bites

Disease	Type of Microbe	Genus of Microbe	Arthropod Vector	Distribution
Plague	Bacterium	*Yersinia pestis*	Fleas	Southeast Asia, Central Asia, South America, western North America
Relapsing fever	Bacterium	*Borrelia* sp.	Lice or ticks	South America, Africa, Asia, western North America
Lyme disease	Bacterium	*Borrelia* sp.	Ticks	Europe, North America, Australia, Japan
Typhus fever (endemic)	Bacterium	*Rickettsia* sp.	Fleas	Worldwide
Typhus fever (epidemic)	Bacterium	*Rickettsia* sp.	Lice	Eastern Europe, Asia, Africa, South America
Eastern equine encephalitis	Virus	Alphavirus	Mosquito (*Culex, Coquillettidia, Aedes*)	North and South America
Japanese encephalitis	Virus	Flavivirus	Mosquito (*Culex*)	Asia, Pacific Islands, Torres Strait of Australia, Papua New Guinea
La Crosse encephalitis	Virus	Bunyavirus	Mosquito (*Ochlerotatus*)	United States
St. Louis encephalitis	Virus	Flavivirus	Mosquito (*Culex*)	North and South America
West Nile encephalitis	Virus	Flavivirus	Mosquito (*Culex*)	Africa, North America, Caribbean, South and Central America, India, Australia, Middle East, Russia, Europe, Southeast Asia
Western equine encephalitis	Virus	Alphavirus	Mosquito (*Aedes*)	North and South America

(continues)

TABLE 8.7 Diseases Transmitted by Arthropod Bites (*continued*)

Disease	Type of Microbe	Genus of Microbe	Arthropod Vector	Distribution
Dengue fever	Virus	Flavivirus (Dengue 1, 2, 3 and 4)	Mosquito (*Aedes*)	India, Southeast Asia, Pacific, Mexico, South America, Caribbean, United States (Texas/Mexico border)
Rift Valley fever	Virus	Phlebovirus	Mosquito (*Aedes*)	Africa, Arabia
Yellow fever	Virus	Flavivirus	Mosquito (*Aedes*)	Tropical South America, Africa
Chikungunya fever	Virus	Alphavirus	Mosquito (*Aedes*)	Africa, Southeast Asia, Philippines
O'nyong-nyong fever	Virus	Alphavirus	Mosquito (*Anopheles*)	Africa
Ross River fever	Virus	Alphavirus	Mosquito (*Culex, Aedes*)	Australia, South Pacific
Venezuelan encephalitis	Virus	Alphavirus	Mosquito (*Culex, Aedes*)	North and South America
Murray Valley or Australian encephalitis	Virus	Flavivirus	Mosquito (*Culex*)	Australia, New Guinea
Barmah Forest fever	Virus	Alphavirus	Mosquito (*Culex, Aedes*)	Australia
California encephalitis	Virus	Bunyavirus	Mosquito (*Aedes*)	United States
Colorado tick fever	Virus	Orbivirus	Tick	United States and Canada
Malaria	Protozoan	*Plasmodium* sp.	Mosquito (*Anopheles*)	Africa, Southwestern Pacific, South America, Southeastern Asia, India
Babesiosis	Protozoan	*Babesia* sp.	Ticks	United States, Europe
American trypanosomiasis (Chagas disease)	Protozoan	*Trypanosoma* sp.	Kissing bug	South and Central America
African trypanosomiasis (sleeping sickness)	Protozoan	*Trypanosoma* sp.	Tsetse flies	West, Central, and East Africa
Leishmaniasis	Protozoan	*Leishmania* sp.	Sand flies	Central and South America, Africa, India, and other parts of Asia, Europe
Filariasis or elephantiasis	Worm	*Wuchereria* sp. *Brugia* sp.	Mosquito (*Culex, Anopheles, Aedes, Mansonia, Coquillettidia*)	Central and South America, Africa, India, and other parts of Asia
Onchocerciasis	Worm	*Onchocerca* sp.	Black flies	Central America, tropical South America, Africa

Ticks are not only significant vectors of disease but also direct sources of disease. Tick paralysis is an example and is characterized by ascending flaccid paralysis resulting from a toxin in tick saliva; the paralysis usually disappears within several days. A 2006 CDC report cited and described a cluster of four cases in Colorado during May 26–31, 2006. Tick populations depend largely on the number of deer in the area; more deer per square kilometer means more ticks.

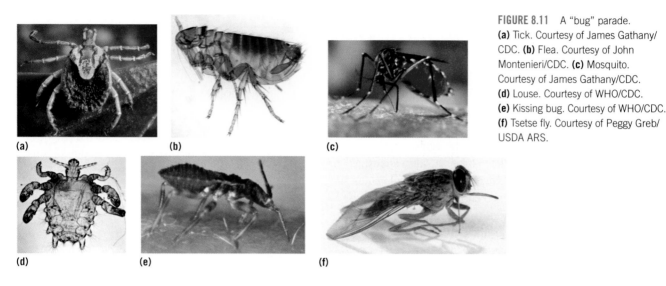

FIGURE 8.11 A "bug" parade. (a) Tick. Courtesy of James Gathany/ CDC. (b) Flea. Courtesy of John Montenieri/CDC. (c) Mosquito. Courtesy of James Gathany/CDC. (d) Louse. Courtesy of WHO/CDC. (e) Kissing bug. Courtesy of WHO/CDC. (f) Tsetse fly. Courtesy of Peggy Greb/ USDA ARS.

Fleas are the biological vectors of the *Yersinia pestis* bacterium, and rats are the reservoirs. Rats are important reservoirs for other microbial diseases, including Lassa fever and, in the southeastern United States, hantavirus pulmonary syndrome. Rats are a serious public health concern around the world (BOX 8.2).

Vector-borne infectious diseases are as numerous and varied as are their vectors, and they are emerging and reemerging throughout the world (Table 8.7 and FIGURE 8.11). Factors responsible include genetic changes in both vectors and pathogens resulting in resistance to insecticides and drugs, public health policy, funding directed toward emergency response, and societal changes. In the first half of the twentieth century, considerable progress was made in the fight against vector-borne diseases. Most of these diseases were brought under control, and by the 1960s their threat, except in Africa was greatly diminished (TABLE 8.8). In fact, malaria was eliminated from many countries of the world. However, no country is immune to the potential threat and spread of vector-borne microbial diseases. A case in point is West Nile virus, which emerged in 1999 in the state of New York. This was the first appearance of the virus in the United States; it had been previously reported only in Africa and Asia.

In 1989, in response to the growing problem of vector-borne diseases, the CDC established what is now known as the Division of Vector-Borne Diseases, presently located in Fort Collins, Colorado. The Division is responsible for information, surveillance, prevention, and control of vector-borne diseases. The Division is charged with the investigation of national and international epidemics of bacterial and viral diseases transmitted to humans by arthropods, primarily mosquitoes, ticks, and fleas. To prevent and control these diseases, biologists in the division work with the three populations involved: the pathogen, the host, and the vector.

Portals of Entry

The next step in the cycle of disease involves access into (or onto) the body through **portals of entry**. Some microbes have a single portal, but others have more than one. Body orifices (openings to the outside), including the mouth, nose, ears, eyes, anus,

BOX 8.2 __Oh, Rats!__

The title of this box is a familiar expression that you have heard or used. Rats transmit infectious diseases by direct contact (biting), serving as reservoirs for arthropod-borne diseases (most notably plague-carrying fleas), and by contamination of food or water with urine and feces. There are many species of rats in the world. The Norway rat is the most common one in the United States; it is also known as the brown rat, sewer rat, and other (nonaffectionate) names. As Shakespeare wrote in Romeo and Juliet (1660) "A rose by any other name would smell as sweet." Poet Gertrude Stein (1922) is responsible for the quotation "A rose is a rose is a rose"; let it now be said that "A rat is a rat is a rat."

The Norway adult rat measures about nine inches long, has a blunt nose, small ears, and heavy-looking body. These rats have an active sex life and produce 4 or 5 litters per year totaling about 20 young. The young become sexually mature at about 3 to 4 months. Norway rats are socially active and live in colonies. They are more than pests and eat just about anything, ruining a large part of the world's food supply. They destroy buildings and household furnishings, attack domestic animals, and gnaw on electrical wires causing fires. Furthermore, they produce about fifty droppings a day in which *Salmonella* bacteria (review bacterial disease), a pathogen for humans, thrive.

Control is based on minimizing their food supply and invoking good measures of sanitation. When a labor strike involving garbage workers hits a large population, mountains of foul-smelling trash bags pile up along the sidewalks and streets, the rat population increases, as does the potential for rat-borne infectious diseases. The area is in a state of emergency. Controlling the rodent population and disposal of trash is a major and ongoing problem in developing countries. Following is a (partial) list of major infectious diseases transmitted to humans by rats and rat fleas:

- Salmonellosis
- Leptospirosis
- Hantavirus pulmonary syndrome
- Rat bite fever
- Murine typhus
- Tularemia
- Eosinophilic meningitis

In the Middle Ages, the rat population was so severe around the world that people lived in fear of the plague and other diseases. Numerous books, poems, and essays have been written about the plague and about rats. The town of Hamelin was particularly besieged and brought in the services of the Pied Piper to rid them of their rats. A part of the delightful poem follows:

Verse 2 (partial)
> *Rats!*
> *They fought the dogs and killed the cats,*
> *And bit the babies in the cradles,*
> *And ate the cheeses out of the vats,*
> *And licked the soup from the cook's own ladles,*
> *Split open the kegs of salted sprats,*
> *Made nests inside men's Sunday hats,*
> *And even spoiled the women's chats. . . .*

Verse 7 (partial)
> *. . . And out of the houses the rats came tumbling.*
> *Great rats, small rats, lean rats, brawny rats,*
> *Brown rats, black rats, grey rats, tawny rats,*
> *Grave old plodders, gay young friskers,*
> *Fathers, mothers, uncles, cousins,*
> *Cocking tails and prickling whiskers,*
> *Families by tens and dozens,*
> *Brothers, sisters, husbands, wives—*
> *Followed the Piper for their lives.*
> *From street to street he piped advancing,*
> *And step for step they followed dancing,*
> *Until they came they followed dancing,*
> *Until they came to the river Weser*
> *Wherein all plunged and perished. . .*

The Pied Piper of Hamelin, Germany: A Children's Story by Robert Browning (1812–1889)

urethra, and vagina, and penetration of the skin make it possible for microbes to gain access. To some extent, human behavior influences the portal of entry; the transmission of the virus that causes AIDS is a case in point. The most common site of

TABLE 8.8 Successful Vector-Borne Disease Control

Disease	Location	Year(s)
Yellow fever	Cuba	1900–1901
Yellow fever	Panama	1904
Yellow fever	Brazil	1932
Anopheles gambiae infestation	Brazil	1938
Anopheles gambiae infestation	Egypt	1942
Louseborne typhus	Italy	1942
Malaria	Sardinia	1946
Yellow fever	Americas	1947–1970
Yellow fever	West Africa	1950–1970
Malaria	Americas	1954–1975
Malaria	Global	1955–1975
Onchocerciasis	West Africa	1974–present
Bancroftian filariasis	South Pacific	1970s
Chagas disease	South America	1991–present

Reproduced from Duane J. Gubler and CDC, *Emerging Infectious Diseases*, 4 (1998): 442–450.

entry for sexually transmitted diseases is the urethra in males and the vagina in females, but the throat and the rectum may also serve for entry. The portal of entry is an important consideration in the outcome of host–parasite interactions. Bubonic plague results from the bite of a plague-infected flea, but if the *Y. pestis* bacteria gains entry into the lungs through the respiratory tract, the result is the more lethal pneumonic plague. Anthrax, also a bacterial disease, is another example. There are three varieties of anthrax; cutaneous anthrax results when the skin is the portal of entry, gastrointestinal anthrax occurs as the result of oral ingestion of the bacteria, and inhalation anthrax is the result of the organisms' entering through the respiratory tract. **TABLE 8.9** and **FIGURE 8.12** summarize the portals of entry by anatomical sites.

Portals of Exit

Once microbes have gained access into the body, whether or not disease results is determined by the interaction between the number of pathogens and their virulence and the

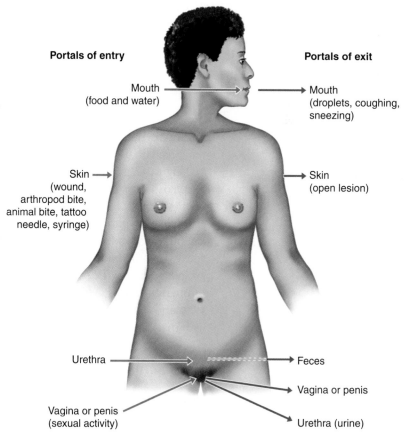

FIGURE 8.12 Portals of entry and portals of exit.

Portals of entry

Mouth (food and water)

Skin (wound, arthropod bite, animal bite, tattoo needle, syringe)

Vagina or penis (sexual activity)

Portals of exit

Mouth (droplets, coughing, sneezing)

Skin (open lesion)

Feces

Vagina or penis

Urethra (urine)

Urethra

TABLE 8.9 Infectious Disease Cycle: Portals of Entry and Exit

	Examples of Disease or Microbe
Portals of entry	
Mucous membranes	
Respiratory tract	*Streptococcus pneumoniae,* tuberculosis, Legionnaires' disease, influenza, hantavirus, common cold
Gastrointestinal tract	Cholera, salmonellosis, *E. coli,* rotavirus, norovirus, hepatitis A, poliomyelitis, guinea worm disease, giardiasis
Urogenital tract	Gonorrhea, chlamydia, AIDS, genital warts genital herpes
Skin (hair follicles, sebaceous glands, wounds, arthropod bites)	Boils, abscesses, cutaneous anthrax, rabies, warts, hookworm, schistosomiasis, malaria
Blood (transfusion, blood products, arthropod bites, placental transfer)	Congenital syphilis, AIDS, German measles, toxoplasmosis, Chagas disease, hepatitis C
Portals of exit	
Respiratory tract	Tuberculosis, Legionnaires' disease, influenza, common cold
Gastrointestinal tract	Cholera, salmonellosis, rotavirus, norovirus, hepatitis B and A, poliomyelitis, hookworm, guinea worm disease
Urogenital tract	Gonorrhea, chlamydia, HIV, schistosomiasis, genital herpes
Skin	Impetigo, boils, abscesses, warts, cold sores, fever blisters, guinea worm disease, candidiasis
Blood (transfusion, blood products, arthropod bites, placental transfer)	Congenital syphilis, toxoplasmosis, HIV, hepatitis B and C, rubella, malaria

immune system of the host. To complete the cycle of infectious disease and to allow the spread of disease into the community, pathogens require a **portal of exit** (Table 8.9 and Figure 8.12). In some cases the portal of exit relates to the area of the body that is infected. This is particularly true for organisms that cause diseases of the respiratory tract (such as colds and influenza). On the other hand, this is not always the case. For example, the spirochete that causes syphilis uses the urogenital tract as the portal of exit but can invade the skin and nervous system. The eggs of some disease-producing worms exit the body in fecal material, survive in soil, and remain infectious for long periods of time. HIV exits the body through semen and vaginal discharges as well as through the blood. Arthropod-borne diseases enter the body through the bites of insects, and these insects also serve as avenues of exit. A mosquito biting an individual with malaria will pick up the parasite.

◼ Nosocomial (Hospital-Acquired) Infections

Nosocomial infections are infections acquired by patients during their hospital stay or during their confinement in other long-term healthcare facilities; infections acquired by hospital personnel are also considered nosocomial. People are hospitalized because they are ill and require treatment beyond what home care can provide. Ironically, while in the hospital they have a 5% to 15% increased risk for contracting an infectious disease. Based on the number of hospital admissions, estimates are that two million nosocomial infections occur each year, resulting in ninety-nine thousand deaths in the United States. These infections account for about 50% of all the major complications of hospitalization.

Hospital Environment as a Source of Nosocomial Infections

What are the factors unique to the hospital environment that place patients and hospital staff at an increased risk for acquiring infections? To begin with, the patient population consists of ill individuals who may have a compromised (weakened) immune system. A weak immune system increases a patient's susceptibility to pathogens and opportunistic microbes, including their own normal flora. Antibiotics are heavily used, but also misused, in hospitals to treat or to prevent infections, fostering the development of antibiotic-resistant strains of bacteria. Drugs to purposely suppress the immune system (as in cancer therapy and organ transplantation), prolonged bed rest, and restrictive diets are necessary components of treatment but are traumatic to the body and counterproductive to the maintenance of a healthy immune system.

Diagnostic and treatment protocols frequently involve extensive surgery and the use of invasive procedures, including the insertion of catheters into the urethra, swallowing of tubes, insertion of needles into veins for intravenous therapy, and insertion of nasal tubes. Thermometers, bedpans, urinals, eating utensils, and night table surfaces are only a few of the many fomites that pose potential risk. Hence, the equipment and devices involved in patient care contribute to transmission. The hospital staff, including physicians, nurses, laboratory technicians, and maintenance workers, may become complacent and unwittingly (and carelessly) transmit microbes from patient to patient; some may be healthy carriers.

All the factors involved in the cycle of infectious diseases are present in a concentrated way in hospitals and in long-term healthcare facilities, establishing these environments as reservoirs of pathogens. A relatively small number of bacterial species are responsible for most nosocomial infections, but these are species common to the environment. Some sites in the body are more prone to nosocomial infections than others; the urinary tract is the most susceptible, followed by surgical sites and the respiratory tract (FIGURE 8.13). Despite the risk of nosocomial infection, be assured that the advances in medicine far outweigh the risks of hospitalization.

FIGURE 8.13 Body site distribution of nosocomial infections. Data from CDC, *National Nosocomial Infection Surveillance.*

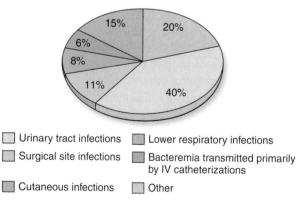

- ☐ Urinary tract infections
- ☐ Surgical site infections
- ☐ Cutaneous infections
- ☐ Lower respiratory infections
- ☐ Bacteremia transmitted primarily by IV catheterizations
- ☐ Other

Control Measures

Nosocomial infections are a serious problem in hospitals and other medical facilities in terms of mortality, morbidity, and financial burden, and every hospital has strategies of prevention and control. The fact that the frequency and spectrum of antibiotic-resistant organisms are on the rise contributes to the problem. About

70% of bacterial nosocomial infections are resistant to at least one antibiotic (**TABLE 8.10** and **BOX 8.3**). All hospitals are required to have an infection control officer and an infection control committee to maintain accreditation by the American Hospital Association. Hospitals spend considerable time and money to minimize the possibility of microbial contamination in all aspects of the hospital environment. The infection control officer is responsible for training of hospital personnel in basic infection control procedures including isolation procedures, proper techniques of disinfection and sterilization, and the surveillance and reporting of cases of infectious diseases in both patients and staff. The infection control officer and the infection control committee are also responsible for insect control, good housekeeping, and safe practices for the disposal of feces, urine, bandages, dressings, and other potentially contaminated materials.

Education emphasizing the importance of the simple act of hand washing is vital. (Semmelweis talked about hand washing 160 years ago!) Numerous studies have demonstrated that this single simple procedure is the most important practice in minimizing nosocomial infections. In some studies shockingly low rates

---

BOX 8.3 _Iraqibacter_

David (last name intentionally omitted) is a young soldier whose military vehicle was hit by an Iraqi pipe bomb throwing him into the air. He suffered numerous injuries, including shrapnel piercing his skin, deep cuts, and his right leg was barely attached. Thanks to the quick attention of a military rescue team, he was airlifted to an army field hospital in Germany. His leg was dangling, his blood pressure was low, and he was suffering from internal bleeding. Despite the poor odds of survival, his life was saved by the experienced medical personnel at the hospital, at the expense of losing his leg at the knee.

It is hard to believe that the worst was yet to come, a scenario that brought him even closer to death and could only be treated by amputation of his left leg, also just below the knee. The second amputation happened, not on the battlefield, not in a war zone, nor in foreign territory, but in a major military hospital in the United States resulting from massive infection by a bacterium—_Acinetobacter baumanii_. It is ironic that people can survive long periods of time without food, massive wounds, loss of limbs, and unimaginable hardships to be felled the unseen—microorganisms. David's ordeal is not uncommon, although the details vary. What happened to David and to the hundreds of other survivors that suffered a similar fate? What's the link to nosocomial infections in hospitals and other long-term medical care facilities?

The answers to these questions are not totally clear. Initially, physicians in the military thought the _Acinetobacter_ were soil organisms carried into the battlefield wounds. It was logical, but disproven; bacteria isolated from the wounds and from the soil were not the same. Actually, wound contamination in the military with this organism was noted back in 2003, and its frequency in returning servicemen led to the name "_Iraqibacter._" Continued sleuthing revealed widespread contamination of field hospitals by servicemen and by equipment along the long route home and, ultimately, to the general population.

A. baumanii is not a "new" organism; it is an opportunistic gram-negative bacillus that is multidrug resistant, sometimes referred to as MDRAB. Its drug resistance is the result of a large section of drug-resistance genes acquired by conjugation with _Legionella pneumophila_.

Table 8.10 indicates that _Acinetobacter baumanii_ is a significant nosocomial infection that some microbiologists fear will continue to increase. It is not a threat to healthy people, but is to senior citizens, those taking antirejection and certain other drugs, children, and surgical patients. The organism has been recovered from a variety of surfaces: respirators, catheters, and other devices used in patient care. It can cause a host of problems including pneumonia, urinary tract infection, bloodstream infections, and infections in other parts of the body.

PART 2 Microbial Disease

TABLE 8.10 Nosocomial Infections Caused by Antibiotic-Resistant Bacteria

Most Common Type of Infection	Pathogen
Bloodstream infections	*Acinetobacter baumannii*
Surgical site infections	*Staphylococcus aureus*
Diarrhea	*Clostridium difficile*
Ventilator-associated pneumonia	*Klebsiella pneumoniae*
Pneumonia, urinary tract infections, and bloodstream infections	*Pseudomonas aeruginosa*
Catheter-associated urinary tract infections	*Escherichia coli*
Catheter-associated urinary tract infections	*Enterococcus faecalis*

(well under 50%) of hand washing by healthcare workers, including physicians and nurses, have been revealed.

Epidemiology of Fear

The fear of epidemics can reach epidemic proportions, but the threat of infectious disease, according to some experts, is out of proportion. *Escherichia coli* outbreaks, avian flu, severe acute respiratory syndrome (SARS), and West Nile virus have captured the public's attention and led to an explosion of television programs, popular books, and a recent (2011) movie, *Contagion*. West Nile virus emerged in the Western Hemisphere for the first time in the summer of 1999 in New York State and caused illness in more than 60 people and the death of 7. By the summer of 2000, infected birds were detected in Massachusetts, Connecticut, and Rhode Island, generating concern among public health officials about an epidemic of fear. As a Massachusetts Department of Health spokeswoman stated, "The message we're trying to get out is to stop people from panicking. West Nile virus is not a major public health threat. It's something people should be aware of, and take precautions, but not let it interrupt their summer."

In his first inaugural address on March 4, 1933, President Franklin Delano Roosevelt spoke eloquently of the danger of fear. His often quoted words were, "So first of all let me assert my firm belief that the only thing we have to fear is fear itself—nameless, unreasoning, unjustified terror which paralyzes needed efforts to convert retreat into advance." His words, focused on World War I, are applicable to the spread of microbial disease. The take-home message is that awareness, surveillance, common sense precautions, and calmness are paramount. Fear can be paralyzing.

Overview

Epidemiologists classify disease as sporadic, endemic, epidemic, or pandemic, depending on its frequency and distribution. These categories are not absolute; a particular disease can slide from one classification to another. Common-source epidemics arise from contact with a single contaminant, resulting in a large number of people becoming ill suddenly; the disease peaks rapidly. Propagated epidemics are characterized by direct person-to-person (horizontal) transmission, a gradual rise in the number of infected individuals, and a slow decline.

A chain of linked factors is required for infectious diseases to exist and to spread through a population. These factors are reservoirs of disease, transmission, portals of entry, and portals of exit. Understanding the characteristics of microbes and the diseases they cause is necessary to break the cycle somewhere along its path. The reservoir is the site where microbes exist in nature and from which they can be spread. Active carriers, healthy carriers, and chronic carriers are reservoirs, as are wild and domestic animals. Nonliving reservoirs include contaminated water, food, soil and surfaces.

Transmission is the bridge between reservoir and portal of entry. Person-to-person contact is the most common type of horizontal direct transmission and allows for the immediate transfer of microbes. Vertical transmission is another type of direct transmission and is categorized by the passage of pathogens from mother to offspring across the placenta, in the birth canal during delivery, or in breast milk. In direct transmission there are no intermediaries. Indirect transmission involves the passage of materials from a reservoir or source to an intermediate agent and then to a host. The intermediate agent can be nonliving or living. Water, food, fomites, and aerosols are significant nonliving vehicles of indirect transmission. Vectors are living organisms that transmit microbes from one host to another. Mechanical vectors passively transfer microbes on their feet or other body parts, whereas biological vectors are required for the multiplication and development of the pathogen within the vector.

Portals of entry are the next consideration in the cycle of disease. Some microbes have a single preferred portal of entry into the body, whereas others have more than one. Body orifices, including the mouth, nose, ears, eyes, anus, urethra, and vagina, are portals of entry; the skin can be penetrated and is another portal of entry.

For the cycle of disease to continue in a population, microbes must exit from the body. In many cases the portals of entry and the portals of exit are the same.

Hospitals and long-term healthcare facilities are hotbeds of infection for patients. Hospital-acquired infections are called nosocomial infections and account for about 50% of all the major complications of hospitalization.

The public should not be paralyzed by the fear of infection. Awareness, surveillance, common sense precautions, and calmness are the best preventive measures.

■ Self-Evaluation

PART I: Choose the single best answer.

1. The breakup of the Soviet Union ushered in an unusually high number of cases of diphtheria from about 1990 to 1995. Which term best characterized the situation?

a. endemic b. epidemic c. pandemic d. herd

2. Morbidity refers to

    a. death rate **b.** illness rate **c.** person-to-person contact
    d. a zombie state

3. Chickenpox in the United States is best described as

    a. sporadic **b.** endemic **c.** zoonotic **d.** pandemic

4. Which one of the arthropods is not a disease vector?

    a. fly **b.** tick **c.** centipede **d.** mosquito

5. Vertical transmission is possible except in the case of

    a. gonorrhea **b.** AIDS **c.** influenza **d.** chickenpox

6. Which term refers to a proportion of immunized individuals in a population?

    a. herd immunity **b.** passive immunity **c.** horizontal transfer **d.** outbreak

7. Ignaz Semmelweis is credited with the practice of

    a. disinfection **b.** vaccination **c.** hand washing **d.** teeth brushing

8. Which inanimate object may be contaminated with microbes and serve in their transmission?

    a. reservoir **b.** fomite **c.** vector **d.** portal

9. Which of the following diseases is not transmitted by a mosquito vector?

    a. West Nile encephalitis **b.** dengue fever **c.** filariasis **d.** leishmaniasis

10. The biological vectors of the bacterium *Yersinia pestis* are

    a. fleas **b.** mosquitoes **c.** tsetse flies **d.** sand flies

PART II: Fill in the blank.

1. A worldwide outbreak of a disease is called a _____ .

2. _____ is a disease transmitted by an arthropod vector.

3. _____ is the only disease that has been eradicated.

4. _____ is an example of indirect contact transmission (fomites) of disease.

5. The body site most susceptible to nosocomial infection is _____ .

6. Typhoid Mary was a healthy _____ of _____, which causes typhoid fever.

7. Hospital-acquired infections are also referred to as _____ infections.

8. Malaria is caused by _____ .

9. R_0 is a measure of _____ .

10. The higher the R_0 value, the _____ .

PART III: Answer the following.

1. Distinguish between vertical and horizontal transmission. Give examples of each.

2. What is meant by zoonoses? What is a common zoonosis in the northeastern United States? Give three examples of zoonoses.

3. Distinguish between biological vectors and mechanical vectors. Give two examples of each.

4. How can vector-borne diseases be prevented? Why are vector-borne diseases a growing problem?

5. Would you consider avian flu or malaria a greater threat to human survival? Explain your answer.

6. Why are patients in a hospital more susceptible to a nosocomial infection than hospital staff?

7. Discuss how rats (rodents) can cause disease. Name two diseases associated with rat populations.

8. Why is it difficult to control rabies in the wild?

9. Explain why hospitals and long-term healthcare facilities are hotbeds of infections for patients.

10. List microbes that are often the contaminants of water and food, resulting in water or foodborne infections.

9

Bacterial Diseases

You need not fear the terror by night
Nor the arrow that flies by day,
Nor the plague that stalks in the darkness

—91st Psalm

■ Preview

A long list of bacterial diseases is presented in this chapter, but it represents only a small number of those that are known. Only the major bacterial diseases of personal and public health significance are presented and are organized according to their route of transmission: foodborne and waterborne, airborne, sexually transmitted, contact, soilborne, and arthropodborne. It should be noted that, in some cases, transmission may be accomplished by more than one route. For example, listeriosis may be foodborne or airborne, and anthrax may be airborne or acquired by direct contact. To the extent possible, statistics indicating the current incidence of these diseases primarily from the United States Centers for Disease Control and Prevention (CDC) is included as is available, but the actual numbers are secondary to the status of the disease. In most cases the numbers vary depending on the source of information and are

Author's photo (TS).

Topics in This Chapter

■ **Foodborne and Waterborne Bacterial Diseases**
  Food Intoxication (Food Poisoning)
  Foodborne and Waterborne Infection

■ **Airborne Bacterial Diseases**
  Upper Respiratory Tract Infections
  Lower Respiratory Tract Infections

■ **Sexually Transmitted Diseases**
  Syphilis
  Gonorrhea
  Chlamydia

■ **Contact Diseases (Other Than STDs)**
  Peptic Ulcers
  Leprosy
  Staphylococcal Infections

■ **Soilborne Diseases**
  Anthrax
  Tetanus
  Leptospirosis

■ **Arthropodborne Diseases**
  Plague
  Ehrlichiosis
  Lyme Disease
  Relapsing Fever
  Rickettsial Diseases

201

underreported as the result of inadequate surveillance, failure to report, misdiagnosis, and underuse of laboratory testing. These factors are of greater significance in developing countries. Attention is paid in varying degrees to the practical considerations of transmission, pathogenesis, prevention, and, to some lesser extent, treatment. Most of these diseases are responsive to antibiotic therapy, although antibiotic resistance is an increasingly greater problem. Preventive vaccines are available in many cases, and others are the subject of active research.

Some of the diseases presented are ones not usually expected to occur in the United States or in other developed countries; however, transmission can happen because of an increase in immigration and in international travel, including eco-travel. Humanitarian purposes, too, require an understanding of public health.

◼ Foodborne and Waterborne Bacterial Diseases

After studying this section you may be overly anxious about the safety of the foods you eat and the source of the water you drink. Pathogens may be lurking in the kitchen, and it is certainly wise to take precautions and to use common sense; it is not necessary, however, to become paranoid! There is protection at two levels: (1) the immune system and (2) a system of surveillance practiced by national, state, and city health departments throughout the country. The U.S. Department of Agriculture, through its Food Safety Inspection Service, is charged with ensuring the safety of the nation's commercial supply of meat, poultry, and egg products. President Clinton launched the National Food Safety Initiative in 1997 to deal with the challenges presented by changes in food supply and distribution. The U.S. Food and Drug Administration (FDA) Food Safety Modernization Act (FSMA) was signed into law by President Obama on January 4, 2011. FIGURE 9.1 illustrates the anatomy of the human digestive system, the usual target for foodborne and waterborne microbes, resulting in gastroenteritis.

Foodborne and waterborne infections have a significant impact on health in the United States and around the world; increasingly, more of the pathogens responsible for these infections are becoming resistant to antibiotics. According to 2011 CDC estimates in the United States there are 1,000 outbreaks, 48 million illnesses, 3,000 deaths, and 128 thousand hospitalizations annually. Many cases are undiagnosed and are sometimes passed off as a "stomach virus" (a nonexistent disease), signifying that these numbers are only the tip of the iceberg. Millions of dollars are spent per year on the recall and disposal of contaminated foods. It is not uncommon for a major food preservation (canning or freezing) company or meat, chicken, or fish processing plant to issue an urgent warning regarding consumption of a particular food item. Even the Gerber baby food company image suffered damage as a result of bacterial contamination of some of its food products. Perhaps the most widely known example in the last twenty years was the 1993 Jack-in-the-Box episode in the state of Washington. Patrons of the fast-food restaurant consumed undercooked hamburgers contaminated with *Escherichia coli* O157:H7. Three young children died, almost seventy children were hospitalized, and a total of 500 people became ill. In February 2007 ConAgra Foods recalled certain brands of peanut butter found to be contaminated with *Salmonella* bacteria; according to

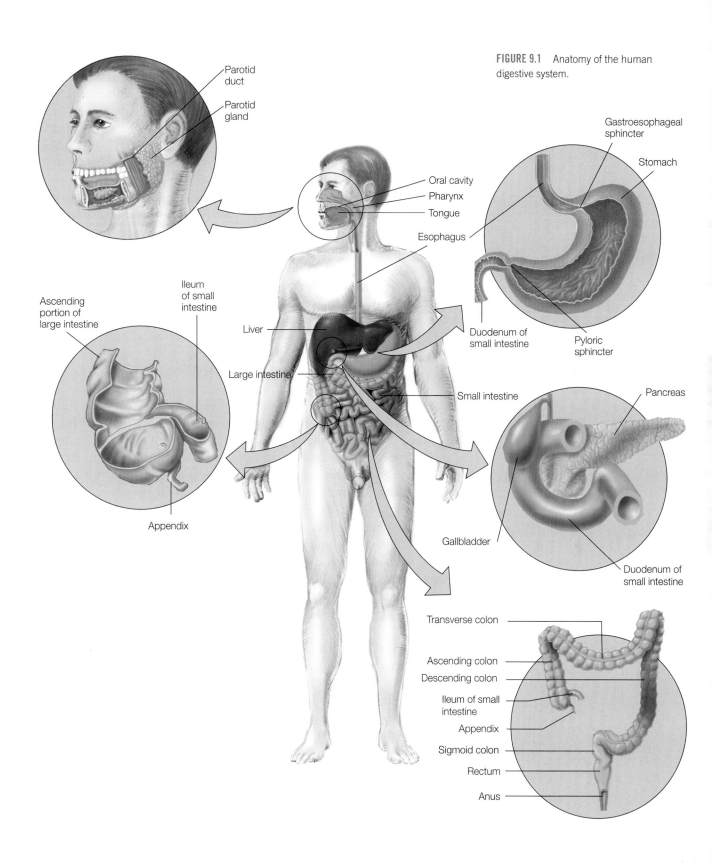

FIGURE 9.1 Anatomy of the human digestive system.

Parotid duct

Parotid gland

Gastroesophageal sphincter

Stomach

Oral cavity

Pharynx

Tongue

Esophagus

Ascending portion of large intestine

Ileum of small intestine

Liver

Duodenum of small intestine

Pyloric sphincter

Large intestine

Small intestine

Pancreas

Appendix

Gallbladder

Duodenum of small intestine

Transverse colon

Ascending colon

Descending colon

Ileum of small intestine

Appendix

Sigmoid colon

Rectum

Anus

a CDC report over 628 people were sickened from consumption of the product in forty-seven states. More than one million pounds of ground beef were recalled in October 2007 by Cargill, Inc. because of possible contamination with *E. coli.* Note that the problem is by no means confined to *E. coli* O157:H7. Popular television shows such as "Dateline" and "20/20" have presented segments relating to the food industry, some of which are shocking.

Travelers to foreign countries need to be particularly careful about ingestion of food and water because the many problems of infrastructure, particularly in developing countries, do not ensure an optimal level of safety. It is easy to forget that ice cubes come from tap water or that thoroughly washing fruits and vegetables is to no avail if the water is contaminated. People who live in an area develop a level of immunity that travelers do not have and hence are less prone to get these diseases. Tourists' plans are frequently curtailed, and sight-seeing becomes limited to the hotel room with an adjoining bathroom. A summary list of major foodborne and waterborne infections is presented in TABLE 9.1. Note that the types of foods and the symptoms are common denominators. "Boil it, cook it, or forget it" is an excellent rule. TABLE 9.2 gives a few simple tips for minimizing the likelihood of infection.

Food Intoxication (Food Poisoning)

A distinction needs to be made between **food intoxication** (also called food poisoning) and **food infection**. In the popular jargon all food-related illnesses are falsely lumped together as food poisoning. Intoxication refers to the ingestion of already-produced bacterial toxins; the actual organisms may, in fact, no longer be present. Foodborne infection is the result of the ingestion of bacterial-contaminated foods and subsequent bacterial growth in the intestinal tract, secretion of a toxin, and, possibly, invasion of the intestinal tract. The toxin is referred to as an **enterotoxin** (a toxin that affects the intestinal tract) and is responsible for the common, and very unpleasant, symptoms of nausea, vomiting, diarrhea, and, possibly, bloody stools and fever. The top ten riskiest foods regulated by the FDA are presented in TABLE 9.3. Nearly 40% of all foodborne outbreaks were linked to the FDA Top Ten.

Botulism

Botulism—the most dangerous food intoxication—is a rare but serious disease caused by *Clostridium botulinum,* a gram-positive anaerobic, spore-forming bacillus commonly found in soil. The neurotoxin produced by this organism is extremely deadly; it interferes with the passage of nerve transmitters (acetylcholine), resulting in muscle paralysis, including the diaphragm (a muscle) and the muscles of the ribs, leading to respiratory paralysis, that is, the inability to breathe. Death occurs within a day or two. Antibiotics are of no value in treatment because the problem is not one of infection but of intoxication. Treatment consists of antitoxin therapy, and in some cases mechanical ventilators are used. The number of cases in the United States is low, and the associated death rate in untreated cases is 70%. The telltale signs of a *C. botulinum*-contaminated can are bulging of the can and offensive odor and taste. The best defense is to heat foods for ten minutes at 90°C, a process that destroys the bacterial toxin.

TABLE 9.1 Foodborne and Waterborne Bacterial Diseases[a]

Disease	Incubation Period	Causative Microbe	Transmission	Symptoms
Food intoxication (food poisoning)				
Botulism	8–24 hr	*Clostridium botulinum*	Contaminated foods, home-canned products	Double vision, blurred vision, drooping eyelids, slurred speech, difficulty swallowing, and muscle weakness that moves down the body
Clostridial food poisoning	8–14 hr	*Clostridium perfringens*	Meat, poultry, beans	Severe cramping, abdominal pain, watery diarrhea
Staphylococcal food poisoning	1–6 hr	*Staphylococcus aureus*	Coleslaw, potato salad, fish, dairy products, cream-filled pastries, spoiled meats	Abdominal cramps, nausea, vomiting, diarrhea
Foodborne (and reptile-borne) infection				
Salmonellosis	2–48 hr	*Salmonella enteritidis* and other species	Undercooked or raw eggs, turtles, chicks, iguanas, reptiles, unpasteurized orange juice, raw alfalfa sprouts	Nausea, vomiting, abdominal cramps
Typhoid fever	7–14 days	*Salmonella typhi*	Fecal–oral; flies, fomites	Blood in feces, fever, delirium, rose-colored spots on the abdomen
Shigellosis	12–72 hr	*Shigella* species	Eggs, shellfish, dairy products, vegetables, water	Diarrhea, dysentery, abdominal cramps, blood in feces
Cholera	24–72 hr	*Vibrio cholerae*	Raw shellfish, water contaminated with fecal material of infected person	Severe diarrhea and dehydration, muscular cramps, wrinkling of the skin
E. coli O157:H7 infection	18–72 hr	*Escherichia coli*	Undercooked hamburgers, radish sprouts, lettuce, spinach	Diarrhea, nausea, abdominal cramps
Campylo-bacterosis	2–10 days	*Campylobacter jejuni*	Poultry, cattle, water, unpasteurized milk	Diarrhea, high fever, bloody stools
Listeriosis	3–70 days (usually within 1 month)	*Listeria monocytogenes*	Vegetables, meats, cheese, ice cream, unpasteurized milk	Sore throat, fever, diarrhea, miscarriage

[a] These diseases are treatable with antibiotic therapy; drug-resistant strains pose a problem in some cases.

Although home-canned string beans, peppers, asparagus, sausage, cured pork and ham, smoked fish, and canned salmon are the most common sources of intoxication, commercially available products can be a source. Four cases of botulism occurred in Florida and Georgia in September 2006; the cause was identified as Bolthouse Farms carrot juice. In July 2007 four cases of botulism—two in Texas and two in India—were reported after ingestion of chili sauce from Castlebury's Food Company. Organic honey can contain spores of *C. botulinum* and should not be fed to children less than one year of age.

TABLE 9.2 The Ten Commandments for Reducing the Risk of Foodborne and Waterborne Infections

- Wash hands frequently, particularly after using the bathroom and after changing a baby's diapers.
- Cook raw beef and poultry products thoroughly.
- Avoid unpasteurized milk and juices and any food made from unpasteurized milk.
- Properly wash food utensils, cutting boards, and raw fruits and vegetables.
- Beware of "double-dippers" (those who dip their cracker, celery, carrot stick, etc., into a spread, take a bite, and then dip the same item into the spread again).
- Avoid eating at disreputable joints.
- Avoid food and drinks sold by street vendors, particularly when traveling abroad.
- Frequently wash kitchen sponges in the dishwasher.
- Do not share eating utensils, water bottles, or other such items.
- Do not swim in contaminated waters.

Ironically, minute doses of botulinum toxin, called **Botox**, have been used to treat a variety of common disorders associated with muscle overactivity, including strabismus (crossed eyes), stuttering, and uncontrolled blinking. Botox is advertised as "from toxin to therapeutic agent." In May 2001 the journal *Neurology* reported on a study of a small number of patients in which it was claimed that treatment with Botox helped sufferers of chronic low back pain. Botox is also used cosmetically by injection into the skin to relieve wrinkling. Unfortunately, unscrupulous practitioners can cause serious problems, as happened to four persons who were injected with an unlicensed botulinum toxin; their symptoms were consistent with those of naturally occurring botulism. The unlicensed toxin prescription was almost three thousand times the estimated human lethal dose.

Clostridium perfringens is associated with **gas gangrene**, a condition that devastates soldiers wounded in battle because of wound contamination with particles of soil containing spores. This organism is also an important cause of

TABLE 9.3 The Top Ten Riskiest Foods Regulated by the FDA

Food	Outbreaks
1. Leafy greens	363
2. Eggs	352
3. Tuna	268
4. Oysters	132
5. Potatoes	108
6. Cheese	83
7. Ice cream	74
8. Tomatoes	31
9. Sprouts	31
10. Berries	25

Source: 2009 report published by the Center for Science in the Public Interest (CSPI), Washington, D.C.

food poisoning (some consider it a food infection). It is commonly associated with spore contamination of meat, poultry, and beans. Recovery usually occurs within twenty-four hours, and there is no specific therapy. In 1993 "wearing of the green" celebrations ended with outbreaks of gastroenteritis associated with contaminated corned beef served at St. Patrick's Day meals in Virginia and Ohio. You may recall that the genus *Clostridium* is a spore former and that spores survive the high temperature used in cooking and germinate as multiplying cells during cooling.

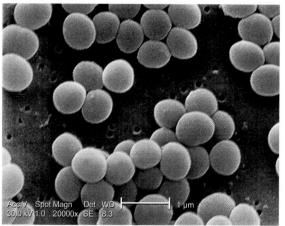

FIGURE 9.2 Colorized scanning electron micrograph of *Staphylococcus aureus.* Courtesy of Matthew J. Arduino, DrPH/CDC.

Staphylococcal Food Poisoning

Staphylococcal food poisoning is the most common type of food poisoning, and many cases remain undiagnosed. You probably have heard of this condition as **"staph food poisoning"** or by an older and incorrect term, "ptomaine poisoning." *Staphylococcus aureus,* a gram-positive coccus, secretes an enterotoxin that produces intestinal tract symptoms, including abdominal cramps, nausea, vomiting, and diarrhea (FIGURE 9.2). The symptoms may be so severe that you wish you were dead, but don't have the luxury. There is hardly anyone who has not suffered from this; all you can do is wait it out and you will feel better within several hours. The list of potential food sources is large and includes coleslaw, potato salad, fish, dairy products, cream-filled pastries, and spoiled meats. How does contamination occur? The main reservoir for *S. aureus* is the nose, but the organism also causes boils, abscesses, or pimples on the skin, and these lesions are the usual source of seeding the staphylococci into food. A few hours under the sun in an improperly iced cooler is all it takes to allow for multiplication of the bacteria and enterotoxin production.

Foodborne and Waterborne Infection

Salmonellosis

How do you like your eggs? If you answered over easy or sunny side up or if you like your Caesar salad dressing made the old-fashioned way with raw eggs, you are at risk for **salmonellosis** caused by *Salmonella enteritidis*-infected eggs. Salmonellosis is a significant public health problem, in that it is a serious and potentially fatal infection that afflicts about 40,000 people each year and causes 400 deaths in the United States. Undoubtedly, these figures can be increased at a factor of three because many cases are not reported. It was once thought that cracked eggs were responsible for the high incidence of salmonella in eggs, but it is now known that seemingly healthy hens may be infected prior to eggshell formation.

Salmonella infection is also associated with contaminated meats, seafood, and unpasteurized fruit juice. Lately, fresh produce has become an increasingly significant source of salmonellosis (and other foodborne microbes), indicating a need for better growing and harvesting practices on the farm. Lettuce, raspberries, tomatoes, cantaloupes, and alfalfa sprouts have made it to the stores contaminated with *Salmonella, E. coli,* or other microbes.

FIGURE 9.3 **(a)** Bearded lizard. Author's photo (RIK). **(b)** *Salmonella enteritidis.* This scanning electron micrograph shows a mixture of small cells with filaments and very large cells that lack filaments. Courtesy of P.J. Guard-Petter, colorization by Stephen Ausmus/USDA Agricultural Research Service (ARS).

(a)

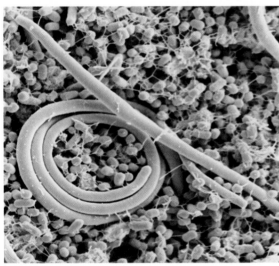

(b)

If you purchase a reptile as a pet you may be buying trouble as well. Iguanas and a variety of lizards, snakes, and turtles are popular pets, but they are carriers of a variety of salmonella species (FIGURE 9.3). In 2008 an outbreak involving 103 cases distributed across thirty-three states was traced back to exposure to turtles. One recommendation of the CDC is that pet store owners advise potential reptile owners of the risk of salmonellosis, particularly in children. Human salmonellosis has been associated with pet treats, such as pig ears, in the United States and Canada for several years. The common denominator is that in all the cases the persons infected had handled salmon or beef dog treats. Children under the age of five, the ones most desirous of having a pet, are the group most likely to acquire infection.

Crows, those big, black birds, have increased in population in recent years, and it is now known that crow feces carry *Salmonella.* (Think about that when a flock flies overhead, and wear a cap!) Remember that salmonellosis is a foodborne infection and requires the presence of bacteria that multiply in the intestinal tract, whereas staphylococcal and clostridial food intoxication, although incorrectly described as food poisoning results from ingestion of already-formed toxins.

Numerous outbreaks of foodborne salmonellosis occurred between 2007 and 2008, including consumption of Veggie Booty (60 cases, mostly in toddlers), Peter Pan peanut butter (628 cases over 47 states), imported cantaloupes from Honduras (dozens of cases and at least 14 hospitalizations), and Banquet pot pies (272 cases in 35 states).

The largest foodborne outbreak in the United States occurred in the spring and throughout the summer of 2008. It started with four cases of salmonellosis in May 2008 and by August spread to about 1,400 people, resulting in 286 hospitalizations and, possibly related, two deaths. The epidemic was over by the end of September. Initially, contaminated tomatoes were thought to be the culprit, but then the blame shifted to jalapeno peppers. In the final analysis the source of contamination has not been confirmed.

Salmonella typhi causes **typhoid fever**. The organism is able to survive in sewage, water, and certain foods, and gains access into the body through the ingestion of fecally contaminated food products; hands contaminated with fecal material, including some imported foods; and, rarely, sexual activities involving the anal area. *Salmonella* is transmitted by flies and fomites. After ingestion the organism invades the lining of the small intestines, causing ulcers and the passage of blood in the feces, accompanied by fever, possibly delirium, and the presence of rose-colored spots on the abdomen as a result of hemorrhaging in the skin.

Once an important disease worldwide, its prevalence has markedly decreased in developed countries as the result of sanitary engineering and sewage treatment facilities, luxuries not available in developing countries. In the early 1900s typhoid fever was a major killer, with over 20 thousand cases occurring annually in the

United States. Approximately 400 to 500 cases are now reported annually to the CDC, about 75% of which are acquired during overseas travel. Some 21.5 million people in less-developed countries become infected with typhoid each year.

About 5% of those infected do not become ill. These healthy carriers of *S. typhi* present a public health problem, although antibiotic therapy usually renders the carrier free of the bacteria. In some cases surgery to remove the gallbladder, a site where the bacilli sometimes take up residence, may be performed to eliminate the carrier state.

The story of **Typhoid Mary** is a classic and significant tale that points out the interface between biology and society (BOX 9.1). An interesting question is whether Mary's human rights were violated.

Shigellosis

Shigellosis, an acute infection of the lining of the intestinal tract caused by various species of *Shigella*, a gram-negative bacillus, is manifest by the usual gastroenteritis symptoms with the possible addition of dysentery, which may be bloody. Eggs, shellfish, dairy products, vegetables, and water are the sources. Transmission is via the fecal–oral route. As in all diseases involving loss of large volumes of water, dehydration is the major cause for concern and can lead to death via circulatory collapse. Treatment involves rehydration, either oral or intravenous, and possibly antibiotics. Estimates are that approximately 18 thousand reported cases appear each year in the United States, but that figure may be as much as twenty times more because of unreported cases. The World Health Organization (WHO) figures indicate that in the developing world approximately 165 million cases and one million deaths occur annually.

Day care centers, because of their very young population, are potentially prone to outbreaks, necessitating that personnel be particularly diligent in handwashing and diapering procedures as measures of both prevention and control. Three outbreaks occurred in 2006 associated with day care centers in Kansas, Kentucky, and Missouri. In most cases the Shigella isolates were multiple-drug resistant. In the three reported outbreaks the median ages were 7, 6, and 4 years, respectively; the case total was 994.

The Battle of Crecy in 1346, in which the English army was defeated by the French, may have been due to, as many historians claim, the English soldiers' suffering from dysentery.

Cholera

Vibrio cholerae is the cause of **cholera**, a gram-negative curved rod (FIGURE 9.4). This organism, like so many other intestinal tract pathogens, secretes an exotoxin that causes diarrhea. As much as 1 liter of fluid can be lost per hour for several hours. A liter is approximately equal to 1 quart, and a human has a total of 5 to 6 liters of blood, which is about 80% water. Unless rehydration is instituted quickly, death results in only a few hours; the mortality rate reaches 70% in untreated cases. The massive dehydration causes the eyes to take on a sunken appearance, the skin becomes dry and wrinkled, and muscular cramps in the legs and arms are a common complaint. Infection is acquired by the ingestion of water or food, including raw shellfish, contaminated by fecal material from *Vibrio*-infected persons.

AUTHOR'S NOTE (RIK)
At about the age of two my daughter became ill and required hospitalization. At the time I was on a sabbatical leave at Georgetown University School of Medicine in Washington, D.C. As crazy as it may sound, my wife and I were relieved when the diagnosis was salmonellosis infection and, therefore, the prognosis was excellent. The likely culprit was a small pet turtle that we had purchased a few weeks earlier. My wife, in a moment of intense relief, flushed the turtle down the toilet before I had the opportunity to culture from it—a foiled epidemiological study!

BOX 9.1 ___Typhoid Mary: A Public Health Dilemma___

Had you accepted an invitation to dinner in the summer of 1906 at the rented vacation home of Charles Henry Warren, a wealthy New York banker, in Oyster Bay, New York, you might have become seriously ill or died from typhoid fever. The culprit would have been the bacterial organism *Salmonella typhi,* harbored unknowingly (at least initially) in the digestive tract of Mary Mallon, the Warrens' cook. Over the next several years Mary became stigmatized as "Typhoid Mary." During that time, as a healthy (asymptomatic) *Salmonella* carrier she was responsible for approximately fifty cases of typhoid fever, at least two of which ended in death. Mary had vowed never to stop working as a cook and refused a gallbladder operation that would have rid her of the bacteria. Consequently, she was forced to spend twenty-six years of her life in relative isolation, shunned by her fellow humans, in a tiny cottage on North Brother Island, a small island in the East River in New York City. Mary, not *Salmonella,* became the culprit. This was hardly the life she envisioned when, as a teenager, she emigrated from Ireland. In a legal sense she was never tried for her "crime."

Mary Mallon was born in 1869 in Ireland and immigrated to the United States in 1883. The first suspicion that

she was a healthy typhoid carrier was during her brief stint as a cook for the Warrens during which, in a 1-week period, 6 of 11 people in the household (Mrs. Warren, 2 daughters, 2 maids, and the gardener) developed typhoid fever. Typhoid and other diseases associated with "filth" were simply "not nice" for the elite of Oyster Bay where the Warrens summered; hence, George Soper, a sanitary engineer, was hired to determine the cause of the outbreak. Soper quickly ruled out soft clams as the source and found a pattern of typhoid in families where Mary had worked as a cook, dating back to 1900 and totaling 26 cases. Soper was unsuccessful in convincing Mary to be tested to determine if she was a *Salmonella* carrier. She proclaimed that her health was excellent. She had no history of typhoid and, hence, saw no reason for compliance.

The case was referred to Herman Biggs, medical officer of the New York City Health Department. In 1907 Biggs authorized police to take Mary against her will to a contagious disease unit in a New York hospital. Examination of her feces revealed a high content of typhoid bacilli, and she was removed to North Brother Island. Three years later she was released by the health commissioner on her promise that she would not cook again, but after about two

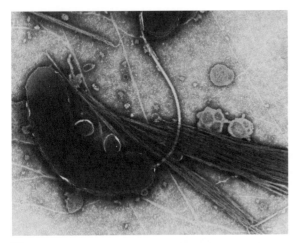

(a)

(b)

FIGURE 9.4 Cholera. **(a)** Transmission electron microscope image of *Vibrio cholerae* that has been negatively stained. Courtesy of Louisa Howard, Tom Kirn, and Ron Taylor, Dartmouth College, Electron Microscope Facility. **(b)** Poverty, overcrowding, and primitive sanitation are a dangerous trio favoring the emergence and dissemination of this waterborne disease. Courtesy of the CDC.

years she returned to cooking, her only employment skill, at a hospital. In 1915 she was sent back to the island for the rest of her life. During her 4 years of freedom she was responsible for another 25 cases of typhoid fever, including 2 deaths. She died on November 11, 1938, not as Mary Mallon but as Typhoid Mary, a public menace blamed for three deaths and about 47 other cases of typhoid fever and with the dubious distinction of being the first healthy *Salmonella* carrier in the United States.

The account of Mary Mallon exposes the dilemma of the public's health and safety versus the rights of the individual. The circumstances of this case and its outcome (Mary Mallon's confinement) can be viewed as an application of objective utilitarianism, a consequence-based moral philosophical position succinctly stated as "the greatest good for the greatest number." Proponents of utilitarianism are clear that all individuals count and count equally, but, nevertheless, the sacrifice of some (Mary, in this case) is justified for the sake of the many (the public). At the time those charged with the public's health exhibited sex and social class prejudice in considering a broad spectrum of factors and in arbitrarily deciding that liberty was an impossible privilege to allow Mary Mallon. In the case of other disease carriers, particularly males, other options were explored.

Do you believe Mary was justly treated? This case serves as a paradigm for today's "Typhoid Mary"—"AIDS Sam" or "Resistant TB Joe." Health officials have not only the right but the obligation to develop public health policies, but, in so doing, they incur the responsibility to provide realistic and appropriate alternatives, including health care, housing, and reemployment training, to those individuals in society who are afflicted. AIDS, in some respects, parallels the Typhoid Mary dilemma.

This poem appeared in *Punch,* a British magazine, in 1909:

> *In U.S.A. (across the brook)*
> *There lives, unless the papers err,*
> *A very curious Irish cook*
> *In whom the strangest things occur;*
> *Beneath her outside's healthy glaze*
> *Masses of microbes seethe and wallow*
> *And everywhere that MARY goes*
> *Infernal epidemics follow.*

Cholera reaches epidemic proportions in developing countries (BOX 9.2). Fortunately, the terrible loss of life from cholera throughout the world can be prevented through **oral rehydration therapy**. These prepackaged mixtures of sugar and salts are available at a very minimal cost, are readily transported to remote areas, and negate the use of intravenous injection and the associated costs that make such therapy prohibitive in countries where it is most needed.

Pandemics of cholera have been documented for hundreds of years, and it seems that no continent has escaped. In 2004 worldwide there were over 100,000 cases resulting in 2,300 deaths. In the United States and in other industrialized countries, because of advances in water and sanitary engineering, cholera is rare. Outbreaks occurring in the United States are usually the result of tourists returning from Africa, Southeast Asia, or South America; in some cases the source has been contaminated seafood brought back to the United States. Occasionally, notices are posted in newspapers citing the closing of shellfish beds and beaches because of fecal pollution and the danger of *V. cholerae* and other fecal pathogens.

The genes for enterotoxin production by cholera vibrios were identified in 1992 by scientists at the University of Maryland. Removal of these genes renders the vibrios nonpathogenic and candidates for a vaccine. Vaccine development has

It was all over in less than a minute on January 20, 2010. The Earth trembled, trees swayed and snapped, buildings shook, houses were devastated, and people scrambled seeking refuge. An estimated 1.5 million Haitians were suddenly left homeless and 300,000 were killed by the hurricane that struck Haiti, an underdeveloped and impoverished nation.

The only good that arose from the rubble of the hurricane disaster was the outpouring of generosity displayed by private citizens and internal agencies around the world. Nevertheless, large numbers of Haitians fled to camps for internally displaced persons, or IDPs, and as so often happens, relatively few dollars trickled to those that needed it the most, namely, the inhabitants of the camps. Hygienic conditions were almost nonexistent and were an embarrassment to the world as well as a violation of human rights. Several months after the hurricane about half of the camps were still without basic needs of water and toilets. People slept under tarpaulins, sheets, cardboard, or out in the open; tents were a limited luxury that few possessed.

The sanitary conditions in the country set the stage for an outbreak of infectious disease that was bound to happen, and it did. In late October 2010, nine months after the hurricane, cholera was confirmed in the country; within only several months it caused thousands to be hospitalized and killed an estimated 5,000 people. The Haitian government feared that there would be over 400,000 cases within a year's time.

This was the first cholera outbreak in Haiti in decades leading to speculation as to the source of cholera microbes. On this point, the scenario becomes contentious and sticky; no country or persons want to be identified as the source of this fecally transmitted pathogen. One explanation is that United Nations peacekeeping troops from Nepal were quartered in an area of Haiti that had a leaky sewage system, and the Nepalese troups carried cholera microbes that were discharged into the sewage. Nepal had recently suffered three waves of cholera that were discharged into the sewage. In this circumstance it would be like throwing kindling wood into a box of red hot coals. The Nepalese government was outraged and denied the allegation despite the fact that the cholera strain isolated from Haitians was much like the one isolated in Nepal. Other epidemiologists posed that the cholera bacteria had been dormant in waters for many years, and perhaps, the breakdown in public health infrastructure resulting from the hurricane was responsible.

The crisis regarding the source of the cholera was defused with the attitude "It is what it is" (one of my wife's favorite sayings), and attention focused on recovery and prevention. The actual source of the cholera microbes was never fully determined.

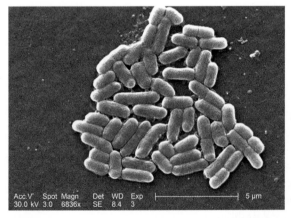

FIGURE 9.5 Colorized scanning electron micrograph of *E. coli* O157:H7. Courtesy of Janice Haney Carr/CDC.

been disappointing because the vaccine failed to protect against many of the cholera strains and is short-lived. A single cholera vaccine is licensed in the United States but is not recommended and is no longer manufactured.

Escherichia coli

E. coli, a gram-negative bacillus, is frequently a news item because of an enterotoxin-producing strain called *E. coli* O157:H7, first recognized in the United States in 1982 (FIGURE 9.5). It is a highly significant cause of foodborne illnesses, and a number of known outbreaks have occurred (TABLE 9.4). What is *E. coli*? The genus is named for Theodor Escherich, who described the organism in the 1880s. The species (*coli*) indicates the location: The organisms are gram-negative bacilli and are a constituent of the normal flora of the large intestines (colons) of humans and animals. Bacteria, particularly *E. coli*, account for much of the weight of fecal material; these normal inhabitants of the

TABLE 9.4 *Escherichia coli* O157:H7 Outbreaks

Year	Outbreak
1994	A multistate outbreak originating in Washington; *E. coli* O157:H7- associated bloody diarrhea and hemolytic uremic syndrome from hamburgers
1994	Japan; a severe outbreak of hemorrhagic colitis and hemolytic uremic syndrome associated with *E. coli* O157:H7
1995	A university outbreak of *E. coli* O157:H7 infections associated with roast beef
1996	Kyoto, Japan; contaminated radish sprouts
1997	Michigan and Virginia; outbreaks of *E. coli* O157:H7 infection associated with eating alfalfa sprouts
1997	A multistate outbreak of *E. coli* O157:H7 infections associated with eating mesclun mix lettuce
2000	Wisconsin; outbreak of *E. coli* O157:H7 infection associated with eating fresh cheese curds
2005	*E. coli* O157:H7 infection associated with drinking water
2006	Multistate outbreak of *E. coli* O157:H7 infections from spinach
2006	Taco Bell; *E. coli*-contaminated lettuce outbreak
2007	Dole bagged salad mix contaminated with *E. coli* O157:H7
2009	Ground beef contaminated with *E. coli* O157:H7. Source of beef was a meat packing plant in Massachusetts
2009	Unpasteurized apple juice contaminated with *E. coli* O157:H7
2011	Germany and other European countries from sprouts contaminated with *E. coli* O104:H4
2012	Canada Diversity Festival from a fruit dessert contaminated with *E. coli* O157:H7
2012	Unpasteurized milk contaminated with *E. coli* O157:H7 consumed from Oregon farm

intestinal tract are beneficial because they produce vitamins that are absorbed into the body. So why the bad rap for *E. coli* O157:H7, just one of the more than 100 strains of *E. coli* in the intestines of cattle and other animals? Somewhere and somehow, *E. coli* picked up the genes responsible for the production of an enterotoxin (Shiga toxin) that produces severe damage, possibly to the point of death. More than half the states in the United States require that isolation of *E. coli* O157:H7 be reported to the state health department for monitoring purposes in an effort to prevent outbreaks.

The consumption of ground beef is particularly dangerous, because contamination of the meat with fecal material may occur during the slaughtering process. The bacteria are mixed throughout large batches of meat obtained from many cattle when it is ground, in contrast to steaks, each of which are from a single source. With a little imagination the expression "One bad apple spoils the barrel" can be applied. The best advice is to avoid rare meat, particularly hamburgers. Many fast-food chain restaurants now specify the cooking time and temperature as a part of their policy and will not serve rare hamburgers. In most cases recovery from *E. coli* infection occurs within five to ten days without specific treatment. In children under five years of age there is a greater risk of death from kidney damage resulting from toxin-produced hemorrhages in the kidney, a condition known as **hemolytic uremic syndrome**.

Germany was under the gun in the spring and summer of 2011 when an outbreak of a new strain of *E. coli*, designated O104:H4, caused an outbreak of gastroenteritis. The strain is similar to O157:H7; it produces the Shiga toxin that can cause kidney, circulatory, and neurological symptoms. Initially, the source of the outbreak was an epidemiologist's nightmare: Cucumbers from Spain were first thought to be the source much to the chagrin of that country, but, ultimately, the cause was pinpointed to contaminated sprouts. How the sprouts became contaminated is not clear. The end result was that in Germany 3,785 became sickened and 45 deaths resulted causing the hospitals to be overwhelmed. Spread to other European countries was limited to 91 confirmed cases and one death.

Note that ground beef and other meats are not the only carriers. The consumption of contaminated fresh spinach was particularly serious in September 2006, resulting in illness in 183 persons in 26 states and one death; 95 of these cases were hospitalized. Shredded lettuce in Taco Bell restaurants in the Northeast was responsible for 71 cases of *E. coli* O157:H7 distributed over 5 states; 53 persons were hospitalized and 8 cases of hemolytic uremic syndrome resulted.

In addition to O157:H7, characterized as enterohemorrhagic *E. coli*, there are several other strains of *E. coli* that exhibit variation in the mechanism by which they cause diarrhea. The **enterotoxigenic *E. coli* strains** are one example. This group is the most common cause of **traveler's diarrhea**.

When you hear that a particular beach or shellfish bed is closed because of fecal pollution, you now know that it is probably because of the presence of large numbers of *E. coli*. This microbe is easy to detect, making it useful as an indicator organism in water quality, and its presence signifies that intestinal pathogenic microbes may be present. Too bad about O157:H7 damaging the reputation of the usually nonpathogenic workhorse *E. coli*, probably the most widely investigated species of life. An *E. coli* culture (nonpathogenic) can be purchased for only several dollars from a variety of biological supply houses and is widely used in laboratories in high schools and colleges.

Campylobacteriosis

Campylobacter is a relatively new cause of intestinal disease in humans, although it has a long history of association with disease in animals, causing abortion and enteritis in sheep and cattle. It is the major cause of bacterial diarrhea in the United States and causes more cases of diarrhea than *Salmonella* and *Shigella* combined; the *Campylobacter* species responsible for this disease is *Campylobacter jejuni*, a gram-negative spiral-shaped bacillus. Poultry, cattle, drinking water, and unpasteurized milk have been identified as potential sources; even one drop of "juice" from raw chicken meat is sufficient to establish infection. (Remember that when you are preparing a barbecue.) Sexual activity involving oral–anal contact is also a mechanism of transmission.

The disease is usually self-limiting and lasts about one week. Fewer than 500 organisms are enough to cause illness. The October 20, 1998 issue of *The New York Times* shocked the public with a report that 70% of chicken meat sampled from supermarkets was contaminated with *Campylobacter*. To make matters worse, many isolates were antibiotic resistant because of the widespread use of antibiotics in feed. On a worldwide basis, *Campylobacter* is the most common cause of gasteroenteritis; about 1,600 cases occur each year in the United States.

Listeriosis

Listeriosis is caused by the gram-positive motile bacillus, *Listeria monocytogenes,* and is an increasingly significant cause of food infection. The organism is distributed worldwide and has been isolated from soil, water, manure, plant materials, and the intestines of healthy animals (including humans, birds, and fish). The disease is associated with the ingestion of *Listeria*-contaminated foods, including vegetables, meats, cheeses, ice cream, raw (unpasteurized) milk, and, in 2007, in pasteurized milk. Processed foods, such as cold cuts, hot dogs, and soft cheeses, may become contaminated after processing and serve as the source of epidemics or threatened epidemics. *Listeria* can grow in refrigerators, contributing to the problem.

Listeriosis is usually mild or subclinical in healthy adults and children and is manifested by nonspecific symptoms, including sore throat, fever, and diarrhea. Infants, the elderly, immunocompromised individuals (e.g., cancer patients receiving chemotherapy and persons with AIDS), and pregnant women are at high risk. Persons with AIDS are approximately 300 times more susceptible than healthy adults, and pregnant women are approximately twenty times more susceptible than nonpregnant women. Newborns can be infected as a result of their mothers' eating contaminated foods containing only a few *Listeria* bacteria and may become acutely ill with meningitis (nervous system involvement); the mortality rate is about 60%, and antibiotics given promptly to pregnant women can often prevent infection of the fetus or newborn.

A wrongful death suit against the Sara Lee Corporation was filed on February 2, 1999, after the death of a woman who consumed *Listeria*-contaminated hot dogs. The U.S. Department of Agriculture reported that maintenance of the company's air-conditioning system and the resulting dust "could have caused widespread contamination of meat products through the plant." Fifteen adults died and three miscarriages occurred in 16 states. In the summer and fall of 2002 in the northeastern United States there was an outbreak of listeriosis involving more than 53 people, most of whom were hospitalized, resulting in 8 deaths and 3 miscarriages or stillbirths. Precooked, sliceable deli meat was the culprit.

A dairy farm described as a "mom and pop" operation had to close down in February 2008 when contaminated milk caused listeriosis in four people. The cases included three men over the age of 70, all of whom died after June 2007, and a miscarriage in a 30-year-old woman. According to the owners, of the 330 cows on the farm about 130 were milked daily and produced 1,000 gallons of milk each day. How the *Listeria* contaminated the milk remains a mystery, because all the farm's milk was pasteurized.

TABLE 9.5 reveals that most outbreaks of listeriosis are associated with meat products and, to some lesser extent, with milk and milk products. However, starting in August 2011, a multistate outbreak of listeriosis was traced to cantaloupes grown at Jensen Farms in Colorado. The cases spread to 28 states resulting in 30 deaths and 146 confirmed cases; it is considered to be the worst foodborne outbreak to have occurred in the United States in 100 years. Those who died ranged from 48 to 96 years.

TABLE 9.5 *Listeria* Recalls and Outbreaks[a]

Food Source	Company	Description
Hot dogs, lunch meat	Thorn Apple Valley	Voluntarily recalled products
Hot dogs, cold cuts	Sara Lee subsidiary	75 cases, including 15 deaths in 16 states; voluntarily recalled product
Chocolate ice cream	Velvet Ice Cream	Voluntarily recalled 244 gallons of ice cream; possible contamination, no reports of illness
Headcheese (a seasoned loaf made of the head meat of a calf or pig in its natural aspic)	Ba Le Meat Processing & Wholesale, Inc.	Voluntarily recalled 2,600 pounds of headcheese; no reports of illness
Beef wieners, hot dogs	Valley Meat Supply	Voluntarily recalled 150 pounds of wieners and frankfurters; no reports of illness
Weisswurst sausage	Alpine Wurst & Meat House	Voluntarily recalled 60 pounds of sausage; no reports of illness
"Hawaiian Winners"	Redondo's Inc.	Voluntarily recalled 9,620 pounds of sausage
Sliced ham, salami, luncheon meat	White Packing Co.	Voluntarily recalled 16,392 pounds of sliced ham, salami, and luncheon meats
Deli meats	Oscar Mayer	Voluntarily recalled 28,313 pounds of deli meat
Hot dogs	Marathon Enterprises	Voluntarily recalled 2.1 million pounds of hot dogs; two deaths
Milk	Whittier Farms	Voluntarily recalled milk
Cantaloupes	Jensen Farms	Voluntarily recalled cantaloupes

[a] In some cases bacteria were presumed to be *Listeria*.

Clostridium difficile

This organism (FIGURE 9.6), a gram-positive spore-forming bacillus closely related to *Clostridium tetani* and *C. botulinum*, is a cause of pseudomembranous colitis. The CDC estimates this organism is responsible for about 13% of nosocomial infections and for close to half of all cases of diarrhea associated with nosocomial infections from 1990 to 2007. The caseload jumped to 40,000 cases in 2007. For some individuals, *C. difficile* colitis symptoms become severe and diarrhea becomes debilitating, resulting in extreme weight loss. The transmission route is fecal–oral; the condition has generally been associated with hospital patients on antibiotic therapy.

Recent evidence shows that the infection has jumped from hospital and hospital-like environments to the community; the emergence of methicillin-resistant *Staphylococcus aureus* is a similar phenomenon. Generally, *C. difficile* is inhibited by the normal flora, but in some individuals on antibiotics the composition of the normal flora is altered to the degree that *C. difficile* may "grow out," causing mild to severe diarrhea and possibly death from inflammation and spread into the bloodstream. While most cases favorably respond to antibiotic therapy, there are cases in which symptoms reappear. Alternative therapies, including probiotics and yogurt, are sometimes successful.

Fecal transplants are practiced by a handful of gastroenterologists in the United States (who have gotten over the "yuck" factor) as a safe, inexpensive treatment option for patients with reoccurring severe *C. difficile* colitis. According to a small series of reports, about 90% of patients are cured. The treatment involves introducing saline-diluted fecal matter from a donor (usually a child or a spouse) into a patient via an enema or through a special catheter that is inserted down the throat, through small intestines, and through the colon. Would you get a fecal transplant? The procedure is more common in Canada and Europe but is not a common practice in the United States because the procedure has not gone through rigorous clinical testing and is not regulated.

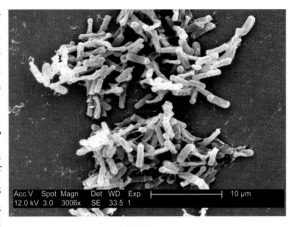

FIGURE 9.6 Colorized scanning electron micrograph of *Clostridium difficile.* Courtesy of Lois S. Wiggs/CDC.

Airborne Bacterial Diseases

Airborne bacterial diseases are respiratory tract infections transmitted by air droplets or by contact with contaminated inanimate objects; in some cases the disease spreads beyond the respiratory tract to other areas of the body. The humidity and temperature in air allow bacteria to remain viable for extended periods of time; hence, air serves as an excellent vehicle for transmission. Air is a passive transmitter because there is no multiplication of the microbes, as is the case with foodborne bacterial infections. Respiratory infections generally respond well to antibiotic therapy, although antibiotic resistance is an increasing problem. Immunization is available for some respiratory infections but may not be practiced in many countries because of problems of availability, cost, and public health infrastructure. The upper respiratory tract includes the tonsils and pharynx, and the lower respiratory tract includes the trachea, larynx (voice box), bronchi, bronchioles, and lungs (FIGURE 9.7). TABLE 9.6 provides a list of major airborne bacterial diseases categorized into **upper** and **lower respiratory tract infections**.

Before the advent of immunization and antibiotic therapy, there were so-called communicable disease hospitals in all communities of the United States and other developed countries of the world to care for those with contagious (also called "infectious" or "communicable") diseases. These hospitals were better equipped to manage the constant threat of bacteria being introduced into the air by their victims through coughing and nasal secretions; young children were especially susceptible to these

FIGURE 9.7 Anatomy of the human respiratory system.

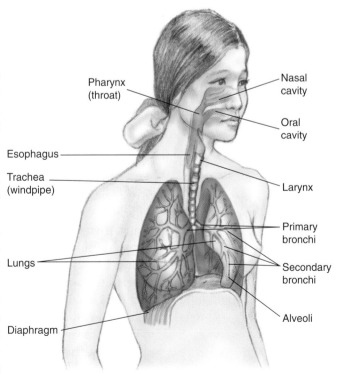

TABLE 9.6 Airborne Bacterial Diseases[a]

Disease	Incubation Period	Causative Microbe	Symptoms
Upper respiratory tract			
Diphtheria	2–4 days	*Corynebacterium diphtheriae*	Respiratory blockage causing suffocation
Pertussis (whooping cough)	7–10 days	*Bordetella pertussis*	Respiratory blockage causing violent and persistent coughing
Streptococcal infections (group A) and sore throat, fever (pharyngitis, tonsillitis strep throat, scarlet fever, flesh-eating strep, toxic shock syndrome, childhood fever)	2–5 days	Streptococci (group A)	Coughing, severe headache, red rash, strawberry tongue, multiorgan failure, destruction or "eating" of the flesh
Meningitis	1–3 days	*Neisseria meningitidis*	Coldlike symptoms, fever, delirium, stiffness of the back and neck
Meningitis	5–6 days	*Haemophilus influenzae* type B	Fever, stiff neck, altered mental status
Lower respiratory tract			
Legionellosis	2–10 days	*Legionella pneumophila*	Fever, muscle pain, cough, pneumonia
Tuberculosis	Several weeks	*Mycobacterium tuberculosis*	Fever, weight loss, cough, fatigue
Tularemia	1–14 days	*Francisella tularensis*	Fever, pleuropneumonitis, respiratory failure, shock

[a]These diseases are treatable with antibiotic therapy; drug-resistant strains pose a problem in some cases.

diseases. High fever, delirium, racking coughs, labored breathing, and horrible rashes were associated symptoms. Little could be done other than supportive therapy. It was a "wait, see, and pray" attitude, and death rates were very high. Frequently, the young and hapless victims were not hospitalized but were confined to their homes; quarantine signs were placed on the doors by local public health officials to discourage visitors. These were the days of the horse-and-buggy doctors, and their selflessness was more than admirable as they sat by the bedsides and could do little more than apply cold towels to the patients burning with raging fevers. In today's world the advancement of sanitation, hygiene, immunization, and antibiotics negates the need for specialized hospitals to care for patients with communicable diseases. Virtually all hospitals are able to invoke "isolation" strategies.

■ Upper Respiratory Tract Infections

Diphtheria

Diphtheria is caused by *Corynebacterium diphtheriae,* a gram-positive bacillus. Infection occurs in the upper part of the respiratory tract, close to the tonsils, causing them to be inflamed. The lymph glands swell resulting in a "bull neck" appearance (FIGURE 9.8). The pathology is the result of the production of a powerful exotoxin that destroys the cells constituting the epithelial lining. These dead cells, along with mucus and scavenger cells, pile up and form a pseudomembrane (false membrane). Its leathery consistency may cause respiratory blockage and death by suffocation, particularly in young children. Before modern medicine, many emergency **tracheotomies** (cutting a hole in the trachea) were performed by a physician on the kitchen table in desperate attempts to save a child's life. Further, the toxin diffuses into the bloodstream, causing widespread damage, particularly to the heart. A milder form of diphtheria limited to the skin can also occur (FIGURE 9.9).

The incidence of diphtheria in the United States is now fewer than fifty cases per year. In fact, many modern physicians, having never seen a case, would have difficulty in making a diagnosis. Worldwide, however, the disease remains a significant problem. In the early to mid-1990s a massive reemergence of diphtheria occurred in the newly independent states of the former Soviet Union caused primarily by the collapse of the healthcare system, resulting in decreased childhood vaccination. It takes a lot of money to maintain an immunization program.

Children in developed countries typically receive their first immunization for diphtheria at about the age of two months. It is actually a triple vaccine, referred to as DPT: D stands for diphtheria, P stands for pertussis, and T stands for tetanus.

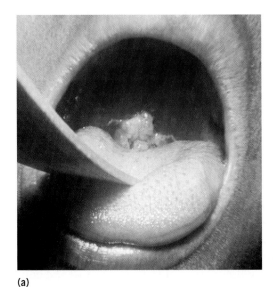

(a)

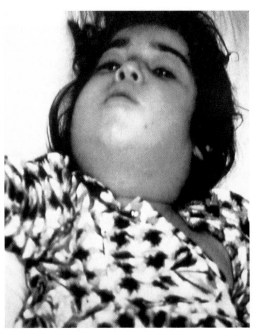

(b)

FIGURE 9.8   Appearance of diphtheria. **(a)** Inflammation of tonsils. © Science VU/Visuals Unlimited, Inc. **(b)** "Bull neck" characteristic. Courtesy of the CDC.

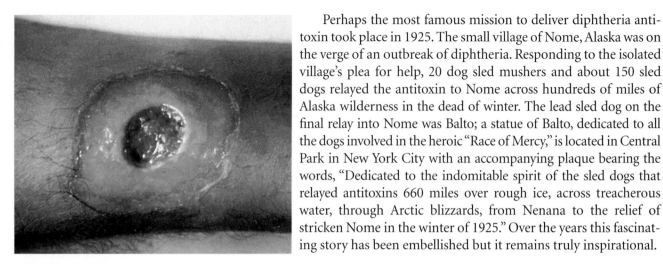

FIGURE 9.9 A milder infection of *Corynebacterium diphtheriae* can cause skin lesions. Courtesy of the CDC.

Perhaps the most famous mission to deliver diphtheria antitoxin took place in 1925. The small village of Nome, Alaska was on the verge of an outbreak of diphtheria. Responding to the isolated village's plea for help, 20 dog sled mushers and about 150 sled dogs relayed the antitoxin to Nome across hundreds of miles of Alaska wilderness in the dead of winter. The lead sled dog on the final relay into Nome was Balto; a statue of Balto, dedicated to all the dogs involved in the heroic "Race of Mercy," is located in Central Park in New York City with an accompanying plaque bearing the words, "Dedicated to the indomitable spirit of the sled dogs that relayed antitoxins 660 miles over rough ice, across treacherous water, through Arctic blizzards, from Nenana to the relief of stricken Nome in the winter of 1925." Over the years this fascinating story has been embellished but it remains truly inspirational.

Pertussis (Whooping Cough)

Whooping cough is a highly infectious and potentially lethal disease caused by the gram-negative coccobacillus *Bordetella pertussis*. The disease is a particular threat to infants and young children under four years of age. Children who have whooping cough are the primary source of the disease; there is no nonhuman reservoir. The bacteria are spread by talking, coughing, sneezing, and laughing. The bacilli bind to ciliated epithelial cells that line the upper respiratory tract and secrete toxins that cause damage to the cells whose function is to clear mucus from the air passages. The net effect is a buildup of a thick glue-like mucus, promoting symptoms of a common cold followed by paroxysms (spasms) of violent, hacking, persistent, and recurrent coughing, usually about 15 to 20 coughs, in an attempt to cough up the mucus. Coughing fits can last up to 10 weeks or more; pertussis sometimes is known as the "100-day cough." These episodes result in a need for oxygen, triggering deep and rapid inspirations throughout the partially obstructed passages that are responsible for the characteristic **"whoop."** The coughing may be so violent as to cause vomiting, hemorrhage, and even brain damage. Whooping cough may last for several weeks or months.

Between 1940 and 1945, before the advent of a pertussis vaccine, whooping cough was rampant. In the United States there were as many as 147,000 cases per year and over 8,000 deaths. In 1991 the pertussis component, consisting of live pertussis bacteria, was modified to the use of killed bacterial cells, referred to as acellular (aP), to minimize side effects. It is usually used in conjunction with the diphtheria and tetanus vaccines; this multiple vaccine is known as DTaP. Pertussis is considered a reemerging disease, and it is not clear why it is on the increase; 27,550 cases occurred in 2010—the greatest level since 1950. In 2012, pertussis cases spiked with 46 states and Washington, D.C. reporting epidemics by August 11, 2012. Wisconsin had the highest number of reported cases. Infants were the most affected, followed by children ages 7 to 10 and adolescents 13 to 14 years of age.

One possible factor is that the original pertussis vaccine was controversial and accused of causing brain damage, although this has never been conclusively proved. Nevertheless, the bad press accounted for fewer pertussis immunizations and a resultant increase in the disease. Two new and improved vaccines were licensed in 2005. Pertussis vaccine does not produce lifelong immunization, nor does recovery from whooping cough, contrary to popular belief. Persistent coughs in older children and adults may well be whooping cough, usually present in a milder form.

Whooping cough is still a devastating disease in nonimmunized populations. The WHO estimates as many as 30 to 50 million cases and 300,000 deaths occur each year, primarily in children. Whooping cough is a leading cause of death worldwide primarily in a nonvaccinated or incompletely vaccinated population.

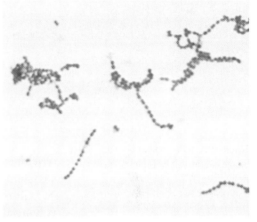

FIGURE 9.10 Chains of *Streptococcus pyogenes*, the bacteria that cause scarlet fever. Courtesy of the CDC.

Whooping cough is endemic in the United States; outbreaks of epidemic proportions occur every three to five years. In 2010 the number of cases rose to a high of 27,550, in sharp contrast to 1,730 cases in 1980, the highest recorded number since the 1950s. Estimates are that over one million undiagnosed cases of whooping cough are present yearly in the United States. An outbreak of pertussis occurred in an Amish community in Kent County, Delaware between September 2004 and February 2005 in which 345 cases resulted primarily in preschool-aged children. Although the Amish religious doctrine allows vaccination, childhood vaccination is low in many Amish communities. These statistics underscore the need to administer vaccination in all populations and to reevaluate the immunization strategy for pertussis. In 2012, the incidence of whooping cough increased among 13 and 14 year olds suggesting that immunity from vaccination was waning. The CDC recommended increased vaccination coverage efforts, especially for pregnant women, adolescents, and adults.

Streptococcal Infections

Streptococci are a large group of gram-positive cocci, the most significant one of which is *Streptococcus pyogenes* (FIGURE 9.10). The particular manifestation of disease is related largely to the specific *Streptococcus* and the toxin it produces. Infections range from generally mild, the most common of which is **"strep throat,"** to a variety of life-threatening conditions. The organisms reside in the nose and throat and are transmitted by respiratory droplets from infected persons or by contact with wounds and sores on the skin.

FIGURE 9.11 Characteristic "strawberry tongue" of a patient with scarlet fever. Children with this disease frequently have an inflamed tongue along with a characteristic red rash. © imagebroker/Alamy Images.

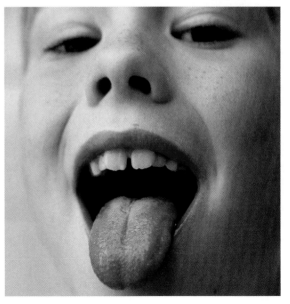

Strep throat, more correctly termed streptococcal pharyngitis or tonsillitis, is a common and usually mild infection predominantly found in children 5 to 15 years of age and is characterized by a red and sore throat, fever, and headache. Definitive diagnosis is by culturing a specimen from a throat swab, or by a rapid test procedure. In some cases, **scarlet fever** results and is caused by a streptococcal strain producing **erythrogenic toxin**. The disease is characterized by a red rash and "strawberry tongue" (FIGURE 9.11).

Some streptococci are invasive and cause serious and life-threatening infections, including streptococcal **toxic shock syndrome,** characterized by multiorgan failure, and **necrotizing fasciitis**, more dramatically called **flesh-eating disease**. The disease is a rarity; it is horrendous and is the result of infection of a minor wound with an invasive strain of streptococci that

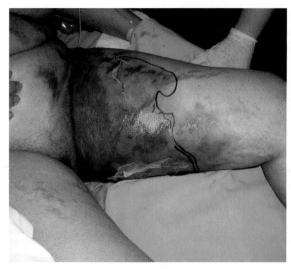

FIGURE 9.12 Necrotizing fasciitis. Image courtesy of Piotr Smuszkiewicz, Department of Anesthesiology, Intensive Therapy and Pain Treatment, University Hospital Przybyszewskiego, Poznan, Poland.

elaborates flesh-destroying enzymes and toxins and can spread through human tissue at a rate of 3 cm per hour (FIGURE 9.12). In 1994 several outbreaks of this rapidly progressing disease, dubbed by the media "flesh-eating bacteria," occurred in England and in the United States; one story was headlined "the bacteria that ate my face." Each year there are about 650 to 800 cases of necrotizing fasciitis caused by *Streptococcus pyogenes*. Another 700 annual cases are caused by other bacteria such as *Staphylococcus aureus, Klebsiella, Clostridium, E. coli,* and *Aeromonas hydrophilia.* The number of annual infections does not appear to be rising. About 1 in 5 die of necrotizing fasciitis. Treatments may include skin graft surgeries and amputations. Hyperbaric oxygen therapy (HBOT) is an emerging therapy in combination with antibiotic treatment. In 2012, several cases describing flesh-eating disease were in the news media (TABLE 9.7). These cases should not be confused with the 2012 reports of cocaine users in Los Angeles and New York City who developed cocaine-related necrotic lesions. The necrosis was caused by levamisole, a contaminant used to cut street cocaine.

Glomerulonephritis, a kidney disease, and **rheumatic fever**, a condition involving the heart and joints, are possible long-term complications of repeated early childhood streptococcal infections. The incidence of these diseases has decreased drastically in the past 25 years as a result of the advent of antibiotics, but, unfortunately, drug-resistant strains are of increasing concern.

TABLE 9.7 2012 High Profile Cases of Necrotizing Fasciitis

Patient	Origin of Infection/Source	Cause	Pre-existing Medical Conditions/Risks	Prognosis/Outcome
24-year-old female, Georgia (Aimee Copeland)	Deep cut injury on left calf when zip-lining along the Little Tallapoosa River. She fell into the river and gashed her leg.	*Aeromonas hydrophila* (waterborne bacteria)	None	Survived; left leg, right foot, both hands, part of torso amputated. Released from hospital July 2, 2012.
36-year-old female, South Carolina (Lana Kuykendall)	Contracted days after giving birth to twins. Large bruise on back of leg.	*Streptococcus pyogenes*	Pregnancy?	Survived, June 2012. No amputations but over 20 skin-grafting operations.
33-year-old female, Michigan (Crystal Spencer)	Boil/abscess on upper right thigh	NA	Obesity; suffered from adult-onset diabetes	Died, August 2012. Death may be related to blood clot, other complications.
Adult male, Alaska (Ruben Pereyra)	Cut on knuckle and got a splinter in left hand via bookshelf injury.	*Streptococcus pyogenes*	NA	Survived, June 2012. Skin grafts on hand and arm. May lose fingers.
Adult male, Los Angeles (Lionel Coates)	NA	NA	NA	Survived, May 2012
Adult female, Los Angeles (Julie Hanna)	Injury to knee, leg, and right foot during karate session	NA	NA	Survived, May 2012

NA = not applicable or known at the time of this writing.

Bacterial Meningitis

Meningitis is an inflammation of the meninges, the three membranes covering the spinal cord and the brain. This condition is potentially serious because microbial invasion of the nervous system is involved. The key to survival is early diagnosis; several hours can make the difference between life and death. Meningitis can be caused by a variety of bacteria, viruses, protozoans, and fungi; this section deals only with bacterial meningitis. Early signs are coldlike symptoms that progress quickly to more definitive complaints, including fever, possibly delirium, and stiffness in the neck and back.

A number of bacteria can infect the meninges, but the two most common species are *Neisseria meningitidis,* a gram-negative diplococcus, and *Haemophilus influenzae* type b, a gram-negative bacillus, the latter frequently referred to as Hib. About 10% of the population are healthy carriers and harbor meningococci in the back of the nose and throat. These healthy carriers, along with infected individuals, spread the cocci between people by respiratory droplets through coughing, sneezing, kissing, and sharing eating utensils. There is no animal reservoir. The meningococci are very fragile outside the body, which limits the incidence of the disease. They enter the nasal pharynx via droplets and may cross the epithelial cell barriers lining the pharynx and invade into the bloodstream, from which they enter the meninges. Between 2003 and 2007, an estimated 4,100 cases and 500 deaths from bacterial meningitis occurred in the U.S.

Hib meningitis occurs primarily in children under five years of age. Once the leading cause of bacterial meningitis in the United States, its incidence has dramatically decreased since the introduction of the Hib vaccine in the early 1990s. Although it is a less serious form of meningitis, it should not be regarded lightly. The Hib vaccine is available in three forms: Hib alone, Hib in combination with DTaP, and Hib in combination with recombinant hepatitis B vaccine. In 2007, a nationwide shortage of vaccine due to the voluntary recall and subsequent production of two Hib vaccines (PedvaxHIB and COMVAX), Hiberix was approved by the FDA as a Hib booster dose for children age 15 months to 4 years in 2009.

Meningococcemia, caused by *Neisseria meningitidis,* is very serious. Fortunately, the infection responds to antibiotic therapy, although some individuals are left with severe, irreversible damage. There is no vaccination in general use for bacterial meningitis. It should be noted that the vaccine, which has been marketed since 2005, carries a risk of severe side effects, including Guillain-Barré syndrome. However, as public health officials point out, the risk of bacterial meningitis causing serious illness in those unvaccinated is about ten times greater than those vaccinated. College students living in dormitories appear to be particularly susceptible to bacterial meningitis. In the last few years episodes have occurred on several college campuses, as exemplified by an outbreak on two campuses of the State University of New York in 2008. College-bound students should be immunized against this potentially deadly disease in view of their future close-quarter living conditions. Between 1,000 and 2,600 cases of bacterial meningitis occur each year in the United States and, despite antibiotics, approximately 11% to 19% develop serious sequelae, necessitating amputation of limbs and toes in an attempt to halt the spread of infection, retardation, and deafness. A vaccine, Menactra®, was licensed in 2005 and is protective against four of the five *Neisseria* meningitis-producing strains.

Lower Respiratory Tract Infections

The lower respiratory tract consists of the larynx (voice box), trachea (windpipe), bronchial tubes, and lungs (Figure 9.7). The tissue of the lungs terminates in millions of air sacs termed alveoli. The lungs are covered by a membrane called the pleura. The terms "**laryngitis**," "**tracheitis**," and "**bronchitis**" refer to their anatomical location; "itis" means inflammation. **Pleurisy** is an inflammation of the pleura, and **pneumonia** is an inflammation of the lungs. Lower respiratory tract infections are caused by a variety of bacteria, viruses, fungi, and protozoans.

Legionnaires' Disease (Legionellosis)

In 1976 the American Legion, a U.S. military veterans' association, held its 58th national convention at the historic Bellevue-Stratford hotel in Philadelphia; this convention was a particularly significant one, because it marked the two hundredth anniversary celebration of the founding of our nation. The occasion was marred by a mysterious pneumonia in almost 200 of the Legionnaires, 31 of whom died. After an exhaustive investigation lasting over a year, it was discovered that the disease was bacterial and was disseminated through the air-conditioning ducts of the hotel. The organism was later identified and termed *Legionella pneumophila*, a gram-negative bacillus (FIGURE 9.13). The publicity that ensued over the next few years was dramatic and inspired musician and composer Bob Dylan in 1981 to write a song about **Legionnaires' disease**.

Isolated and unreported cases of legionellosis occur in the general population; symptoms of the disease are fever, muscle pain, cough, and pneumonia. These are symptoms shared with other respiratory infections, making definitive diagnosis difficult. The infections can be acquired by persons of any age, but middle-aged and older persons are the most susceptible, as are cigarette smokers and those with chronic lung diseases. When death occurs, it is usually attributable to shock and kidney failure.

Transmission does not involve person-to-person contact. Inhalation of aerosols that come from a water source contaminated with *L. pneumophila* is the mode of transmission. Air-conditioning cooling towers, spas, decorative fountains (e.g., in malls and hotel lobbies), vegetable mists in markets, showers, and whirlpools have all been implicated, and outbreaks continue to occur throughout the world (TABLE 9.8). The organisms grow well in the warm and stagnant waters afforded by these environments and are then aerosolized. Ironically, hospital outbreaks have occurred on frequent occasions because the temperature in the hot water lines is kept relatively low for safety reasons, allowing microbial multiplication. Furthermore, this microbe is more resistant than most to disinfection by chlorine. Fortunately, home and automobile air-conditioning units have never been implicated. Healthcare providers should be aware that legionellosis has been associated with gardening and the use of potting soil as has been shown in Australia, Japan, the United States, and, more recently in 2006, in The Netherlands.

Somewhere between 8,000 and 18,000 cases of Legionnaires' disease are hospitalized each year. The reason for the wide disparity in numbers is that many cases are not diagnosed or are misdiagnosed. Infection can be acquired any time of the year, but the

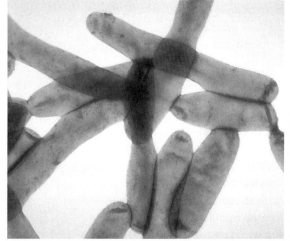

FIGURE 9.13 Transmission electron micrograph of *Legionella pneumophila*, the causative agent of legionellosis or Legionnaires' disease. Courtesy of Margaret Williams, PhD; Claressa Lucas, PhD; Tatiana Travis, BS/CDC.

TABLE 9.8 Legionnaires' Disease Outbreaks

Year	Location	Source	Outcome
1976	American Legion Convention, Philadelphia	Air conditioning ducts in a hotel	182 taken ill, 29 deaths
1989	Stafford Hospital, England	Unknown	98 taken ill, 10 deaths
1998	University of Washington Medical Center	Water-cooling tower	3 deaths in 5 months
1999	The Netherlands	Decorative water display at a bulb flower show	79 taken ill, 7 deaths
1999	Havre de Grace Hospital and Hartford Memorial Hospital, Connecticut	Water supply	4 deaths
1999	U.S. Naval Academy air-conditioning system	*L. pneumophila* found during a routine check	No illness
2001	Murcia, Spain	Hospital cooling towers	450 confirmed, 6 deaths
2002	United Kingdom	Cooling tower	7 deaths
2003–2004	Northern France	Cooling plant	86 confirmed, 18 deaths
2007	Russia	Hot water supply	Over 175 hospitalized
2008–2009	Scotland	Potting soil	Cluster of cases

summer and early months of the fall are the seasons when most cases occur. Prevention of Legionnaires' disease can be accomplished by improvements in the design and maintenance of cooling towers, whirlpools, spas, and other sources of warm, stagnant waters in which there is the potential for aerosolization.

Pontiac fever, first reported in Pontiac, Michigan, is also caused by *L. pneumophila*. The symptoms resemble those of a variety of other respiratory illnesses and last a few days. No deaths from Pontiac fever have been reported.

Tuberculosis

Mycobacterium tuberculosis (tubercle bacillus) is the causative agent of **tuberculosis**, a disease originating in the Middle East about 8,000 years ago. Evidence suggests that human tuberculosis was acquired from cattle and has affected civilization through the centuries. Tuberculosis is the leading cause of deaths resulting from microbes worldwide. Estimates are that about 9 million people are infected, the majority in the developing world; that 1.4 million people die of tuberculosis annually; and that about eight million new cases develop annually at the rate of one case per second. The WHO estimates that almost one billion people will become infected by 2020 if control measures are not improved. In 2010 in the United States there were 11,182 cases, representing a 3.1% decline from the previous year's rate. Tuberculosis is a current and dangerous epidemic made worse by the mushrooming occurrence of multiple drug-resistant and extensively drug resistant strains.

Tularemia

Francisella tularensis, a gram-negative bacillus, is the causative agent of tularemia, also called rabbit fever. It is one of the most virulent of all bacteria; inhalation of as few as ten organisms is sufficient to establish disease. Tularemia is a zoonotic disease; its natural reservoir consists of small mammals, including mice, voles, rabbits, and squirrels. Humans acquire tularemia through inhalation, contact with infected animal tissues, ingestion of contaminated food or water, and, most commonly, arthropod bites. There is no evidence of person-to-person transmission. Approximately 200 cases occur each year in the United States. A vaccine is being used to protect U.S. military laboratory workers; its use for other high-risk groups has been hampered by several obstacles. A new live vaccine strain is under investigation at the U.S. Army Medical Research Institute of Infectious Diseases (USAMRIID).

The highly infectious nature of *F. tularensis* and the ease of its dissemination by aerosolization renders it a potential weapon in the arsenal of biological warfare. It is considered a class A agent. Inhalation would result in severe respiratory illness, including pneumonia. Tularemia is treatable with several antibiotics.

■ Sexually Transmitted Diseases

Unlike most other microbial diseases, the **sexually transmitted diseases (STDs), formerly called venereal diseases** or **VD,** carry with them the stigma of "human wickedness" because they are associated with personal sexual behavior. Syphilis and gonorrhea may be passed from mother to child in utero. Ironically, these miserable diseases, recognized in the 1600s, were named after Venus, the Roman goddess of love. Before the germ theory of disease it was believed that some

FIGURE 9.14 **(a)** Anatomy of the human genitourinary system and **(b)** male and **(c)** female reproductive systems (see next page).

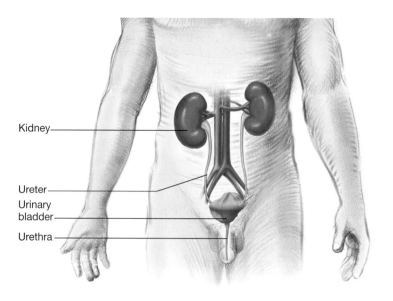

Kidney

Ureter
Urinary
bladder
Urethra

(a)

females passed onto their male partners a "noxious substance" that resulted in VD (a flagrant example of sexism).

Microbes that cause STDs are directly transmitted from the warm, moist mucous membranes of one individual to the warm, moist membranes of another individual during sexual practices, including vaginal and anal intercourse and oral sex. Semen and vaginal discharges are the major sources. The male and female genitourinary tracts are diagrammed in FIGURE 9.14. Promiscuous individuals are

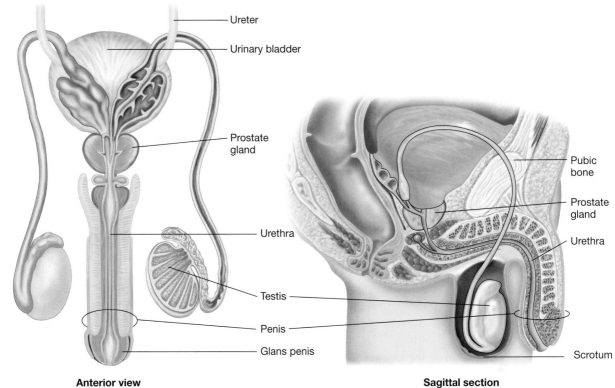

(b)   **Anterior view**   **Sagittal section**

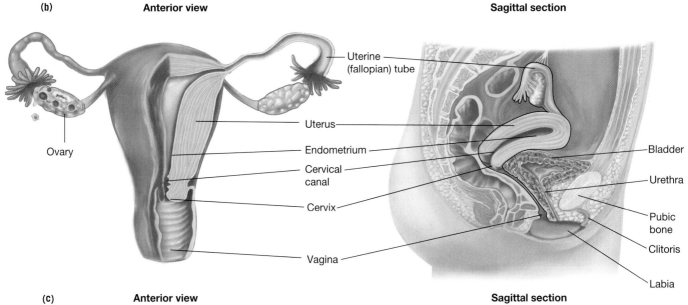

(c)   **Anterior view**   **Sagittal section**

TABLE 9.9 Sexually Transmitted Bacterial Diseases[a]			
Disease	Incubation Period	Causative Microbe	Symptoms
Syphilis	16–28 days (possibly as long as 10 weeks)	*Treponema pallidum*	Painless sores on the penis or cervix, rashes on the palms and soles, and eventual paralysis and insanity
Gonorrhea	2–7 days	*Neisseria gonorrhoeae*	Burning urination, cervical and urethral infection, abdominal pain, sterility
Chlamydia	2–6 weeks	*Chlamydia trachomatis*	Abdominal pregnancy, infertility, inflammation of the testicles
Lymphogranuloma venereum	1–4 weeks	*Chlamydia trachomatis*	Sores on the penis and vagina that develop into painful buboes
Chancroid	3–5 days	*Haemophilus ducreyi*	Painful ulcers on the penis, labia, or clitoris

[a]These diseases are treatable with antibiotic therapy; drug-resistant strains pose a problem in some cases.

particularly prone to STDs because of their sexual behavior. Unfortunately, one STD is not exclusive of the others, so individuals may have more than one STD disease simultaneously. TABLE 9.9 summarizes STDs.

The advent of public education programs, particularly during World War II, and the appearance of penicillin in 1943 led to a dramatic decrease in the incidence of syphilis and gonorrhea; predictions were that they would largely become diseases of the past. Unfortunately, these predictions did not come true. In recent times STDs have become a serious public health problem, particularly because of the emergence of antibiotic-resistant strains. The current concern with AIDS, a viral infection, overshadows the continuing problem of other STDs. Vaccines are not available; prevention focuses on abstinence, monogamy, and safe sex.

Syphilis

Girolamo Fracastoro, a poet and physician, wrote a poem in 1530 about a shepherd boy named Syphylis who unknowingly offended Apollo, the sun god. Apollo vented his wrath on the boy by inflicting him with a "loathsome" disease. Ultimately, this new disease became known as *syphilis sive morbus gallicus.* The causative agent of **syphilis** is the spirochete-shaped bacterium *Treponema pallidum* (FIGURE 9.15). It frequently coinfects with pathogens that cause other STDs, including AIDS. Humans are the only known reservoirs. In 2010 in the United States there were over 45,834 cases. A total of 87% of primary and secondary syphilis cases were males. Higher number of cases were recorded in males in the 20- to 29-year-old and 45 to 54-year-old age groups. Men who have sex with men account for 67% of cases in the U.S. However, syphilis among men having sex with women continues to be a problem. The disease is sometimes referred to as "the great imitator" because, in its later stages, it mimics other diseases.

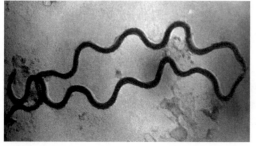

FIGURE 9.15 A photomicrograph of two spirochete-shaped *Treponema pallidum* bacteria. Courtesy of Joyce Ayers/CDC.

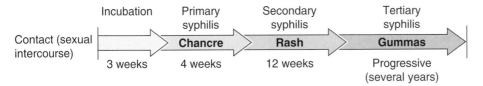

| Incubation | Primary syphilis | Secondary syphilis | Tertiary syphilis |

Contact (sexual intercourse) → **Chancre** → **Rash** → **Gummas**

3 weeks | 4 weeks | 12 weeks | Progressive (several years)

FIGURE 9.16 Stages and time line of syphilis. Times are approximate and subject to considerable individual variation.

Syphilis progresses through a series of three stages: **primary**, **secondary**, and **tertiary** (FIGURES 9.16 and 9.17). The primary stage is manifested by the appearance of painless **chancres** (sores) on the penis or on the cervix, which may be undetected. These chancres shed **treponemes** continuously. The chancres disappear in about four to six weeks and give false hope of recovery to those infected. In untreated individuals, over approximately five years, symptoms of a rash, particularly on the palms and soles, appear and disappear. This secondary stage is systemic, meaning that the bacteria multiply and spread throughout the body. About one-third of those untreated progress to tertiary syphilis, an advanced stage that develops over the next forty years, during which numerous organs and tissues, particularly those of the cardiovascular and nervous systems, show degenerative changes. **Neurosyphilis** is particularly disabling and can result in a shuffling walk (tabes), paralysis (paresis), and insanity. Before the antibiotic era individuals with tertiary syphilis constituted a significant portion of the population of mental institutions. A poem by an anonymous author describes the progression of syphilis (BOX 9.3). In the poem the term "**gummas**" refers to tumorlike lesions that develop during the tertiary stage; these gummas destroy nerve or skin tissue.

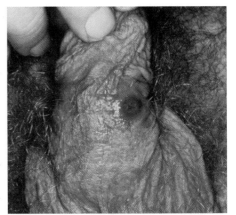

(a)

Although sexual contact is the usual means of transmission of syphilis, the organism can be passed in saliva and poses a threat to dentists, dental hygienists, and "deep kissers." Of particular concern is that infants can acquire **congenital syphilis** as the result of spirochetes that pass across the placenta from mother to baby. From 2009 to 2010 the number of cases decreased from 429 to 377 in the United States. These infants develop serious problems, including **saber shins**, a condition in which the shinbone develops abnormally.

Considering the high incidence of syphilis on a worldwide basis, it is not surprising that there have been intensive efforts to develop a vaccine; thus far, efforts have been futile. This has been particularly disappointing because a genetic map of *T. pallidum* was completed in1998 opening up new avenues of vaccine development. Campaigns aimed at educating the public to practice safe sex remain vital; the military still plays an important role in this effort.

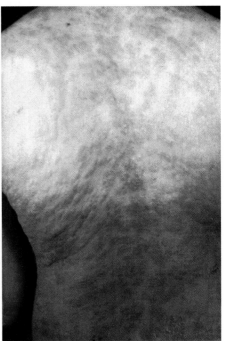

(b)

FIGURE 9.17 Characteristic lesions of syphilis. **(a)** Chancre of primary syphilis on penis. Courtesy of Dr. Gavin Hart/CDC and of Dr. N. J. Fiumara. **(b)** Lesions of secondary syphilis on the back; these lesions may occur on other body surfaces. Courtesy of Dr. Gavin Hart/CDC.

Gonorrhea

Gonorrhea, sometimes referred to as "clap" or "drip," is caused by *Neisseria gonorrhoeae,* a gram-negative diplococcus. Like syphilis, the only reservoir is the human. The number of cases of this disease increased from 301,174 in 2009 to 309,342 in 2010 in the United States, but this is an underestimated figure because only about half are reported to the CDC. In 2010 the rate of infection was 100.8 per 100,000 persons, with the highest rates in sexually active teenagers, young adults, and African Americans. *N. gonorrhoeae* may be transmitted via vaginal, oral, or anal sex. Pili enable the organisms to attach to cells lining the urethra,

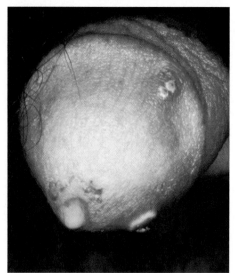

FIGURE 9.18 Male penis with discharge due to _Neisseria gonorrhoeae_ infection. Courtesy of the CDC.

allowing them to hang on during the passage of urine. Gonorrhea is the second largest STD in the United States (after chlamydia).

A major problem of controlling gonorrhea is that carriers often remain without symptoms after infection but continue to transmit the disease for over ten years. As many as 10% of males and 30% of females may not show symptoms. In males the disease is characterized by a pus-containing drip from the penis and burning during urination (FIGURE 9.18). In females the disease may remain hidden, and the cervix and urethra are the most common sites of infection. Furthermore, _N. gonorrhoeae_ is now on the list of "superbugs"; antibiotic resistance. From 2001 to 2006, the number of antibiotic resistance isolates from patients jumped from less than 1% to 13.8%, and subsequently to 23.5% in 2009.

Pelvic inflammatory disease occurs in about 50% of untreated females and is characterized by abdominal pain and, possibly, sterility. _N. gonorrhoeae_ can be transmitted into the eyes of newborns during delivery, causing corneal damage and possibly blindness, a condition referred to as **ophthalmia neonatorum**. Erythromycin or another antibiotic ointment is placed into the eyes of newborns as required by law in most states.

Chlamydia

Chlamydia, caused by _Chlamydia trachomatis_, a gram-negative coccobacillus, is the most common STD, as well as the most commonly reported nationally notifiable diease. Most people are not even aware of its existence; it is sometimes referred to as "the silent epidemic." There were 1,244,180 cases reported in 2009 in the United States. This disease is particularly prevalent in young adults and teenagers. Even those with the disease may be unaware that they are infected because approximately 70% of infected females and 30% of infected males lack symptoms. Females are particularly vulnerable to complications, including pelvic inflammatory disease (as in gonorrhea), abnormal pregnancy, and infertility. Males are subject to inflammation of the testicles and infertility. The silent nature of the infection is a major obstacle in prevention; education and alertness to this disease is a major preventive measure. The CDC recommends sexually active females 25 years old and younger need testing every year.

As in gonorrhea, newborns whose mothers are infected have a 50-50 chance of developing an eye-threatening condition; therefore, antibiotics are administered at birth. Chlamydia is the easiest STD to treat, and a single dose of antibiotic usually cures the disease in one week.

A silent epidemic of chlamydia presented a serious problem in the United Kingdom in 2005. It was the most common STD affecting over 10% of sexually active young men and women. In reality, the figures are probably very much higher, because many cases are not reported. The pharmaceutical company, Boots, with locations spread across the UK, inaugurated a program in 2005 to encourage young people to be screened and to seek treatment if necessary (FIGURE 9.19). More than 908,488 chlamydia tests were done in England from January 1 to September 20, 2011, of which 7.4% tested positive.

AUTHOR'S NOTE (RIK)
From 1956 to 1958 I was a young army officer stationed in a medical laboratory outside Tokyo. Gonorrhea was rampant among the military personnel, most of whom were eighteen to twenty years old. The diagnosis was simple. Look for the presence of a drip from the penis, and place a drop of the discharge on a glass slide. The slide was then Gram-stained and examined under a microscope for the presence of *Neisseria* organisms. The examination lasted less than five minutes, and, fortunately, penicillin was the cure. I took my turn in the VD clinic, behind a curtain, checking one young man after another.

FIGURE 9.19 Chlamydia is rampant in the United Kingdom, prompting Boots, a pharmaceutical company with retail stores throughout the UK, to sponsor an awareness program. Banners like this are prominently displayed in drug stores. Author's photo (RIK).

Lymphogranuloma Venereum

Lymphogranuloma venereum, an infection of the lymphatic system, is another STD caused by any one of three strains of *Chlamydia*; they are not the same strains that cause genital chlamydia. Lymphogranuloma is a chronic and long-term condition manifested by genital papules (bumps), ulcers, and swelling of lymph glands in the genital area. Those who practice anal sex are particularly prone to rectal ulcers. In untreated cases, the external genitalia appear larger leading to an elephantiasis appearance resulting from obstruction of the flow of lymph. Estimates are that a few hundred cases appear each year in the United States but the actual number is not known. It is more common in Central and South America. Lymphogranuloma, like genital chlamydia, is treatable by the use of antibiotics.

Contact Diseases (Other Than STDs)

There are a number of contact diseases, only three of which are discussed here (TABLE 9.10).

Peptic Ulcers

Students sometimes complain that the stress of exams, papers, (boring) lectures, and all other facets of academic life are "giving them **ulcers**." It is true that student life is stressful (so is a professor's!), but it won't give you ulcers. The

TABLE 9.10 Contact Diseases (Other Than STDs)[a]

Disease	Incubation Period	Causative Microbe	Transmission	Symptoms
Leprosy	3–6 yr	*Mycobacterium leprae*	Skin contact; air droplets	Development of lepromas; neurological damage
Staphylococcal skin infections	3–4 days	*Staphylococcus aureus*	Skin contact	Pimples, abscesses, and peeling on skin; systemic symptoms in toxic shock syndrome
Peptic ulcers	Unknown	*Helicobacter pylori*	Unknown, but probably person to person, possibly food and contaminated water	Burning or gnawing in epigastrium, particularly when stomach is empty; possibly bleeding; nausea, vomiting

[a] These diseases are treatable with antibiotic therapy; drug-resistant strains pose a problem in some cases.

popular myth that ulcers are directly associated with stress is no longer relevant. Take a look at TABLE 9.11, and you may be surprised to learn that those engaged in stressful occupations are no more at risk of getting ulcers than those in occupations considered nonstressful. Ulcers were also once thought to be the result of eating spicy foods. Treatment focused on hospitalization, bed rest, bland diet, and consuming large quantities of milk, cream, and antacids to coat the lining of the stomach and the duodenum (the first part of the small intestine). In some cases surgery was performed to remove the diseased area. But relief was only temporary.

In 1982 two Australian physicians, J. Robin Warren and Barry Marshall, claimed that *Helicobacter pylori* (a comma-shaped bacterium), not stress or diet, was the cause of peptic ulcers. This idea knocked the medical establishment off its feet. Marshall went to the extreme of drinking cultures of *H. pylori* to prove his claim to a skeptical medical community, and, sure enough, he developed ulcers. Finally, in 1994 a National Institutes of Health Consensus Development Conference affirmed a strong association between *H. pylori* and peptic ulcers and recommended antibiotics for treatment.

How does *H. pylori* survive in the harsh acidic environment of the stomach and in the duodenum—environments hostile to most microbes? The answer is a remarkable example of biological adaptation. The bacterium produces the enzyme **urease**, which breaks down the compound **urea**, a metabolic product of the body, into carbon dioxide and ammonia, an alkaline compound that neutralizes the acidity. This clever survival strategy protects *H. pylori*.

The mechanism of transmission of *H. pylori* is not clear, nor is there an explanation why some persons infected with the bacterium develop ulcers but others do not. Direct person-to-person contact appears to be the most plausible route;

TABLE 9.11 Stress and Ulcers: A Myth[a,b]

Ten Most Stressful Jobs in the United States in 2011	Ten Least Stressful Jobs
1. Commercial Airline Pilot	1. Audiologist
2. Public Relations Executive	2. Dietitian
3. Senior Corporate Executive	3. Software Engineer
4. Photojournalist	4. Computer Programmer
5. Newscaster	5. Dental Hygienist
6. Advertising Account Executive	6. Speech Pathologist
7. Architect	7. Philosopher
8. Stockbroker	8. Mathematician
9. Emergency Medical Technician	9. Occupational Therapist
10. Real Estate Agent	10. Chiropractor

[a] Do you believe that stressful jobs lead to ulcers? Which of these people listed above are most likely to have an ulcer? The answer may surprise you. All the workers on this list are just as likely to get an ulcer as any others you can imagine. An airline pilot's stomach is no more likely to be riddled with ulcers from the stress of dealing with stressed passengers than a dietician's stomach. Although stress and diet can irritate an ulcer, they do not cause it. Ulcers are caused by the bacterium *H. pylori* and can be cured with a 1- or 2-week course of antibiotics, even in people who have had ulcers for years.

[b] According to CareerCast.com.

humans are the primary reservoir. There is also the possibility that food and water are involved in transmission.

The diagnosis of peptic ulcers is based on several methods, including the presence of antibodies specific for *H. pylori* and the culturing of tissue, obtained by biopsy, to demonstrate the presence of the bacterium. A test, called the **breath test,** is about 94% to 98% accurate and is based on the organism's production of urease. For this test the patient drinks a preparation containing isotopically labeled urea. If present, *H. pylori* breaks down the urea, resulting in the formation of carbon dioxide, which is exhaled. Measurement of the labeled carbon dioxide in the breath determines the presence or absence of *H. pylori*.

H. pylori infection can be cured in two to three weeks with appropriate antibiotics and other medications. The recurrence rate after antibiotic treatment is about 6%, compared with 80% when only antacids are used. The bottom line is that ulcers are a bacterial infection and not caused by stress.

H. pylori infection is common throughout the world. It is estimated that 70% of the population in developing countries and 30% to 40% of the population of the United States and other developed nations are infected with *H. pylori* but only about 10% develop ulcers during their lifetime. Infection typically occurs during childhood and may persist lifelong unless treated. Studies have indicated that long-term *H. pylori* infection is associated with the development of gastric cancer, the second most common cancer worldwide.

Leprosy

Among all those diseases carrying a social stigma, **leprosy** is at the top of the list (BOX 9.4). Historical accounts of "affliction" point to leprosy; through the ages the

BOX 9.4   Politics of the Leprosy Bacillus

In August 2001 I (RIK) visited the Leprosy Museum at St. Jørgens Hospital in Bergen, Norway. This might appear to be an unusual activity; most tourists are attracted to the other museums and cultural sites that Norway has to offer. (In fact, not to my surprise, I was the only tourist there!) The hospital was founded around 1411 and stayed in operation continuously until 1946, providing a home for thousands of Norwegian lepers. Leprosy was common in the western parts of Norway into the late nineteenth century.

While visiting the museum I met Sigurd Sandmo, its curator, and spent several hours with him over a two-day period. Sigurd is a medical historian with a vast amount of knowledge. I was so impressed with this young man that I invited him to write the following piece about leprosy:

"Today Norway is a fully developed country and among the wealthiest nations in the world. But 150 years ago one would consider Norway to be a poor outpost in Europe, with many of the problems to be found only in developing countries today. One of the most conspicuous problems was leprosy, a disease that was looked upon with great interest and concern by Norway's physicians and authorities. In the second half of the nineteenth century, the city of Bergen had the highest concentration of leprosy patients in Europe, and the city was an international center for leprosy research.

For most other European countries, where the disease had ceased to be a problem around 1500 leprosy had during the nineteenth century more or less become a subject for missionaries. In the reports from work among African and Asian lepers, the Levitical meaning of the disease was now reproduced. An almost forgotten medieval disease had renewed its religious and racial actuality. In Europe only Norway and

Iceland experienced a second bloom of leprosy in the nineteenth century. The Norwegian distribution of leprosy was rather local, and the disease was mainly found among poor fishermen and peasants in the western parts of the country. With an incidence of more than 3% of the population in some districts, this was a shame for the young Norwegian state. In the 1870s a Swedish physician referred to the problem as "the waves of shame on the shores of Norway."

The national authorities offered the medical circles in Bergen both economic freedom and political influence to have the problem solved. In return they expected visible results. From our modern point of view, we may see several scientific breakthroughs. In 1847 Daniel C. Danielssen published the first symptomatology of leprosy, which represents the birth of modern leprosy research. In 1856 Ove G. Hoegh founded the Norwegian Leprosy registry, probably the first national disease registry in the world. But the value of these events was hard to recognize at the time. Much more spectacular was Armauer Hansen's discovery of the leprosy bacillus in 1873. Hansen's publication from 1874, Preliminary Contributions to the Characteristics of Leprosy, is the earliest description of a microorganism as the cause of a chronic disease.

Hansen's discovery represented the kind of breakthrough the public health authorities were waiting for. It was the microscope against the Old Testament and the scientist's rationalism against the Levitical myth. And ever since, health workers have used the story of the discovery to fight the myth where it has survived. Hansen has become a symbol of dignity, humanism, and a scientific approach toward patients who are still suffering from Hansen's disease. When we come across

disease conjured up dreadful images of rejection and exclusion from society, disfiguring skin lesions, and fingers and toes falling off (FIGURE 9.20a). People suffering from the disease were required to carry and ring bells and to shout "unclean" as they approached others (Figure 9.20b). Fear in the minds of healthy individuals led to terrible tales of cruelty. The story of a Belgian priest, Father Damien, is one of great dedication and compassion. In 1870 Damien established a leper colony on the island of Molokai in Hawaii as a refuge for those with the disease. He spent his life there and ultimately died of leprosy.

Hansen stamps from Thailand or a Hansen monument in the Vietnamese jungle, we easily recognize the impact Hansen's discovery once had. The authorities, the mob, the church, the peasants, and the patients could all see how Hansen's scholarly approach challenged the 2,000-year-old stigma.

However, Hansen's discovery had immediate political implications too. At the time, Norway did not have any leprosy legislation. Admission to all hospitals was voluntary, as the common opinion was that the disease was either inheritable or a nonspecific condition caused by bad living conditions. During the first years after the discovery, Hansen's theory of a contagium vivum was met with a certain amount of skepticism from his colleagues, both in Norway and abroad. When Hansen presented his proposal for a new act requiring isolation of leprosy patients in Norway, many of his colleagues felt that this was an inhumane law, branding the patients as criminals. Hansen defended himself as a pragmatic scientist: How can we act humanely toward persons who have a contagious disease? His opponents read Hansen's contributions to the discussion with regret. However, the law was passed in 1885 and served as a model and inspiration for leprosy legislation and public action in disease colonies in several other countries, such as Greece, Hawaii, and Japan. Today the remains of colonies in Spinalonga, Greece, and in Molokai are sad sights for visitors. In Japan surviving patients now receive compensation for their sufferings caused by harsh legislation.

In the discussions concerning the public health work and legislation in Norway in the 1870s and 1880s, we can see how Hansen's work gradually became politicized. The discovery itself was internationally accepted during the 1880s, but many of Hansen's opponents and colleagues saw that the political answer to the leper question was complicated and that the new legislation in a way revitalized the old stigma. In one of Norway's medical journals, the discussion went on for several years. After years of heated debate with irritated colleagues, Hansen might have considered whether his fundamentalistic positivism and rationalism were, after all, suitable for discussing public health work.

Several elements from the discussion on leprosy legislation in the 1880s were repeated with regard to another disease 100 years later in Norway, as in most other countries; the AIDS discussions of the 1980s were surprisingly similar. Again, the possibility of establishing disease colonies was discussed and society's rights and responsibilities with respect to infected individuals were presented as a question of yes or no. And, again, the evidence of contagion was used by different groups with different interests to justify their political and religious opinions. Once again, scientific discoveries of microorganisms turned out to be powerful but unpredictable political weapons.

We often think of biological and medical discoveries as nonpolitical statements. And when we look at the single participant, we are usually right. But like the discovery of the leprosy bacillus, scientific breakthroughs are usually caused by priority given by political actors, and scientific results will often take on a political dimension once they leave the laboratory. Both the discovery of the leprosy bacillus and the discovery of human immunodeficiency virus remind us how vulnerable and risky this dimension can be."

Source: Sigurd Sandmo, curator, Leprosy Museum, St. Jørgens Hospital, Bergen, Norway.

Today, the term "leper," because of its stigma, is not socially acceptable, and the disease is now known as **Hansen's disease**, after Gerhard Henrik Armauer Hansen (Figure 9.20c), the physician who first observed *Mycobacterium leprae*. *M. leprae* is the causative agent of Hansen's disease; note that it belongs to the same genus, *Mycobacterium*, as the tubercle bacillus. The disease is transmitted by skin contact and respiratory droplets over a long period of time; it is not considered to be highly infectious. The incubation time is long, and it takes about three to six years of close contact with an individual with the disease for

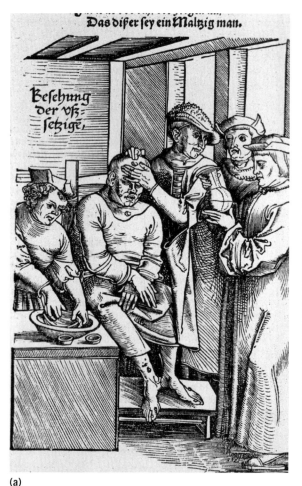

(a)

(b)

FIGURE 9.20 **(a)** A patient with leprosy, in this medieval drawing, is examined by a physician, who is consulting other physicians. Courtesy of National Library of Medicine. **(b)** A historic painting from the 14th century showing a female leper with missing limbs ringing a bell to announce her presence. © The British Library/age footstock.

it to be acquired. Children appear to be more susceptible. The disease is characterized by a variety of physical manifestations (Figures 9.20 d and e), including disfiguring tumorlike skin lesions (**lepromas**) and neurological damage to the cooler peripheral areas of the body, such as the hands, feet, face, and earlobes. This may lead to curling of the fingers (described as "claw hand"), thickening of the earlobes, collapse of the nose, and possibly blindness. The individual may lose the ability to perceive pain in the fingers and toes, and accidental burns may occur, resulting in serious deformities. Fortunately, drugs are now available to arrest the disease, and patients with Hansen's disease can look forward to a normal life.

It is estimated that there are over 500 cases in New York City and well over 6,500 cases in the United States. The cases are typically new immigrants from Central and South America and Asia. The national Hansen's Disease Program in Baton Rouge, Louisiana is the United States' main care and research facility and operates under the auspices of the U.S. Department of Human Health and Services. There were once more than 400 residents at the original facility in Carville, Louisiana, which dated to the 1890s. Closing the hospital caused great anxiety among patients who had spent their lives there. These residents were referred to

(c)

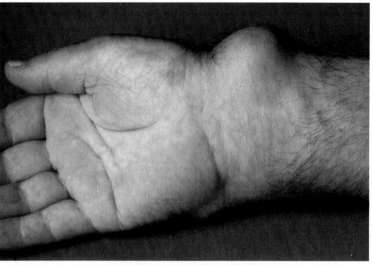

(d)

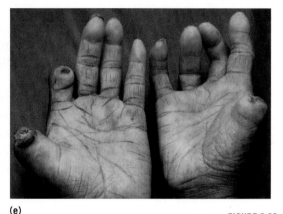

(e)

specialized outpatient facilities throughout the country and were offered financial subsidies by the government. The leprosy center in Molokai has also been closed. Persons having completed or under appropriate antibiotic therapy are considered free of active infection. The WHO reports that the global registered prevalence of Hansen's disease early in 2010 was close to 192,246 cases and the number of new cases annually continues to decline.

Staphylococcal Infections

Bacteria of the genus *Staphylococcus,* frequently called "staph," a gram-positive group of cocci, of which there are over 30 types that cause a multitude of infections, are normal flora of the skin, mouth, nose, and throat; the skin is the largest organ in the human body. Usually, these organisms present no problem. However, when the skin is broken, as can result from a wound, burn, or other circumstance, its normal integrity as a barrier is lowered making possible the invasion of opportunistic strains of staphylococci. Certain strains of *S. aureus* are the most frequent staphylococci involved in infections. Nosocomial infections are a major

AUTHOR'S NOTE (RIK)

I visited the hospital for Hansen's disease in Carville, Louisiana in 1962. The atmosphere in this modern hospital, located in a parklike setting, was remarkably upbeat, as was the attitude of the people with the disease. I chose not to include some of the cases with advanced disfigurement.

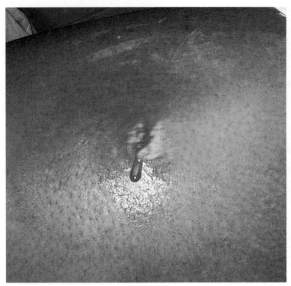

FIGURE 9.21 Staphylococcal infection. Note discharge of pus. Courtesy of Bruno Coignard, MD, and Jeff Hageman, MHS/CDC.

FIGURE 9.22 NCAA educational poster on community-acquired MRSA infections. Courtesy of the CDC.

Practice good hygiene: Do not share personal items, such as towels or razors. Wash your hands frequently. Shower immediately after every practice and game. Use clean towels each time you shower. Launder clothes and towels after each use. Your health matters.

problem, and methicillin-resistant staphylococci are the leading cause of nosocomial infections in the United States.

Staphylococci are frequent causes of localized skin infections, occurring as pimples, abscesses, and inflamed lesions filled with a core of pus (FIGURE 9.21). Abscesses can progress to produce **boils**, which can develop into **carbuncles**. Carbuncles are larger and deeper lesions and may, in fact, reach baseball size. These lesions are extremely painful and are dangerous because they can progress into **systemic** (bloodborne and widespread) **infections** throughout the body. A relatively common manifestation of staphylococcal (and streptococcal) infection is **impetigo**. This is a superficial infection of the skin that usually occurs around the mouth in the form of blisters that ooze a yellowish liquid. More frequently, impetigo occurs in very young children, particularly after a runny nose, which sets up irritation in the surrounding tissue. Impetigo is annoying but not particularly serious, other than the fact that it is spread easily from child to child. Impetigo can also be caused by streptococci. **Scalded skin syndrome** is also caused by staphylococci and tends to be more prevalent in children with infection of the stem of the umbilical cord. In these cases the skin can become blistery as a result of the production of an **exfoliative toxin** that peels away the skin to expose a red layer.

Nosocomial (hospital acquired infections) are a major problem, and methicillin-resistant *Staphylococcus aureus* (MRSA) is the leading cause of nosocomial infections in the United States. Everyone is at risk. MRSA is widespread and is no longer only hospital acquired. MRSA infections are defined as **community acquired** (CA-MRSA) if the person infected with MRSA was not hospitalized within two years before the date of the MRSA infection. These infections are usually skin infections that appear in the form of boils or sores. About 30% of the people carry MRSA in their noses, but show no signs and symptoms.

Community-acquired MRSA infections are associated with locations such as athletic facilities, dormitories, military barracks, correctional facilities, and day care centers. Media has focused attention on community-acquired MRSA strains that cause outbreaks among professional, college, and high school athletes participating in highly physical sports (person-to-person contact) such as football, wrestling, rugby, and soccer. Athletes with preexisting cuts and abrasions that shared bars of soap or towels with an MRSA-infected person or carrier were associated with MRSA infections. This is another good reason not to share hygiene or personal products! Today athletic locker rooms and health clubs are filled with posters educating the public about MRSA and personal hygiene, including the sharing of personal items and equipment (FIGURE 9.22).

Toxic shock syndrome came to light in the late 1970s. A major outbreak occurred in 1980 and caused considerable fear, particularly among menstruating females who used a certain brand of tampon. The outbreak led to a recall of the tampons in question. Some strains of *S. aureus* secrete an exotoxin that produces high fever, nausea, vomiting, peeling of the skin (particularly on the palms and the soles), and a dangerous drop in blood pressure that leads to life-threatening shock. The infection carries about a 3% fatality rate. Toxic shock syndrome can also occur in men and in nonmenstruating females and may be associated with surgical wound infections. It can also be caused by streptococci.

Current statistics regarding the incidence of toxic shock syndrome are not available, and an active surveillance is overdue, but there have been no reported outbreaks. About 1,000 cases were reported in 1981, but by 1997 the incidence had been reduced to about 100 cases. However, for reasons that are not known the CDC reported an 18% increase in the caseload from 2000 to 2004.

■ Soilborne Diseases

Anthrax, tetanus, and leptospirosis are a few of the most common soilborne bacterial diseases (TABLE 9.12). Anthrax and tetanus are spore-forming bacilli and survive in soil for many years. Bacterial spores are extremely resistant to many environmental stresses.

■ Anthrax

Anthrax, a potentially deadly disease caused by *Bacillus anthracis*, a gram-positive bacillus, is considered by security experts as the most probable weapon for biological warfare because their spores can be readily spread by missiles and bombs (FIGURE 9.23).

TABLE 9.12 Soilborne Bacterial Diseases[a]

Disease	Incubation Period	Causative Microbe	Transmission	Symptoms
Anthrax	1–15 days	*Bacillus anthracis*	Contact with endospores in soil and in occupations involving handling of wool, hides, and meats	Cutaneous form produces skin lesion, headache, nausea, fever
Tetanus	4 days to several weeks	*Clostridium tetani*	Contact with spores from soil, animal bites, gunshot wounds	Lockjaw, muscle stiffness, and spasms due to production of tetanospasmin
Leptospirosis	7–13 days	*Leptospira interrogans*	Urine, soil, or water contaminated with urine from dogs, cats, sheep, rats, mice	Flulike with possible complications in liver and kidney

[a]These diseases are treatable with antibiotic therapy; drug-resistant strains pose a problem in some cases.

FIGURE 9.23 **(a)** *B. anthracis* bacteria (intertwined chains of bacilli) and refractile endospores. Courtesy of Larry Stauffer, Oregon State Public Health Laboratory/CDC. **(b)** *B. anthracis* spread on plate. Note mucoid characteristic of growth due to capsule formation. Courtesy of the CDC.

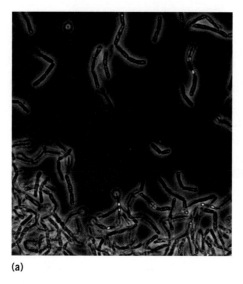

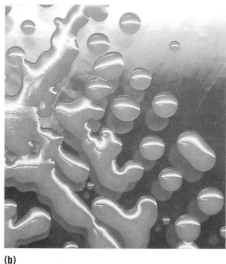

(a)   (b)

Anthrax is primarily found in large, warm-blooded animals, mostly sheep, cattle, and goats, and is acquired by the ingestion of spores during grazing. Infection can also occur in humans and is most prevalent in agricultural regions of the world. There are three varieties of anthrax, which vary by the route of transmission and subsequent symptoms.

Inhalation anthrax is an occupational hazard for humans exposed to contaminated dead animals and animal parts (e.g., tanners and sheep shearers), who, in the course of their work, are most likely to inhale spores; this condition is also called woolsorter's disease. Inhalation anthrax is the most severe form of the disease and is the greatest threat with the deadliest consequences in terms of biological warfare. The incubation period is one to six days. Early symptoms are coldlike and progress to severe breathing problems within several days, followed by death one to two days later.

A second form of anthrax is **cutaneous anthrax**, also acquired by contact with anthrax bacilli or spores via wool, hides, leather, or hair products (FIGURE 9.24).

FIGURE 9.24 A cutaneous anthrax lesion. Courtesy of the CDC.

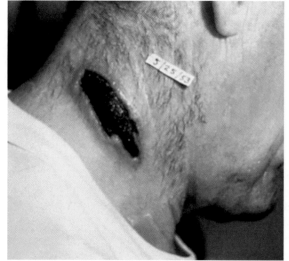

About 20% of untreated cases result in death; death is rare with appropriate antibiotic therapy. A case of cutaneous anthrax was identified in a musician and a family member in Connecticut, presumably from animal skins used to make drums.

A third form, **gastrointestinal anthrax**, results from the ingestion of inadequately cooked meat contaminated with *B. anthracis,* leading to acute inflammation of the gastrointestinal tract characterized by abdominal pain, vomiting of blood, and severe diarrhea. Approximately 25% to 60% of untreated cases result in death.

Antibiotics are effective against all three forms of anthrax, but early intervention is necessary. A vaccine is available for animal workers, laboratory or remediation workers because their occupation classifies them as a high-risk group. The use of anthrax vaccine in military personnel is a controversial issue that is being addressed. Person-to-person transmission does not occur.

Tetanus

Tetanus, also called **lockjaw**, conjures up terrible images and is usually associated with stepping on a rusty nail. Tetanus is a noncommunicable disease since it does not require human or animal-to-human contact. It is acquired by exposure to spores of *Clostridium tetani,* a gram-positive bacillus. The tetanus organism produces **tetanospasmin**, a neurotoxin considered to be the second most deadly bacterial toxin; estimates are that as little as 100 micrograms can kill an adult. (The botulinum toxin is considered the most potent bacterial toxin.)

Tetanus bacilli are present worldwide and are abundant in soil, manure, and dust. They are normal inhabitants of the intestinal tracts in horses and cattle and, more rarely, in humans. The fact that these organisms produce spores and the hardiness of the spores explain their longevity and abundance in the soil. Tetanus develops when spores gain access into the body through wounds. These spores then germinate into bacilli and multiply in deep wounds that do not tend to bleed a lot (puncture wounds). Hence, in the "stepping on a rusty nail" scenario, the nail produces a relatively deep puncture wound, which in itself is of little concern; "rusty" implies that the nail has been in the soil for a long time. The crevices in the nail provide for the adherence of minute amounts of soil that potentially contain large numbers of spores. Gunshot wounds, animal bites, knife wounds, and wounds caused by the prongs of a pitchfork all serve as grisly mechanisms of puncture wounds, possibly allowing the entrance of tetanus spores. The incubation period is anywhere from three days to three weeks but is usually about eight days.

C. tetani, unlike most bacteria dealt with in this chapter, has no invasive ability. During growth, tetanus toxin is produced that interferes with the relaxation phase of muscle contraction, resulting in uncontrollable muscular contraction. Stiffness in the jaw (lockjaw) is an early symptom resulting from the contraction of facial muscles; as more toxin is produced the neurotoxin continues to spread, leading to contraction in other muscles, particularly in the limbs, stomach, and neck. Tetanus survivors report excruciating pain resulting from the spasms of muscle contractions, which can be strong enough to break bones. The extreme contractions in the back and rib muscles cause the body to arch severely to the extent that only the victim's head and heel are in contact with the surface, a position referred to as **opisthotonos** (FIGURE 9.25).

Death occurs by suffocation as a result of contraction of the muscles involved in breathing. Imagine taking a very deep breath and not being able to blow out to take another breath. In the United States schoolchildren are required to be immunized against tetanus, as are children in day care facilities. The number of tetanus cases in the United States is now about 50 to 100 cases per year, a sharp decline from the 500 cases per year in the 1940s, and 5 deaths. The occurrence of tetanus in developed countries is primarily due to absent or inadequate immunization. The tetanus vaccine does not provide lifelong protection and needs to be repeated at about ten-year intervals. Several vaccines are available to prevent tetanus in children, adolescents, and adults including Tdap (against tetanus, diphtheria and pertussis) and Td (against tetanus and diphtheria).

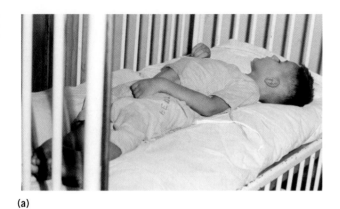

(a)

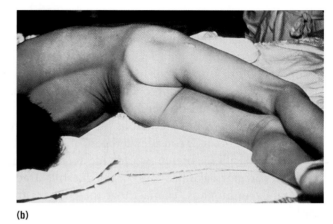

(b)

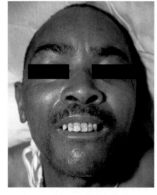

(c)

FIGURE 9.25 Tetanus. The tetanus toxin causes muscle contraction and rigidity. (**a** and **b**) Opisthotonos (a. Author's photo (RIK); b. Courtesy of CDC.) (**c**) Risus sardonicus (sardonic grin) or facial tetany. Courtesy of Dr. Thomas F. Sellers, Emory University/CDC.

AUTHOR'S NOTE (RIK)

The photograph in Figure 9.25a was taken in Central America. The boy was approximately ten years old and was brought into the hospital near death; there was no possible intervention. He died several hours later, a horrible outcome from, presumably, a minor wound. Immunization would have prevented this.

Tragically, **neonatal (newborn) tetanus** (NT) is a common manifestation of tetanus in the first month of life in developing countries, and the course of this disease follows the same pattern as tetanus. Symptoms occur within two weeks after birth; the infant fails to suck properly on the mother's breast, becomes irritable, and has convulsions. The disease is prevalent in the poorest nations of the world and occurs when an infant is delivered under unsanitary conditions to a nonimmunized mother.

Imagine a scenario in which a nonimmunized woman delivers a baby on the dirt floor of a small hut in a remote village in some impoverished area. An untrained birth attendant, unaware of the need for sanitary conditions, cuts the umbilical cord with an unclean razor blade, thereby introducing tetanus. In some traditions the baby's umbilical stump is treated with cow dung, ash, mustard oil, or other unsterilized "folk" products.

Neonatal tetanus remains a killer in the developing world. In 2002 about 180,000 newborns died of the disease, but only one of the deaths was in the United States. Neonatal tetanus is rare in countries with tetanus vaccine programs. Prevention is centered on vaccination of all women of childbearing age (because their protection is passed to their newborns), improved delivery and postdelivery practices, and education.

Neonatal tetanus is a severe, often fatal disease caused by a toxin of *Clostridium tetani*. In March 1998, an NT case was reported by the Missoula City-County Health Department (MCCHD) and the Montana Department of Health and Human Services (MDHHS) in an unvaccinated mother after the application of nonsterile clay to the umbilical cord of a nine-day-old newborn. Medical history revealed that the mother had not been vaccinated. For home umbilical cord care, a "Healthy and Beauty Clay" powder provided by a midwife was used. The preparation was an unsterilized bentonite clay product from (of all places) Death Valley, California. The parents reported a 10-hour history of an inability of the baby to nurse and difficulty in opening her jaw. They also noticed a foul-smelling discharge from the baby's umbilical cord during the preceding one to two days.

Culture from the umbilical cord grew several anaerobic (*Clostridium perfringens, C. sporogenes*) and aerobic (*Staphylococcus* and *Bacillus*) bacterial species.

The baby was treated with tetanus immune globulin (tetanus antibodies) and penicillin and at one point required a respirator. She was ultimately discharged with no apparent neurological sequelae.

Leptospirosis

Leptospirosis, also known as **swamp fever**, is caused by the spirochete *Leptospira interrogans.* These spirochetes are harbored in nonhuman hosts, including dogs, rodents, and a variety of wild animals, and is spread through their urine. Rats appear to be the most significant source of disease, probably because there is greater opportunity for contact with rat urine than with that of other animals. In fact, there is concern that inner-city residents may be at particular risk for leptospirosis, along with farmers, sewer workers, and workers in other occupations that may be exposed to infected rat urine. Additionally, bodies of water, riverbanks, and vegetation may become contaminated as a result of urine runoff from surrounding soil. The spirochetes penetrate the human skin, enter the bloodstream, and rapidly invade virtually all organs. The infection can also be acquired by swallowing water from sources where rats may have urinated. It is not transmitted from human to human. Outbreaks among triathlon participants have been reported. A triathlon is a multisport involving the completion of three continuous endurance events, typically swimming, biking, and running components. Swimming in natural waters is the first leg of the race. (Neither author has ever considered participating in a triathlon but is accustomed to marathon writing as in preparation of this text.)

Leptospirosis can be a serious and potentially fatal disease, with symptoms including jaundice, fever, headache, nausea, and chills, and may require hospitalization. Diagnosis is based on blood tests. It is treatable with antibiotics such as penicillin and doxycycline, which should be given early in the course of disease.

Leptospirosis is most common in tropical countries. It is the most widespread zoonotic illness and has been classified as an emerging and infectious disease by the WHO and CDC. Approximately 100 to 200 cases per year are reported in the United States, with about 50% occurring in Hawaii. The actual number is considered to be much higher because many cases are not reported. The risk of acquiring leptospirosis can be greatly reduced by not swimming or wading in water that might be contaminated with animal urine and staying away from infected animals.

Arthropodborne Diseases

Arthropods play a major role in the transmission of microbial diseases. Vector control is vitally important in breaking the complicated transmission cycle of these diseases. Numerous species of fleas, mosquitoes, flies, ticks, and lice are responsible for a variety of devastating diseases that have plagued humankind for thousands of years and have influenced the course of civilization. TABLE 9.13 summarizes the arthropodborne bacterial diseases presented in this section.

TABLE 9.13 Arthropodborne Bacterial Diseases[a]

Disease	Incubation Period	Causative Microbe	Transmission	Symptoms
Plague		*Yersinia pestis*	Fleas	Hemorrhages under skin (Black Death), fever, buboes
Bubonic	2–6 days			
Pneumonic	2–3 days			Pneumonia
Septicemic	1–6 days			Pneumonia
Lyme disease	3 days to 1 month	*Borrelia burgdorferi*	Ticks	Possibly expanding bull's-eye rash, flulike symptoms (headache, fatigue, fever, muscle pain); in later stages, arthritislike neurological impairment, heart inflammation
Relapsing fever		*Borrelia recurrentis*		High fever, headaches, muscle pain, weakness
Endemic	5–10 days		Ticks	
Epidemic	5–10 days		Body lice	
Rocky Mountain spotted fever	2–6 days	*Rickettsia rickettsii*	Ticks	Rash that begins on days 2–4, appears noticeably on wrists and ankles and then spreads upward
Epidemic typhus fever	1–2 weeks	*Rickettsia prowazekii*	Body lice	Severe headache, high fever, rash, on about the 5th day muscle pain, and chills
Endemic typhus fever	1–2 weeks	*Rickettsia typhi*	Fleas	Severe headache, high fever, cough, nausea, vomiting
Ehrlichiosis	7–10 days	*Ehrlichia chaffeensis* and *E. canis*	Ticks	Like a "bad flu" with fever, chills, headache, muscle pain, nausea; rash only rarely present

[a] These diseases are treatable with antibiotic therapy; drug-resistant strains pose a problem in some cases.

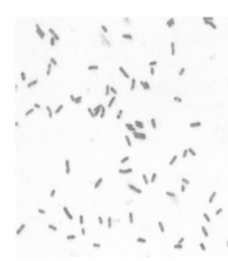

FIGURE 9.26 *Yersinia pestis* as single bacilli. Courtesy of Larry Stauffer, Oregon State Public Health Laboratory/CDC.

Plague

Plague, also known as the **Black Death**, conjures up terrible images of disease throughout the course of history. The term "plague" is frequently used in a general sense to describe an explosive outbreak with a high death rate. More strictly, however, plague refers to "the plague," caused by the gram-negative bacillus *Yersinia pestis,* one of the most virulent bacteria known (FIGURE 9.26). All it takes is a single bacillus to establish infection.

Medieval Europe was beleaguered with *Y. pestis* from 1347 to 1351, when homes were inhabited by fleas carrying the plague bacillus. In the short span of only four years one-third of the population, estimated at twenty-five million people, were ravaged and died of the Black Death. The dead were piled up and fed upon by rats, resulting in even more infected rats whose fleas spread the disease as the plague swept through populations. At the peak of the epidemic in the 1300s as many as 800 people died each day in Paris, cutting the population in half. It was once believed that carrying sweet-smelling herbs and flowers and holding them to the nose protected against the poisonous vapor of plague. FIGURE 9.27 depicts a person in protective clothing; the bird-like mask contains the plant material. *Y. pestis* was used as a weapon of biological warfare by the Japanese army during World War II. A major plague epidemic that killed over twelve million people began in Asia in 1890 and was carried to San Francisco by rat-infested ships at the turn of the nineteenth century.

Plague is considered a reemerging infection because of its increase throughout the world. Plague resurfaced in 1994 in Surat, India, about 400 miles from Bombay, and resulted in the infection of several hundred people and about 50 deaths. In the fourteenth century the Black Death required seventeen years to cross the trade routes from China to Iceland. Today it takes only about ten hours to fly from Bombay to London's Heathrow Airport. This is less than the incubation time for an infected person to make the trip without any symptoms of illness. If treatment is started early enough, the patient should be isolated and contacts should be notified. Antibiotic therapy, typically streptomycin or gentamycin, should begin as soon as possible.

In the United States sporadic cases occur, particularly in the Southwest, because of contact with infected rodents. For more than three decades before 1965 only a few cases per year occurred in the United States, but this rate of incidence gradually increased to 30 to 40 cases in the mid-1980s before decreasing again (though not to pre-1965 levels). According to the CDC 107 cases (or an average of 10 to 15 cases per year) of plague in humans were reported during 1990 to 2005. In 2006, 13 cases of human plague occurred with 7 deaths in New Mexico, 3 in Colorado, 2 in California, and 1 in Texas. The 2006 increase was the highest number of cases in 12 years and is thought to be climate associated, leading to increased reproduction and survival rates among rodents and fleas. The WHO reports 1,000 to 3,000 cases of plague every year.

Plague is a zoonosis; over 200 species of mammals, primarily rodents (including gophers, ground squirrels, mice, and wild rats), serve as reservoirs. Rat fleas are the usual vectors, and transmission occurs primarily from infected rats to other animals, including humans, as a result of contact with a dead plague-infected animal or, more usually, being bitten by an infected flea. In **bubonic plague**, the best-known form of the disease, the bacilli localize in the lymph nodes, particularly in the nodes of the groin, armpits, and neck causing the nodes to swell to the size of eggs. These enlarged nodes, known as **buboes**, are hard, red, and painful (FIGURE 9.28). The bacilli invade the bloodstream, liver, lungs, and other sites. Hemorrhages occur under the skin, and the dried blood turns black, hence the term "Black Death." The mortality rate in untreated cases is over 50%. Disease control, as in all zoonotic diseases, centers on interrupting the transmission between reservoirs and humans. Bubonic plague,

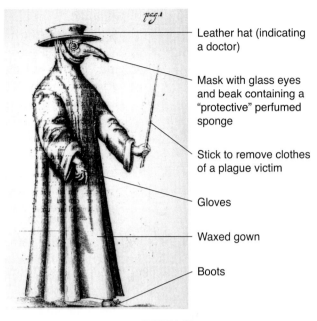

Leather hat (indicating a doctor)

Mask with glass eyes and beak containing a "protective" perfumed sponge

Stick to remove clothes of a plague victim

Gloves

Waxed gown

Boots

FIGURE 9.27 Costume that was once thought to be protective against plague. © National Library of Medicine.

FIGURE 9.28 Bubo in the armpit of a person with the plague. Courtesy of the CDC.

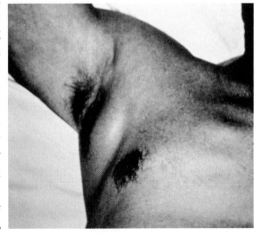

although it may be severe enough to kill its victims, is not normally infectious from person to person.

A second form of the plague, known as **pneumonic plague**, occurs when individuals with the bubonic form develop pneumonia and transmit the bacteria to others by coughing and through saliva. The bacteria invade the victims' lungs, which become filled with a frothy, bloody fluid. Pneumonic plague approaches 100% fatality without early detection and treatment. A vaccine is available but offers protection for only a few months and is only given to persons at high risk. **Septicemic plague** is a third form of the disease; it results from the spread of infection from the lungs to other parts of the body, but it can also be acquired by direct contact of contaminated hands, food, or objects with the mucous membranes of the nose or throat. This form of plague is considered to be 100% fatal.

Ironically, plague, a devastating disease with a long and terrible history, has given rise to the children's rhyme and playground game, "Ring Around the Rosie," that has been around since the 1800s and maybe even before. The actual wording varies by country and region, but it is characterized by children standing in a circle chanting the words. The American version is as follows:

> *Ring around the rosy,*
> *A pocketful of posies.*
> *Ashes, ashes,*
> *We all fall down!*

The significance of these words is open to interpretation, but many historians link the rhyme to the Great Plague. The first line refers to the rosy rash, an alleged symptom of the plague, posies refers to herbs carried as protection, ashes (or, in other versions, hush-a, a-shoo, or a-tishoo) refers to the victims' sneezing fits or to the burning of the bodies, and, finally, the last line is the outcome. A modern version might read as follows:

> *Symptoms of serious illness,*
> *Flowers to ward off the stench.*
> *We're burning the corpses,*
> *We all drop dead!*

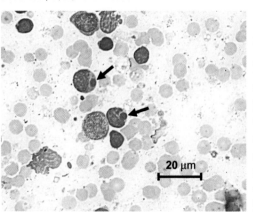

FIGURE 9.29 *Ehrlichia,* the rickettsia that cause ehrlichiosis, hide in white blood cells. The *Ehrlichia*-infected cells in the figure are stained purple. Reproduced from Williams CV, Van Steenhouse JL, Bradley JM, Hancock SI, Hegarty BC, Breitschwerdt EB. Naturally occurring *Ehrlichia chaffeensis* infection in two prosimian primate species: ring-tailed lemurs (*Lemur catta*) and ruffed lemurs (*Varecia variegata*). *Emerg Infect Dis.* 2002 December; 8(12): 1497–1500.

20 μm

■ Ehrlichiosis

Ehrlichiosis is an emerging tickborne infection, similar to Lyme disease, caused by a gram-negative intracellular bacillus (FIGURE 9.29). It was first reported in humans in 1986, although the disease has been known for years to exist in horses, dogs, cattle, and other mammals. Some biologists have suggested that the ticks may be moving from dogs to humans. There were approximately 200 cases in 2002, but in 2008, there were 961 cases occurring in most of the eastern half of the United States. The majority of cases were reported in Oklahoma, Missouri, and Arkansas. Deer hunters should be diligent to check for ticks and be aware of and respond to the symptoms.

Property owners should discourage deer from coming on to their property by using deer-resistant plants and eliminating bird feeders.

There are two forms of the disease, **human monocytic ehrlichiosis** and **human anaplasmosis**, formerly known as human granulocytic ehrlichiosis, both of which respond well to antibiotic therapy; about 5% of untreated cases result in death. Although the two appear to be the same disease, symptomatically they are caused by different species. Human monocytic ehrlichiosis is caused by *Ehrlichia chaffeensis*; the particular species that causes anaplasmosis is ascribed to *Anaplasma phagocytophilum*. The disease is characterized by fever, headache, general malaise, and, in some cases, a rash. The lone star tick is the vector of human monocytic ehrlichiosis, and the dog and deer tick (the same tick that transmits Lyme disease) transmit anaplasmosis. In 2008 the CDC cited a case in which an individual acquired anaplasmosis through blood transfusion.

Lyme Disease

The residents of the small New England town of Old Lyme, Connecticut are probably not thrilled that a zoonotic disease bears its name because it was first described in their community in 1975. This tickborne disease is now present throughout the United States, particularly in the Northeast, the upper Midwest, and the Pacific Coast, and its incidence has dramatically increased for at least the last 15 years in the United States (FIGURE 9.30). It is the most common vector-borne disease in the United States; there were close to 30,000 confirmed cases of Lyme disease in 2009 and another "probable" 8,500 cases. It is also found in Europe, Australia, the former Soviet states, China, and Japan.

The biology of Lyme disease is particularly complex, because five closely interrelated organisms must be present at appropriate times: (1) the spirochete, *Borrelia burgdorferi*; (2) the tick that serves as the vector (the particular species is

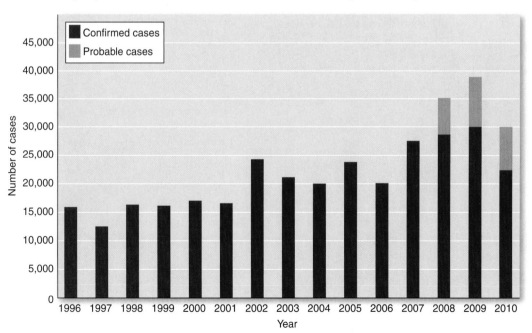

FIGURE 9.30 Number of reported cases of Lyme disease in the United States, 1996 to 2010. Lyme disease has been on the increase since it was first discovered in Old Lyme, Connecticut. Data from the CDC.

a function of the geographical area); (3) deer that serve as hosts for the maturation of tick eggs to the adult stage; (4) mice that allow the further development of the tick; and (5) humans or other final hosts. The cycle requires two years to complete. It is important to understand the life cycle so as to target mechanisms that break the transmission cycle (FIGURE 9.31).

Adult ticks feed and mate on large mammals, particularly deer in the fall and early spring. Female ticks become engorged with blood, fall off the deer, and lay their eggs on the ground; by summer the eggs hatch into **larvae**. The larvae feed on small mammals and birds in the summer and early fall and are inactive until the next spring, when they molt into **nymphs**. The nymphs feed on small rodents and other small mammals and birds in the late spring and summer and mature into adults in the fall, thereby completing the two-year cycle. The larval and nymph stages become infected with the spirochetes as they take a blood meal from their infected intermediate hosts, particularly white-footed mice; the spirochetes remain in the tick during the tick's developmental stages. The nymph is the major source of transmission to humans and other animals. Two or more days of feeding are required by the nymphs, during which the ticks transmit the spirochetes.

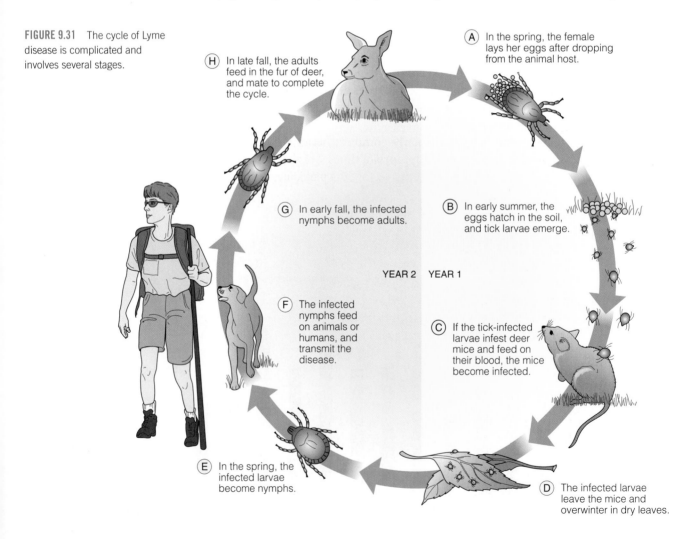

FIGURE 9.31 The cycle of Lyme disease is complicated and involves several stages.

H In late fall, the adults feed in the fur of deer, and mate to complete the cycle.

A In the spring, the female lays her eggs after dropping from the animal host.

G In early fall, the infected nymphs become adults.

B In early summer, the eggs hatch in the soil, and tick larvae emerge.

YEAR 2   YEAR 1

F The infected nymphs feed on animals or humans, and transmit the disease.

C If the tick-infected larvae infest deer mice and feed on their blood, the mice become infected.

E In the spring, the infected larvae become nymphs.

D The infected larvae leave the mice and overwinter in dry leaves.

The virtually undetectable poppy seed size of the nymphs allows them to escape being removed. Adults (FIGURE 9.32) are about the size of sesame seeds.

In its early stages the disease causes flulike symptoms, including fever, headache, and muscle and joint pain. Some individuals have a spreading "bull's-eye" skin rash (FIGURE 9.33). These are the "lucky" people because the vague flulike symptoms apply to many diseases, making the early diagnosis of Lyme disease difficult. Unfortunately, the rash may not occur until one month or more after the tick bite, or it may never occur; other symptoms, including arthritis and numbness, may occur years later. Diagnosis includes history of tick exposure, symptoms, and blood tests to look for the presence of the antibodies to the spirochetes. A not unusual story illustrating the problems of Lyme disease diagnosis is that of a man who lived with a faulty diagnosis of multiple sclerosis for nine years before being correctly diagnosed with Lyme disease. Infected individuals treated early with appropriate antibiotics usually recover completely and quickly. Those diagnosed with late and chronic Lyme disease may experience arthritis-like symptoms and neurological difficulties.

FIGURE 9.32 A deer tick, the vector of Lyme disease. Courtesy of James Gathany/CDC.

Intervention is based on avoiding tick-infested areas, particularly in May through July, when ticks are active and looking for a blood meal. Hunters should take note of this and exercise the proper preventive measures. Measures of protection include use of an insect repellent containing DEET, appropriate clothing, personal inspection for and removal of ticks after outdoor activity, and removal of leaves. Ticks live on the tips of grasses and shrubs, so it is important to clear brush and tall grasses near houses, particularly if there are young children in the area. No vaccine is currently available against Lyme disease. "Lymerix®," introduced in 1998, was the first Lyme disease vaccine to be approved by the U.S. FDA. Disappointingly, the product was withdrawn by the manufacturer in 2002, presumably due to poor sales, but also numerous outbreaks of severe arthritic-like conditions did surface. A research group at the U.S. Department of Energy's Brookhaven laboratory has already been issued a patent for a novel vaccine consisting of proteins that trigger an immune response again the *Borrelia* spirochete at two stages of its complicated life cycle. Remarkably, these proteins can be made in *E. coli* by genetic engineering.

Relapsing Fever

The two varieties of relapsing fever, caused by the spirochete *Borrelia recurrentis,* are **endemic relapsing fever** and **epidemic relapsing fever**. The spirochetes undergo antigenic switching, an important mechanism of virulence for avoiding host antibody defenses, to escape immunological detection. This phenomenon accounts for the recurrence of symptoms and is the basis for the name relapsing fever. Relapses occur approximately every two to four days; in the endemic form there may be three or four

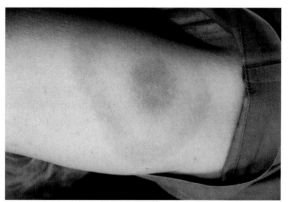

FIGURE 9.33 Bull's-eye rash of Lyme disease. About one month or more after the tick bite this rash may appear, but some patients never develop a rash. Courtesy of James Gathany/CDC.

recurrences, whereas in the epidemic form there are typically only two or three relapses. However, there have been reports of over ten relapses.

The endemic form is transmitted by ticks and occurs sporadically in most areas of the world. About twenty-five cases occur each year in the western half of the United States; the mortality rate is less than 5%. The epidemic form is the more serious and is transmitted directly from human to human by body lice that bite and cause itching; infection results from scratching the skin, thereby crushing the spirochete-infected lice into the skin. The epidemic variety has a mortality rate of approximately 30% to 70% in untreated cases. This disease currently exists in eastern and central Africa, China, and the Peruvian Andes and is associated with overcrowding (as in refugee camps), war, poverty, and other conditions that lead to social breakdown. The best forms of prevention are the delousing of people and their clothing and improved hygienic conditions; the implementation of these measures is dehumanizing and difficult and presents serious or even overwhelming difficulties in catastrophic conditions.

Rickettsial Diseases

Rickettsiae are atypical bacteria in the sense that, like viruses, they are obligate intracellular parasites and require eucaryotic host cells. They are carried by arthropods (with a single exception), including ticks, body lice, and fleas. The symptoms generally include a rash, which varies depending on the particular rickettsial infection, and flulike symptoms. Antibiotic therapy is effective, and further methods of control are directed at minimizing the contact between the vector and humans.

Rocky Mountain Spotted Fever

The tickborne disease **Rocky Mountain spotted fever** is misnamed, because it is more common in the southeastern part of the United States than in the Rocky Mountains, where it was first reported. It is the most severe tickborne rickettsial disease in the United States. The causative organism, *Rickettsia rickettsii,* is transferred by a tick feeding on a human or other mammalian host. As in all tickborne diseases, early removal of the tick is vital, because transmission occurs during the feeding cycle. Estimates are that the death rate is higher than 6% when treatment is delayed for more than three days after the symptoms first appear; if treatment is started earlier the death rate is 1.3%. Up to 20% of those infected but not treated die; the elderly population is particularly vulnerable. From 1981 to 1996 Rocky Mountain spotted fever was present in almost all states in the United States. Approximately 1,500 cases of this disease occurred in the United States in 2004.

Typhus Fever

There are two varieties of typhus (not to be confused with typhoid) fever: endemic and epidemic. **Endemic typhus**, also known as **murine typhus**, is caused by *Rickettsia typhi* and is transmitted by fleas found on mice and rats. Those whose jobs or residences are in rat-infested areas are particularly at risk for endemic typhus. Rodent control is primary in breaking the disease cycle.

Epidemic typhus is caused by *Rickettsia prowazekii* and is transmitted by the human body louse; the human is the only reservoir. The rickettsiae are named in tribute to Howard Ricketts and Stanislaus von Prowazek, both of whom died from epidemic typhus. As the louse feeds it defecates; the bite causes an itch, and as the individual scratches the rickettsiae and the louse's feces are inoculated into the bite. Typhus is truly a "lousy" disease. The infected louse dies usually within a few weeks, and its entire life cycle (from egg to adult) can take place on the clothing of an infected person. No epidemics of this disease have occurred since 1922, but there have been sporadic cases, with flying squirrels serving as possible reservoirs for the disease. Like relapsing fever, also a louseborne disease, conditions of war, crowding, unsanitary conditions, and little opportunity to bathe or to wash clothing are conducive to this disease. As with all vector-borne diseases, environmental control of the vector is important.

Overview

A variety of diseases caused by bacteria is presented and organized on the basis of their mode of transmission, namely, foodborne and waterborne, airborne, sexually transmitted, contact, soil, and arthropodborne. Some bacterial diseases are diseases of antiquity, but others are new, emerging, or reemerging. Some bacteria have caused major epidemics and pandemics throughout the centuries, and their death tolls have influenced the course of history.

The information presented makes it clear that, despite advances in hygiene, immunization, and drug therapy in the twentieth century, humans all over the world remain threatened by pathogenic bacteria. Foodborne and waterborne intoxications and infections are characterized by gastroenteritis and are a huge burden manifested by morbidity and mortality, particularly in less-developed countries with an inadequate public health infrastructure. Despite the watchdog efforts of the U.S. Department of Agriculture, they continue to have a significant effect on the health of Americans.

Airborne bacteria cause diseases that are categorized into upper and lower respiratory tract infections. Immunization and antibiotic therapy have been instrumental in reducing the morbidity and mortality in those countries that are able to implement these practices.

STDs caused by bacteria have been overshadowed by the pandemic of AIDS, a virally caused disease. STDs were on the decline after the appearance of penicillin in the 1940s, but the sexual revolution in the 1960s resulted in an increase in their prevalence. Contact diseases (other than those transmitted sexually) remain a problem; the necessity for direct contact limits their spread. The relatively recent discovery that most gastrointestinal ulcers are caused by bacteria, and not by stress as previously believed, is a major breakthrough in the treatment of ulcers. Direct person-to-person contact is thought to be the most plausible route of transmission.

Pathogenic bacteria are present in the soil; those that are spore formers survive for very long periods of time. Circumstances leading to their transmission into the body result in infection. Arthropods are vectors in the transmission of several bacterial diseases, including those caused by spirochetes and rickettsiae, that are responsible for major epidemics.

■ Self-Evaluation

PART I: Choose the single best answer.

1. Families with pet iguanas or turtles are particularly at risk for
   a. *Campylobacter* infection **b.** listeriosis **c.** salmonellosis **d.** shigellosis
2. One of the most potent known toxins is associated with
   a. *Salmonella* **b.** botulism **c.** staphylococci **d.** cholera
3. Oral rehydration therapy is associated with
   a. *Salmonella* **b.** meningitis **c.** cholera **d.** Legionnaires' disease
4. The beautiful fountain in a shopping mall may be a source of
   a. listeriosis **b.** leptospirosis **c.** Legionnaires' disease **d.** leprosy
5. Which of the following toxins is a neurotoxin?
   a. botulinum toxin **b.** diphtheria toxin **c.** cholera toxin **d.** more than one of the preceding
6. Which one of the following is an upper respiratory tract infection?
   a. diphtheria **b.** tetanus **c.** legionellosis **d.** tuberculosis
7. Which of the following is not a location associated with community-acquired MRSA?
   a. football fields **b.** hospitals **c.** dormitories **d.** weight rooms
8. Leprosy is caused by
   a. *Staphylococcus* **b.** *Mycobacterium* **c.** *Bacillus* **d.** *Yersinia*
9. Which of the following is not associated with a tick bite?
   a. Lyme disease **b.** ehrlichiosis **c.** relapsing fever **d.** anthrax
10. Which of the following bacteria do not have vaccines available to prevent disease caused by them?
    a. *Staphylococcus aureus* **b.** *Bordetella pertussis* **c.** *Bacillus anthracis* **d.** *Haemophilus influenzae*

PART II: Match the descriptions with the disease from the lists below. A letter may be used once, more than once, or not at all.

Soilborne and Arthropodborne Diseases

1. rickettsial disease transmitted by lice
2. possibly transmitted by dogs and cats
3. strong candidate for biological warfare
4. named after town in Connecticut
   a. leptospirosis **b.** anthrax **c.** Lyme disease **d.** epidemic typhus **e.** endemic typhus

Airborne Infections

5. can cause obstructive membrane in pharynx
6. characterized by severe cough
7. associated with Philadelphia
   a. Legionnaires' disease **b.** diphtheria **c.** meningitis **d.** pertussis

Sexually Transmitted and Other Contact Diseases.

    8. syphilis

    9. plague

   10. gonorrhea

       a. "great imitator" **b.** toxic shock syndrome **c.** buboes **d.** clap or drip
       e. Hansen's disease

PART III: Answer the following.

    1. Distinguish between food poisoning (intoxication) and food infection. Name two diseases in each category.

    2. Explain the following terms:

       a. oral rehydration therapy

       b. opisthotonos

       c. epidemic relapsing fever

       d. pelvic inflammatory disease

    3. You are about to travel to a foreign country where sanitation is at a lower level than you are accustomed to. What precautions will you take during your stay?

    4. List measures you can use to prevent tickborne diseases.

    5. What is meningitis? List the bacteria that can cause meningitis.

    6. List three bacterial pathogens found in the soil and what diseases they cause. How can these diseases be prevented?

    7. Describe the three stages of syphilis. How is it treated?

    8. Explain why *H. pylori* can withstand the harsh pH of stomach acid.

    9. Explain what it means if you call someone a "typhoid Mary."

   10. List at least four bacterial pathogens that are multidrug resistant. Explain why there is cause for concern if a pathogen is antibiotic resistance.

CONVENT OF
MARY IMMACULATE
—— (1878) ——
Built by the Sisters of the Holy Names
of Jesus and Mary, a Canadian Order which
first established a school here in 1868.
Designed by William Kerr of Ireland, of
Romanesque style, with dormered, mansard roofs
and central tower. In the Spanish-American
War the Sisters offered their services as
nurses and the Convent to the Navy as a
hospital, and rendered devoted service to
the wounded and yellow fever victims.
F-76   FLORIDA BOARD OF PARKS AND HISTORIC MEMORIALS   1962

Viral and Prion Diseases

An efficient virus kills its host. A clever virus stays with it.

—James Lovelock

Prions are "almost immortal."

—Paul Brown

■ Preview

This chapter presents the major viral and prion diseases of humans. The field of virology changes quickly. Special effort was made to update the epidemiology statistics of the viral diseases presented here. The main resource used to do this was the Centers for Disease Control and Prevention (CDC) data and statistics information. Emerging viruses, including Ebola virus, HIV (human immunodeficiency virus), Lassa virus, and hantavirus, to mention only a few, seem to have appeared suddenly from nowhere and threaten our existence. Viruses are assigned to a particular family based on their shape, nucleic acid content, size, and other properties; those that cause human disease fall into fifteen RNA and seven DNA virus families.

For each viral disease presented, particular attention is paid to its mechanism of transmission and other factors involved in the microbial

Photo © Jason O. Watson/historical-markers.org/Alamy.

cycle of disease: pathogenicity, incubation time, treatment, and (in some cases) immunization. With many bacterial diseases the affected area of the body frequently correlates with the mechanism of transmission and the route of entry of the bacteria into the body. For example, foodborne bacteria primarily infect the digestive system, and airborne bacteria primarily infect the respiratory tract. In the case of viruses there may be little connection between the route of entry and the particular organs and tissues of the body involved. The measles and chickenpox viruses, for example, are airborne and enter the body via the respiratory route, but the skin is their major target. Rabies is acquired through a break in the skin from the bite of an infected animal, but the virus attacks the nervous system. In many cases there may be more than one route of entry and/or transmission, making many viral diseases difficult to pigeonhole.

Immunization is available against some viral diseases but not against others. Keep in mind that antibiotics are not effective against viruses, but there are some antiviral agents (not antibiotics) that are used in specific cases.

As you study each of these viral diseases, recall the five stages of viral replication: adsorption, penetration, replication, assembly, and release. Remember that these stages occur in infected cells, causing damage to the host, which may be anywhere from subclinical to mild to severe to fatal. Frequently, the diagnosis of a particular viral disease is based on the clinical symptoms, which may be readily apparent, such as swollen jaw in mumps and the characteristic rash of measles or chickenpox. Laboratory diagnosis, however, based on demonstration of viral antigens or antibodies and culture of the specific virus, may be the only way to definitively diagnose many viral diseases. Because of the reporting time necessary and the expense involved, these procedures are not always done routinely. Frequently, it is assumed that the condition will simply run its course.

Prions are infectious proteins that cause a group of diseases of the brain and nervous system in humans and animals called transmissible spongiform encephalopathies (TSEs). TSEs are rare, incurable neurodegenerative fatal diseases. Creutzfeldt-Jakob disease (CJD) and variant CJD (caused by eating prion-contaminated beef products) are the most common TSEs in humans.

Foodborne and Waterborne Viral Diseases

The major foodborne and waterborne viral diseases contribute to the overall public health problem of diseases and cause significant morbidity and mortality. The Centers for Disease Control and Prevention (CDC) estimated foodborne diseases (bacterial, parasitic, and viral) in the United States at an annual average of 47.8 million cases, including 128,000 hospitalizations and 3,000 deaths in 2011. The CDC and the U.S. Environmental Protection Agency (EPA) completed studies that determined an average of 16.4 million cases of gastroenteritis were associated with contaminated public drinking water in the U.S. (TABLE 10.1).

Gastroenteritis

Gastroenteritis is characterized by stomach and abdominal pain, diarrhea, vomiting, and abdominal cramps as a result of inflammation of the stomach and intestinal tract. The term "**stomach flu**" is popular but meaningless. An exact

TABLE 10.1 Foodborne and Waterborne Viral Diseases

Disease	Incubation Period	Virus(es)	Symptoms	Immunization and Comments
Gastroenteritis	2–10 days	Noroviruses, rotaviruses, adenoviruses	Stomach ache, diarrhea, dehydration, vomiting	RotaTeq® and Rotarix® vaccines available against rotavirus infections.
Hepatitis	3–8 weeks	Hepatitis A virus, hepatitis E virus	Jaundice, abnormal liver function in tests	Vaccine available against hepatitis A virus for those at risk
Poliomyelitis	7–14 days	Poliovirus	Usually asymptomatic but can cause lifelong paralysis	Vaccine available since 1950s; disease close to eradication

diagnosis of gastroenteritis is difficult because these symptoms are caused by a variety of viruses, bacteria, and protozoans. The symptoms usually last a few days, and recovery is uncomplicated. Dehydration is a potential problem, particularly in infants and in the elderly, because of fluid loss resulting from excessive diarrhea and vomiting. Transmission is by the fecal–oral route. Raw or undercooked shellfish from contaminated waters are also a potential threat. Outbreaks, following a one- to two-day incubation period, are most likely to occur in schools, child care centers, cruise ships, dormitories, nursing homes, military barracks, refugee camps, and disaster relief camps with their close quarters and shared toilet facilities.

Children under the age of five are particularly vulnerable to infection by a group of RNA viruses known as **rotaviruses.** In fact, it is the most common cause of seasonal viral gastroenteritis in this age group; it is sometimes called "winter diarrhea." The CDC estimates that in the United States rotavirus infections account for twenty to sixty deaths in children younger than 5 years of age, 200,000 emergency room visits, about 55,000 to 70,000 hospitalizations, and as many as 400,000 visits to physicians annually. Globally, rotaviruses cause about 600,000 deaths per year in the less-developed countries of southern Asia and sub-Saharan Africa. In 1998, the U.S. Food and Drug Administration (FDA) licensed a live attenuated oral vaccine against rotavirus, **RotaShield®**, but it was withdrawn after only a little more than a year on the market because of reports of bowel blockages in a small number of vaccinated children. Two new rotavirus vaccines, Rotarix® and RotaTeq® were made available worldwide in 2006. Both **vaccines** can be used to vaccinate infants from the age of 6 weeks. Immunity lasts over 2 years, peaking at a time when young children are most vulnerable to rotavirus infections.

Noroviruses (also known as **Norwalk** and **Norwalk-like viruses** named after Norwalk, Ohio) are frequent causes of gastroenteritis or "winter vomiting disease" in older children and adults. People living or gathering together can be easily infected for several reasons: Fewer than ten viral particles are required for transmission, the viruses can persist in the environment, and the viruses continue to be shed after recovery. Cruise ships, some of which carry over 3,000 passengers and 1,000 crew, are natural hotspots for noroviruses. The short incubation period was more than enough time for 304 passengers and crew on the cruise ship *Queen Elizabeth 2*, which departed from

New York on January 2007 on a worldwide 106-day voyage, to be clinging to the rails or "heading for the head" as they suffered from the nausea, vomiting, and diarrhea associated with norovirus. Cruise ships sailing from a foreign port that participate in the CDC's Vessel Sanitation Program report gastroenteritis cases confirmed by the ship's medical staff to the CDC before entering a U.S. port. You don't need to ride the waves to become debilitated by this highly contagious virus. Over the past few years, norovirus outbreaks have been associated with smorgasbords, food service workers, hospitals, even celebrants at family reunions. Unfortunately, there are many different strains of norovirus, which means it is difficult to develop immunity.

Hepatitis A and E

Hepatitis is an inflammation of the liver and is commonly caused by one of five viruses designated hepatitis A, B, C, D, and E. **Hepatitis A virus**, like the other hepatitis viruses (with the exception of hepatitis B virus), is an RNA virus. Jaundice, a yellowish discoloration of the skin and eyes caused by increased levels of bilirubin in the blood, frequently occurs in hepatitis virus infections, along with abnormal liver function test results (FIGURE 10.1). Hepatitis B, C, and D are presented later based on their route of transmission.

Hepatitis A, formerly called infectious hepatitis, is usually a mild and self-limiting disease with an abrupt onset. Its average incubation time is approximately thirty days, but the incubation period may be as long as two months. Unlike other hepatitis infections, recovery is usually complete without chronic infection. The virus is found in feces, and transmission is by the fecal–oral route, most frequently resulting from contamination of food and drinking water. Close personal contact, including oral sex, contributes to transmission. Approximately 50% of cases are subclinical. The CDC estimated that approximately 21,000 new hepatitis A infections occurred in the United States in 2009. Rates of hepatitis A infections have steadily declined in the U.S. Good personal hygiene, emphasizing hand washing and good sanitation, is the most effective means of control.

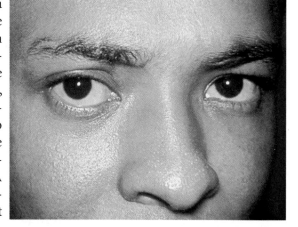

FIGURE 10.1 Jaundice. Courtesy of Dr. Thomas Fe. Sellers/Emory University/CDC.

Diagnosis is based on laboratory detection of antibodies in the blood serum; there is no specific treatment. The CDC recommends that all children in the U.S. be vaccinated against hepatitis A virus between their first and second birthdays. Additionally, it is recommended for use in high-risk individuals, including those traveling to foreign countries where the disease is endemic (FIGURE 10.2). Injections of blood products rich in hepatitis A virus antibodies are also used before and after exposure under defined conditions.

Hepatitis E virus, like hepatitis A virus, is transmitted by the fecal–oral route. It is uncommon in the United States but is endemic in Mexico, Africa, Central America, India, and Asia. There is currently no vaccine available for hepatitis E virus.

Poliomyelitis

Two generations back, the word **poliomyelitis** (usually shortened to **polio**), also known as **infantile paralysis**, struck fear into the hearts of parents the world over.

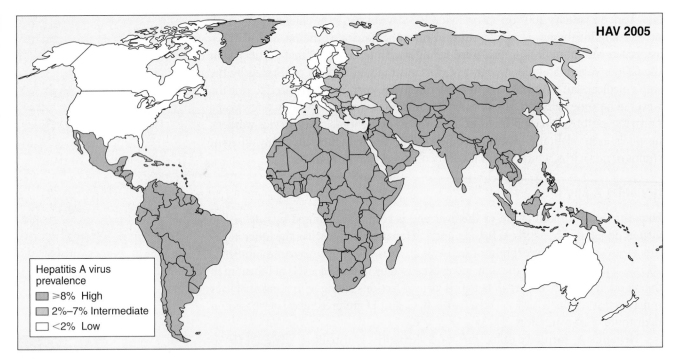

HAV 2005

Hepatitis A virus
prevalence

- ≥8% High
- 2%–7% Intermediate
- <2% Low

FIGURE 10.2 Prevalence of Hepatitis A worldwide. Adapted from CDC. *Health Information for International Travel 2008.* Elsevier, 2007.

AUTHOR'S NOTE (RIK)

In the summer of 2006 I visited the Jonas Salk Institute in La Jolla, California, to talk with Peter Salk, MD, about his father, Jonas. The two main cement laboratory buildings facing the Pacific enclosed a spacious courtyard. The inscription leading up to the courtyard states, *Hope lies in dreams, in imagination, and in the courage of those who dare to make dreams into reality.* —Jonas Salk

Now polio is expected soon to join smallpox, the first eradicated disease, in extinction. Immunization in most parts of the world has greatly reduced the incidence of the disease. The last case of naturally occurring polio in the United States was in 1979. Since that time, there have been a few cases of unvaccinated immune compromised children who contracted vaccine-derived polio in the U.S. In 2010 there were only 1,349 cases of (wild-type) polio in the four countries in which the disease remains endemic (Nigeria, Pakistan, India, and Afghanistan). All this has taken place since the introduction of the first poliovirus vaccine, the **Salk vaccine**, in the 1950s, followed a few years later by the **Sabin vaccine**. The Salk vaccine uses dead viruses of all three polio virus strains. The Sabin vaccine is composed of live but attenuated (weakened) polio viruses. In the United States and in other countries the Salk vaccine was replaced by the Sabin vaccine; in some countries the Sabin vaccine continues to be used.

Poliomyelitis is a highly transmissible infectious disease. Most cases of polio are asymptomatic, and only a small number result in paralysis, as happened with President Franklin Roosevelt (FIGURE 10.3). Transmission is from person to person by direct fecal–oral contact, by indirect contact with infectious saliva or fecal material, or by contaminated sewage or water. The virus, an RNA virus, replicates in the tonsils and Peyer's patches (lymph nodes in the walls of the intestine near the colon), and it may invade the blood and be disseminated to the nervous system. Replication in spinal cord nerve cells causes **paralytic poliomyelitis**, resulting in severely deformed limbs (FIGURE 10.4). Paralysis occurs in about seven to twenty-one days from the time of initial infection; in about 2% to 10% of cases death may result.

Bulbar poliomyelitis is an extremely dangerous form of polio that can paralyze the respiratory muscles, causing difficulty in swallowing and breathing. In the past, patients spent long periods, perhaps their entire lives, in **iron lungs** (FIGURE 10.5),

PART 2 Microbial Disease

(a)

(b)

FIGURE 10.3 **(a)** and **(b)** Franklin Delano Roosevelt fell victim to polio in 1921 at age thirty-nine, leaving this future president (1933–1945) a paraplegic. Throughout his four years as governor of New York and his twelve years as president of the United States, his legs were in heavy metal braces. This courageous and highly visible individual lent his name and influence to the fund-raising campaigns of the National Foundation for Infantile Paralysis and to the March of Dimes to combat polio. Courtesy of Franklin D. Roosevelt Presidential Library and Museum.

FIGURE 10.4 The crippling effects of polio. A man with a weak, withered leg. Courtesy of NIP/Barbara Rice/CDC.

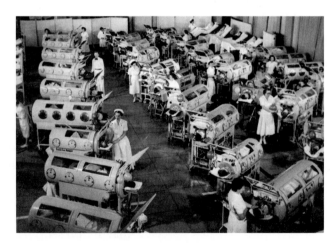

FIGURE 10.5 Life in an iron lung in 1953. Many victims of polio had to spend their lives in an iron lung because they could not breathe on their own. Courtesy Rancho Los Amigos National Rehabilitation Center, L.A. County Department Health Services. Used with permission.

which are airtight metal tanks that enclose the entire body, except for the head, and force respiration. These tanks have a diameter slightly larger than that of an oil drum and a length approximately equal to that of two drums. In the 1950s iron lungs were replaced by positive pressure ventilators, but a few are still used by about 30 people in the United States. Electrical failures were a nightmare; attendants had to mechanically pump the bellows of an iron lung to keep a patient breathing. Dianne Odell died in 2008 at the age of 61 as the result of a power failure in her home after having lived in an iron lung for 51 years. She was one of the nation's oldest survivor of polio to be living in an iron lung.

After exposure to poliovirus more than 90% of susceptible contacts become infected and acquire lifetime immunity, but only to the particular viral type (of which there are three) that caused the infection. Humans are the only reservoir for poliovirus, as is the case with smallpox and measles, making possible the eradication of these diseases.

■ Airborne Viral Diseases

The primary source of airborne diseases is respiratory droplets from an infected person to a susceptible person. Hence, having an "infectious laugh" may not be a desirable trait. A summary of the major airborne viral diseases is presented in TABLE 10.2.

■ Respiratory Syncytial Virus

The RNA virus **respiratory syncytial virus (RSV)** is highly contagious and **endemic** worldwide. The virus is the common cause of two serious respiratory diseases—bronchiolitis and pneumonia—and is particularly life threatening to infants under six months of age with preexisting lung or heart conditions. In fact, it is the most widespread cause of respiratory infection in this age group, and by the age of two most children have had RSV infections. RSV has an incubation period of four to five days. The symptoms of the infection are nonspecific, which makes diagnosis difficult; they include fever, runny nose, ear infection, and pharyngitis. The virus is so named because in the respiratory tissues, a syncytium—a network of large, abnormal cells with multiple nuclei—may be present. The disease can progress to serious lower respiratory tract infection, including obstructed airways.

It is estimated that approximately 125,000 children in the U.S. are hospitalized each year, but the virus rarely causes death. The mortality rate is less than 1% with less than 500 deaths per year attributed to RSV. Virtually all children have had one RSV infection by their third birthday. In older children and in adults, the infection is manifested as a common cold. Outbreaks are a threat in pediatric wards and in nurseries, and the results can be devastating. Frequent and careful hand washing decreases the transmission of RSV. Treatment is largely supportive; oxygen therapy may be necessary in some cases. The antiviral drug **ribavirin** may be administered as an aerosol. There is no vaccine.

TABLE 10.2 Airborne Viral Diseases

Disease	Inoculation Period	Genome Type	Symptoms	Immunization and Comments
Influenza	1–2 days	ssRNA (8 pieces)	Chills, fever, muscle aches, sneezing, sore throat	Annual flu vaccine for senior citizens and others at high risk
Measles (rubeola)	10–21 days	ssRNA	Coldlike, Koplik's spots; rash	MMR vaccine
Mumps	10–20 days	ssRNA	Swollen jaw	MMR vaccine
German measles (rubella)	12–32 days	ssRNA	Fever, rash	MMR vaccine, prenatal transmission
Respiratory syncytial virus	4–5 days	ssRNA	Coldlike	No vaccine; life threatening; most prevalent cause of respiratory infection under six months of age
Common cold	1–3 days	ssRNA (rhinoviruses) ssRNA (coronaviruses) ssDNA (adenoviruses)	Coldlike	No vaccine
Chickenpox and shingles	10–23 days (chickenpox); few to many years (shingles)	dsDNA	Chickenpox: rash; shingles: rash and intense pain along nerves	Vaccine now available[a]; prenatal transmission Shingles vaccine (Zostavax®)
Hantavirus	1–3 days	ssRNA	Influenza-like	No vaccine; transmitted by aerosolized fecal material, saliva, or urine from infected rodents; high fatality rate
Parainfluenza virus	2–7 days	ssRNA	Croup, bronchitis, and pneumonia; especially in infants and children	No vaccine
Severe acute respiratory syndrome (SARS)	2–10 days	ssRNA	Fever, headache, cough, body aches	No vaccine

ds, double stranded; MMR, measles-mumps-rubella; ss, single stranded.

[a]A "combination" vaccine called MMRV, which contains both chickenpox and MMR vaccine, is available to children under the age of 12.

Common Cold

The **common cold** is the most frequent infection, with the highest loss of workdays and school days, establishing an economic burden. Hence, although not considered life threatening, a cold is "nothing to sneeze at" (pun intended!). Next time you are in a drugstore or a supermarket, take a look at the dozens of remedies marketed to relieve the symptoms of the common cold. Manufacturers of these remedies are ever wary to avoid use of the term "cure" for fear of legal liability, but they come very close to the edge in their claims.

The symptoms of sneezing, coughing, sore throat, stiffness, and that general "blah" feeling are all too familiar. Colds are popularly described as "summer colds," "head colds," "chest colds," and "winter colds." These expressions have no medical basis, but they do indicate a seasonality and hint at the variety of symptoms

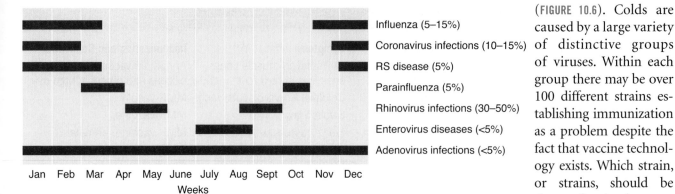

Influenza (5–15%)
Coronavirus infections (10–15%)
RS disease (5%)
Parainfluenza (5%)
Rhinovirus infections (30–50%)
Enterovirus diseases (<5%)
Adenovirus infections (<5%)

Jan Feb Mar Apr May June July Aug Sept Oct Nov Dec
Weeks

FIGURE 10.6 The seasonal variations of viral respiratory diseases and their annual percentages in the United States.

(FIGURE 10.6). Colds are caused by a large variety of distinctive groups of viruses. Within each group there may be over 100 different strains establishing immunization as a problem despite the fact that vaccine technology exists. Which strain, or strains, should be chosen for development of a vaccine? Although a common denominator might exist within all strains, one has not been identified.

Cold viruses are transmitted not only by respiratory droplets but also, to a large extent, by hands or **fomites**, including doorknobs, faucets, furniture, and toys. The incubation period is about one to three days. Some individuals with colds carry a handkerchief or tissues into which nasal and cough secretions and saliva are discharged. Others sneeze or cough directly into their hands and have no reluctance in offering their hand for a friendly and infectious handshake. Three-ply tissues were introduced in the 1990s; the moisture-activated middle ply was impregnated with citric acid and sodium lauryl sulphate designed to kill the cold virus. Citric acid is a flavoring agent in soft drinks and sodium lauryl sulfate is a **surfactant** commonly found in many shampoos, shaving foams, toothpastes, and detergents. However, for a variety of reasons, including cost, efficiency, and the abrasiveness of the tissue, these tissues were not popular with the public but did not totally disappear from store shelves. The "antiviral" Kleenex® is still sold today, advertised to kill 99% of cold and flu viruses, including H1N1 influenza virus (FIGURE 10.7).

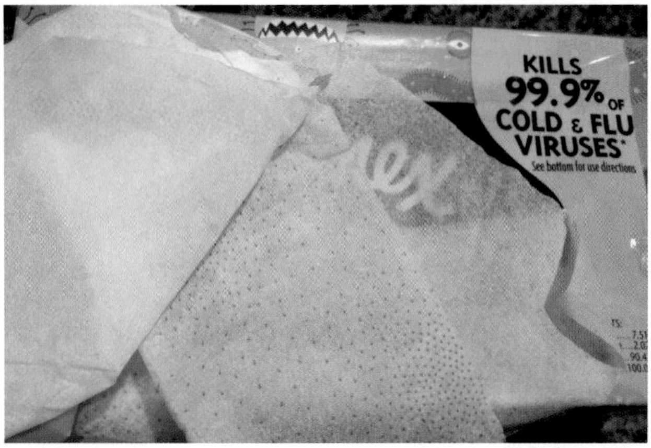

As mentioned, there is a large variety of viruses responsible for the common cold, but about half of all colds are caused by two groups, the **rhinoviruses** and the **coronaviruses**; both groups are RNA viruses. Most of the cold viruses possess mechanisms of adhesion that allow them to attach to the nasal pharynx and avoid being entirely eliminated during vigorous coughing and sneezing. This permits replication and spread to neighboring cells. Those viruses that escape become airborne.

FIGURE 10.7 Three-ply anti-viral tissues with a special middle layer containing blue dots of chemicals that kills cold and flu viruses upon contact with mucus. Author's photo (TS).

Diagnosis of a cold is symptom based. The illness is generally mild and self-limiting, although a nuisance. Laboratory tests are not necessary or worthwhile, except for epidemiological purposes.

■ Measles (Rubeola), Mumps, and German Measles (Rubella)

Measles, mumps, and German measles are considered childhood diseases, and each is caused by a specific RNA virus species in the Paramyxoviridae family. In 1968, a **measles-mumps-rubella** vaccine, called **MMR**, was introduced. MMR consists of a mixture of live, attenuated (weakened) measles, mumps, and German measles viruses. The vaccine has drastically reduced the incidence of these diseases. The low incidence does not mean people are unlikely to contract the diseases and therefore do not need to be vaccinated. The public needs to be aware of this, because a false sense of security can lead to failure to immunize, resulting in a nonimmunized and vulnerable population. If vaccination was halted and the population's herd immunity waned, many new cases would result. Although the triple vaccine is composed of live, attenuated viruses, complications are rare, and the risks of acquiring one or more of these diseases is much greater than are the risks of complications. The vaccine is given at twelve to fifteen months of age and again at four to six years of age. The three diseases have no specific treatment, and prevention by immunization is the key.

Measles

Measles is distinct from what is commonly known as German measles, discussed below. The **measles** virus causes one of the most infectious viral diseases; estimates are that there is more than a 98% chance of becoming infected if exposed directly to someone with measles. The mechanism of transmission is by respiratory droplets, and the disease is fostered by overcrowding, low levels of immunity in the population, malnutrition, and inadequate medical care. Humans are the only reservoir for this disease (as is true of smallpox, chickenpox, and mumps), and therefore it could be eradicated. The symptoms of measles are cold-like, with the development of characteristic **Koplik's spots** in the mouth early in the disease, followed by a red rash (**FIGURE 10.8**) that starts on the face and spreads to the extremities and over most of the body. The disease is usually mild and self-limiting, but one in 500 children with measles develops potentially serious and even fatal complications, including pneumonia, ear infections, brain damage, and seizures.

Worldwide, measles is a significant disease and is a frequent cause of death, particularly in developing countries, despite the availability of an effective vaccine. In developing countries and in populations lacking immunization, more than 15% of children who contract measles die of the disease or its complications.

Malnourished children are particularly vulnerable. The Measles Initiative Program, launched in 2001, had as one of its goals to cut the measles death rate by 95% by the year 2015. The program led to measles immunization campaigns in Africa. Before the program's launch the estimated death rate around the world in 1999 was 873,000. Between 2000 and 2007 the death rate was drastically reduced to 197,000. More than 1 billion children in more than 80 countries have been vaccinated. The program is a partnership between the American Red Cross, the CDC, WHO, the United Nation's Children's Fund, and the United Nations Fund.

The acquired immunity resulting from having had the disease is considered to be lifelong for survivors, whereas the MMR

FIGURE 10.8 Measles (rubeola). **(a)** Characteristic diffuse red rash covering the face and shoulders. Courtesy of CDC. **(b)** Koplik's spots inside the mouth. Courtesy of CDC.

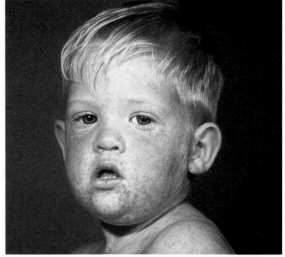
(a)

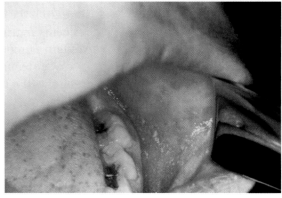
(b)

Early report

Ileal-lymphoid-nodular hyperplasia, non-specific colitis, and pervasive developmental disorder in children

A J Wakefield, S H Murch, A Anthony, J Linnell, D M Casson, M Malik, M Berelowitz, A P Dhillon, M A Thomson, P Harvey, A Valentine, S E Davies, J A Walker-Smith

Summary

Background We investigated a consecutive series of children with chronic enterocolitis and regressive developmental disorder.

Methods 12 children (mean age 6 years [range 3–10], 11 boys) were referred to a paediatric gastroenterology unit with a history of normal development followed by loss of acquired skills, including language, together with diarrhoea and abdominal pain. Children underwent gastroenterological, neurological, and developmental assessment and review of developmental records. Ileocolonoscopy and biopsy sampling, magnetic-resonance imaging (MRI), electroencephalography (EEG), and lumbar puncture were done under sedation. Barium follow-through radiography was done where possible. Biochemical, haematological, and immunological profiles were examined.

Findings Onset of behavioural symptoms was associated, by the parents, with measles, mumps, and rubella vaccination in eight of the 12 children, with measles infection in one child, and otitis media in another. All 12 children had intestinal abnormalities, ranging from lymphoid nodular hyperplasia to aphthoid ulceration. Histology showed patchy chronic inflammation in the colon in 11 children and reactive ileal lymphoid hyperplasia in seven, but no granulomas. Behavioural disorders included autism (nine), disintegrative psychosis (one), and possible postviral or vaccinal encephalitis (two). There were no focal neurological abnormalities and MRI and EEG tests were normal. Abnormal laboratory results were significantly raised urinary methylmalonic acid compared with age-matched controls (p=0.003), low haemoglobin in four children, and a low serum IgA in four children.

Interpretation We identified associated gastrointestinal disease and developmental regression in a group of previously normal children, which was generally associated in time with possible environmental triggers.

Lancet 1998; **351:** 637–41

See Commentary page

Inflammatory Bowel Disease Study Group, University Departments of Medicine and Histopathology (A J Wakefield FRCS, A Anthony MB, J Linnell PhD, A P Dhillon MRCPath, S E Davies MRCPath) **and the University Departments of Paediatric Gastroenterology** (S H Murch MB, D M Casson MRCP, M Malik MRCP, M A Thomson FRCP, J A Walker-Smith FRCP), **Child and Adolescent Psychiatry** (M Berelowitz FRCPsych), **Neurology** (P Harvey FRCP), **and Radiology** (A Valentine FRCR), **Royal Free Hospital and School of Medicine, London NW3 2QG, UK**

Correspondence to: Dr A J Wakefield

THE LANCET • Vol 351 • February 28, 1998 637

Introduction

We saw several children who, after a period of apparent normality, lost acquired skills, including communication. They all had gastrointestinal symptoms, including abdominal pain, diarrhoea, and bloating and, in some cases, food intolerance. We describe the clinical findings, and gastrointestinal features of these children.

Patients and methods

12 children, consecutively referred to the department of paediatric gastroenterology with a history of a pervasive developmental disorder with loss of acquired skills and intestinal symptoms (diarrhoea, abdominal pain, bloating and food intolerance), were investigated. All children were admitted to the ward for 1 week, accompanied by their parents.

Clinical investigations

We took histories including details of immunisations and exposure to infectious diseases, and assessed the children. In 11 cases, the history was obtained by the senior clinician (JW-S). Neurological and psychiatric assessments were done by consultant staff (PH, MB) with HMS-4 criteria.[1] Developmental histories included a review of prospective developmental records from parents, health visitors, and general practitioners. Four children did not undergo psychiatric assessment in hospital; all had been assessed professionally elsewhere, so these assessments were used as the basis for their behavioural diagnosis.

After bowel preparation, ileocolonoscopy was performed by SHM or MAT under sedation with midazolam and pethidine. Paired frozen and formalin-fixed mucosal biopsy samples were taken from the terminal ileum; ascending, transverse, descending, and sigmoid colons, and from the rectum. The procedure was recorded by video or still images, and were compared with images of the previous seven consecutive paediatric colonoscopies (four normal colonoscopies and three on children with ulcerative colitis), in which the physician reported normal appearances in the terminal ileum. Barium follow-through radiography was possible in some cases.

Also under sedation, cerebral magnetic-resonance imaging (MRI), electroencephalography (EEG) including visual, brain stem auditory, and sensory evoked potentials (where compliance made these possible), and lumbar puncture were done.

Laboratory investigations

Thyroid function, serum long-chain fatty acids, and cerebrospinal-fluid lactate were measured to exclude known causes of childhood neurodegenerative disease. Urinary methylmalonic acid was measured in random urine samples from eight of the 12 children and 14 age-matched and sex-matched normal controls, by a modification of a technique described previously.[2] Chromatograms were scanned digitally on computer, to analyse the methylmalonic-acid zones from cases and controls. Urinary methylmalonic-acid concentrations in patients and controls were compared by a two-sample *t* test. Urinary creatinine was estimated by routine spectrophotometric assay.

Children were screened for antiendomysial antibodies and boys were screened for fragile-X if this had not been done

---

vaccination affords approximately a 20-year protection. There are "susceptibles" who slip through the regimen of immunization because of a false sense of security that these diseases are gone, a proposed link to autism, or because of the burden of healthcare costs, particularly in less-developed countries. Therefore, outbreaks of measles will continue to occur as happened in Fiji during February to May 2006, resulting in 132 cases.

The United States is no exception. From January 1 to April 25, 2008, 64 cases were reported to the CDC, the highest number since 2001. Fifty-four of these cases were acquired outside the United States. San Diego was hit by a measles outbreak in February 2008 in which there were 11 confirmed cases in children ranging in age from 10 months to 9 years; all the children were not immunized because of their parents' fear of autism.

In 1998, Dr. Andrew Wakefield, a British gastroenterologist was the lead researcher and main author of a study published in *The Lancet* that described a new autism syndrome speculated to be triggered by MMR vaccination. The study lowered parental confidence in vaccination programs and created a MMR vaccination crisis in the United Kingdom and sparked questions about vaccine in North America. Evidence mounted that there was no increased autism risk associated with the MMR vaccine.

The paper was retracted 12 years later (**FIGURE 10.9**). Britain's General Medical Council ruled that the data presented by Dr. Wakefield was fraudulent, and he was found guilty of unethical medical practice and scientific misconduct. Despite the retraction, parents and autism advocacy groups continue to support Dr. Wakefield. Dr. Paul Offit, Chief of Infectious Diseases at Children's Hospital of Philadelphia, said: "This retraction by *The Lancet* came far too late. It's very easy to scare people; it's very hard to unscare them."

Because of effective MMR vaccination, measles in the U.S. declined to an average number of seventy cases per year from 2001 to 2008. However, by July 9, 2008, the measles case number had increased to 127 spread across fifteen states and continued to climb. The common denominators were no measles vaccine, frequently due to religious beliefs, overseas exposure, and home schooling. By mid-June 2011, the U.S. had reports of 150 cases. Many of the cases were unvaccinated residents returning to the U.S. after being exposed to measles while traveling in Western Europe, Asia, and Africa. As of April 19, 2011, more than 10,000 measles cases were reported in Europe. In Nigeria there were 29,871 cases and the Democratic Republic of Congo had 16,000 suspected measles cases by April 22, 2011. This demonstrates that travelers need to make sure their vaccinations are up-to-date! These outbreaks underscore the necessity to maintain a high level of immunity in the population as a preventative measure.

Mumps

The hallmark of mumps is a large swelling on one or both sides of the face, resulting from infection of the **parotid gland**, one of three pairs of salivary glands, which is

FIGURE 10.9 The *Lancet* retracted this controversial paper 12 years after it was published. The author suggested a link between the MMR vaccine and autism. Many UK parents lost confidence in vaccination because of this research study. Author's photo (TS).

located at the junction of the upper and lower jaw (FIGURE 10.10). Humans are the only natural hosts for this disease, which has an incubation period of ten to twenty days. In temperate climates the disease usually occurs in late winter and early spring, most commonly in children under the age of fifteen years. Many children are asymptomatic, that is, they show no symptoms but nevertheless are rendered immune.

Symptoms other than swelling of the parotid glands (which may not occur) include fever, nasal discharge, and muscle pain. The virus can spread to other structures, including the testes, ovaries, meninges, heart, and kidneys. Complications are unlikely, but there has been some concern that in young males, inflammation of the testes (**orchitis**) could lead to sterility. This has not proved to be the case, although this temporary complication is painful. Rarely, permanent deafness occurs and is usually confined to one ear.

Before the MMR vaccine, estimates are that 50% of the children in the United States became infected with mumps. The decline, due to the vaccine, is dramatic; 2,000 cases occurred in 1964, whereas 291 cases occurred in 2005. To the surprise of public health officials, outbreaks of mumps occurred in college students in 1990 despite them having had a second dose of the vaccine as recommended. In 2006, the U.S. experienced multistate outbreaks involving 6,584 mumps cases, affecting mainly midwestern college students. These outbreaks were dubbed the "two dose vaccine failure." The largest U.S. mumps outbreak since 2006 occurred beginning in July 2009. The outbreak was traced to an 11-year-old boy who returned from the United Kingdom in which there was an ongoing mumps outbreak.

In 2011, the Global Alliance for Vaccines and Immunization (GAVI) Alliance Board decided to fund a rubella campaign in countries of most need, introducing a measles-rubella (MR) vaccine in many more countries.

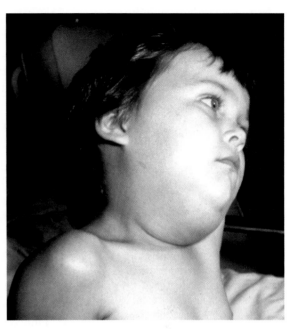

FIGURE 10.10 Mumps. This child has swollen parotid glands. Courtesy of NIP/Barbara Rice/CDC.

Rubella (German Measles)

Of the several viral diseases that cause a rash, rubella is the mildest and has an incubation period of twelve to thirty-two days. In fact, for generations, **rubella**, also called "three-day measles," caused by an RNA virus, was thought to be a mild form of measles (rubeola). In 1829 a German physician recognized that these were two different diseases; hence, the term "German measles" came into use. The disease is endemic worldwide and is highly infectious; it is spread largely through respiratory secretions, but urine from infected individuals can also transmit rubella. Some individuals are asymptomatic and transmit the virus without knowing it. The characteristic rash starts on the face and progresses down the trunk and to the extremities; it resolves in about three days (FIGURE 10.11).

In itself, rubella is not regarded as a serious disease. However, prenatal transmission can occur, particularly if the mother is infected during the first trimester of pregnancy, even if she is asymptomatic. Major consequences include cardiac lesions, deafness, ocular lesions resulting in blindness, mental and physical challenges, and glaucoma. Approximately 15% of those exposed may escape infection during their childhood years, posing a risk that females may enter their childbearing years without having had measles. Therefore it is particularly important that females be immunized against rubella. Immunization has been very successful in decreasing the incidence, leading to the CDC declaring in 2004 that rubella had been eliminated from the United States.

▧ Chickenpox (Varicella) and Shingles (Herpes Zoster)

The expression "brothers under the skin" aptly describes chickenpox and shingles, two seemingly different diseases that cause skin eruptions. Historically, they were thought to be caused by different viruses, but it has long been established that they are manifestations of the same DNA virus. The virus, therefore, is called **varicella-zoster virus**. Chickenpox is a disease usually associated with childhood; shingles typically occurs after about the age of sixty in some individuals who had chickenpox in their earlier years, but it has been reported in children as young as eight years.

Humans are the only hosts for chickenpox and shingles. Chickenpox has an incubation period of ten to twenty days and is transmitted by airborne droplets or by contact with the fluid in the blisterlike skin lesions (**vesicles**) that develop. The disease is highly contagious, especially before the emergence of the rash. Early signs include fever, headache, and generalized aches and pains, followed in a few days by an itchy rash with fluid-filled vesicles appearing on the scalp, face, trunk, and extremities (FIGURE 10.12). More than 100 vesicles can appear on the body at any one time; usually, adults with chickenpox develop more vesicles than do children. The vesicles tend to occur in a succession of crops over a two- to four-day period, and an individual may have a combination of newly emerging and old, "crusty" vesicles.

Before 1995 close to four million American children (under the age of fifteen years) per year developed chickenpox, resulting in about 7,400 hospitalizations and more than fifty deaths. In 1995 the FDA approved a live, attenuated chickenpox vaccine that dramatically reduced chickenpox deaths. In 2006 the CDC

FIGURE 10.11 Rubella (German measles) strikes both adults and children and is very dangerous to infants. **(a)** Teenager with rash on his chest. © PR. PH. Franceschini/CHRI/Science Photo Library. **(b)** Infant born with rubella. Courtesy of CDC. **(c)** Newborn with thickening of the lens of the eye that causes blindness. Courtesy of CDC.

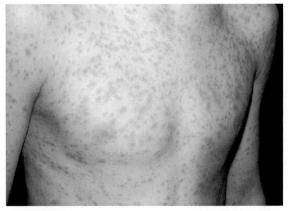

(a)

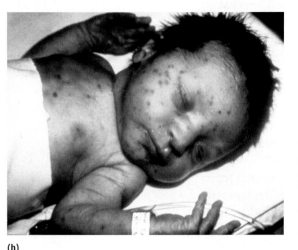

(b)

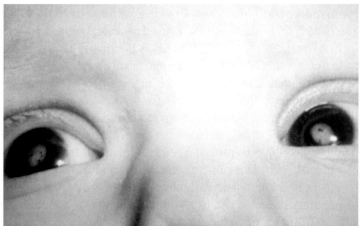

(c)

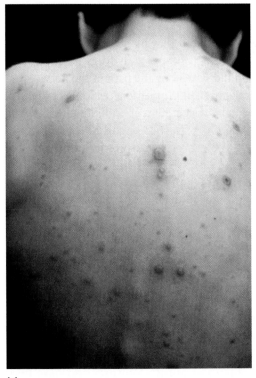

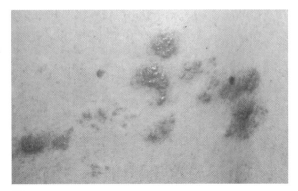

(b)

FIGURE 10.12 Chickenpox and shingles. These two diseases are caused by varicella-zoster virus and are manifested by characteristic rashes. **(a)** Chickenpox. Courtesy of CDC. **(b)** Shingles. This painful condition is caused by the presence of the virus along sensory nerves, where they replicate and produce painful skin eruptions along the path of the nerve. © Stephen VanHorn/ShutterStock, Inc.

(a)

recommended a booster shot for children ages four to six. The disease exists worldwide, and nearly all unvaccinated children are infected during their early years. Chickenpox, with rare exceptions, confers lifelong immunity, eliminating the need for vaccination. Like measles, chickenpox can cross the placenta and cause serious fetal damage. In 2010, an MMRV (measles, mumps, rubella, varicella) vaccine became available as a childhood vaccine.

FIGURE 10.13 In 2011, some parents who feared the chickenpox vaccine purchased lollipops prelicked by children suffering from chickenpox to try to infect their children the "natural" way during "pox parties." Health and legal authorities warned parents against this practice. Author's photo (TS).

Some parents purposely promote "chickenpox parties" to get their children naturally infected with varicella zoster virus. Their belief is that something "natural" (the disease) is better than something artificial (the vaccine) or that immunity from the disease will be more permanent than that from the vaccine. In 2011, state officials cracked down on a Tennessee woman whose Facebook page "Find a Pox Party Near You" was advertising the sale of lollipops licked by children with chickenpox (FIGURE 10.13). It is a federal crime to send "diseases" or viruses across state lines. Chickenpox is not spread through oral secretions. A child licking one of the tainted pops will not get chickenpox. Consequently, the child could pick up bacteria or other viruses, spreading other diseases.

Shingles is not life threatening, but it is miserable with painful complications occurring in some individuals (Figure 10.12). About one in five adults with a history of chickenpox will develop shingles. The condition is described as causing one of the most intense pains of any disease, which is not surprising because the virus infects nerve fibers. During the initial chickenpox infection, viruses take up residence, in a latent state, in collections of nerve cells (**ganglia**) located along the spinal column or in

nerves supplying the face. The spinal nerve, which girdles the trunk at about the belt line, is frequently involved, resulting in this area being a common site for shingles. These viruses remain in a latent state for years but may be triggered into an active replicating state by a variety of agents, including x rays, certain drugs, and immunodeficiency, causing a painful eruption on the skin along the path of the nerve.

An individual with shingles can transmit chickenpox to a person who has not had the disease. More rarely, shingles can be triggered in an individual with a history of chickenpox upon exposure to a person with chickenpox. The good news is that Zostovax®, a vaccine consisting of live varicella viruses, was licensed by the FDA in 2006. The vaccine, manufactured by Merck & Co. Inc., was field tested in 38,000 individuals in the United States ages 60 and older and reduced the incidence of shingles by about 50%; in those 60 to 69 the incidence decreased by 64%. The CDC recommends anyone 60 years or older should get the Zostovax vaccine regardless if they have had chickenpox or not to prevent shingles.

Hantavirus Pulmonary Syndrome

The first known U.S. outbreak of **hantavirus pulmonary syndrome (HPS)** occurred in 1993 in the Four Corners area (where New Mexico, Colorado, Utah, and Arizona meet), with twenty-four cases of a severe influenza-like respiratory illness complicated by respiratory failure and, in some cases, death. The case fatality rate approximates 6% to 8%. Hantaviruses are RNA viruses and have been present in Africa since the 1930s but were associated with hemorrhagic (bleeding) symptoms. Many of the hantaviruses attack the kidneys, but the Sin Nombre strain discussed here attacks the lungs.

The fatality rate for HPS is at least 60%. The disease primarily strikes young, healthy adults, and death occurs in only several days after infection because of accumulation of fluid in the lungs, which interferes with oxygen diffusion. Oxygen therapy is frequently required in an attempt to avert death.

The HPS virus is carried by rodents, especially the deer mouse (FIGURE 10.14) and the cotton rat, both inhabitants of the Southwest. Because of the explosive nature of HPS, the mouse carrier has been referred to as "the mouse that roared." Transmission of the virus to humans is the result of exposure to dried and aerosolized fecal material, saliva, or urine from infected rodents. It is also possible to contract the disease when fresh or contaminated rodent droppings get into the skin, eyes, food, or water or as the result of a rodent bite. The virus cannot be transmitted from human to human.

Prevention is based on avoiding contact with rodents and their excreta. Because the virus is primarily spread by airborne contaminated droplets, entering barns or other buildings that are frequently closed for extended periods may be hazardous to your health, as is disturbing rodent-infested structures. Diagnosis of hantavirus disease is difficult, because the early flulike signs are characteristic of a variety of viral and bacterial diseases. Definitive diagnosis requires laboratory procedures.

Since the first case in 1993, a total of 465 cases of HPS has been reported in the United States in thirty states. Two cases of

FIGURE 10.14 A deer mouse. This mouse species is a vector of HPS; it sheds the virus in urine, feces, and saliva. Courtesy of James Gathany/CDC.

HPS occurred in 2004 in West Virginia; in both cases the Monongahela hantavirus, distinct from the Sin Nombre strain of the Four Corners area, was the causative agent. The first case involved a 32-year-old wildlife sciences graduate student suffering from a laboratory-confirmed strain of Hantavirus. The patient was hospitalized, continued to deteriorate, and died. The patient's coworkers reported that he had spent the previous month trapping small animals and handling mice daily in conjunction with his studies. All members of the team reported carelessness about washing their hands and not having worn gloves while handling the animals or the excreta. The group also reported frequent rodent bites on their hands. The second patient was a 41-year-old man who had spent a long weekend at a log cabin with his family in July 2004. Upon arriving, the cabin smelled of rodent urine, and two live mice were discovered. The patient killed the mice, disposed of their remains, and cleaned the nearby trashcans; gloves were not worn. Symptoms developed two days later, leading to hospitalization. The patient showed improvement after five days of hospitalization and recovered slowly over the next few months. This case underscores the necessity of being alert to avoid exposure to mice and mouse droppings. In the summer of 2012, two Californians who had been hiking in Yosemite National Park were diagnosed with hantavirus pulmonary syndrome. One died. The California Department of Public Health worked with Yosemite Park workers to increase routine measures to reduce the risk of hantavirus exposure to park visitors. Efforts included regular inspection and cleaning of cabins, rodent-proofing buildings, maintaining sanitation levels to discourage rodent infestations, and public education.

Severe Acute Respiratory Syndrome

Between November 2, 2002, and July 3, 2003, it appeared that the world was on the brink of a pandemic of severe acute respiratory syndrome (SARS) caused by SARS–CoV, an RNA coronavirus. Over that relatively short period of time, there were approximately 8,096 cases and 774 deaths, establishing about a 10% mortality rate as reported by the WHO. The outbreak first occurred in a rural area in Guangdong Province in China in 2002, but the People's Republic of China covered up the extent of the problem, thus delaying public health intervention. It wasn't until February 2003 when an American businessman flying from China to Singapore became ill with respiratory distress, causing the plane to land in Hanoi, Vietnam, where the person died. This event was widely publicized, and the world was now aware of the cover up in China and a threatening pandemic. In short order the disease spread rapidly across continents into some thirty-seven countries (TABLE 10.3). Images of deserted streets and subways and people wearing face masks were common; thermal checkpoints for the detection of passengers with elevated temperatures were recommended by WHO at selected airports with international flights—all in an effort to halt the further march of the disease (FIGURE 10.15).

FIGURE 10.15 Thermal imaging screens for SARS screening at Taiwan Taoyuan Airport. © Reuters/Romeo Ranoco Landov.

TABLE 10.3 Probable Cases of SARS by Country, November 2002 Through July 2003

Country or Region	Cases	Deaths	Fatality (%)
People's Republic of China[a]	5,327	349	6.6
Hong Kong[a]	1,755	299	17
Canada	432	44	10
Taiwan[a]	346[b]	37	11
Singapore	238	33	14
Vietnam	63	5	8
United States	27	0	0
Philippines	14	2	14
Germany	9	0	0
Mongolia	9	0	0
Thailand	9	2	22
France	7	1	14
Malaysia	5	2	40
Sweden	5	0	0
Italy	4	0	0
Guatemala	1	0	0
United Kingdom	4	0	0
India	3	0	0
Republic of Korea	3	0	0
Indonesia	2	0	0
South Africa	1	1	100
Macau[a]	1	0	0
Kuwait	1	0	0
Republic of Ireland	1	0	0
Romania	1	0	0
Russian Federation	1	0	0
Spain	1	0	0
Switzerland	1	0	0
Total	8,271	775	9.1

[a]Figures for the People's Republic of China exclude the Special Administrative Regions (Macau SAR, Hong Kong SAR) and the Republic of China (Taiwan), which were reported separately by the WHO.

[b]Since July 11, 2003, 325 Taiwanese cases have been "discarded." Laboratory information was insufficient or incomplete for 135 discarded cases; 101 of these patients died.

Adapted from World Health Organization. *Epidemic and Pandemic Alert and Response.* April 21, 2004.

SARS is spread by close personal contact. Respiratory droplets propelled within about a meter distance from an infected person onto the mucous membranes of the eyes, mouth, and nose of another person can establish infection. Transferring the virus by touching SARS-CoV-contaminated fomites and with transference onto the mucous membranes can also establish infection. Diagnosis can be difficult because of the appearance of the usual influenza-like symptoms—cough, fever, myalgia, sore throat, and shortness of breath—within usually a two- to ten-day window. As is true of viral diseases in general, there is no

PART 2 Microbial Disease

specific treatment, limiting the care of those with SARS to supportive treatment. In the United States only eight people, all of whom had traveled to countries with SARS, had laboratory-confirmed evidence of the disease.

Much remains unknown about SARS and its epidemiology. Continued surveillance by the WHO, CDC, and other public health interests is imperative as documented by the rapid dissemination of the virus and its high morbidity and mortality rates. Further, the socioeconomic spin-offs were tremendous in this near pandemic, including ethnic and racial stereotyping and tourism. The SARS epidemic and its sociological and ethnic impact serves as an excellent example of the broad parameters of public health.

Smallpox

The early signs of smallpox include high fever, fatigue, headaches, and backaches, followed in a few days by a rash concentrated on the extremities (FIGURE 10.16). The rash is characterized by flat red lesions that become filled with pus and crust over in the second week; the lesions break out at the same time. (The rash of chickenpox appears in waves.) The crusts dry up and fall off, leaving deeply pitted scars, particularly on the face. The case fatality rate is about 30%.

Smallpox is transmitted directly from person to person by virus-infected saliva droplets expelled from an infected individual onto a mucosal surface of another. **Variola** (smallpox virus)–containing clothing and bedding can also spread the virus. The first week of illness is the most infectious time because of the high numbers of viruses or viral particles in the saliva.

The eradication of smallpox stands as a public health triumph of the twentieth century. Generations to come will never know the horrors of this disease. The

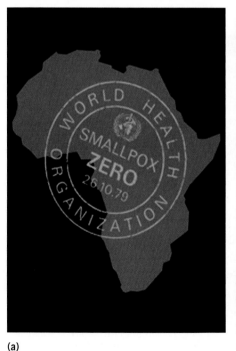

(a)

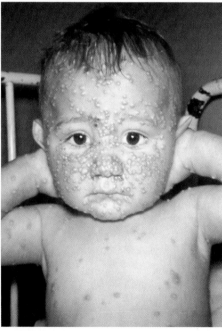

(b)

FIGURE 10.16 **(a)** Poster celebrating the eradication of smallpox. Courtesy of WHO. Used with permission. **(b)** Brazilian boy with mild form of smallpox in 1966. Courtesy of Dr. Noble/CDC.

unique characteristics of smallpox and of the smallpox virus that led to success in smallpox eradication and its establishment as the criterion by which to evaluate other diseases as candidates for eradication are as follows:

- It is a disease only of humans; there are no natural reservoirs or biological vectors.
- The infection is easily diagnosed because of a characteristic rash.
- The duration and intensity of infectiousness is limited.
- Recovery establishes permanent immunity.
- A safe, effective, inexpensive, easily administered, stable (even in tropical climates), one-dose vaccine is available.
- Vaccination confers long-lasting, possible lifetime immunity.
- Vaccination often results in a permanent and recognizable scar, allowing for detection of immune versus nonimmune individuals in a population.

Sexually Transmitted Viral Diseases (STDs)

There are three major sexually transmitted viral diseases: genital herpes, genital warts, and AIDS (TABLE 10.4). Genital herpes and genital warts are discussed in this chapter.

Genital Herpes

There are a variety of herpes viruses, and they are all "bad actors." Herpes is a highly infectious disease caused by **herpes simplex virus (HSV)**, a DNA virus, of which there are two closely related types. HSV type 1 (HSV-1) is usually associated with painful sores around the mouth and lips, called cold sores or fever blisters, and, occasionally, on the throat and the tongue. HSV type 2 (HSV-2) is the major cause of

TABLE 10.4 Sexually Transmitted Viral Diseases (STDs)

Disease	Incubation Period	Genome Type	Symptoms	Comments
Genital herpes	4–10 days	dsDNA	Painful and itchy sores on the penis or on the labia, vagina, or cervix; can be asymptomatic	Lifetime infection with recurrent lesions
AIDS	Up to 12 years, maybe more	ssRNA	Weight loss, tiredness, loss of appetite, repeated infections	Always fatal, but longevity has been substantially improved; causes severe immunosuppression
Human papillomavirus and genital warts	1–6 months	dsDNA	Presence of warts; can be asymptomatic	Can lead to cervical cancer and other genital cancers

ds, double stranded; ss, single stranded.

genital herpes and manifests as painful and itchy sores on the penis in males or on the labia, vagina, and cervix in females. HSV-1 and HSV-2 have an incubation period of 4 to 10 days. The disease is usually mild, but initial infection with HSV-1 and HSV-2 can be accompanied by high fever and large numbers of painful sores. Prompt treatment with antiviral drugs is somewhat effective. The disease can be severe in individuals with AIDS or other immunosuppressive conditions.

Once acquired, infection is lifelong, and episodes of recurrent, painful genital ulcers occur. **Latency**, a period of "taking up residence and hiding," is characteristic of herpes viruses in general, including varicella-zoster virus. The virus' genetic material "hides out" in the host's chromosomes and is replicated along with the genetic material of the host. This strategy allows for evasion of the "foreign" genetic material by the immune system of the host. The viral genetic material can be triggered into an active replicating state by a variety of factors, including stress, fever, sunlight, colds, and menstruation. In many respects latency is a state of peaceful coexistence like that described for bacterial prophages. Generally, within the first year of infection four or five symptomatic recurrences take place.

Transmission occurs through direct contact with herpes sores from one person to another or from one part of the body to another. The herpes lesions are highly infectious until the sores heal within about two to four weeks. Frequently, the virus is transmitted by asymptomatic individuals who have no knowledge of their disease; it is important to realize that these individuals, as well as those who are aware that they are harboring the virus in a latent state, can transmit the disease.

Pregnant HSV-2–infected women are at risk for miscarriage during the course of their pregnancy. In those carrying full-term infants, a cesarean section is performed to minimize the risk of infection to the newborn during passage through an HSV-2–infected birth canal. Finally, HSV-2–infected females are at a greater risk of developing cervical cancer than uninfected females.

Estimates are that 45 million people in the United States, aged 12 and older, are infected with genital herpes. The disease is more common in females (one of four) than in males (one of five) and is more common in blacks (45.9%) than in whites (17.6%). The disease is diagnosed by visual inspection or by culturing tissue from herpes sores to determine the presence of the virus. There is no cure, but antiviral medications may help reduce the duration and severity of occurrences. There is no vaccine, and prevention is based on safe sex practices.

Genital Warts

Human papillomaviruses (HPVs) are responsible for **common warts**, which are benign, painless elevated growths that occur most frequently on the fingers; **plantar warts**, which are deep and painful warts on the soles of the feet; and **genital warts**, which are fleshy growths in the genital areas of men and women.

Genital warts (also called venereal warts) are now one of the most common sexually transmitted diseases in the world; in the United States an estimated 20 million people are infected with HPV. The disease is on the increase, with as many as one million new cases in the United States each year; more than 50% of sexually active men and women will become infected with genital HPV during their lifetime. There are more than 100 types of HPV, of which more than

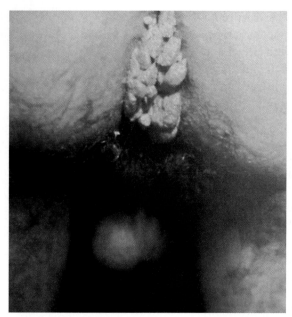

FIGURE 10.17 Genital warts located in the anal region. Courtesy of Dr. Weisner/CDC.

one-third are sexually transmitted and invade genital tissue. Genital warts are highly contagious, and about two-thirds of people whose sexual partners have genital warts develop warts within three months.

In females warts can occur on the outside and inside of the vagina, on the **cervix** (the opening to the uterus), and around the anus. In males warts may occur on the shaft or head of the penis, on the scrotum, and around the anus. Oral sex with a person with genital warts can result in warts in the mouth or throat. The warts may disappear without treatment, but they can develop into large fleshy growths that resemble pieces of cauliflower (FIGURE 10.17).

A major concern is the link between some strains of HPV and cancer, including cervical, vulvar, throat, and anal cancer and, more rarely, cancer of the penis. The best preventative is to avoid direct contact with the virus. Recent data indicate that 70% of HPV-related cervical cancers are caused by HPV types 16 and 18. Types 6 and 11 can also cause cancer but are more associated as responsible for 90% of genital warts. About 12,000 American women are diagnosed with cervical cancer and about 3,700 die on an annual basis; cervical cancer is the fifth leading cause of cancer. About 400 men a year are diagnosed with HPV-associated penile cancer. CDC studies reveal that more than 80% of American females will harbor at least one strain of HPV by the age of fifty. These facts are very disturbing, but the good news is that two vaccines were approved for use in the U.S. Gardasil® (Merck) a vaccine that prevents the aforementioned four types, was approved for females and males ages 9 to 26 by the FDA in 2006. Cervarix® (GlaxoSmithKline), a bivalent vaccine against HPV types 16 and 18 was approved for females 9 to 26 in 2009. The vaccines are safe and are given as shots of three required doses. Vaccination is expected to greatly decrease the incidence of HPV-related cervical cancer.

Implementation of the use of the vaccine was highly controversial because of the ethical questions raised. For example, in 2007 the governor of Texas issued an executive order mandating that Gardasil be given to all girls entering the sixth grade as of September 2008; the Texas legislation overruled the order several months later, barring mandatory vaccination until at least 2011. In 2006, legislators in at least 41 states and Washington, D.C. introduced legislation to require the HPV vaccine, fund research for the vaccine, or educate the public about the vaccine. To date, only two bills have passed that mandate the vaccination of girls: Virginia and Washington, D.C. Major issues are the age or grade of the girls, with most states favoring entrance into sixth grade, and the role of health insurance companies in covering the cost of immunization; most pending legislation allows for parents to opt their daughters out.

The CDC recommends that girls age 11 or 12 be **routinely** vaccinated but it can be given to girls starting at age 9, the goal being to vaccinate before girls are sexually active. A catch-up vaccination for females ages 13 to 26 is also recommended. In October 2011, the CDC Advisory Committee on Immunization Practices recommended to include that boys ages 13 to 21 be routinely vaccinated with the Gardasil vaccine. It is not mandated but it is a public choice; for public health reasons, the vaccine is encouraged. The recommendations come at a time when a study correlating HPV and heart disease in women was published in the *Journal of the American College of Cardiology*. Could the vaccine prevent cancer and heart disease too? The answer is that more rigorous studies are needed to determine this.

Contact Diseases (Other Than STDs) and Bloodborne Viral Diseases

Several viral contact diseases (TABLE 10.5) currently of considerable significance in the United States and around the world are presented in this chapter.

Infectious Mononucleosis

Mononucleosis, often called "mono" or "kissing disease," is a frequent and unwelcome guest on college campuses. The term "mononucleosis" refers to monocytes, one of five categories of white blood cells. By adulthood, antibodies against the causative virus are present in most people. As many as fifty of every 100,000 Americans have symptoms of **infectious mononucleosis**, primarily in the 15- to 30-year-old age group. Most cases of mononucleosis are caused by **Epstein-Barr virus (EBV)**, a DNA virus. The virus infects and replicates in salivary glands and is transmitted by saliva and mucus during kissing, coughing, and sneezing. The incubation period is approximately four to seven weeks after exposure. An infected person poses little risk to household members or college roommates, assuming there is no direct contact with the infected person's saliva.

Symptoms are vague, making diagnosis difficult. In fact, in many people the disease is asymptomatic or so mild they are not even aware they are infected. The early symptoms include "not feeling great," headache, fatigue, and loss of

TABLE 10.5 Contact Viral Diseases Other Than STDs and Bloodborne Viral Diseases

Disease	Incubation Period	Genome Type	Symptoms	Comments
Infectious mononucleosis	4–7 weeks	dsDNA	Headache, loss of appetite, fever, sore throat, swollen glands	"Kissing disease" (caused by EBV); ages 15–30 years most affected
Hepatitis B	45–180 days	dsDNA	May be asymptomatic Nausea, vomiting, appetite loss, jaundice	Blood, blood associated products, sexual practices, AIDS
Hepatitis C	2–22 weeks	ssRNA	Asymptomatic	Blood, blood products, no vaccine
Hepatitis D		ssRNA		Can't infect without presence of HBV, severe liver disease, defective virus
Rabies	1–2 months	ssRNA	Encephalitis, anxiety, drooling, agitation, "hydrophobia"	Almost always fatal if not treated; immunization available; zoonotic with many wild animal reservoirs
Ebola hemorrhagic fever	4–16 days	ssRNA	Massive hemorrhage	

ds, double stranded; ss, single stranded.

appetite, followed by the later triad of fever, sore throat, and swollen glands, particularly in the neck. Enlargement of the lymph nodes and spleen is a more serious complication.

Diagnosis can be definitively established only by laboratory tests based on the detection of antibodies and abnormal white blood cells. There is no specific treatment, and symptoms generally disappear in four to six weeks. EBV, like other herpesviruses, remains latent in the body for life and is held in check by the immune system, although recurrences can occur. It is now known that a condition called **chronic fatigue syndrome** is not caused by EBV, as was formerly suspected.

EBV is the cause of **Burkitt's lymphoma**, a malignant cancer of the jaw and abdomen that occurs in children in central and western Africa. Why this lymphoma is found predominantly in Africa remains a mystery, although there is speculation that having malaria predisposes a person to this condition. New research led by James Lawson at the University of New South Wales (Australia) in 2012 identified DNA sequences of HPV 16 and 18 and EBV normal and cancerous prostate tissues. Further research is needed to determine a causal link between these two viruses and prostate cancer.

Hepatitis B, C, and D Viruses

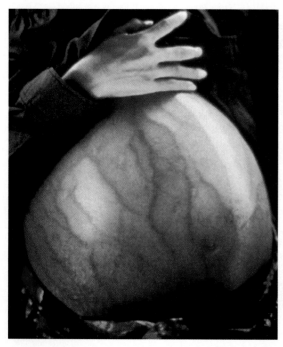

FIGURE 10.18 Hepatitis B. This woman has a hepatoma (liver cancer) resulting from chronic HBV infection. Courtesy of CDC. Used with permission of Patricia F. Walkers, M.D., DTM & H, Health Partners, Center for International Health & International Travel Clinic, St. Paul, Minnesota.

As previously mentioned, there are other hepatitis viruses in addition to hepatitis A and E viruses, and these are transmitted primarily by blood and blood products and by sexual practices. Infected pregnant women can transmit the viruses to their infants. Hepatitis B and hepatitis C, along with other diseases, are frequently found in persons with AIDS. In fact, hepatitis B is sometimes referred to as an "AIDS twin" because of their similarities of transmission.

Hepatitis B is 50 to 100 more times infectious than HIV; an individual pricked with a needle from an individual who has both hepatitis B and AIDS has a 40% chance of acquiring hepatitis B but only a 0.5% chance of acquiring AIDS. Both hepatitis B and hepatitis C are transmissible by blood and blood products, can be asymptomatic or be manifested as chronic diseases, and result in serious lifelong liver problems (FIGURE 10.18), possibly necessitating a liver transplant.

The term "hepatitis" refers to inflammation of the liver. Although the clinical symptoms caused by all the hepatitis viruses are somewhat similar, their biology, circumstances of transmission and infection, and consequences of infection are different. Laboratory tests are required for a definitive diagnosis of hepatitis.

Hepatitis B virus (HBV), originally known as **serum hepatitis**, has a long incubation time, ranging from 45 to 180 days, with an average of about 75 days. HBV is a common cause of cancer, second only to tobacco. In the Far East and in sub-Saharan Africa, where HBV is widespread, liver cancer is a common cause of death. HBV is also an occupational hazard for health professionals, including physicians, medical and dental hygienists, dentists, and others who have contact with blood. These individuals may serve as sources of HBV to the population at large. Infection can also be acquired by sharing needles, tattooing, or body art, acupuncture, and ear piercing.

According to a 2008 report of the WHO, two billion people around the world are infected with HBV, 350 million live with chronic infection, and over 620,000 die each year; these alarming figures establish HBV as a major public health problem and the most dangerous of all hepatitis viruses. Estimates are that each year in the United States 140,000 to 320,000 people of all ages get hepatitis B, 700,000 to 1.4 million have chronic HBV infection, and approximately 1,800 die annually from complications caused by HBV.

Although a vaccine was introduced in the early 1980s, the incidence of disease reduction was disappointing. An improved vaccine made from a component of the virus proved to be a more effective and safer vaccine, resulting in the 1992 recommendation by the American Academy of Pediatrics of universal immunization for newborns. In the same year the WHO set a goal for all countries to implement HBV into national routine infant immunization programs by 2007, and by 2006, 84% of the 193 countries had conformed. Older children should also be vaccinated, particularly adolescents if they are sexually active. The vaccine has reduced the incidence of acute hepatitis B in children and adolescents by over 95% and by 75% in all age groups. One of the problems is that the three-shot vaccination strategy is expensive and beyond the reach of poor countries.

Hepatitis C virus (HCV) was first identified in 1988 and was responsible for almost all cases of blood transfusion–transmitted (non-A, non-B) hepatitis. Hepatitis C is a chronic bloodborne infection with an incubation period of 2 to 22 weeks; it may be subclinical or mild, but approximately 50% of the cases progress to chronic hepatitis. The first reliable screening tests for HCV were not implemented until 1992; therefore, many people may be infected with HCV but unaware that they have the disease, particularly considering that it can take 20 years for the appearance of symptoms. An estimated four million persons in the United States are infected; 130,000 children are infected primarily due to intrauterine transfer; and 16,000 new cases and 10,000 to 20,000 deaths occur each year. Intravenous drug users constitute the vast majority of hepatitis C victims, and as many as 300,000 may have contracted it from blood transfusions before screening tests were available. As a result, in 1988 the U.S. Department of Health and Human Services announced that people who received transfusions before 1992 from donors who later proved to be positive for hepatitis C would be notified. Hepatitis C is the major reason for liver transplants in the United States.

Keep in mind that there is no immunization against hepatitis C. Drugs, including alpha interferon and ribavirin, are available for the treatment of HCV, but, unfortunately, the results are disappointing. Clinical studies indicate that a combination of interferon and ribavirin clears the virus from about 40% of patients, whereas only 20% to 30% are helped by interferon alone. Researchers are actively looking for new drugs against HCV; a major problem is that the virus does not grow in laboratory animals. In 2011, two new protease inhibitors, boceprevir and telaprevir, were approved for hepatitis C treatment. Data is still emerging; however, evidence is mounting that these two new drugs are substantially improving the cure rate in HCV patients and may benefit individuals coinfected with HIV.

Hepatitis D virus (HDV), an RNA virus, also called hepatitis delta virus, is a factor in severe liver disease. It has a close relationship with HBV, and both may be transmitted together, as a coinfection. HDV can also be transmitted independently if HBV is present. In fact, without HBV, HDV cannot replicate and is called a "defective virus." The coinfection of HBV and HDV results in a greater likelihood of liver disease than does HBV alone; furthermore, the HDV combination carries a 20% mortality rate, the highest rate of all the hepatitis viruses. HDV is rare in developed countries and is primarily associated with intravenous drug use.

Rabies

Death is almost a certainty in untreated cases of **rabies**, a disease formerly known as **hydrophobia**; few survive this virus-caused encephalitis. There is no treatment once symptoms appear, but thanks to Louis Pasteur, a French chemist, a rabies vaccine is available for both prevention and early treatment (before the development of symptoms) of this disease. Two types of rabies are described, namely "furious" and "dumb," both of which cause death. About 80% of rabies cases are of the furious type and characterized by brain dysfunction; the animals display aggressiveness, excitability, and (not always) foaming at the mouth—the "mad dog" image. "Dumb," or paralytic, rabies primarily involves the spinal cord; the animals display weak limbs and are unable to raise their heads and/or make sounds because of paralysis of the neck and throat muscles.

Rabies virus is an RNA virus; the disease is zoonotic, with a worldwide distribution in diverse wild animal reservoirs, including coyotes, skunks, cats, bats, foxes, mongooses, and raccoons, as well as domestic animals. Although control measures and immunization have substantially reduced the incidence of domestic animals as vectors in the United States and in other countries, dogs remain the major source of rabies in Asia, Africa, and Latin America, causing thousands of deaths each year. The annual worldwide human death toll is approximately 55,000, of which 34,000 occur in Asia and 24,000 in Africa. At least 150 people die of rabies every day. India has the highest rate in the world attributed to the large number of stray dogs. So many cases occurred in Beijing that the country introduced a "one dog per family" policy.

In the United States there has been a dramatic shift over the past 40 years in the principal wildlife carrier. In 1966 foxes were the major reservoirs, followed several years later by skunks; since about 1993 raccoons have emerged as the principal reservoir (FIGURE 10.19). The raccoons frequent neighborhoods in search of food and may bite dogs and cats. Only about 5% to 15% of individuals bitten by rabid animals develop the disease. Seventy-nine human cases occurred in 2006 and deaths due to rabies have been reduced to one to two per year.

The bat is a rabies vector, and most human rabies cases in the United States are caused by bat bites. Bats have acquired a bad but unearned reputation perhaps, at least in part, due to the novel *Dracula* by Bram Stoker, published in 1897, in which bats were depicted as evil, blood-sucking creatures capable of turning their victims into horrible life forms. However, the novel contained no reference to a connection between bats and rabies. Bats have been dramatized by the

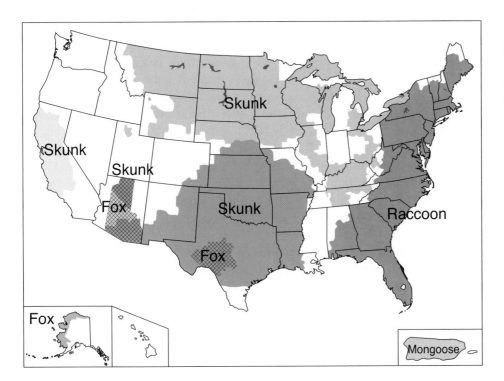

FIGURE 10.19 The distribution of the major reservoirs of rabies in the United States in 2005. Adapted from the CDC, Animals and Rabies, Rabies: Just for Kids, http://www.cdc.gov/ncidod/dvrd/kidsrabies/Animals/animals.htm.

escapades of Batman as he wheels around in his Batmobile. Expressions such as "bats in the belfry," "batty," "blind as a bat," and "crazy as a bat" are part of our jargon and indicate our fear of bats. But let's clear up a few misconceptions: Bats are not blind, most species do not suck blood (an exception being vampire bats of South America), and they do not particularly like to "get in your hair." Actually, they are remarkable creatures (some say "beautiful," but that is too much) and play an important role in ecosystems around the world by eating insects, including agricultural pests.

Despite rabies vaccination in domestic pets, wild animals that carry the disease and unvaccinated pets remain a constant threat. A rabies scare occurred in June 2008 when a dog imported from Iraq under the "Operation Baghdad Pups," an affiliate of the International Society for the Prevention of Cruelty to Animals organization, was confirmed to have rabies. Twenty-three other dogs in the shipment were exposed to the rabid dog and shipped to 16 states, but none became ill. Thirteen of the 28 persons who had contact with the dogs were administered postexposure treatment. Other similar cases have occurred in conjunction with other programs. The CDC is responsible for implementation of federal regulations governing the importation of dogs (and other animals).

The downside of organ transplantation is that it introduces opportunity for the transmission of disease. Two examples make this clear. In June 2004 three organ recipients died from rabies transmitted from the organs of an infected donor; the donor had been bitten by bats. In February 2005 three persons contracted rabies as the result of organ and corneal transplants.

Jeanna Giese made history after contracting rabies at age fifteen. Jeanna had picked up a bat and suffered a minor bite on her finger. She received no medical attention and 21 days later developed the symptoms of rabies. It was too late for rabies vaccine therapy because she had already developed an antibody response to the virus. Her doctors, aware that death from rabies is associated with brain dysfunction, decided on an experimental treatment known as the **Milwaukee Protocol**. Giese was put into a drug-induced coma using a unique combination of antiexcitatory and antiviral medications. The coma allowed her brain to recover and allowed time for her immune system to produce antibodies to the rabies virus. After six days and signs that her immune system was responding, she was brought out of the coma. Jeanna is the first person known to have recovered from symptomatic rabies without receiving the rabies vaccine after exposure; six more individuals survived rabies to date (**FIGURE 10.20**). Before Jeanna's unusual case, seven other rabies survivors had received prophylaxis. Of the 7, only 4 made a full recovery.

The survivors after Jeanna's unusual case did not receive postexposure vaccination but were treated by a modified Milwaukee Protocol administered to Jeanna, the most recent survivor being eight-year-old Precious Reynolds of Willow Creek, California. In 2011, Precious was scratched by a rabid feral cat and developed rabies symptoms. She was in the hospital for 53 days. She made a full recovery aside from a slight limp in her left leg. A total of 41 attempts using the Milwaukee Protocol have been made worldwide. Five of the 41 patients survived, including Precious who had the best recovery to date.

The most common mode of rabies transmission is through the bite of a rabies-infected animal. In addition to a bite, the virus can also be transmitted, although rarely, into the eyes, nose, and respiratory tract. The early (prodromal) symptoms of rabies are nonspecific and flulike, and there may be pain and tingling at the site of the bite. In the secondary (illness) stage there are symptoms of anxiety, confusion, agitation, delirium, abnormal behavior, hallucination, drooling, and hydrophobia. Persons with rabies are tortured by thirst and, ironically, are revolted by water. The mere sight of water results in uncontrollable spasms in the muscles of the mouth and pharynx, leading to spitting and choking.

FIGURE 10.20 Jeanna Geise, the first rabies survivor who did not receive postexposure vaccination. © Morry Gash/AP Photo.

Before the work of Pasteur the public's fear of hydrophobia and the prevailing attitude that the disease could be transmitted through the saliva or breath of rabies victims was so overwhelming that violent modes of death, including suffocation, were inflicted upon those who had suffered bites of a rabid animal. Such events must have been quite frequent, because in 1810 a bill in France was written in the following terms: "It is forbidden under pain of death, to strangle, suffocate, bleed to death, or in any other way, murder individuals suffering from rabies, hydrophobia, or any disease causing fits, convulsions, furious and dangerous madness." Early treatments were terrible. **Cauterization** was the most frequently used method; if the wounds were somewhat deep, it was recommended to use long, sharp needles and to push them well in, even if the wound was on the face. Another practice involved the sprinkling of gunpowder over the wound and setting a match to it.

The incubation time for the onset of symptoms is related to the extent of the wound and its closeness to the brain. The incubation time for wounds on the hands is about eight weeks, whereas for wounds on the face it is about five weeks. The average incubation period is one to two months, with extremes of one week and more than one year. Initially, the virus multiplies in the muscle and connective tissue at the site of the bite; then it migrates along nerves to the central nervous system.

The treatment of rabies has come a long way since Pasteur's first treatment in 1885 of a boy who had been bitten by a rabid dog; his second success was even more dramatic, because about six days had elapsed before immunization (BOX 10.1). The vaccine currently in use consists of inactivated virus cultured in tissue culture. Treatment consists of one dose of immunoglobulin (blood serum from individuals containing high levels of antibodies against rabies) and a first dose of the vaccine given as soon as possible after exposure, with the remaining four doses given on days 3, 7, 14, and 28, respectively. The injections are relatively painless and are given in the arm in the same way that other immunizations are given. Earlier versions of the vaccine were administered into the abdomen and were painful.

Whether the individual needs to be vaccinated is carefully evaluated and depends on the circumstances. Laboratory tests are available to determine whether the animal does, in fact, have rabies. If the bite is from a domestic animal, the animal can be quarantined and observed for symptoms. Vaccinations are advised if a wild animal is involved and the animal cannot be captured. The development of the immunofluorescent antibody test in 1958 allows for immediate determination of whether an animal suspected of being rabid is actually rabid; based on the test outcome, a decision as to whether the exposed person needs to be given the vaccine can be made. It may have occurred to you that, for most diseases, immunization is usually given as a preventive measure and not as a treatment; if it has, you are correct (and very astute). However, the situation with rabies is somewhat unusual in that postexposure vaccination, as a treatment, is effective because of the long incubation time generally associated with rabies. On the other hand, one case of human rabies developed in only 10 days after the person had been bitten. After an animal bite, the wound should be thoroughly cleansed. Preexposure vaccination is recommended for those in high-risk groups, including veterinarians, animal handlers, and laboratory personnel engaged in work with rabies.

A study carried out by researchers at the CDC collaborating with the Peruvian Ministry of Health determined that people living in two remote Amazon communities (Santa Marta and Truenococha) who were repeatedly exposed to rabies virus survived without vaccination. Blood samples were taken from 63 people, and 11% of them had neutralizing antibodies against rabies virus. Researchers were unable to determine if any of the people in the study experienced any symptoms of rabies. Over the past 20 years, outbreaks of fatal human rabies caused by vampire bat bites occurred regularly. The results open the door to the idea that a small percentage of remote Peruvians encountered enough exposure to the vampire bat rabies virus to evoke natural immunity but not enough to kill them. This counters the traditional belief that 100% of individuals exposed to rabies virus died unless they sought post-exposure vaccination.

BOX 10.1 Pasteur and the Development of the Rabies Vaccine

When one thinks of the famous names associated with the history of science, the name Louis Pasteur ranks among the greatest. Pasteur's phenomenal success in diverse fields of research is without parallel, but his development of the rabies vaccine was his crowning glory. Nobel laureate Selman A. Waksman, discoverer of streptomycin, the first antibiotic effective against tuberculosis, wrote in *The New York Times* on February 5, 1950, "Pasteur was not only the great scientist who was largely responsible for the creation of the science of microbiology, he was its high priest, preaching and fighting for the recognition of its importance in health and in human welfare."

In the early 1870s, at about the age of fifty, Pasteur began to conduct research related to human infection. By this time he was well renowned and respected by the scientific community in France and beyond for his work on crystallography, fermentation, diseases of wine and beer, spontaneous generation, silkworm disease, cholera, and anthrax. It is not possible to cite with certainty the event or events that resulted in Pasteur's entry into rabies research. One story describes how, as a young boy, Pasteur had witnessed and heard screams of a victim bitten by a rabid wolf; the victim was undergoing cauterization of the wounds with a red hot iron.

Before the mid-1880s Pasteur succeeded in cultivating the virus of rabies in the brainstem of rabbits. He then attenuated the virus (weakened its virulence) by suspending a fragment of the rabbit virus-infected brainstem from a thread in a sterilized vial. As the brainstem preparation dried, virus virulence gradually decreased, and by fourteen days these crude preparations were totally without virulence. Dogs that were immunized with this crude vaccine and experimentally bitten by rabid dogs or had the active rabies virus applied directly onto their brains did not develop rabies. Pressure was immediately brought to bear on Pasteur to use the vaccine on humans, but Pasteur resisted the temptation, realizing that the vaccine was not yet ready for a trial on human subjects.

Monday, July 6, 1885 was a momentous day for Louis Pasteur and for the history of medicine. It was on that day that Joseph Meister, a nine-year-old boy, was brought to Pasteur's laboratory by his mother. On his way to school, two days earlier, Joseph had been severely attacked by a rabid dog and suffered fourteen wounds on his hands, legs, and thighs. The child's wounds had been treated with carbolic acid 12 hours after the incident by a local physician, who advised that Joseph be brought to Paris to be seen by Pasteur. The boy's mother pleaded to Pasteur for help for her doomed son. Pasteur's reputation was on the spot. Here was this mother pleading with him to save her child, but he worried that his vaccine, which had been tried only on dogs, would not be effective. Pasteur relented and administered the vaccine to his first human subject. Young Meister received thirteen inoculations of Pasteur's rabies vaccine over the next several days and survived his ordeal.

Jean-Baptiste Jupille, age fifteen, was the second patient treated by Pasteur, under less favorable circumstances.

Ebola Virus

Following are vivid excerpts about **Ebola hemorrhagic fever** from *The Coming Plague*, by Laurie Garrett (1994, The Penguin Group):

> "And he was bleeding. His nose bled, his gums bled, and there was blood in his diarrhea and vomit. . . . They pumped Antoine full of antibiotics, chloroquine, vitamins, and intravenous fluid to offset his dehydration. Nothing worked. . . . The horror was magnified by the behavior of many patients whose minds seemed to snap. Some tore off their clothing and ran out of the hospital, screaming incoherently. . . . Some, the huts of the infected, were burned by hysterical neighbors."

Jupille was in a group of six shepherd boys who were attacked by a rabid dog. One of the many biographers of Pasteur describes the event:

> "The children ran away shrieking, but the eldest of them, Jupille, bravely turned back in order to protect the flight of his comrades. Armed with his whip, he confronted the infuriated animal which flew at him and seized his left arm. Jupille wrestled the dog to the ground and succeeded in kneeling on him, forcing his jaws open in order that he might disengage his left hand, and in so doing, his right hand was seriously bitten in its turn; finally, having been able to get hold of the animal by the neck, Jupille called his little brother to pick up his whip which had fallen in the struggle and secured the animal's jaws with the whip. He then took his wooden shoe, with which he battered the dog's head."

Source: R. Vallery-Radot, *La vie de Pasteur* (Hachette et Cie, Paris, France, 1900).

Jupille was brought to Pasteur for treatment six days after being bitten, in contrast to the two-day interval in the Meister case. Pasteur was particularly reluctant to treat Jupille because of the extended time that had elapsed since the attack. Nevertheless, Pasteur again put his reputation on the line and injected Jupille with the vaccine. Jupille did not develop rabies, and the boy's act of bravery was commemorated with an impressive statue depicting the struggle between a boy and a rabid dog, which stands today on the grounds of the Pasteur Institute (FIGURE B10.1).

FIGURE B10.1 Statue of Jean-Baptiste Jupille at the Pasteur Institute in Paris. Jupille was attacked by a rabid dog while attempting to protect younger children from the animal. He was the second person to receive the vaccine. Author's photo (RIK).

The following is an excerpt from an article in *Newsweek* published on May 22, 1995:

> "Then, as the virus starts replicating in earnest, the victim's capillaries clog with dead blood cells, causing the skin to bruise, blister, and eventually dissolve like wet paper. By the sixth day, blood flows freely through the eyes, ears, and nose, and the sufferer starts vomiting the black sludge of his disintegrating internal tissues. Death usually follows by day nine."

These descriptions are enough to make you break out in a sweat. It is no wonder that near panic resulted in the United States in 1989, 1990, and again in 1996 when Ebola virus was introduced into primate quarantine facilities in Pennsylvania, Texas, and Virginia from monkeys imported from the Philippines. Several individuals developed antibodies, but no human cases of disease were identified. In 2008, another

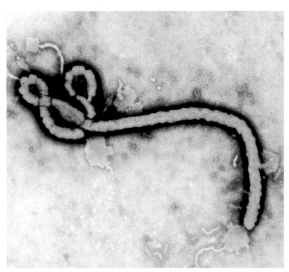

FIGURE 10.21 Colorized transmission electron micrograph of the Ebola virus. Courtesy of Cynthia Goldsmith/CDC.

scare—the first known infection of Ebola-Reston virus in pigs on a farm in the Philippines—occurred. Six workers from a pig farm and slaughterhouse developed antibodies but did not get sick.

Ebola virus, an RNA virus, the cause of Ebola hemorrhagic fever, is named after the Ebola River in the African nation of Zaire (now the Democratic Republic of the Congo), where it was first detected in 1976 (**FIGURE 10.21**). The virus causes sporadic outbreaks of severe infection with a high fatality rate in humans and in monkeys and chimpanzees. The disease is considered to be zoonotic, although the reservoir of the virus is not definitely known.

Symptoms, including fever, chills, muscle aches, headache, stomach pain, sore throat, and abdominal pain, appear after an incubation period of 4 to 16 days. The blood fails to clot, resulting in massive hemorrhage throughout the body, both externally and internally. Transmission requires close contact with an individual ill with the disease. Sexual contact may also transmit Ebola because the virus may remain infective in genital secretions for a brief time. Humans are "incidental" hosts—they do not "carry" the virus.

An outbreak of Ebola hemorrhagic fever originated in a hospital in 1995 in Kikwit, Democratic Republic of the Congo. Kikwit is 240 miles east of Kinshasa, the capital of that nation. The epidemic raged through the hospital staff as they cared for and cleansed the bodies of the ill and of those who succumbed to the disease. Syringes were reused and isolation procedures were not carried out, as hospital personnel walked freely among the patients and unknowingly spread the disease from one to another. In poor countries families play an intimate and crucial role in caring for hospitalized relatives. Following centuries of tradition, the dying were washed along with other practices that increased contact between the living and the dead. As it later became known, a major mechanism of transmission is direct—from person to person—transmission. As the families left the Kikwit hospital and returned to their villages, little did they know that Ebola was with them. According to the WHO, there were 315 cases during this outbreak, of which 250 died—an overwhelming 81% fatality rate. In 2007 and 2008 to 2009 outbreaks of Ebola reemerged in the Democratic Republic of the Congo for reasons that are not known, documenting the theme of "new, emerging, and reemerging." Of the 264 cases in 2007, 187 resulted in death (71%). There were 32 cases in the outbreak that began in 2008, and of these, 15 (47%) died. In May 2011, a single fatal case of Ebola hemorrhagic fever caused by an Ebola-Sudan strain occurred in the Luwero district in Uganda. The disease stopped in its tracks. In August 2012, Ebola virus resurfaced in western Uganda. Over 17 individuals died and over 300 potential contacts were under surveillance.

Definitive diagnosis of Ebola virus infection is based on the detection of antibodies or of Ebola virus genetic material. Ebola has a mortality rate of 50% to 80%, and there is no specific treatment available against Ebola virus infection.

Control measures in a community or hospital environment center on isolation of those with the disease and avoidance of direct contact with infected persons and their blood or secretions and with the bodies of deceased patients. Barrier nursing precautions are called for, including the use of gowns, goggles, masks, and gloves by physicians, nurses, and other healthcare workers, the restriction of visitors, and proper disposal of wastes and corpses. Researchers and other laboratory workers handling Ebola virus and other viruses that cause hemorrhagic fevers must work in biosafety level four cabinets (FIGURE 10.22) as a matter of protection. The CDC has established biosafety levels one through four with reference to work practices, safety equipment, and facilities to minimize the consequences of escape of infectious microbes to laboratory personnel and to the environment. Level four is reserved for the "hot" microbes—extremely dangerous and exotic microbes with a potential to cause life-threatening disease and for which there is no vaccine or treatment.

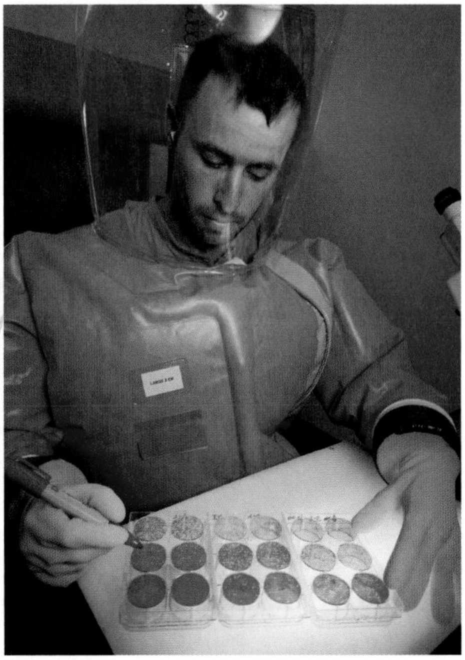

Arthropodborne Diseases

A variety of viral diseases are transmitted by arthropods (TABLE 10.6). (Arthropod vectors are also involved in some diseases caused by protozoans and worms.) Arthropodborne viruses are called arboviruses. With a single exception, that of Colorado tick fever, they are transmitted by mosquitos.

Dengue Fever

Dengue fever, also called **breakbone fever**, can result from four dengue viruses. The vectors are the *Aedes aegypti* and, rarely, *Aedes albopictus*. The recent appearance of *A. albopictus* in the United States is of great concern because this species is aggressive, and its biting habits could spread dengue fever.

Dengue fever is caused by an RNA virus and is usually self-limiting; although the disease can be debilitating, recovery occurs in about ten days. **Dengue hemorrhagic fever** is a serious and potentially fatal infection caused by a dengue virus

FIGURE 10.22 Biosafety cabinet. Highly virulent microbes and viruses must be handled in biosafety cabinets to protect laboratory personnel and the environment. This special pathogens laboratory worker is counting viral plaques within a fixed monolayer of cells in a BSL-4 laboratory at the CDC. Courtesy of Dr. Scott Smith/CDC.

TABLE 10.6 Arthropodborne Viral Diseases

Disease	Genome Type	Symptoms	Comments
Arboviral encephalitis	ssRNA	Meningitis, coma, tremors, convulsions, paralysis, brain damage	EEE, WEE, St. Louis encephalitis, West Nile virus. La Crosse encephalitis (LEV) (all mosquito-borne), tickborne Colorado fever
Yellow fever	ssRNA	Fever, bloody nose, black vomiting, jaundice	Immunization by live attenuated virus; relatively long-lasting immunity; used for those at high risk
Dengue fever	ssRNA	Flulike symptoms	Dengue hemorrhagic fever can develop; *Aedes aegypti* and *Aedes albopictus*
Dengue hemorrhagic fever	ssRNA	Hemorrhages in skin, gums, and other areas	Potentially fatal

ds, double-stranded; EEE, eastern equine encephalitis; ss, single-stranded; WEE, western equine encephalitis.

strain different from the one causing the initial infection. The condition is manifested by hemorrhages occurring in the skin, gums, and other areas within the body. Shock may develop, requiring immediate treatment to prevent death, and even with intensive supportive measures as many as 40% of those infected may die. There is no immunization against any of the four dengue virus strains. Because the early characteristics of the disease are flu-like, a definitive diagnosis is difficult and requires laboratory confirmation, either through virus isolation or through detection of virus-specific antibodies in the blood.

Dengue fever epidemics date back to the early 1700s in Asia, Africa, and North America and continued through the decades with caseloads of 40 million people and several hundred thousand deaths. In March 2008 an epidemic occurred in Rio de Janeiro, resulting in close to twenty-four thousand cases including thirty deaths; other outbreaks continue to occur.

There is no vaccine against dengue fever. As in all arthropod-borne diseases, insect control is the best prevention. Travel-associated dengue fever and small outbreaks do occur in the continental United States; most dengue cases in U.S. citizens occur as endemic transmission among residents in some of the U.S. territories. Dengue fever was placed on the CDC *Morbidity and Mortality Weekly* (*MMWR*) list of "reportable diseases" January 22, 2010, requiring laboratories and heathcare providers to report all U.S. cases of dengue fever to the CDC.

Yellow Fever

Yellow fever, also called yellow jack, is caused by an RNA virus. The disease once had a widespread distribution, but mosquito control measures have resulted in elimination in many countries, including the United States. In 1793 a yellow fever epidemic devastated Philadelphia, America's early capital city, causing 2,000 deaths. The largest epidemic in the United States was in New Orleans in 1853, resulting in a death toll of 7,849; New Orleans was also the site of the last epidemic in 1905. The disease is still present in areas of South America, Central America, and Africa. In jungle areas monkeys serve as reservoirs, and the incidence of disease is highest in these areas because mosquitoes bite both monkeys and humans. The vector is *A. aegypti*, a mosquito that has jumped the species to humans; it bites by day and breeds in standing water, such as in old tire casings. The disease is manifested by fever, bloody nose, headache, nausea, muscle pain, (black) vomiting, and jaundice. In about a week the infected individual is either dead or in the process of recovery.

Immunization against yellow fever has been available since 1950 with live, attenuated viruses. Protection is relatively long-lasting, and immunization is required for travelers to certain destinations.

The history of the Panama Canal is linked with yellow fever and malaria (FIGURE 10.23). The construction of the canal was an engineering triumph thwarted for years by the yellow fever virus and by a protozoan, the malaria parasite (BOX 10.2). Hence, its construction was also a triumph over microbes.

FIGURE 10.23 The Panama Canal. Construction of the Panama Canal was possible only when yellow fever and malaria were conquered. Courtesy of Edwin P. Ewing, Jr./CDC.

The construction of the Panama Canal linking the Atlantic and Pacific Oceans was a monumental undertaking over a thirty-five-year span. In 1914 the canal's waterways were finally opened to traffic. The canal is a winding waterway stretching miles through a series of locks that raise and lower ships feet. The obstacles encountered in completing this massive project were tremendous and seemingly impossible to overcome. There were political, financial, and engineering problems of gigantic proportions, but one by one they were resolved. As it turned out, the task of moving thousands of tons of earth was overshadowed by a more daunting problem—the microbial enemies that could not be seen, namely, the virus of yellow fever and the protozoan of malaria. The challenge was waging and winning the war against their carriers—mosquitoes.

In 1879 France's Ferdinand de Lesseps, having successfully completed the Suez Canal in Egypt ten years earlier, embarked on building the Panama Canal. A French resident of Panama warned de Lesseps, "If you try to build the Canal, there will not be trees enough on the Isthmus to make crosses for the graves of your laborers." His prediction, unfortunately, proved to be correct. De Lesseps had underestimated the power of mosquitoes, malaria, and yellow fever. The workers were plagued by these natural enemies and the project was abandoned after a ten-year battle.

David McCullough vividly describes yellow fever in his book, *The Path between the Seas:*

"In those places where it was most common—Panama, Havana, Veracruz—it was the stranger, the newcomer, who suffered worst, while the native was often untouched. Wherever or whenever it struck, it spread panic of a kind that could all but paralyze a community. . . .

As with malaria, the patient was seized first by fits of shivering, high fever, and insatiable thirst. But there were savage headaches as well, and severe pains in the back and legs. The patient would become desperately restless. Then, in another day or so, the trouble would appear to subside and the patient would begin to turn yellow, noticeably in the face and in the eyes.

In the terminal stages the patient would spit up mouthfuls of dark blood, the infamous, terrifying vómito negro, *black vomit. The end usually came swiftly after that."*

Source: D. McCullough, *The Path between the Seas* (Simon & Schuster, New York, 1977).

In 1903 the United States acquired the Panama Canal during the presidency of Theodore Roosevelt. In 1904 William C. Gorgas, an American military physician, went to Panama. Epidemics were in full force at the time, but in less than a year Gorgas cleared the area of disease by draining the swamps to deprive mosquitoes of breeding places. Gorgas implemented other sanitary measures, and it was his success in controlling the mosquito population that led to the completion of the canal.

You are going to have the fever, Yellow eyes!
In about ten days from now
Iron bands will clamp your brow;
Your tongue resembles curdled cream,
A rusty streak the center seam;
Your mouth will taste of untold things,
With claws and horns and fins and wings;
Your head will weigh a ton or more,
And forty gales within it roar!

—Excerpted from James Stanley Gilbert, *Panama Patchwork* (The Trow Press, New York, 1909).

Encephalitis-Causing Arboviruses

Common encephalitis-causing arboviruses in the United States are **West Nile fever, eastern equine encephalitis, western equine encephalitis, St. Louis encephalitis,** and **Colorado tick fever.** These are all RNA viruses. Note that Colorado tick fever is the most common tickborne viral fever in the United States. All other arboviruses are carried by mosquitoes. Humans do not serve as reservoir hosts; these viruses cycle between wild animals, primarily wild birds and mosquitoes, with humans and horses serving as dead-end hosts (FIGURE 10.24). When the virus invades the spinal cord, the **meninges** (wrappings around the spinal cord and brain) become inflamed, possibly resulting in coma, convulsions, tremors, paralysis, memory deficits, and permanent brain damage.

In areas where the disease may be a problem, sentinel animals (caged rabbits or chickens) are left in mosquito-infested areas; blood is drawn from these animals on a periodic basis and tested for the presence of specific antibodies, which indicate the presence of viruses. Also, mosquitoes are collected and tested to determine whether they are carrying encephalitis viruses.

West Nile Fever

It appears that mosquitoes have a vengeance against New York. In late August 1999, on the heels of a St. Louis encephalitis outbreak, New Yorkers were plagued by yet another arbovirus, **West Nile virus**, first described in the 1930s in Africa. This virus is carried by the same mosquito (*Culex* sp.) that transmits St. Louis encephalitis. In a period of only several weeks 56 cases, including 7 deaths, had been reported in the New York City area. Mosquito control measures, including aerial spraying, were quickly implemented.

West Nile fever is a new and emerging disease in the United States, with its origins in Egypt and Israel. It is not certain how the virus arrived in New York, but experts believe that it probably came in a bird or mosquito imported on a jet aircraft, another example of the pitfalls of technology.

FIGURE 10.24 Cycle of transmission of arthropodborne viral encephalitis. EEE, eastern equine encephalitis; WEE, western equine encephalitis.

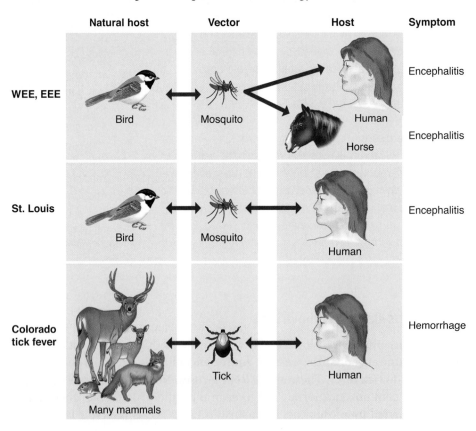

PART 2 Microbial Disease

Outbreaks of illness and deaths occurred in horses and in hundreds of crows in the area by 2000. At the Bronx Zoo several rare birds died and others became ill, including "a trumpeter swan doing the backstroke." A variety of mammals including skunks, chipmunks, cats, dogs, and horses are also susceptible. Mosquito spraying abatement measures, along with the onset of cooler temperatures, were effective in breaking the transmission cycle.

In a matter of only several years the disease spread across the country and into Canada, Mexico, and the Caribbean. A CDC surveillance report summarizing West Nile virus covering January 1 to October 16, 2007 cited 42 states with a total of 3,022 cases of human West Nile virus illness; 76 cases were fatal. Additionally, 1,924 birds (primarily crows, jays, and magpies) died as a result of West Nile infection in 34 states and in New York City. As of 2011, human or animal cases of West Nile encephalitis have been reported in 48 states.

Between January and August 21, 2012, the highest number of West Nile encephalitis cases were reported to the CDC since the disease was first detected in 1999—tallying 1,118 cases. Of these cases, 56% caused encephalitis resulting in 41 deaths and 44% were classified as a non-neurovasive illness. About 75% of the cases were reported from five states: Texas, Mississippi, Louisiana, South Dakota, and Oklahoma. In Dallas, Texas (the epicenter of the West Nile encephalitis epidemic), a "state of emergency" was declared and the entire city was sprayed with mosquito insecticide by airplanes. It was the first time in 50 years that aerial spraying occurred in Dallas. Most cases occur in elderly; usually in the summer, when humans, mosquitoes and migratory birds are in close proximity outdoors. Elderly patients are affected most severely. Worldwide, most cases are children and young adults.

Colorado Tick Fever

Colorado tick fever is transmitted by the bite of an infected tick and is prevalent in mountain forest environments at altitudes of 4,000 to 10,000 feet. Most cases occur between the months of April through July.

St. Louis Encephalitis

St. Louis encephalitis is found throughout most of the United States. Birds are the reservoirs. The disease can be asymptomatic, but most cases display mild illness characterized by headache and fever. Severe illness sometimes results in high fever and neurological symptoms.

La Crosse Encephalitis

La Crosse encephalitis is named after La Crosse, Wisconsin where the first case occurred in 1963. Historically, most cases of La Crosse encephalitis were reported from the upper midwestern states of Wisconsin, Minnesota, Iowa, Illinois, Ohio, and Indiana. More recently cases have been reported in the mid-Atlantic and southeastern states. Most individuals infected by La Crosse encephalitis virus are asymptomatic. About 80 cases are reported each year in the United States. Most cases of severe La Crosse encephalitis occur in children under the age of sixteen who are bitten by eastern tree hole mosquitoes (*Aedes triseriatus*) in outdoor recreational areas. *Aedes triseriatus* lay their eggs in tree holes or man-made containers and typically bite during the day.

Eastern Equine Encephalitis

Eastern equine encephalitis occurs in the eastern half of the United States in humans, horses, and in some birds. On the average there are five deaths per year. The states of Florida, Georgia, Massachusetts, and New Jersey appear to be the most prone to eastern equine encephalitis. Contrary to popular opinion, infected horses are a "dead end" and not a significant risk factor, because the amount of virus in their blood is not sufficient to infect mosquitoes.

Western Equine Encephalitis

Western equine encephalitis is endemic to the western half of the United States. The disease is primarily subclinical and rarely symptomatic but can cause severe neurological damage to the point of death. The mortality rate is 4%.

Prion Diseases

Prions are infectious proteins that cause a number of neurodegenerative diseases in animals and humans called **transmissible spongiform encephalopathies** (TSEs). The first human TSE discovered was **kuru**. TSEs are characterized by long incubation periods of several years but once symptoms begin, the condition progresses rapidly and is fatal. There is no cure. Treatment is supportive. The prions do not evoke any host specific immune response and are unusually resistant to disinfection. The prion disease that created the most awareness is bovine spongiform encephalopathy (BSE).

Mad Cow Disease (MCD)

It all began in 1986 when an outbreak of BSE, better known as mad cow disease (MCD), in the United Kingdom frightened meat eaters around the world and raised the questions, "Can it happen here? Is it transmissible to humans?" In the United States the MCD scare was fueled by an Oprah Winfrey show about the disease and Oprah's statement, "I will never eat another hamburger." Television news shows wasted no time in showing pastures populated by mad cows drooling, stumbling, and unable to rise to their feet.

BSE is characterized by spongy degeneration of the brain accompanied by severe and fatal neurological damage. Cows suffering from BSE are referred to as "downer" cows. A **downer cow** is unable or unwilling to stand for greater than twelve hours. Other symptoms include:

- Changes in temperament such as nervousness and aggression towards other cattle or humans
- Kicking when being milked
- Incoordination
- Difficulty walking
- Incoordination
- Head shyness (hanging head low)
- High stepping hind leg gait
- Skin tremors
- Weight loss despite having a good appetite

Since the 1986 initial reports of BSE, according to the CDC, over 180 thousand cases have been confirmed in the United Kingdom. The disease spread to cattle in other European countries, primarily France, Switzerland, Portugal, and Ireland (BOX 10.3). More than five million cattle were slaughtered in the United Kingdom to halt the epidemic. Ten infected cows have been reported in the United States since 2008. All of these cases were imported from infected cows imported from Canada to the United States.

There are other forms of TSEs in addition to BSE, also characterized by spongy deterioration of the brain. Scrapie is a TSE of sheep and is manifested by animals scraping against trees, fencing, and whatever else might be available. TSEs that resemble BSE are also found in household cats, deer, elk, and mink, as well as in other ruminants (Box 10.3 and BOX 10.4).

Human TSEs

Creutzfeldt-Jakob disease (CJD), the model for the human TSEs, is transmitted in three ways. Sporadic cases occur throughout the world at a rate about one case per one million people, accounting for 85% to 90% of CJD cases. Another 5% to 10% of CJD cases are due to hereditary predisposition associated with gene mutation. Fewer than 5% of CJD cases are **iatrogenic**, meaning they are transmitted via contaminated surgical equipment, corneal transplants, or natural human growth hormone. The prototype human TSE was kuru, which was epidemic in Fore natives of New Guinea in the 1950s and 1960s. Kuru was spread by an ancient cannibalistic practice in which, as a sign of respect, Fore people, particularly females, ate the brains of deceased relatives, thereby infecting themselves with prions. This ritual has been abolished, leading to the disappearance of kuru.

In March 1996 a new form of CJD appeared. CJD develops at an average age of 65 years and has a mean duration of illness of about four and a half months,

AUTHOR'S NOTE (TS)

It was Thanksgiving week during my first year as an assistant professor at the University of Wisconsin Oshkosh. I noticed that there were hardly any males in my classes that week. I was confused as I didn't notice a drop in female attendance as I would expect before a holiday. Naively, I asked colleagues about my class attendance observations, and the response was: "Students are deer hunting during the week of Thanksgiving." Deer hunting is a tradition that is deeply ingrained in the state of Wisconsin. Faculty also warned me that there are drops in attendance at Monday night classes because of "Monday night football." Packer fans are strong and loyal through thick and thin; no matter if they are winning or losing, I was relieved to learn that I would never have to teach a Monday night class.

BOX 10.4 Where the Deer and the Antelope Play

Deer hunting is an age-old tradition, dating back to tens of thousands of years ago. According to the 2001 *National Survey of Fishing, Hunting, and Wildlife-Associated Recreation* (FHWAR) deer hunting is the most popular form of hunting in the U.S. There were 10.3 million deer hunters in 2001. Nearly 1 in every 20 Americans and 8 in 10 hunters hunted deer in 2001, and their hunting-related expenditures while seeking deer totaled nearly $10.7 billion. At least 50% of hunters in all but a few states hunt deer.

The deer population is plagued by a new disease. Symptoms of the disease were first observed in mule deer grazing on northern Colorado wildlife research land in 1967. The sick animals had poor hair coats and appeared emaciated, or starving and "wasting away." Other symptoms of the sick mule deer were blank facial expressions, excessive drooling and thirst, frequent urination, teeth grinding, nervousness, holding head in a lowered position, and sluggish behavior, and sick deer isolated themselves from the herd. In 1978, researchers named it chronic wasting disease (CWD) and classified it as a TSE. To date, CWD has been identified in sixteen states: Colorado, Illinois, Kansas, Missouri, Wisconsin, Minnesota, New York, North Dakota, Utah, West Virginia, Nebraska, South Dakota, Oklahoma, New Mexico, Michigan, and Wyoming. It has also been identified in deer and farmed elk in the providences of Saskatchewan and Alberta, Canada.

While human prion diseases are very rare, the incidence of CWD can be over 30% in wild deer populations and as high as 100% in captive deer herds. CWD prions are found in saliva, feces, blood, and urine of the infected animal. The stability of the prions makes it difficult to remove from the environment (e.g., soil, watering holes, and birthing sites). Transmission likely occurs through direct animal-to-animal contact or indirectly through contaminated water, carcasses, and food sources.

The catchphrase for returning home with deer and elk is "no skull, no backbone." This is based on the idea that the prions are concentrated in the nervous tissues, spinal cord, and antlers of the animal. Removing them would reduce exposure to prions. Hunters should take precautions by not eating brain, spinal cord, eyes, spleen, tonsils, or lymph nodes of deer. Proper field dressing removes most if not all of these body parts (FIGURE B10.4).

FIGURE B10.4 Hunters field dressing a deer. © Farlap/Alamy.

Since eating BSE-contaminated meat can cause vCJD in humans, what about eating CWD-contaminated meat? Between 1993 and 1999, three men who participated in wild game feasts in northern Wisconsin died of unknown degenerative neurologic illnesses. A CDC investigation confirmed one death to be caused by CJD. To date there is no proof of an association between CWD and CJD. Research is ongoing to determine if there is a species barrier that protects humans from CWD. CWD is worrisome because it is spreading through wild deer populations and could possibly make the species leap to other animals in the wild or in pastures, including livestock, squirrels, chipmunks, or even humans as has happened with other diseases previously mentioned.

whereas the new form, variant CJD (vCJD), affects individuals at an average age of 29 years and has a duration of illness of about 14 months (TABLE 10.7). Further, vCJD is transmissible through prion-contaminated beef. By November 2001, 106 cases had been identified in the United Kingdom, four cases had occurred in France, and one case had occurred in Ireland.

TABLE 10.7 Comparison of CJD and vCJD

Disease	Average Age at Infection (yr)	Duration of Illness (mo)	Transmission
CJD	65	4.5	Sporadic (cause unknown), 85%–90%; genetic mutation, 5%–10%; iatrogenic, <5%
vCJD	29	14	Prion-contaminated beef, 100%

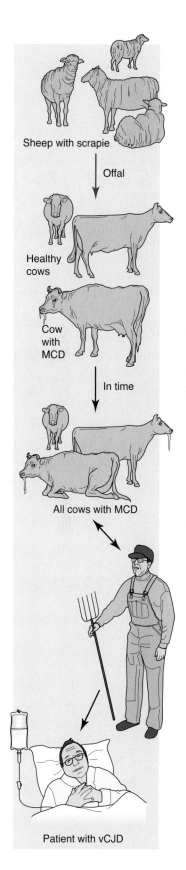

Sheep with scrapie

Offal

Healthy cows

Cow with MCD

In time

All cows with MCD

Patient with vCJD

Disturbingly, on April 24, 2012, for the first time in six years, a United States Department of Agriculture (USDA)–confirmed case of BSE was reported in a cow in central California; tissue samples revealed "atypical BSE." This marked the fourth discovery of BSE in United States cattle, the first occurring in December 2002, followed by cases in 2005 and 2006, respectively. On the positive side, the finding indicates not only that the safeguards in place are effective but also that there is the need for constant vigilance.

BSE-vCJD Link

The most plausible explanation for vCJD is the contamination of beef products by central nervous system (brain and spinal cord) tissue. The cluster of vCJD cases is most likely due to the same agent that causes BSE. Support for this hypothesis is based on several lines of evidence, including an association between these two TSEs in time and place, a resemblance of pathological features in the brains of monkeys inoculated with either vCJD or BSE, and the nearly identical and distinctive distribution and pathology of the infectious agent in the brains of mice injected with infected cow BSE tissues or with human vCJD tissue.

Assuming that BSE and vCJD are caused by the same prion strain, how did all of this come about? The events that brought us to this stage are illustrated in (FIGURE 10.25). The widely held view is that the outbreak of BSE in cattle in 1986 in the United Kingdom was the result of the species leap of prions from sheep to cattle when prion-infected animal proteins were used as cattle feed to "beef up" milk production.

The term "animal protein" sounds innocuous, but animal protein is a concoction of ground-up carcasses of sheep and cattle, including their brains, spleens, thymus glands, tonsils, and intestines, a mixture called offal. In essence, cattle were eating the remains of other cattle, including "downers" (sick and dead animals found in the fields). Upton Sinclair's 1906 novel, *The Jungle*, describes conditions in "Packingtown" and the inclusion of downers in meats packaged for human consumption, a practice that continued for almost another 100 years. In 1989 the use of offal from cattle was banned in the United Kingdom. Bovine milk and milk products are not considered to pose any risk for transmission of TSEs.

FIGURE 10.25 Transmission of spongiform encephalopathies.

But why did this outbreak of BSE suddenly occur in 1986? What changed on the farm or enroute from farm to table? The most plausible explanation has to do with a change that occurred in the **rendering process** used in the production of food products for cattle. Rendering is a process somewhat like the making of stew, in which all ingredients are put into a pot and boiled. In the late 1970s the rendering process was altered by the elimination of the solvent-steam treatment phase. Solvents and steam reduce the infectivity of prions and, hence, the transmissibility of TSEs. The upshot, allegedly, is that the scrapie prion, in the absence of the solvent-steam treatment, survived the rendering process and contaminated the dietary supplement, which was then distributed throughout the United Kingdom, resulting in infected herds.

Experiments to study the rendering process and its effect on the inactivation of prions confuse the situation, because none of the different rendering processes completely inactivates prion infectivity. Paul Brown, a career investigator of TSEs at the National Institute of Neurological Disorders and Stroke, suggests that even though the number of infectious scrapie particles may have been only modestly reduced by the solvent-steam treatment, the reduction may have been sufficient to make the difference if, in fact, the level of infectivity was at the borderline of the number of prions necessary to cause BSE in cattle that are fed offal. This might explain why the United States, and other countries that similarly modified the rendering process at about the same time as the United Kingdom, did not experience an outbreak of BSE. But why did the outbreak occur in the United Kingdom? Because the ratio of sheep to cattle is lower in the United States than in the United Kingdom, so is the number of prions (from scrapie-infected sheep), resulting in less than the number of prions required for infectivity. The explanation is confusing but logical.

■ Overview

Humans serve as hosts for a variety of viral diseases. West Nile and Ebola virus infections are regarded as "new" diseases, whereas rabies and plague are diseases of antiquity. The viral diseases discussed in this chapter are categorized on the basis of their primary mode of transmission: foodborne and waterborne, airborne, sexually transmitted, contact, and arthropodborne. Each of these diseases is described with attention to factors involved in the microbial cycle and pathogenicity.

On the basis of complementary receptor molecules on the surface of viruses and host cells, viruses target a specific type of host cell or, in some cases, a variety of cell types. Some viruses cause asymptomatic or mild infection, whereas others are almost always fatal; some are zoonoses, and some exhibit latency. Fever, muscle aches, and respiratory distress are common symptoms and are sometimes too vague to make a definitive diagnosis. Skin rashes accompany some viral infections and are significant in establishing a clinical diagnosis. Several viruses cross the placenta or are acquired by newborns during delivery and cause serious damage to the newborn.

Prions are infectious proteins that cause TSEs in humans and animals. TSEs are characterized by long incubation periods (years) but once symptoms start, the disease progresses quickly and is always fatal.

PART I: Choose the single best answer.

1. Foodborne hepatitis is most commonly the result of
   a. hepatitis virus B, C, and D **b.** hepatitis virus A and E **c.** hepatitis A
   d. hepatitis E

2. HSV-1 causes_____, and HSV-2 causes_____.
   a. cold sores; genital herpes **b.** fever blisters; cold sores **c.** fever blisters;
   shingles **d.** chickenpox; genital herpes

3. Congenital herpes is usually acquired through
   a. contaminated hands **b.** other babies **c.** placental transfer
   d. an infected birth canal

4. Chickenpox and shingles are caused by
   a. varicella-zoster virus **b.** different strains of varicella-zoster virus
   c. herpes-simplex and herpes zoster **d.** contaminated foods

5. Which of the following pathogens has the longest incubation period?
   a. scrapie agent **b.** rabies virus **c.** norovirus **d.** RSV

6. Which of the following pathogens is the hardest to kill using disinfectants?
   a. herpes simplex virus **b.** West Nile virus **c.** CWD prions **d.** Ebola virus

7. This viral pathogen was added to the list of CDC *Morbidity and
   Mortality Weekly* (*MMWR*) list of "reportable diseases" in 2010.
   a. West Nile encephalitis virus **b.** dengue fever virus **c.** herpes simplex
   virus **d.** norovirus

8. Which HPV vaccine can be used to prevent genital warts in boys and girls?
   a. Cervarix **b.** MMR **c.** Gardasil **d.** smallpox vaccine

9. Who was the lead author on journal article (that was retracted)
   correlating autism and the MMR vaccine?
   a. Walter Reed **b.** Darin Reiger **c.** Andrew Wakefield **d.** C. J. Peters

10. Which coronavirus caused a pandemic between November 2002 and
    July 2003?
    a. HIV **b.** hepatitis C virus **c.** SARS-CoV **d.** CWD

PART II: Match the statement on the left with the disease (or microbe) on the right
by placing the correct letter in the blanks. Not all letters are used.

_____ **1.** kissing disease	**a.** St. Louis encephalitis	
_____ **2.** most frequent encephalitis in United States	**b.** rabies	
_____ **3.** Hantavirus	**c.** CWD	
_____ **4.** bloodborne	**d.** RSV	
_____ **5.** common under 6 months of age	**e.** humans are incidental hosts	
	f. herpes simplex virus	
	g. infectious mononucleosis	
_____ **6.** TSE of deer	**h.** hepatitis C	
_____ **7.** Yellow fever vector	**i.** Colorado tick fever	

_____ **8.** Bats are vectors   **j.** mosquito

_____ **9.** Cold sores   **k.** transmitted by rodents

_____**10.** Ebola   **l.** hepatitis A

PART III: Answer the following.

1. Discuss the reasons why the WHO program was able to eradicate smallpox.
2. Discuss why it would be difficult, if not impossible to eradicate rabies virus.
3. "One bad apple spoils the barrel" is a popular expression. Discuss this expression and how it applies to vCJD.
4. Explain why postexposure rabies vaccination is able to prevent rabies.
5. Discuss the inherent problems associated with "chickenpox parties."
6. Why has there been an increase in measles cases in the United Kingdom and the United States?
7. List at least five different ways in which viruses can be transmitted from person to person.
8. Discuss why HPV vaccines are controversial.
9. What is a zoonotic disease? List at least five viral zoonotic diseases.
10. How can viral diseases transmitted by mosquitoes be prevented? List at least five viral diseases transmitted by mosquitoes.

Protozoan, Helminthic, and Fungal Diseases

If a man take no thought about what is distant, he will find sorrow at home.

—Confucius

■ Preview

Malaria, leishmaniasis, sleeping sickness, and giardiasis are examples of protozoan diseases. Ascariasis, river blindness, and tapeworm infection are examples of diseases caused by helminths (worms). Histoplasmosis, ring worm, and candidiasis are fungal diseases. Why are these diseases lumped together in this chapter? Protozoans are microscopic and unicellular microbes, whereas helminths are macroscopic and multicellular. Some fungi are microscopic and unicellular, while others are multicellular and macroscopic. Protozoans, helminths, and fungi are eucaryotic cells, and this is the characteristic that links them together. The helminths are not microbes but are included because some cause infection (sometimes referred to as infestations) with accompanying symptoms that challenge the immune system. Many parasitic protozoans, helminths, and fungi exhibit complicated life cycles

Author's photo (RIK).

involving more than one host. Some biologists reserve the term "parasite" for the multicellular parasites—the worms. In this chapter the term "parasite" has been is used in an inclusive sense and refer not only to worms but to all microbes.

■ Biology of Protozoans

Protozoans are eucaryotic microorganisms. The term "protozoan" is derived from the Greek *protos* meaning first, and *zoon*, meaning animal. There are predicted to be at least 66,110 species and their distribution is worldwide. They are found in fresh water, marine habitats, mud, drainage ditches, water-filled tires, in the guts of termites, and in soil.

Protozoans' unicellular characteristic requires that each cell bear the total burden of staying alive on its own; there is no cellular specialization and differentiation allowing for sharing of functions, as is the case in multicellular organisms. Like all eucaryotes (and as distinct from procaryotes), protozoans have a cell membrane, a membrane-bound nucleus, and other membrane-bound organelles within the cytoplasm. Some species have a protective rigid cover, the **pellicle**, outside the cell membrane; freshwater species continually take in water and eliminate it by contractile vacuoles that push water out. Most protozoans are heterotrophic and aerobic. Many ingest food particles by phagocytosis; the particles are then enclosed within food vacuoles in which digestion takes place.

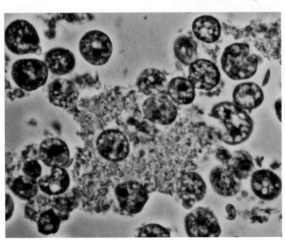

FIGURE 11.1 A photomicrograph showing an *Entamoeba histolytica* cyst. © Eric V. Grave/Photo Researchers, Inc.

Many parasitic protozoans have complicated life cycles involving more than one host and, in some cases, survival in nature outside of a host. **Encystation** allows for outside survival and is a process resulting in a dormant resting stage **cyst** surrounded by a thick capsule (FIGURE 11.1). (In some helminthic life cycles transmission between hosts takes place by ingestion of cysts.) **Excystation** to the **trophozoite** form (the reproductive and feeding stage) occurs after ingestion of the cyst by a host.

Asexual reproduction by binary fission is the most common mode of reproduction in protozoans. A primitive form of sexual reproduction, called **conjugation**, occurs in some protozoans and allows for the direct exchange of genetic material during physical contact of mating pairs.

Relatively few protozoans are human pathogens; some are pathogens for other animals. Most protozoans are free living and constitute a sizable portion of **plankton**, a primary food source for many aquatic organisms and the base of food chains. Although protozoan (and helminthic) diseases are more common in the tropics, they are also significant in temperate zones, particularly with increased immigration and tourism.

Classification of the protozoans is complicated and controversial. There is limited agreement between protozoologists on a universally acceptable classification, but from a practical and medical point of view, four groups are defined and are classified on the basis of their mechanism of locomotion: **sarcodina**, **mastigophora**, **ciliata**, and **sporozoa** (TABLE 11.1 and FIGURE 11.2).

TABLE 11.1 Classification of the Protozoans

Group	Locomotion	Disease(s)
Sarcodina	Ameboid-like	Amebic dysentery
Mastigophora	Flagella	Chagas disease, African sleeping sickness, giardiasis, trichomoniasis, leishmaniasis
Ciliata	Cilia	Balantidiasis
Sporozoa	No locomotion	Malaria, babesiosis

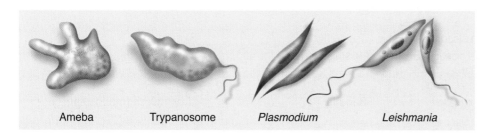

Ameba   Trypanosome   *Plasmodium*   *Leishmania*

FIGURE 11.2 Some protozoan parasites.

Protozoan Diseases

Protozoan diseases are organized by their mode of transmission, following the pattern for bacterial diseases and viral diseases (TABLE 11.2).

Foodborne and Waterborne Diseases

Giardiasis

Consider this hypothetical scenario. A group of Boy Scouts returns from a hiking trip in the Rocky Mountains tired, hungry, and dirty but in good health and in good humor. Over the next ten to twelve days disaster strikes: Several of the scouts fall ill with cramps, nausea, and diarrhea. The diagnosis is likely to be **giardiasis**, a common protozoan intestinal infection in the United States acquired by drinking contaminated water. If you are a hiker or a camper, beware of drinking from "pristine" mountain streams and other bodies of water. The days are long gone when you could drink directly from clear, cool, refreshing lakes and streams without running the risk of acquiring giardiasis, an infection caused by the flagellated protozoan *Giardia lamblia* (FIGURE 11.3). The intestinal disease caused by this parasite has been dubbed "hiker's diarrhea" and "beaver fever" after one of the protozoan's animal hosts.

Over the past fifteen years giardiasis has been recognized as one of the most common waterborne human diseases in the United States, and it is the most commonly diagnosed intestinal parasite in public health laboratories. The parasite has been isolated from humans and a variety of animals, including dogs, coyotes, cats, cattle, horses, birds, and the eponymous beavers—all of which serve as reservoirs and pass cysts into water.

After excystation of the cysts to the vegetative and multiplying trophozoites, attachment to the lining of the intestine occurs by means of **sucking disks**. The parasites feed on mucous secretions and undergo reproduction, resulting in large

TABLE 11.2 Major Protozoan Diseases

Organism	Disease	Transmission	Principal Site(s)
Foodborne and waterborne			
Giardia	Giardiasis	Water, direct contact	Intestinal tract
Entamoeba	Amebiasis	Water, food, direct contact	Intestinal tract
Cryptosporidium	Sporozoa	Water	Intestinal tract
Toxoplasma	Toxoplasmosis	Food; contact with cat feces resulting in fecal–oral transmission	Brain, heart, lungs; possible transfer to fetus transmission
Arthropodborne			
Trypanosoma brucei	African sleeping sickness	Tsetse fly	Blood
Trypanosoma cruzi	Chagas disease (South American trypanosomiasis)	Kissing bug	Heart
Leishmania	Leishmaniasis	Sand fly	Liver, spleen, mucocutaneous membranes, skin
Plasmodium	Malaria	Mosquito	Blood
Babesia	Babesiosis	Tick	Blood
Sexually transmitted			
Trichomonas	Trichomoniasis	Sexual contact	Urogenital tract

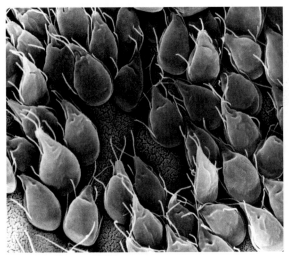

FIGURE 11.3 Scanning electron micrograph of the small intestinal surface covered by *Giardia lamblia* trophozoites. Courtesy of Dr. Stan Erlandsen/CDC.

populations. Typical symptoms include diarrhea, abdominal pain, and large amounts of gas. Although the diarrhea may be extensive, it is not bloody because the protozoans do not usually invade deeply into the lining of the intestine. Symptoms usually appear about two weeks after infection and may last six weeks or longer.

Transmission occurs by swallowing water from contaminated swimming pools, water theme parks, lakes, rivers, shallow wells, springs, and even municipal water supply systems. The parasite is resistant to the chlorine concentration used in municipal systems, making control of giardiasis a major problem. Adding to the problem is that *Giardia* has a very low infectious dose—as few as ten cysts are enough to establish giardiasis. The parasite can be transmitted from contaminated environmental surfaces and person-to-person contact by the fecal–oral route, as might occur in day care centers. *Giardia* epidemics have broken out in childcare centers, with as many as 70% of the children infected with parasites. The parasites may be accidentally acquired from contaminated toys and diaper-changing tables. An infected person may shed 1 to 10 billion cysts daily in their feces (poop) for several months. Good hand-washing practices are essential. Fecally contaminated food is another source. It is imperative that uncooked vegetables and fruits be thoroughly washed in uncontaminated water.

The finding of cysts or trophozoites is diagnostic for giardiasis, but this is problematic because the organisms may be shed intermittently, necessitating the examination of stool specimens over several days. Prior to the development of tests based on molecular biology, a person exhibiting the symptoms of giardiasis despite negative multiple stool examinations may undergo the Entero-Test®, which is an alternative diagnostic procedure. The person swallows a lead sinker in a gelatin capsule attached to a string, and after four hours the string is withdrawn and the capsule is examined for the presence of trophozoites. New tests using probes for the detection of cysts in stool samples are available. A number of drugs are effective in treating this disease.

Giardiasis is a worldwide disease. Nearly 33% of people in developing countries have had giardiasis. It is the most common intestinal parasitic disease afflicting humans in the United States. Seasonal peaks coincide with the summer recreational season. Anyone can be infected with *Giardia*!

Cryptosporidiosis

The largest outbreak of waterborne disease in the history of the United States occurred in Milwaukee in 1993. The source of the outbreak was the city's water supply system contaminated with human fecal material containing *Cryptosporidium parvum*. The episode resulted in about 100 deaths and 400,000 illnesses.

C. parvum is found in a variety of mammals, birds, and reptiles. Infectious **oocysts** are discharged into the water in fecal material, and their ingestion initiates infection. The oocysts undergo excystation to **sporozoites** that penetrate the intestinal cells where multiplication results in a new batch of oocysts (FIGURE 11.4). The release of these oocysts into the environment completes the life cycle. In patients with AIDS and in other immunocompromised hosts, potentially life-threatening diarrhea often occurs. Up to twenty-five bowel movements occur per day, resulting in a loss of over 10 liters of fluid and severe dehydration. Other symptoms include weakness, fever, nausea, and abdominal pain. The incubation period ranges from several days to a few weeks. Outbreaks of cryptosporidiosis are difficult to prevent because oocysts are not inactivated by the usual doses of chlorination and can survive for days in treated water. As few as ten oocysts are sufficient to cause disease. Further, field tests are not reliable; in tests of water samples intentionally seeded with oocysts, only about 10% were detected.

Cryptosporidium parasites can be found throughout the world and in every region of the United States. The incidence of cryptosporidiosis in the United States has increased from 1 case per 100,000 people in 1999 to greater than 3 cases per 100,000 people in 2008. The reason for this increase is unknown. Eighteen outbreaks occurred in 2006, and all involved public recreational water use. An outbreak of cryptosporidiosis occurred in 2009 in a summer camp in North Carolina; 40 cases were identified. The appearance of clean water does not, in itself, guarantee its safety, and the continued occurrence of outbreaks indicates the need for improved strategies of disinfection and constant vigilance.

Amebiasis

You might be fascinated by the movements of amebas in a drop of pond water examined under the microscope. Unfortunately, these "cute" amebas have disease-producing cousins. **Amebiasis**

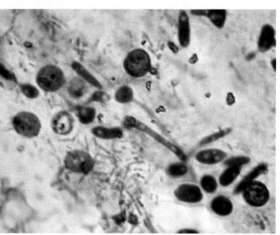

FIGURE 11.4 *Cryptosporidium* oocysts (red), yeast stained blue-green. Three-step stool examination for cryptosporidiosis in 10 homosexual men with protracted watery diarrhea. © *J Infect Dis.* 1983 May; 147(5): 824–8/Dr. Pearl Ma, Dir., Microbiology, St. Vincent's Hospital, New York/CDC.

is primarily caused by the protozoan *Entamoeba histolytica* and occurs worldwide, especially in regions with poor sanitation. In the United States amebiasis is most prevalent in immigrants from developing countries, in people who have traveled to developing countries, in people who live in institutions where it is difficult to maintain good sanitary conditions, and in those who practice oral–anal sex.

The life cycle is initiated by the ingestion of cysts and excystation in the intestinal tract (FIGURE 11.5). Four trophozoites emerge from each cyst, move into the large intestine, and attach to the wall where they mature, multiply, and feed. About 90% of infected individuals remain asymptomatic or suffer from only mild disease characterized by diarrhea and stomach pain. Severe disease can result, however, accompanied by fever, bloody stools, and stomach pain due to destruction of the lining of the large intestine. Symptoms usually occur about one to four weeks after infection and include dysentery (bloody, mucous-filled stools),

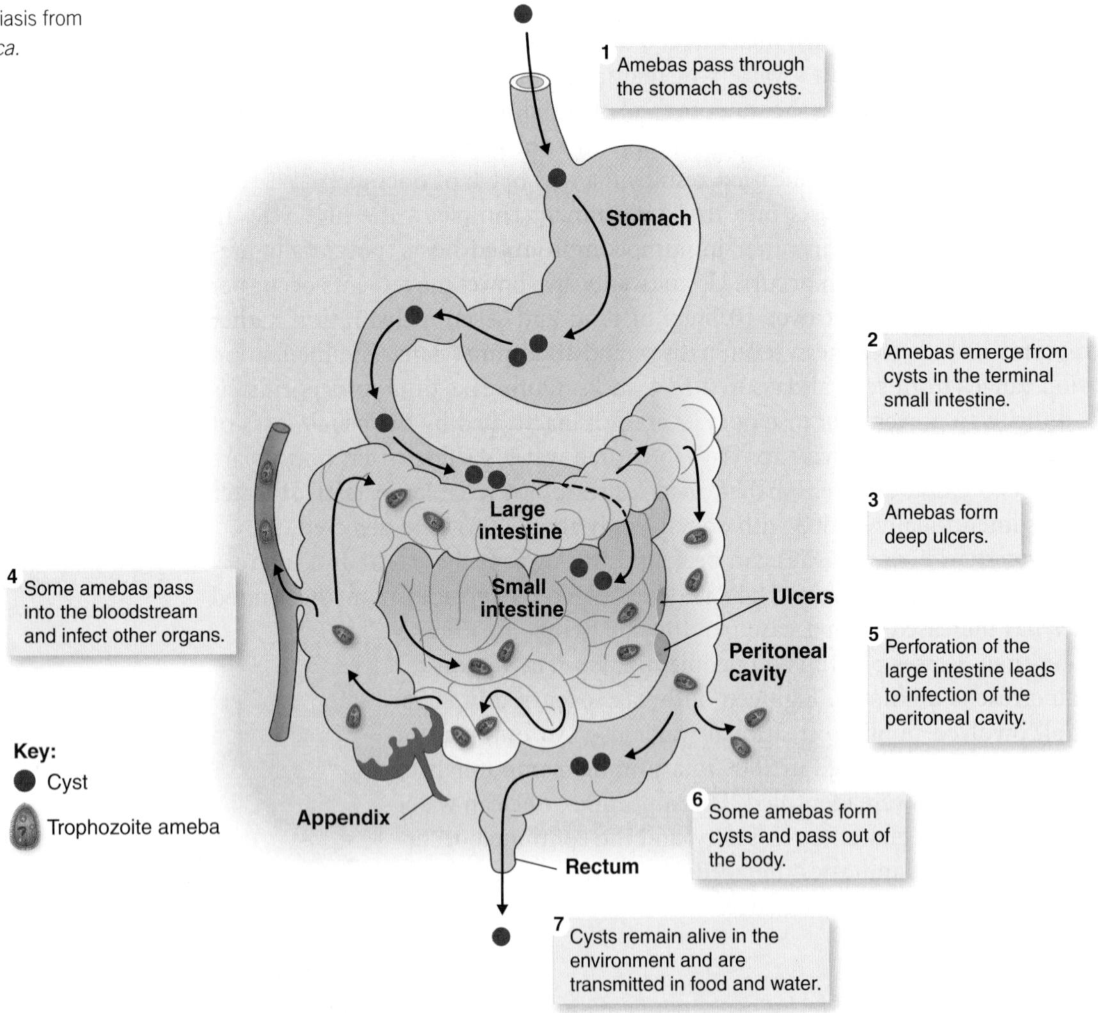

FIGURE 11.5 Amebiasis from *Entamoeba histolytica*.

1 Amebas pass through the stomach as cysts.

Stomach

2 Amebas emerge from cysts in the terminal small intestine.

3 Amebas form deep ulcers.

Large intestine

Small intestine

Ulcers

4 Some amebas pass into the bloodstream and infect other organs.

Peritoneal cavity

5 Perforation of the large intestine leads to infection of the peritoneal cavity.

Key:
● Cyst
🦠 Trophozoite ameba

Appendix

6 Some amebas form cysts and pass out of the body.

Rectum

7 Cysts remain alive in the environment and are transmitted in food and water.

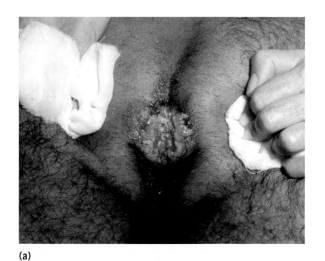

(a)

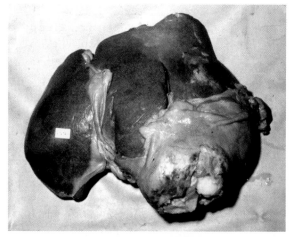

(b)

FIGURE 11.6 **(a)** Perianal cutaneous amebiasis. The amebas have caused erosion of the tissues around the anus. Author's photo (RIK). **(b)** Amebic abscesses in the liver. Amebas may invade from the intestine and produce lesions outside the intestine. Hepatic (liver) amebiasis, a serious disease, results from invasion of the liver by trophozoites. Author's photo (RIK).

weight loss, fever, and fatigue. The dysentery may be severe enough to cause the rectum to extrude through the anus (prolapsed rectum). In some cases amebas invade the kidneys, skin, brain, spleen, and liver (FIGURE 11.6). Invasion of the liver can result in amebic hepatitis and liver abscesses. Several drugs are available for the treatment of amebiasis.

Amebiasis is one of the most common parasitic diseases worldwide, and has a 10% fatality rate. There may be as many as fifty million cases per year, with an estimated mortality of over 100,000 deaths. In the United States amebiasis is relatively rare and frequently remains undiagnosed, indicating that there may be more cases of amebiasis than are reported. Fecal–oral amebic infections are more common in countries where sanitation is poor, and as many as 10% to 50% of the population in tropical countries may be infected with *E. histolytica*.

Diagnosis of amebiasis is based on finding trophozoites or cysts in an examination of fecal smears. It can be difficult to distinguish nonpathogenic intestinal amebas from those that are pathogenic. Other, more definitive diagnostic laboratory tests are available to help in diagnosis.

As in the case of giardiasis and cryptosporidiosis, chlorination of water supplies may fail to kill amebic cysts. Practicing good personal hygiene, including thorough hand washing after using the toilet and before handling food, minimizes the risk. Individuals traveling to countries in which poor sanitary conditions prevail need to be particularly careful about their source of water, fruits, and uncooked vegetables; these items may have been washed with contaminated water.

"Brain-eating amebas" refer to *Naegleria fowleri*, an ameba that has caused several recent deaths. A few have been associated with the use of neti pots, devices that are filled with water that is forced into the nose and used to irrigate the sinuses. In these fatal cases tap water containing amebas was used. The microbe travels up the olfactory nerve and into the brain feeding on brain cells. The condition is almost always fatal. Diving, water skiing, or other scenarios that propel water into the nose can also result in the form of amebiasis. On the other hand, *Naegleria* is of no consequence if swallowed in drinking water.

Toxoplasmosis

Toxoplasmosis is caused by the parasite *Toxoplasma gondii* and has a worldwide distribution. The disease is a zoonosis and occurs in over 200 species of birds and mammals; members of the feline family (FIGURE 11.7), both domestic and wild, serve as the primary reservoir and host. Toxoplasmosis can be acquired by humans after the accidental ingestion of oocysts present in cat feces or by eating meat that contains cysts. Exposure to infective oocysts from cat feces usually occurs when people are careless in their hand-washing habits and ingest minute amounts of cat feces containing oocysts in cat litter, sandboxes, or garden soil in which cats have defecated. In moist surroundings the oocysts are capable of remaining viable and infective for several months.

FIGURE 11.7 Cats play an important role in the spread of toxoplasmosis. They become infected by eating infected rodents, birds, or other small animals. The parasite is then passed in the cat's feces. Kittens and young cats can shed millions of parasites in their feces for as long as 3 weeks after infection. These are Russian Blue cats (Sasha and Leo), a very intelligent breed. Author's photo (TS).

Additionally, humans may acquire toxoplasmosis through the ingestion of cysts in pigs, sheep, cattle, and poultry that have picked up oocysts in the soil. The oocysts develop into cysts in the tissue of these animals and may then be consumed by humans. Studies have indicated that pork and lamb are frequently contaminated with cysts. Eating raw or undercooked meats is a common cause of toxoplasmosis; cysts are killed by heating above 60°C.

There is about a 30% chance that the disease can be transmitted during pregnancy to the fetus, resulting in stillbirth, neonatal death, or in serious fetal defects, including brain damage, convulsions, and retinal damage leading to blindness. Pregnant women should not change a cat's litter box and should minimize the touching of cats, because the possibility of picking up oocysts on the hands and transferring them into the mouth exists. The complicated life cycle of *Toxoplasma* is illustrated in FIGURE 11.8.

Most cases of toxoplasmosis are asymptomatic or produce only mild symptoms, including sore throat, low-grade fever, and lymph node enlargement making it difficult to assess the burden of the disease. In those individuals with AIDS or with immune systems weakened from other causes, however, toxoplasmosis is frequently a rapidly fatal disease resulting from massive invasion of the parasites into the brain. Diagnosis of toxoplasmosis is difficult because the disease resembles infectious mononucleosis. Definitive diagnosis can be accomplished by isolating the parasites, by identifying them in samples of infected tissues, or by detecting the presence of antibodies. Drugs are available to treat the disease and may need to be taken for as long as one year to prevent recurrent infection. About 22.5% of the population twelve years and older in the United States have been infected with *Toxoplasma*. In various places throughout the world as high as 95% of the population is infected. Infection is usually highest in locations that have hot, humid, climates and lower altitudes.

Arthropodborne Diseases

Trypanosomiasis

African trypanosomiasis, also known as **sleeping sickness**, is an ancient disease and one of many that have plagued Africa for millennia. There are two types of

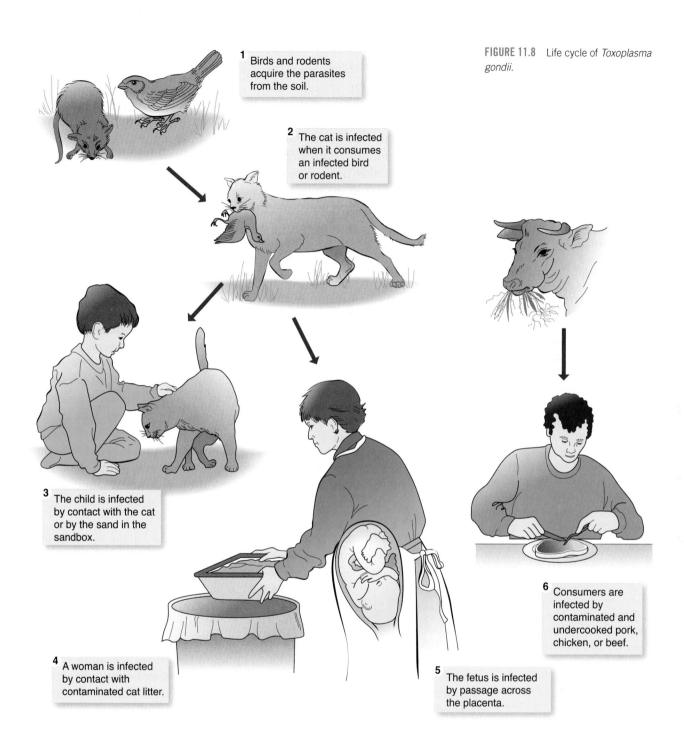

1 Birds and rodents acquire the parasites from the soil.

2 The cat is infected when it consumes an infected bird or rodent.

3 The child is infected by contact with the cat or by the sand in the sandbox.

4 A woman is infected by contact with contaminated cat litter.

5 The fetus is infected by passage across the placenta.

6 Consumers are infected by contaminated and undercooked pork, chicken, or beef.

FIGURE 11.8 Life cycle of *Toxoplasma gondii.*

African trypanosomiasis named for the area of Africa in which they are present. West African trypanosomiasis is caused by *Trypanosoma brucei gambiense,* found in the rainforests of western and central Africa. The East African variety, caused by *Trypanosoma brucei rhodesiense,* is found in the upland savanna of East Africa. Both species are transmitted by the bite of infected tsetse flies, which are large and aggressive flies that inflict painful bites. The geographical distribution of the two varieties of African trypanosomiasis is a reflection of the different ecological niches occupied by the tsetse fly vectors.

The cycle of trypanosomiasis starts when a tsetse fly takes a blood meal from an infected reservoir host: wild animals (e.g., hyenas, lions, and wild pigs), domestic animals (e.g., goats, cows, and dogs), and humans are reservoirs. The trypanosomes multiply and migrate from the gut of the fly to the salivary glands, where further development takes place. As many as 50,000 parasites can be injected by a single fly bite, far in excess of the approximately 500 that are required to establish infection. After the bite a red chancre (sore) develops at the site, and from there the parasites move into the blood, spinal fluid, lymph nodes, and brain. Early symptoms include fever, fatigue, swollen lymph nodes, and aching muscles and joints. As the disease progresses over a few months to a few years, the trypanosomes invade the brain and cause personality changes, progressive confusion, difficulty in walking, and altered sleep patterns, which become worse with time. Sleeping for long periods of the day (hence, the name "sleeping sickness"), accompanied by insomnia at night, is common. If the disease is left untreated, death occurs within a few months to several years after infection.

Trypanosomes (like the influenza virus) exhibit antigenic variation—they change their "coats" as a defensive virulence strategy resulting in evasion of the host's immune defense mechanisms. This property interferes with vaccine development.

Estimates are that a few hundred new cases of East African trypanosomiasis are reported to the World Health Organization (WHO) each year. Over 95% of the cases of human infection occur in Tanzania, Uganda, Malawi, and Zambia. In the United States there is one case on average per year, usually a traveler on safari in East Africa.

West African trypanosomiasis is on the rise in areas of Sudan. In 1997 a team of epidemiologists examined almost 1,400 persons in 16 villages and reported the presence of the disease in every village surveyed. It is estimated that 10,000 new cases occur each year. Over 95% of the cases of human infection are found in Democratic Republic of Congo, Angola, Sudan, Central African Republic, Chad, and northern Uganda.

Trypanosoma cruzi is the causative agent of **American trypanosomiasis**, also called **Chagas disease**. The life cycle is similar to that of the trypanosomes that cause sleeping sickness. Wild animals, including rodents, opossums, and armadillos, serve as reservoirs for *T. cruzi*. The vector is a **triatomid insect**, commonly referred to as a "kissing bug," that lives in thatched roofs, cracks in the mud walls, and dark places of adobe huts, close to humans (FIGURE 11.9). The vector harbors the trypanosomes in its hindgut and has the nasty habit of defecating as it bites, causing itching and scratching, resulting in inoculation of the trypanosomes into the skin. The insect bites at night. Trypanosomiasis transmission is not limited to an arthropod vector. It can also be spread by blood transfusion and by organ transplantation. Trypanosomiasis should become less of a problem since, in 2006, the U.S. Food and Drug Administration approved a new test for the detection of

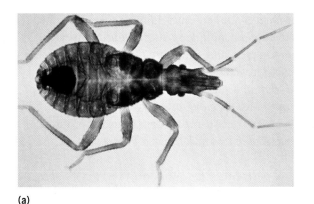

(a)

(b)

FIGURE 11.9 (a) A triatomid "kissing bug," the vector of *Trypanosoma cruzi*. Courtesy of WHO Geneva, Switzerland/CDC. (b) Thatched-roof huts. This type of structure is an ideal habitat for triatomids. © Lucian Coman/ShutterStock, Inc.

antibodies against *T. cruzi*. The CDC estimates that 300,000 people living in the United States are infected with *T. cruzi*. Most people with Chagas disease in the United States acquired their infections in endemic countries.

Chagas disease occurs in Mexico, and Central and South America where an estimated 8 to 11 million people are infected, many of whom are unaware of their infection. The parasite causes slow, widespread tissue damage, particularly to the heart, causing it to enlarge and impairing its function, leading to heart failure. In about 1% to 2% of the cases the eyelid and the area around the eye become swollen, resulting in Romaña's sign (FIGURE 11.10). Death occurs within 30 years of infection if the disease is untreated. In addition to the usual diagnostic methods, **xenodiagnosis** is sometimes used; in this technique kissing bugs known to be free of the trypanosomes are allowed to feed on the individual suspected of having Chagas disease. After a few weeks the insects are examined for the presence of the parasites.

The following case study was described in *"Kiss of Death—Chagas' Disease in the Americas"* by Joseph William Bastien:

> *"Bertha (pseudonym) lives in La Paz, Bolivia, and her medical history provides insight into the effects of Chagas. She suffers from chronic heart ailments from Chagas disease.*
>
> *As a child living in the 1930s, she was bitten by vinchucas ("kissing bugs") and infected with T. cruzi when she lived in Tupiza, a small rural village in Bolivia. She later married and bore four daughters. In 1960, she moved to La Paz after her husband abandoned the family. She made a living sewing for wealthy people, but in 1974 she was diagnosed with Chagas disease.*
>
> *She tells the story of her life coping with Chagas. Until she was forty-four she was healthy, going up and down the hills of La Paz to do her sewing. In 1974 she felt fatigue. She began to get a swollen throat and spit blood. She didn't know what it was, she had no idea it had to do with the vinchucas bites years before. She would get tired, fatigued, and experience*

AUTHOR'S NOTE (RIK)
The photo in Figure 11.9b is of a row of traditional huts in Botswana, Africa, with a high incidence of Chagas disease and kissing bugs. In a similar village, over fifty bugs were isolated from a wooden bed frame, the only piece of "furniture" in the one room hut. The bugs were examined for the presence of trypanosomes, and most of the insects were positive.

FIGURE 11.10 Romaña's sign. Courtesy of WHO/TDR; used with permission.

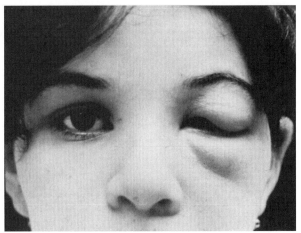

dizzy fainting spells. She continued to do her sewing though she sometimes would faint while she was working. The fainting spells continued for a year; the next year her fainting got more severe and she eventually suffered a stroke. Her children took her to a doctor, Dr. Jauregui, who hospitalized her. She underwent testing, xenodiagnosis, that indicated she had Chagas disease. X-rays showed that she didn't suffer from cardiomegaly (an enlarged heart), but that she probably had lesions in her heart's electrical system. These were caused by T. cruzi amastigotes being encysted in her cardiac tissue. This condition can be fatal.

Dr. Jauregui implanted a pacemaker in 1980 when Bertha's heart rhythm worsened. The pacemaker keeps the heart rhythm constant and Bertha's condition improved. She was able to resume her seamstress work, although she suffered minor fatigue as she climbed the streets of La Paz at 12,000 feet." (Courtesy of the University of Utah Press.)

Chagas disease was endemic in Uruguay because blood was not screened before transfusion and kissing bugs were present in 80% of households. Through an intensive program of vector (kissing bug) reduction by both indoor and outdoor spraying and by replacing thatched roofs with metal, infection rates fell to below 0.1%, except in a single area. This was a decline of 80% in only sixteen years. Uruguay was declared free of Chagas disease in 1997, a great triumph of public health measures and an inspiration to other countries that the burden of Chagas disease could be lifted.

Leishmaniasis

Leishmaniasis is caused by several species of *Leishmania* and is endemic in many tropical and subtropical areas. Transmission occurs by the bite of female **sand flies** that become infected while taking a blood meal from wild and domestic animal reservoir hosts (FIGURE 11.11). In the Mediterranean region, including North Africa, jackals and dogs are important reservoirs, and dogs are significant reservoirs in Brazil. In India there are no animal reservoirs, and sand flies transmit the disease from human to human. Donor blood is not universally screened for the presence of leishmania, thereby contributing to the problem.

When an infected sand fly bites a human the parasites enter the skin and are engulfed by cells called **macrophages**. These cells are phagocytes and are of prime importance to the host immune system. The parasites multiply and are released when a macrophage bursts. The life cycle is illustrated in FIGURE 11.12.

There are three human manifestations of leishmaniasis, determined by the particular parasite species, the geographical location, and the host immune response. **Visceral leishmaniasis**, also called **kala-azar**, is the most severe form of the disease, with close to a 100% mortality rate within two to three years if not treated. About 90% of visceral leishmaniasis cases occur in parts of India, Bangladesh, Nepal, Sudan, Ethiopia, and Brazil. The parasites invade the liver and spleen and cause characteristic symptoms, including irregular bouts of fever, weakness, weight loss,

FIGURE 11.11 *Phlebotomus dubosci,* a sand fly vector of *Leishmania* parasites, taking a blood meal through human skin. Courtesy of James Gathany/CDC.

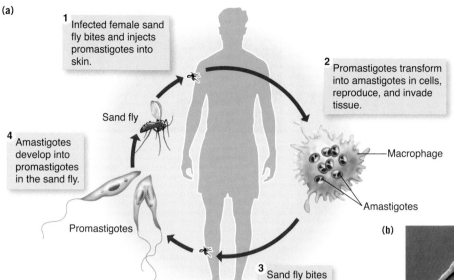

(a)

1 Infected female sand fly bites and injects promastigotes into skin.

2 Promastigotes transform into amastigotes in cells, reproduce, and invade tissue.

Sand fly

Macrophage

Amastigotes

4 Amastigotes develop into promastigotes in the sand fly.

Promastigotes

3 Sand fly bites and ingests amastigotes.

(b)

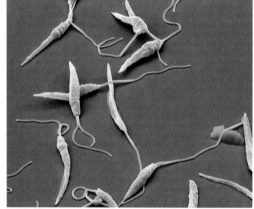

anemia, and protrusion of the abdomen due to the swelling of the spleen and liver. The other two varieties of leishmaniasis are more localized and are seldom fatal. In **mucocutaneous leishmaniasis**, also called **espundia**, the parasites invade the skin and mucous membranes, causing destruction of the nose, mouth, and throat. **Cutaneous leishmaniasis** (FIGURE 11.13), also called Baghdad boil and tropical boil, results in mild to disfiguring skin lesions, primarily on exposed parts of the body, particularly the face, arms, and legs. Typically, only a few *Leishmania* lesions appear on an individual, but there may well be over 100. About 90% of cutaneous leishmaniasis cases occur in parts of Afghanistan, Algeria, Iran, Saudi Arabia, Syria, Brazil, Colombia, Peru, and Bolivia.

FIGURE 11.12 **(a)** Life cycle of *Leishmania.* **(b)** A colorized scanning electron micrograph of *Leishmania.* © Eye of Science/Photo Researchers, Inc.

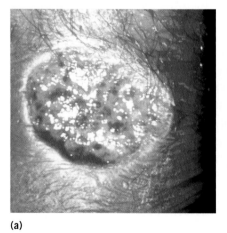

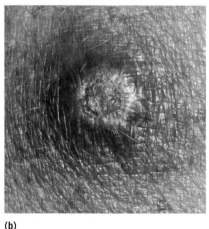

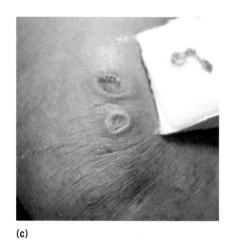

(a)

(b)

(c)

FIGURE 11.13 Cutaneous leishmaniasis. A variety of skin lesions are caused by infection with certain species of *Leishmania.* They are described as dry **(a)** or moist and diffuse (Courtesy of CDC), **(b)** raised (© Dr. Morley Read/Science Photo Library), or **(c)** ulcerated (Author's photo, RIK).

(a)

(b)

People with Oriental sores rarely get **kala-azar**, encouraging the practice of self-vaccination on inconspicuous areas of the body. Some parents purposely infect their children with *Leishmania* strains that cause Oriental sores as prevention against getting kala-azar. Although this practice may prevent cutaneous lesions, it does not always prevent kala-azar.

Estimates are that approximately 12 million people in the tropics and sub-tropics suffer from leishmaniasis. Further, 1.5 million new cases of the cutaneous form and a half a million new cases of the visceral form are recognized each year. This disease is one of the most common infectious diseases among soldiers on duty in Iraq and Afghanistan. Military personnel serving in Afghanistan or Kuwait where the disease is also endemic are at risk as well (FIGURE 11.14).

Malaria

Ask almost anyone to name a tropical disease, and the chances are they will name **malaria**. The symptoms of malaria were described in Chinese medical writings dating back to 2700 B.C. Despite many years and many dollars spent on trying to control this disease, it remains a scourge on humankind. It is the world's biggest burden of tropical disease and, with the exception of tuberculosis, kills more people than any other microbial disease. The term "malaria" is derived from the Italian words *mala* (bad) and *aria* (air). In the 1940s the spraying of the insecticide dichlorodiphenyltrichloroethane (DDT) almost eradicated malaria, but it has since returned with a vengeance as a result of the emergence of DDT-resistant, other insecticide-resistant, and antimalarial drug-resistant strains. The disease is not limited to Africa, although the burden of malaria falls heavily on that continent (FIGURE 11.15). Gro Harlem Brundtland, a previous director general of WHO, stated, "Malaria is hurting the living standards of Africans today and is also preventing the improvement of living standards for future generations. This is an unnecessary and preventable handicap on the continent's economic development." Studies have indicated that Africa's gross domestic product would now be substantially higher if malaria had been eliminated years ago. According to the 2011 WHO *World Malaria Report*, international funding to control malaria

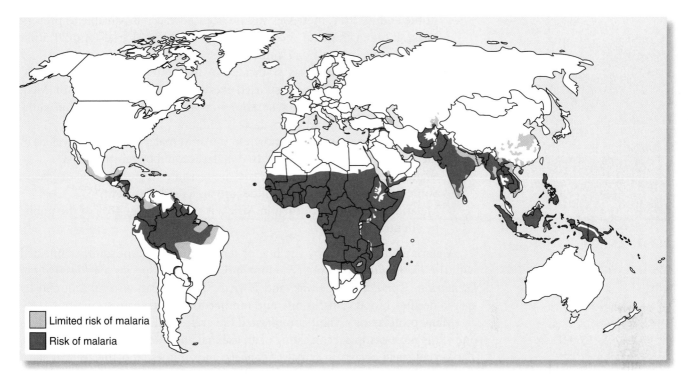

Limited risk of malaria
Risk of malaria

FIGURE 11.15 Estimated malaria risk by country—2010. Adapted from WHO Global Health Observatory Map Gallery.

continues to increase and significant progress has been made toward controlling malaria. Here are some facts about malaria:

- An estimated 3.3 billion people living in 109 countries or territories were at risk of contracting malaria in 2010.

- Malaria is the fifth leading cause of death from infectious diseases worldwide (after respiratory infections, HIV/AIDS, diarrheal diseases, and tuberculosis).

- Financing provided for an increase of 3% to 50% of households owning at least one insecticide-treated net in 2011.

- Household surveys indicate that 96% of individuals with access to an insecticide-treated net use it.

- The number of diagnostic tests manufactured is increasing and the percentage of cases reported that have been confirmed by testing has increased from 67% in 2005 to 76% in 2010.

- Many cases are still treated without parasitological diagnosis.

- In 2010, there were an estimated 216 million malaria cases worldwide.
- Approximately 81% or 174 million of these cases occurred in Africa.
- There were 650,000 malaria deaths in 2010; 91% of these deaths were in Africa and 85% of cases were children under five years of age.
- The number of cases reported has decreased by 50% from 2000 to 2010 in forty-three of ninety-nine countries in which malaria transmission is ongoing.

- At the end of 2011, there was still no licensed vaccine available to prevent malaria. One candidate vaccine is in phase 3 clinical trials and 20 other projects are in phase 1 or 2 clinical trials.
- Even though malaria was eradicated in the United States in the 1950s, 1,500 cases of malaria are reported every year in the United States. Most cases are first- and second-generation immigrants traveling to home countries to visit relatives where malaria is endemic.
- Two species of mosquitoes known to transmit malaria are still prevalent in the United States, posing a risk that malaria could be reintroduced.
- Many strains of the malaria parasite are resistant to antimalarial drugs.
- Some malaria-carrying mosquitoes are insecticide resistant.
- Malaria is a financial burden in Africa; it eats up about 40% of the public health budget.

Malaria is transmitted by the bite of female *Anopheles* mosquitoes infected with the *Plasmodium* parasite. The four most common species are *Plasmodium falciparum, P. malariae, P. ovale,* and *P. vivax.* Malaria can also be acquired by shared needles, blood transfusions, and mother-to-fetus transmission.

Many protozoans exhibit complicated life cycles, but the malaria life cycle is one of the most complex, consisting of an **asexual** and a **sexual** stage (FIGURE 11.16). The sexual stage occurs in female *Anopheles* mosquitoes, and the asexual stage occurs in the liver and red blood cells of the infected individual. When a plasmodium-infected mosquito bites an individual, the infective forms of the parasite,

FIGURE 11.16 Life cycle of the malaria parasite.

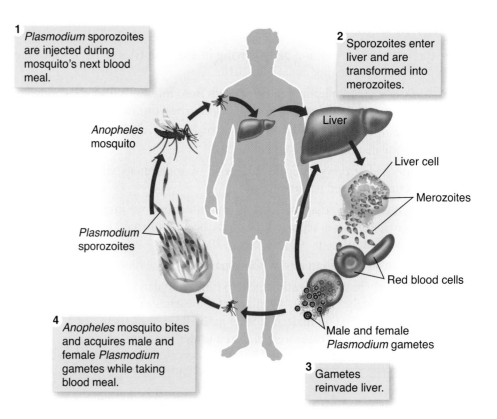

1 *Plasmodium* sporozoites are injected during mosquito's next blood meal.

2 Sporozoites enter liver and are transformed into merozoites.

Anopheles mosquito

Liver

Liver cell

Merozoites

Plasmodium sporozoites

Red blood cells

4 *Anopheles* mosquito bites and acquires male and female *Plasmodium* gametes while taking blood meal.

Male and female *Plasmodium* gametes

3 Gametes reinvade liver.

called **sporozoites**, pass from the salivary glands of the mosquito into the person's bloodstream. Within about an hour the sporozoites travel to the person's liver, where asexual multiplication takes place, resulting in forms known as **merozoites** that eventually leave the liver. A single sporozoite can produce thousands of merozoites. The merozoites enter **erythrocytes** (red blood cells), where further asexual multiplication occurs. The blood cells lyse, or burst, releasing more merozoites into the bloodstream, where they may infect other erythrocytes. If the mosquito bites while the plasmodium parasites are in the blood, the parasites are ingested and develop into male and female **gametes** (sex cells), initiating the sexual cycle. This results in sporozoites that migrate into the mosquitoes' salivary glands. The cycle is now complete.

After a *Plasmodium*-infected mosquito bites a person symptoms usually appear in two to four weeks but may not appear for one year. While the parasites are in the liver the infected individual usually experiences no symptoms. In the symptomatic stage the person experiences the characteristic shaking chills and fevers associated with malaria. The symptoms are caused by the destruction of large numbers of red blood cells accompanying the release of merozoites and their toxins into the bloodstream. This event is frequently synchronized during an attack, and many new merozoites are released at one time. It is as though the parasites can tell time! In the case of *P. malariae* the chills and fever occur about every 72 hours, whereas in the other three species these symptoms occur about every 48 hours. Several episodes of chills and fever constitute an attack (paroxysm); between attacks the individual feels normal. Muscle aches, fatigue, diarrhea, and nausea may accompany the chills and fever. *P. falciparum* is the most dangerous of the four species, because it can cause **cerebral** (brain) malaria. Pregnant women have an increased susceptibility, resulting in lower birth weights, decreasing the chance of the baby's survival. This form of malaria is fatal for 1% to 2% of those infected.

Malaria can cause anemia if large numbers of red blood cells are lysed. The anemia, if severe, contributes to maternal deaths. Further, pregnant women who are infected with malaria and are positive for HIV are more likely to infect their unborn child with the virus.

Before the early 1900s malaria was endemic in the United States, primarily in the Southeast. Massive mosquito abatement programs and a concomitant reduction in the number of human carriers were effective; the disease is no longer endemic. *Plasmodium* parasites and infected mosquitoes have been known to survive flights from countries where malaria is present and can transmit the disease in the vicinity of the airport, a phenomenon known as **airport malaria**. In 2005, 6 out of 30 U.S. students returning from East Africa as part of an educational program came down with malaria symptoms in a United Kingdom airport. They had been on a flight from Nairobi to London before their flight to the United States. This is a reminder to travelers to take malaria prophylaxis.

The diagnosis of malaria is often based on the detection of malaria parasites in red blood cells during microscopic examination of a drop of blood smeared onto a slide (**FIGURE 11.17**). Rapid test kits that provide results within 2 to 15 minutes detect antigens derived from the malaria parasites. The rapid kits are a useful

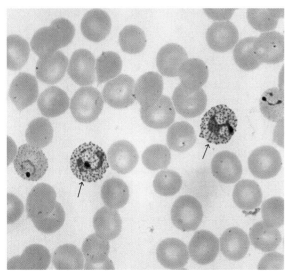

FIGURE 11.17 Blood smear showing a malaria parasite. *Plasmodium* parasites infect red blood cells. Arrows indicate the cells in the center of the photo that are infected with the parasite. Courtesy of Dr. Mae Melvin/CDC.

alternative in situations where microscopy is not available. Malaria is a curable disease if promptly diagnosed and treated with appropriate drugs.

Several drugs are available for treatment and prevention; the particular drug prescribed depends on the circumstances, including the malaria species identified, severity of disease, and age of the individual (TABLE 11.3). Drug resistance is a serious problem, and international travelers and travel clinics need to be aware of this and appropriately advised. During January to March 2000 two U.S. citizens died from malaria as the result of receiving malaria-preventive drugs to which the parasite in the area of their travels was resistant.

In several areas of Africa the use of bed nets regularly treated with recommended insecticides reduced the incidence of child deaths by about 30%. Mosquito abatement programs in malaria-endemic countries are vital (FIGURE 11.18). World Malaria Day occurs on April 25th of each year. The WHO instituted it in 2007 as a day to recognize the control of malaria. Malaria control, frequently in the form of campaigns, has involved numerous agencies at the local, national, and international levels since the 1950s. An eradication program in 1950 substantially decreased malaria incidence and deaths but fell short in eradicating this disease as acknowledged by the world health assembly. But the battle was not over and global partnerships were established including the Roll Back Malaria program. Large amounts of money became available from several sources including the World Bank and the United States President's Malaria Initiative (PMI) started in 1995. The latter was originally designed to focus on fifteen carefully chosen target countries in sub-Saharan Africa and designed to cut malaria deaths in half; the initiative was later modified with a broader goal and will soon be evaluated. Only time will tell. PMI supports four initiatives: insecticide-treated mosquito nets, indoor insecticide spraying, treatment programs for pregnant women, and prompt use of approved therapies for those that have malaria.

The world eagerly awaits a vaccine against malaria and it appears that one may soon arrive. There have been many efforts and many contenders over the

TABLE 11.3 Drugs Used to Treat or Prevent (Prophylaxis) Malaria	
Drug	**Use**
Atovaquone-proguanil combination	Treatment or prophylaxis
Artemether-lumefantrine combination	Treatment or prophylaxis
Mefloquine	Prophylaxis
Chloroquine phosphate	Prophylaxis
Deoxycycline	Prophylaxis
Mefloquine	Prophylaxis
Primaquine	Prophylaxis
Hydroxychloroquine sulfate	Prophylaxis

Source: *CDC Health Information for International Travel 2012: The Yellow Book*, Oxford University Press.

years. It is not easy to make a vaccine against malaria; one of the problems is that the *Plasmodium* parasite changes its shape as it moves along from the blood to liver and back again exposing different protein surfaces as antigens somewhat analogous to the way trypanosomes "change their coats." Preliminary results from the trial of a vaccine indicates that it protected about 41% of the children; the vaccine, produced by GlaxoSmithKline is known as RTS,S and has been under development for more than twenty-five years.

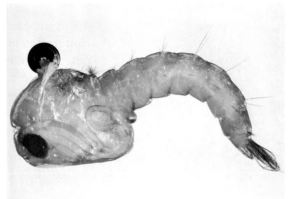

FIGURE 11.18 *Anopheles* pupa. The adult anopheles mosquitoes are the known vector for malaria. Courtesy of Harry Weinburgh/CDC.

Babesiosis

Babesiosis, sometimes described as a malaria-like parasite, is caused by the protozoan parasites *Babesia microti* and *Babesia divergens*. **Babesiosis**, or "the Babe" as some biologists call it, is transmitted by ticks. The ticks become infected by feeding on infected vertebrate hosts (rodents, cattle, and wild animals). The parasites multiply and develop in the ticks and are transmitted to the next vertebrate host, where they invade the red blood cells. As in malaria, the parasite undergoes cycles of multiplication and reinvasion of red blood cells. Based on its similarity to malaria, babesiosis can be easily misdiagnosed as malaria in areas where malaria is endemic. Cases may be asymptomatic, but when symptoms occur they include severe headache, high fever, and muscle pain. Because red blood cells are destroyed, anemia and jaundice can also appear. The symptoms may last for several weeks, and in some cases a prolonged carrier state may develop. Almost all untreated cases of *B. divergens* babesiosis are fatal, whereas cases of *B. microti* babesiosis are seldom fatal in healthy persons. Avoiding tick bites is the best means of protection. Babesiosis is distributed worldwide; in the United States it has been most frequently identified in the Northeast and Midwest.

This section has presented several arthropod-borne protozoan diseases. Although they are most common in tropical and semitropical areas, they remain a constant threat to other areas of the world, including the United States. Measures of control center around programs designed to control populations of the arthropod vectors involved and include spraying and avoidance of standing water in which mosquitoes breed.

Sexually Transmitted Diseases

Trichomonas vaginalis causes **trichomoniasis**. It is estimated that as many as 3.7 million people have the infection in the United States and close to 180 million cases occur annually worldwide, making this disease one of the most common sexually transmitted diseases. Rarely, it is also transmitted by contaminated towels and articles of clothing. The human urogenital tract is the reservoir, and many infected persons, both female and male, are asymptomatic carriers. The organism has no cyst stage and therefore cannot survive for long periods outside the host. It frequently accompanies infection with the bacterial diseases chlamydia and gonorrhea.

In females symptoms include intense itching, urinary frequency, pain during urination, and vaginal discharge. In males the disease may be characterized by pain during urination, inflammation of the urethra, and a thin milky discharge, but most infections are asymptomatic. Diagnosis is accomplished by direct microscopic examination of the discharge and searching for the presence of actively swimming trichomonads. The infection is successfully treated by the oral administration of appropriate drugs, and it is recommended that both sex partners be treated simultaneously to prevent "ping-pong" reinfection.

■ Biology of Helminths

The worms crawl in,
the worms crawl out,
in your belly,
and out your snout.
        —Anonymous

This may sound like a cute little nursery rhyme, but in reality it describes a gruesome picture "and opens up a can of worms." Live worms of several species have been known to crawl out the mouth, nose, and umbilicus (belly button) and to be discharged into toilet bowls or onto the ground from the anuses of infected individuals, particularly in impoverished countries where poor sanitation is frequent. As is the case with bacteria and viruses, it should be emphasized that most helminths in nature are not parasites, but there are a number of worms that cause disease in humans, in other animals, and in plants.

The idea of a worm, a eucaryotic, multicellular animal, slithering around within your body is repugnant (FIGURE 11.19). All the bacteria, viruses, and protozoans discussed thus far are subcellular, procaryotic, or eucaryotic single-celled microscopic biological agents. They may, in fact, be more serious but are not necessarily viewed as loathsome *because* they are not visible. In other words, what you cannot see may not be so bad, but looking at worms is another story!

The worms are divided into two morphological categories, **roundworms** (nematodes) and **flatworms**. Worms are at the systems level of biological organization and possess circulatory, nervous, reproductive, excretory, and digestive systems. Depending on the species, some systems may be lacking or may be rudimentary. The parasitic mode of existence of worms eliminates, or at least greatly minimizes, the need for a sophisticated digestive system, nervous system, and method of locomotion. They are surrounded by absorbable nutrients already digested, have little to react to, have no place to go in the quest for food, and are passively transferred from host to host. On the other hand, worms have complex reproductive systems ensuring the production of very large quantities of fertilized eggs and maximizing the continuance of the species. In some species males and females are **dioecious** (the two sexes have different reproductive organs), whereas in others single animals are

FIGURE 11.19 A handful of *Ascaris lumbricoides* worms, which had been passed by a child in Kenya, Africa. Courtesy of James Gathany/CDC.

hermaphroditic and produce both sperm and eggs. Cross-fertilization or self-fertilization may occur, depending on the species. In some cases their life cycles are very complex and include both larval and adult forms as well as multiple intermediate hosts.

Helminthic Diseases

Helminthic diseases are major problems in tropical countries. Infection with worms may result in death, but more frequently worm diseases are chronic and debilitating and may lower the quality of life in an entire community. In some areas, primarily in developing countries, as many as 90% of the population have at least one type of worm; they are almost like normal fauna. Their presence in the United States and in other temperate areas is becoming of increasing concern because of immigration, travel, and AIDS.

Helminthic diseases are organized by their mode of transmission (TABLE 11.4), as are bacterial, viral, and protozoan diseases. The mode of transmission varies with the particular species and may include food and water, arthropods, and direct contact.

Foodborne and Waterborne Diseases

Ascariasis

Ascariasis is caused by *Ascaris lumbricoides,* large roundworms that resemble earthworms and set up housekeeping in the small intestine. Females can grow to over 1 foot (30.5 cm) in length; adult males are 8 to 10 inches long.

TABLE 11.4 Major Helminthic Diseases

Organism	Classification	Disease	Transmission[a]	Principle Site(s)
Foodborne and waterborne				
Ascaris	Roundworm	Ascariasis	Food, water	Intestinal tract, lungs
Trichuris	Roundworm	Whipworm	Food, water	Intestinal tract
Trichinella	Roundworm	Trichinellosis	Pork consumption	Intestinal tract
Dracunculus	Roundworm	Guinea worm	Water	Skin
Taenia saginata	Flatworm	Beef tapeworm	Beef consumption	Intestinal tract
Taenia solium	Flatworm	Pork tapeworm	Pork consumption	Intestinal tract
Vectorborne				
Wuchereria	Roundworm	Filariasis	Mosquitoes	Lymph vessels
Onchocerca	Roundworm	River blindness	Blackflies	Skin, eyes
Contact				
Ancylostoma/Necator	Roundworm	Hookworm	Contact	Intestinal tract, lungs, lymph
Enterobius	Roundworm	Pinworm	Contact	Intestinal tract
Schistosoma	Flatworm	Schistosomiasis	Water contact	Skin, bladder, blood vessels
Strongyloides	Roundworm	Strongyloidiasis	Contact	Small intestine, skin, bladder, small blood vessels

[a]Transmission may occur by more than one mechanism.

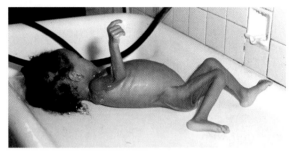

FIGURE 11.20 Malnutrition associated with ascariasis. Lack of food or a poor diet results in malnutrition, a problem particularly present in less-developed countries. Most people with malnutrition are infected with one or more species of worms. The swollen belly of the child in the photo is characteristic of malnutrition accompanied by worm infection. Author's photo (RIK).

Infective eggs are passed onto the soil in the feces of infected individuals. People become infected when they ingest food or water contaminated with the eggs. The eggs are very hardy; they resist drying and thrive in warm, moist soils. Once in the intestine, the eggs hatch into larvae (immature worms), which penetrate the intestinal wall and begin their journey through the body. They enter blood vessels and are carried along by the blood flow into the heart and then into the capillaries of the lungs, migrate up the respiratory tree, and enter the back of the throat. They are swallowed and returned to the small intestine, where they mature into male and female worms over a period of two to three months. Female worms produce 200,000 or more eggs per day, and adults live approximately two years.

Many people experience no symptoms when infected with *Ascaris,* but in some people damage to the lungs may occur as the worms journey through the lung tissue, causing a cough and the threat of secondary bacterial infection. Rarely, adult worms may be coughed up or block the pharynx, causing suffocation. Further malnutrition may result from a large number of worms feeding on intestinal contents (FIGURE 11.20). When the worm burden is high, intestinal obstruction and perforation of the intestinal tract may occur, leading to death (FIGURE 11.21).

Estimates are that worldwide as much as 25% of the population is infected with *Ascaris,* particularly in tropical and subtropical areas, as a result of poor sanitation and hygiene. In some rural areas of the southeastern United States up to 50% of the children are infected and are burdened with five to ten worms. Individuals practicing poor hygiene may suffer repeated bouts of infection.

The presence of *Ascaris* eggs or worms in the feces is diagnostic; an adult worm may be coughed up or passed in fecal material. A number of drugs are effective to rid the body of adult worms; they work by temporarily relaxing the worms, resulting in their passage through the anus.

Whipworm Infection

Trichuris trichiura resembles a whip, hence its English name "whipworm." Whipworm is morphologically distinct from *Ascaris* but has a similar life cycle. Eggs passed in fecal material onto moist, warm soil become infective about three to six weeks after ingestion. The eggs hatch in the small intestine and release larvae that mature, migrate to the large intestine, and take up residence. After sexual maturation and fertilization the females begin to lay eggs 60 to 70 days after infection and produce as many as 5,000 eggs, which can be discharged onto the soil, per day. The life span of the adults is about four to seven years.

In most cases the infection produces no symptoms, but it can cause abdominal pain, diarrhea, and rectal prolapse, particularly in small children. Identification of whipworm eggs in feces is proof of infection. Drugs are available for treatment. Trichuriasis is the third most common roundworm disease of humans and is more frequent in the tropics and subtropics. It is estimated that nearly 300 million people, including some in the southeastern United States, carry these worms.

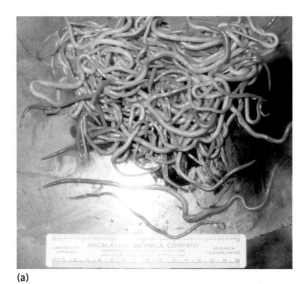

(a)

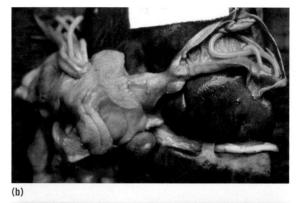

(b)

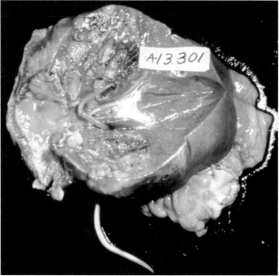

(c)

FIGURE 11.21 Intestinal obstruction due to ascarid worms.
(a) A tangle of ascarid worms caused an intestinal blockage,
resulting in death. **(b)** Worms invaded the gallbladder.
(c) Worms invaded the liver. Author's photos (RIK).

Trichinellosis

Beware of eating raw or undercooked pork, because it may be infected with *Trichinella spiralis* and cause **trichinellosis**. For that matter, if you like to eat wild game, including bear, boar, walrus, or seal, make sure the meat is well cooked. (Remember that the next time you order a walrus sandwich!) This parasitic nematode has a direct life cycle meaning that there is no intermediate host.

When pork or other meat that contains trichinella cysts is ingested, digestive juices dissolve the hard covering of the cysts and worms emerge in the small intestine (**FIGURE 11.22**). Maturation occurs in one to two days, and, after mating, adult females lay eggs that develop into larvae. Larvae migrate through the blood and lymph vessels and are transported to muscles, including the eye, tongue, diaphragm, and chewing muscles, within which they form cysts. The human host is a dead end for trichinellosis (barring that the person is consumed by a cannibal). Infected rats and other rodents are fed upon by meat-eating animals, including pigs, and thus play a major role as reservoirs in maintaining trichinellosis.

Early symptoms of trichinellosis occur within only one to two days after infection and include nausea, diarrhea, vomiting, fatigue, and fever, followed by later symptoms of headaches, chills, aching muscles, and itchy skin. If the burden of worms is heavy, the individual may experience cardiac and respiratory problems; in severe cases death may occur. Mild cases may never be diagnosed and are limited to flulike symptoms.

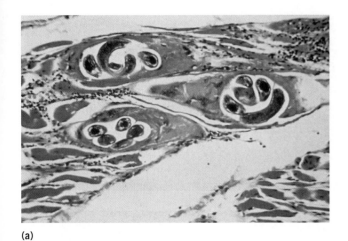

(a)

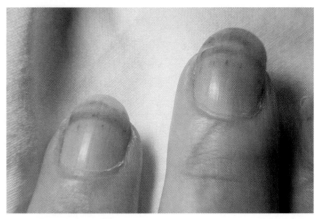

(b)

FIGURE 11.22 (a) *Trichinella* cysts within human muscle tissue. Courtesy of the CDC. (b) Splinter hemorrhages under the fingers of a patient suffering from trichinellosis. Courtesy of Dr. Thomas F. Sellers/Emory University/CDC.

The incidence of trichinellosis has markedly declined through the years. Worldwide, an estimated 10,000 cases of trichinellosis occurs each year. According to the CDC, in the late 1940s an average of 400 cases and 10 to 15 deaths occurred in the United States each year; from 1997 through 2001 the number declined to 72 cases, of which 31 were associated with eating wild game. From 2002 to 2007, eleven cases were reported to the CDC each year on average. The decline reflects a combination of factors, the primary one of which is the passage of laws prohibiting the feeding of **offal** (the organs and trimmings of butchered animals), raw meat, and uncooked garbage to pigs. It is therefore highly unlikely that trichinellosis will be acquired from pigs raised in the United States. Awareness on the part of the public regarding the danger of eating raw or undercooked pork or wild game has also contributed to the decline of trichinellosis. Infection occurs worldwide but is mostly found in areas where raw or undercooked pork and pork products (ham and sausage) are consumed. A muscle biopsy and a blood test can reveal the presence of trichinellosis. Several drugs with limited effectiveness are available to treat this infection.

Dracunculiasis (Guinea Worm Disease)

Dracunculiasis is caused by the parasitic roundworm *Dracunculus medinensis* (also called guinea worm and the fiery serpent) and is present in poor communities in Africa that lack access to safe drinking water. The life cycle is initiated by drinking water contaminated with dracunculus larvae; small **copepods** ("water fleas") feed on these larvae (FIGURE 11.23). Copepods are small crustaceans in the phylum *Arthropoda*. When someone ingests water containing infected copepods, the copepods are digested but not the worm larvae, which penetrate the stomach and intestinal wall and enter the abdominal cavity. Over the next year the larvae mature into adult worms, copulation takes place, the male worms die, and the females migrate toward the skin surface, most commonly the foot or the leg. A painful blister that eventually ruptures occurs at the site. In many cases persons immerse the blistered area in water in an attempt to gain some relief from the pain. The temperature change can induce the blister to open, exposing the adult worm. The gravid ("pregnant" with internal eggs or larvae) female releases a

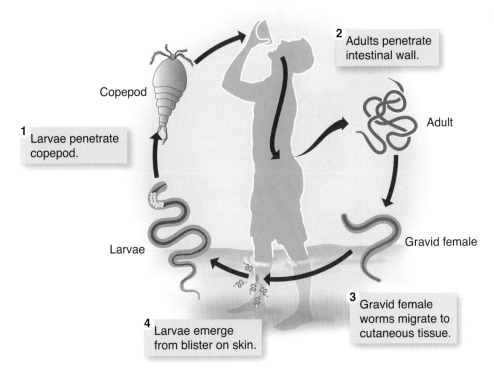

FIGURE 11.23 Life cycle of guinea worms.

2 Adults penetrate intestinal wall.

Copepod

Adult

1 Larvae penetrate copepod.

Larvae

Gravid female

4 Larvae emerge from blister on skin.

3 Gravid female worms migrate to cutaneous tissue.

milky white fluid that contains millions of larvae into the water, a process that continues each time the worm comes in contact with water. The larvae are ingested by copepods and mature in two weeks (and after two molts) into infective larvae, completing the cycle.

Symptoms of infection occur about one year after the ingestion of water contaminated with infected water fleas. A few days to several hours before rupture of the blister and emergence of the worm, fever, pain, and swelling in the area occur. After the rupture the one-meter-long worm begins to enzymatically bore its way out through the skin, cutting the flesh during its migration and causing excruciating and debilitating pain (FIGURE 11.24). The worm is the diameter of a paper clip or a piece of thin spaghetti so it cannot be pulled out. The risk is too great that part of it will be left inside the leg, leading to a potentially life-threatening bacterial infection. The most you can do is coil the worm around a small stick as it emerges from the blister and hope the agony will end in two to three weeks and not drag on for as long as three months.

Diagnosis is made clinically and does not require laboratory confirmation. Difficulty in walking, loss of appetite, and exhaustion are common symptoms. There is no antihelminthic medication available to prevent or end infection. Surgical removal before the formation of a blister is possible, assuming medical care is available. Healing at the blister site takes place in about eight weeks, but secondary bacterial infection and permanent crippling and disability may result.

In addition to the human misery associated with dracunculiasis, the disease poses an economic and financial burden for families. While the worm is emerging an infected person is often unable to work or go about daily activities for a few

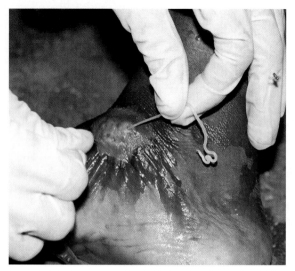

(a)

(b)

FIGURE 11.24 Guinea worm, the "fiery serpent." **(a)** After a female worm matures, it induces formation of a blister at the site from which it will emerge. **(b)** A worm is removed from a young girl's foot, a painful process that can take several weeks. Photos courtesy of the Carter Center.

months. The inability to farm and to harvest results in heavy crop losses, and children are often kept out of school to work in the fields.

Once the larvae are ingested, guinea worm disease cannot be cured. Prevention is based on teaching villagers in affected communities, many of whom are illiterate, to filter their water through cloth or nylon filters to rid the drinking water of copepods. Additionally, people need to avoid seeking relief from the burning sensation of emerging worms by immersing their feet or other body areas in reservoirs of drinking water.

Guinea worm disease is on a path toward eradication. Former President Jimmy Carter took the initiative in 1986 in marshaling global resources against dracunculiasis and was knighted in Mali, Africa in April 1998 in recognition of his efforts. A coalition of organizations, initiated by the Global 2000 Program of the Carter Center and the CDC, in concert with the WHO and UNICEF, the World Bank, numerous bilateral and multilateral agencies, private corporations, and foundations, is working toward the eradication. In May 2000 at the fifty-third World Health Assembly (the highest governing body of the WHO), the Bill and Melinda Gates Foundation awarded a grant of $28.5 million to accelerate the eradication of dracunculiasis. The eradication campaign reduced the number of reported cases by 97%, from an estimated 3.2 million annually in 1986 in twenty countries, to about 9,585 reported in 2007 in five African countries and to 1,747 in 2010. WHO statistics report that from January 1 to September 30, 2011 there were 1008 confirmed cases from 461 villages: South Sudan, 982; Mali, 10; Ethiopia, 8; Chad, 8. Asia is free of dracunculiasis. The collaboration of the public and private sectors can serve as a model for eradication of other diseases. Progress toward eradication is slow, steady, and ongoing, but the goal is in sight. Asia is now reported to be free of the parasite. Education about filtering drinking water and avoidance of entering water when worms are emerging, treating water sources with compounds to kill copepods, and the provision of clean water are the necessary steps to eradicate guinea worm. Further, the provision of clean drinking water is essential.

Tapeworms (Cysticercosis/Taeniosis)

The two most common species of tapeworm are *Taenia saginata* (the beef tapeworm) and *Taenia solium* (the pork tapeworm). Their life cycles are similar and are initiated by the ingestion of undercooked beef or pork containing encysted larvae (FIGURE 11.25). The beef tapeworm is the larger of the two and may reach 20 to 25 feet in length, whereas the pork tapeworm reaches approximately 15 to 20 feet in length. An individual with a tapeworm might have a 20-foot animal dangling into their intestinal tract! The renowned parasitologist Asa Chandler had a tapeworm he named Homer.

Infection is acquired by eating meat (muscle) containing larvae that have been encysted into forms known as **cysticerci** (FIGURE 11.26). Upon ingestion these larvae are digested, except for the **scolex** (the head), which attaches to the intestinal wall by suckers. The scolex produces compartment-like segments known as **proglottids**, each of which is a reproductive bag containing both testes and ovaries. The tapeworm body continues to lengthen as new proglottids are produced, and the worms reach adult size. Each worm has anywhere from 1,000 to 2,000 proglottids, which mature and contain 80,000 to 100,000 eggs per proglottid. Proglottids detach from the worm and migrate to the anus, and approximately six per day are passed in the stool. The eggs can survive for long periods in the environment; cattle and other herbivorous (plant-eating) animals become

FIGURE 11.25 A butcher shop stall in a less-developed country. The unrefrigerated pieces of meat are open to the environment and contamination. © Regien Paassen/ShutterStock, Inc.

FIGURE 11.26 Tapeworm cysticerci. **(a)** A tapeworm larva-contaminated piece of beef in a butcher shop in Costa Rica. **(b)** Tapeworm cysticerci resemble small bladders. Author's photos (RIK).

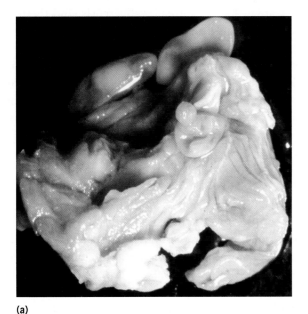

(a)

(b)

infected by grazing on contaminated vegetation. Larvae hatch from the eggs, penetrate through the intestinal wall, and become encysted as cysticerci in the muscle of the animal. When humans (or other animals) ingest these cysticerci, the cycle is completed.

Both *T. saginata* and *T. solium* produce only mild abdominal symptoms. Laboratory diagnosis is made to identify eggs and proglottids in the feces. It takes about three months after infection for adult tapeworms to develop; therefore, laboratory confirmation is not initially possible. Eggs of *T. saginata* and *T. solium* cannot be differentiated microscopically, but examination of proglottids and, rarely, the scolex allows species identification.

Both species are worldwide in distribution. *T. solium* is more common in poor communities where people may live in close contact with pigs and eat undercooked pork. Medication is available for the treatment of tapeworms. In the United States the Department of Agriculture is responsible for the inspection of beef that is intended for human consumption.

A serious disease called **cysticercosis** can result from the ingestion of pork tapeworm eggs (as opposed to ingesting cysticerci). These eggs are passed in the feces of an infected person and may contaminate food, water, or surfaces. Those infected with pork tapeworm can reinfect themselves by poor sanitary habits. The tapeworm eggs hatch inside the stomach, penetrate the intestine, and travel through the bloodstream; ultimately, they may cause cysticerci in the muscles, brain, or eyes. Lumps may be present under the skin, and infection in the eyes may cause swelling or detachment of the retina.

Neurocysticercosis (cysticerci in the brain or spinal cord) is serious and may cause death. It is considered to be the most frequent preventable cause of epilepsy in developing countries. More than 80% of people in the world who are affected by epilepsy live in developing countries. Seizures and headaches are the most common symptoms and can occur months, or even years, after infection. Antiparasitic drugs are available to treat cysticercosis, and their use depends on the number of brain lesions judged to be present and the individual's symptoms.

Other species of tapeworms can be acquired by humans. Fish tapeworm infections are common in Scandinavia, Russia, and other areas of the world, including the Great Lakes area of the United States. The disease is acquired by humans through the ingestion of raw or undercooked contaminated fish. Sushi eaters beware (BOX 11.1)!

Arthropodborne Diseases

Lymphatic Filariasis (Elephantiasis)

FIGURE 11.27 shows a dramatic example of **elephantiasis**. The term "elephantitis" is not correct, but, as illustrated, the grotesque swelling in the legs is elephant-like, and the texture and the appearance of the skin resemble the hide of an elephant.

What causes elephantiasis and brings about this deformity? The disease is caused by tiny adult threadlike worms, called **filarial worms**, which block the **lymphatic vessels** and cause the accumulation of large amounts of lymphatic fluid in the limbs. The lymphatic vessels function in the body's fluid balance and

BOX 11.1 _ Sushi Eaters Beware

Sushi lovers the world over enthusiastically devour the tasty and exquisitely presented little cakes of rice topped with a filet of raw fish. Sushi bars can be found in just about all large cities as well as small ones. Platters of these delicacies are found on buffet tables at weddings, anniversary parties, and other festive occasions.

The making of sushi dates back to at least the 1600s; it was introduced in Southeast Asia as a method of preservation of fish. The fish were packed with rice, which fermented and produced lactic acid, resulting in their being pickled. Why is sushi so popular? By no means is it inexpensive; a single little tidbit will cost you anywhere from $2.50 to $6.00—even more if you crave the exotic. The vinegar in sushi rice is claimed to have antibacterial properties, reduce fatigue, and reduce blood pressure. Nutritionists recommend two to three servings of fish per week, because fish is an excellent source of omega-3 fatty acids thought to be effective in the control of heart disease, rheumatoid arthritis, cancer prevention, and maybe even depression. In short, sushi, according to sushi lovers, is a health food.

What about the dangers of acquiring worms from eating raw fish? Statistics indicate that about 2,000 Japanese acquire worms each year from sushi; statistics are not kept in the United States. *Diphyllobothrium latum* is a fish tapeworm that can reach 25 feet in length in a person's intestinal tract. More commonly, *Anisakis simplex,* a slender half-inch-long worm, may be found in sushi. You can even buy a key chain with this worm embedded in plastic at the museum store in the Parasitological Museum in Tokyo! The presence of these worms causes abdominal pain that is frequently misdiagnosed as appendicitis, resulting in numerous unnecessary appendectomies. Adding to its miserable reputation, the worm excretes waste from its face and has a borer-like tooth that enables it to drill holes into the intestinal wall. Fortunately, the human is a dead-end host, and the worm dies after several days. On the other hand, if the *Anisakis* worm is lucky enough to be swallowed by a seal, it is able to complete its life cycle.

So the bottom line is this: The more likely risk is in handling the fish. The best sushi restaurants routinely freeze their fish before its appearance on cakes of rice, which kills the worms, affording a measure of protection to potential sushi eaters. Sushi chefs appear to be meticulous and frequently wash their hands, counters, and utensils, but choose your sushi restaurant carefully (as you should any restaurant), and then go ahead and order whatever strikes your fish fancy. You might choose the shiromi (whitefish), hamachi (yellowtail), uni (sea urchin), or ikura (salmon roe). The biggest danger is that you might ingest too much wasabi and feel like you are choking to death! For the uninitiated, a minute amount of wasabi is used on the sushi. It is even more potent than curry and will cause you to gasp, bring tears to your eyes, and burn your lips. Enjoy!

return **lymph** back into the circulation. Hence, blockage of these vessels results in a "backup" of lymph, causing swelling, particularly in the legs. The affected areas may swell to several times their normal size. In males the scrotum and penis may be involved, and scrotal swelling can be so severe that a wheelbarrow is used to support the scrotum to allow limited walking. In females the breasts may become considerably enlarged. Individuals with filariasis are prone to bacterial infections.

Wuchereria bancrofti and *Brugia malayi* are the two major species responsible for lymphatic filariasis. The cycle of infection is initiated by the bite of mosquitoes carrying filarial larvae. The larvae migrate into the lymphatic vessels where, within about a year, they grow into adult worms with a life span of up to several years. The adult worms mate, and the females release into the blood millions of microfilariae that may be ingested by mosquitoes taking a blood meal. The mosquitoes transmit the microfilariae to other humans, completing the cycle.

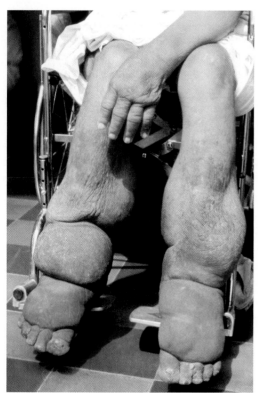

FIGURE 11.27 Elephantiasis. Infection with filarial worms can block lymphatic vessels, causing severe edema (swelling). Author's photo (RIK).

About half of the people with lymphatic filariasis have overt clinical disease. The early symptoms are chills, fever, and inflammation of lymph nodes and lymphatic vessels; some individuals remain asymptomatic and may never develop elephantiasis. The remaining people harbor infections with millions of microfilariae and dozens of adult worms in their bodies but with internal damage undetected and untreated. Despite the elephant-like deformities, the disease is not usually fatal. The WHO ranks filariasis as the second leading cause of permanent and long-term disability worldwide. The disease carries a social stigma because of its extreme disfigurement, and individuals are frequently shunned to the point their families neglect them. Their disability may cause them to be unable to work.

Examination of blood samples allows the identification of microfilariae. In patients with elephantiasis, swollen limbs are wrapped in compressive bandages in an effort to force out lymph thereby reducing discomfort and swelling. Lymphatic filariasis is most common in tropical and subtropical areas and is fostered by poor sanitation and population growth factors, which create increased breeding areas for mosquitoes. (Population growth is a factor in emerging and reemerging infections.) Filariasis affects 120 million people in 81 countries where it is endemic, including numerous countries in Africa, Southeast Asia, and the western Pacific and at least seven countries in the Americas. About 65% of cases occur in Southeast Asia and 30% in Africa. Further estimates are that debilitating genital swelling occurs in 25 million men and that elephantiasis of the leg occurs in over 15 million persons. Further, an additional 900 million people are estimated to be at risk for filariasis; approximately one-third live in India. In 1993 the International Task Force for Disease Eradication evaluated 94 infectious diseases for the feasibility of eradication; lymphatic filariasis is one of only six diseases considered potentially eradicable. As a result of an effective single-dose treatment, global elimination of lymphatic filariasis is now possible. A number of agencies in the public and private sector are collaborating to reduce the burden of filariasis.

Onchocerciasis (River Blindness)

Onchocerciasis, also called **river blindness**, is the world's second leading cause of blindness; it exists in many parts of Africa and Central America and is caused by *Onchocerca volvulus*, a parasitic worm. Onchocerciasis can devastate entire communities. In many small villages nearly all villagers past the age of forty are blind, presenting a serious obstacle to socioeconomic development (**FIGURE 11.28**).

The disease vector is a black fly, which abounds in fertile riverbanks and breeds in fast-flowing rivers and streams. When the fly bites, larval worms are deposited under the skin. The larvae mature into adult worms about a year later and live for as long as 14 years. They produce millions of microfilariae (microscopic larvae) that migrate throughout the body, causing rashes, intense itching, depigmentation of the skin, and nodules that form around the adult worms. Some migrate into the eyes and invade the cornea, causing blindness. These manifestations begin to occur one to three years after injection of the larvae by the flies. The cycle is perpetuated when a biting black fly ingests microfilariae from the blood of an infected person while taking a blood meal and then injects larvae into a new host (FIGURE 11.29).

Diagnosis is difficult, and there is no blood test to detect the presence of microfilariae. The presence of nodules under the skin is diagnostic, and their removal by minor surgery reduces the number of microfilariae produced (FIGURE 11.30). Several anthelmintic drugs are available; ivermectin is particularly effective because it kills migrating microfilariae. In practice, ivermectin is the only drug suitable for treatment, because other drugs cause severe reactions.

The WHO estimates that 123 million people worldwide, of whom 96% are in Africa, are at risk for onchocerciasis. Over 6.5 million suffer from severe itching and skin disease, about a half a million are blind or severely visually impaired due to the disease. The disease is endemic in thirty-seven countries, of which thirty are in sub-Saharan Africa and six are in the Americas.

Until the 1980s control focused on the use of chemicals to kill immature black flies in rivers. A renewed and major effort to eliminate river blindness was initiated in 1987 when the drug **ivermectin** was licensed by Merck & Co. and provided free of charge for treating the disease. The Carter Center, the Lions Club Sight First Project, the CDC, and Merck are collaborating in an effort to curb river blindness.

FIGURE 11.28 A statue at the WHO headquarters in Switzerland, *A Little Child Shall Lead*, commemorates the battle against onchocerciasis. In some African villages a child leading a blind person is a common sight. Author's photo (RIK).

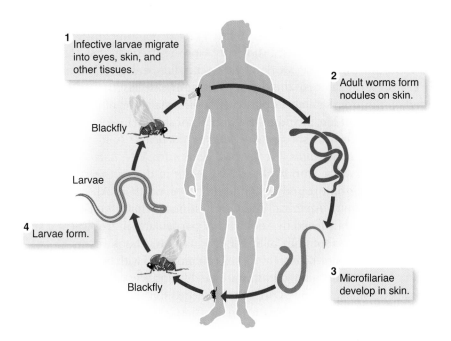

1 Infective larvae migrate into eyes, skin, and other tissues.

2 Adult worms form nodules on skin.

Blackfly

Larvae

4 Larvae form.

Blackfly

3 Microfilariae develop in skin.

FIGURE 11.29 Life cycle of *Onchocerca volvulus*.

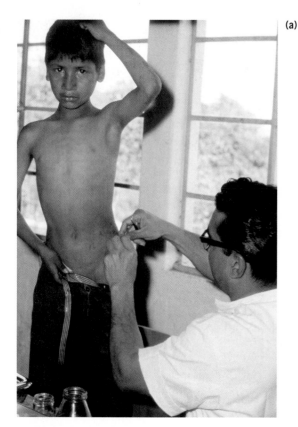

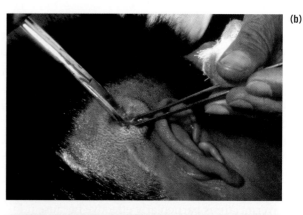

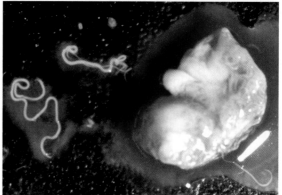

FIGURE 11.30 Onchocerciasis intervention. A bite by an infected black fly causes the formation of a nodule containing the larval form of the worm. **(a)** A healthcare worker examining nodules. **(b)** Surgical removal of worms from nodules. **(c)** Appearance of worms. Author's photos (RIK).

AUTHOR'S NOTE (RIK)
The photos in Figure 11.30 were taken on a field trip in Costa Rica. Figure 11.30b demonstrates nodules under the skin, which were easily removed by minor surgery. I took my turn with the scalpel.

Direct Contact Diseases

Hookworm Disease

In the United States rural areas of the South were, at one time, burdened with **hookworm disease**, a disease that insidiously saps the strength of those parasitized. Thanks to the pioneering work of the Rockefeller Sanitary Commission in the early 1900s, the incidence of this major parasitic disease has markedly declined as the result of improved sanitation, particularly the installation of indoor plumbing, and a better way of life, including the wearing of shoes. Hookworm disease is caused by two roundworm species, *Ancylostoma duodenale* (Old World hookworm) and *Necator americanus* (New World hookworm). *N. americanus* is the most common hookworm worldwide and is present in the southeastern United States. *A. duodenale* is only found in southern Europe, northern Africa, northern Asia, and areas of South America. The CDC estimates that about 20% of the world's population is burdened by hookworms, particularly in tropical and subtropical climates where about one billion individuals carry hookworms.

Infection occurs by direct contact with soil contaminated with hookworm larvae while walking barefoot, touching soil with bare hands, or accidentally swallowing bits of contaminated soil. Eggs require warm, moist, shaded soil and hatch in one to two days into barely visible larvae. After five to ten days the larvae reach the infective stage and penetrate the skin; they are carried to the lungs, where they are coughed up and swallowed, reaching the intestinal tract in about a week. In the small intestine the

(a)

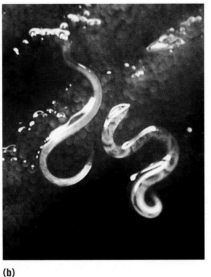

(b)

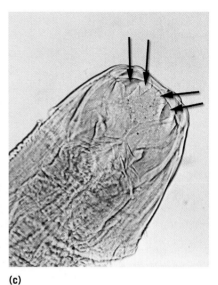

(c)

FIGURE 11.31 **(a)** Hookworm in wall of intestine. Author's photo (RIK). **(b)** Hookworm attached to intestinal mucosa. Courtesy of CDC. **(c)** Mouth parts of a hookworm. The hookworm uses its teeth (indicated by the arrows) to grasp the intestinal wall. Courtesy of Dr. Mae Melvin/CDC.

larvae develop into half-inch-long adult worms. The mature adult worms live and mate attached to the wall of the small intestine and produce thousands of eggs, which are passed in the feces (FIGURE 11.31). The eggs ultimately reach the soil, completing the life cycle. The cycle takes about five to six weeks from larval invasion to egg deposition by female worms (FIGURE 11.32). Person-to-person contact is not possible, because development of the larvae must be in soil as part of the life cycle.

The adult worms have mouth parts, called **biting plates**, by which they attach and suck blood, causing anemia, protein deficiency, and fatigue; in children, continual

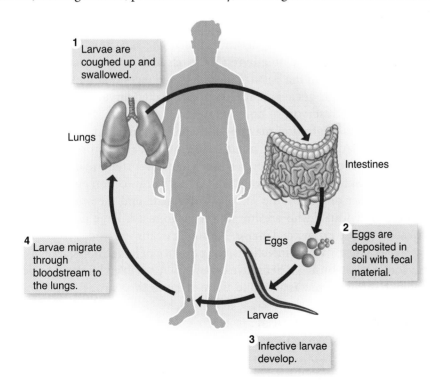

1 Larvae are coughed up and swallowed.

Lungs

Intestines

4 Larvae migrate through bloodstream to the lungs.

Eggs

2 Eggs are deposited in soil with fecal material.

Larvae

3 Infective larvae develop.

FIGURE 11.32 Life cycle of hookworms.

infection can cause retarded growth and mental development as the direct result of blood loss. Serious infection can be fatal in infants and causes children to be malnourished. The severity of symptoms appears to be directly related to the hookworm burden. Fewer than 25 worms do not usually cause disease, whereas 500 to 1,000 worms result in severe and often fatal damage. According to the WHO, hookworm disease is the leading cause of anemia and protein malnutrition, afflicting an estimated 740 million people in developing nations of the tropics. Diagnosis of hookworm is confirmed by the presence of eggs in a fecal sample. Medication is available for treatment.

Strongyloidiasis

Strongyloidiasis is caused by the nematode *Strongyloides stercoralis,* sometimes called the threadworm because of its minute size. Around 30 to 100 million people worldwide are infected with this parasite. *S. stercoralis* is unique in that the females produce eggs by **parthenogenesis**, a process that does not require fertilization by males (tough luck, boys). Infective larvae in the soil penetrate the skin and are sequentially transported by blood to the heart, lungs, and bronchial tree and into the pharynx, where they are coughed up and swallowed, finally reaching the small intestine. There, they mature into adult female worms and produce eggs. Some eggs develop into larvae that penetrate the small intestine or the skin around the anus and repeat the cycle through the heart, lungs, bronchial tree, and pharynx and to the small intestine where they mature into adult worms (**FIGURE 11.33**). This repeated cycle results in internal **autoinfection**, a process unique among worms in humans. Autoinfection accounts for persistent infections for over 20 years in infected persons who have moved to areas where the parasite is not present. Further, autoinfection can result in hyperinfection, characterized by the dissemination of the parasite within the body.

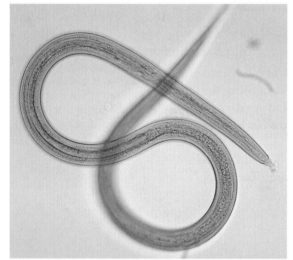

FIGURE 11.33 A *Strongyloides* larva. Courtesy of Dr. Mae Melvin/CDC.

The free-living cycle is an alternative strategy. It results from the passage of eggs in the feces that develop into free-living worms that produce large numbers of infected larvae. The free-living cycle is an important reservoir of human infection.

Strongyloides infection may be mild and go unnoticed, but it can be severe, resulting in nausea, vomiting, anemia, weight loss, and chronic bloody diarrhea. In 1958 a Japanese biologist infected himself by the intradermal injection of larvae to follow the course of the disease. The severity of infection appears to be related to the number of worms, the site of involvement, and the immune status of the host. In individuals who are immunocompromised, large numbers of invasive larvae can cause extensive damage, leading to death.

As with many worm infections, the presence of larvae in the feces and duodenal contents establishes a definitive diagnosis. A reliable blood test to detect antibodies to *Strongyloides* is available as are drugs to treat strongyloidiasis.

Enterobiasis (Pinworm Disease)

Pinworm disease is the most common helminthic disease in the United States, with an estimated forty million people infected according to the CDC. *Enterobius vermicularis,* about the length of a staple, is the culprit.

The life cycle is a simple one. The adult male and female worms live and mate in the intestine. At night, while the infected person sleeps, egg-bearing females exit through the anus, deposit as many as 15,000 eggs around the skin in the anal area, and then die. Egg laying is triggered by the lower temperature outside the body. Infective larvae develop within the eggs in about four hours. The eggs are swallowed and hatch in the intestine, releasing larvae that mature into adults, and become established in the intestine. The interval from ingestion of larvae to egg-producing females is about one month; the life span of the adults is about two months.

Self-infection is a common mode of transmission, particularly in children. Pinworms cause an intolerable itch, resulting in intense scratching. Eggs cling to the fingers, are lodged under the fingernails, and are then swallowed, reinitiating infection. Further, an infected person can transfer eggs through direct contact, on bedclothes, and even by inhalation of airborne eggs.

Anal itching, restlessness, and loss of appetite are common and relatively mild symptoms; the disease is rarely debilitating. School-age children have the highest rate of infection, and in some childcare centers and other institutional settings over 50% of children are infected. If one member of a family has pinworm, there is a good chance that other family members are infected.

Diagnosis is based on the "Scotch tape test"; a piece of tape is applied to the anal region during the night or early morning hours. If eggs are present, they stick to the tape and are identified under a microscope (FIGURE 11.34). Treatment involves a two-dose course of an anthelmintic drug; the second dose should be given two

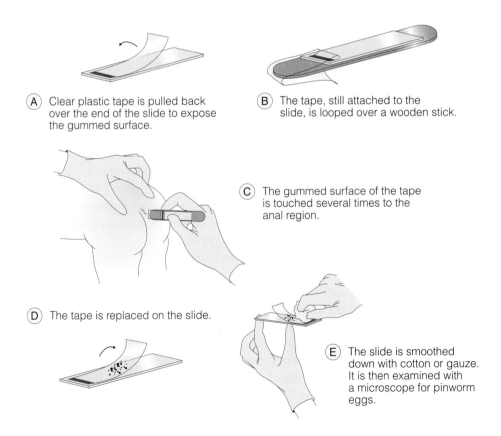

(A) Clear plastic tape is pulled back over the end of the slide to expose the gummed surface.

(B) The tape, still attached to the slide, is looped over a wooden stick.

(C) The gummed surface of the tape is touched several times to the anal region.

(D) The tape is replaced on the slide.

(E) The slide is smoothed down with cotton or gauze. It is then examined with a microscope for pinworm eggs.

FIGURE 11.34 Diagnosing pinworm disease. The transparent tape technique is used in the diagnosis of pinworm disease.

(a)

(b)

FIGURE 11.35 Schistosomiasis. Larval forms called cercariae are present in water and penetrate the skin, initiating an infection. **(a)** Two boys wading in water despite a warning sign. Courtesy of CDC. **(b)** A boy with a swollen belly resulting from schistosomiasis. Courtesy of CDC.

weeks after the first. In 2009, over 300 million children were dewormed. It is usually wise to treat all members of a family to break the cycle. According to the CDC, cleaning and vacuuming the entire house and washing sheets every day are probably not necessary or effective.

Schistosomiasis

More than 207 million people have **schistosomiasis** in seventy-four countries of Africa, South America, the Caribbean, the Middle East, and the Far East. The incidence of this disease is second only to malaria. It is caused by *Schistosoma japonicum, Schistosoma haematobium,* or *Schistosoma mansoni;* these worms are frequently called **blood flukes**. Certain species of snails serve as intermediate hosts in these geographical regions; they release immature, free-swimming microscopic forms, called **cercariae**, resembling miniature tadpoles that can survive in the water for about 48 hours. Exposure to water containing cercariae allows their penetration into the skin (**FIGURE 11.35**).

Penetration is followed by migration into blood vessels and, in about a week, entrance into the liver, where the organisms achieve sexual maturity. The adults have evolved a clever strategy of coating themselves with antigens of the host, thereby escaping detection by the host's immune system. Female worms reside in a groove in the body of the larger male, and the two remain entwined as a (happy) mating pair for up to 10 years. The pair migrate into small blood venules (veins) of the intestinal tract or of the bladder, depending on the particular schistosome species. Ancient Egyptian writings describe males with blood in their urine as "menstruating"; possibly, this was the result of schistosomes in their urinary tract. The worms feed on blood, and the females produce eggs within several weeks; the eggs are discharged by infected persons defecating or urinating into the water. The eggs hatch into actively motile larvae that penetrate the snails and develop into cercariae, completing the cycle.

Itchiness develops at the penetration site within several days and may be followed by fever, chills, cough, and muscle aches within the next several weeks, but some people experience no symptoms during this early stage. The major consequence of schistosomiasis is the damage caused to the liver, intestine, or bladder. The eggs cause allergic reactions. More rarely, eggs may move into the brain or spinal cord, causing seizures, paralysis, or inflammation of the spinal cord. The adult worms can also damage the spleen, liver, lungs, intestine, and bladder, causing the abdomen to become distended (FIGURE 11.35b). The adult worms escape detection by the immune system and can survive for years.

Diagnosis is confirmed by the presence of eggs in the urine or stool. Safe and effective drugs are available for treatment, but they are relatively expensive. Prevention through improved sanitation and control of the snail vectors is a more efficient and cost-effective approach. In some areas of Africa the introduction of a species of small fish that feeds on snails (biological control) is reducing the cercarial population.

Technological development has often been cited as a causative factor of new and emerging infections. An unfortunate and unpredictable consequence of the building of the Aswan Dam in Egypt in the 1960s was the significant increase in schistosomiasis because of the favorable conditions created for the intermediate snail hosts. In the United States the required intermediate snail host is not present so schistosomiasis does not occur.

A condition known as **swimmer's itch**, or **cercarial dermatitis**, occurs sporadically and unpredictably in the United States, particularly in the northern lakes and in Mexico and Canada. Swimmer's itch is due to penetration of the skin by cercariae from eggs of water birds infected with the avian species of schistosomes; humans are accidental hosts. The body's immune system destroys the cercariae, resulting in the release of allergenic products that cause a rash and itching.

Transmission of Worms From Pets

Pet cats, dogs, reptiles, and birds (but not goldfish) that share our homes and sometimes beds are potential sources of worm and other infections. Simple measures of personal hygiene, hand washing in particular, and veterinary care to maximize healthy pets minimize transmission from pets to people (BOX 11.2).

Biology of Fungi (Not To Be Confused with "Fun Guy")

Fungi are a category of eucaryotic microbes; some are unicellular and others are multicellular. They have distinctive characteristics as compared to other microbes; they are ubiquitous. Fungi are a varied group as is apparent in the many kinds of mushrooms that seemingly spring onto your lawn overnight, or look at a pair of old sneakers thrown into your closet or observe a piece of vegetable or fruit (FIGURE 11.36) that you forgot to take out of your backpack a few weeks ago and you will likely see a fuzzy moldy growth.

Most would agree that dogs are man's best friend (**FIGURE B11.2**). Dogs have been domesticated for thousands of years and have been used as sheep and cattle herders and as guard dogs. They conjure up visions of St. Bernards trudging through deep snow in an effort to rescue stranded mountaineers, Dalmatians accompanying fire engines, Lassies playing with children and alerting household members of a fire or other danger. They are known for their gentleness, loyalty, and affection. No matter how stressful a dog owner's day may be, upon returning home the owner will be relieved by the sight of an excited, happy animal anxious to bestow kisses. A pat on the head, a soothing "good dog," or a treat is all that is asked for in return. Dogs, of course, can be pests at times, such as when they demand to "go out" after having just "gone out."

More significantly, dogs (and other pets) can be a source of disease. Fortunately, most microbial diseases, including worms, that dogs acquire cannot be transmitted to humans, and, fortunately for dogs, most human microbial diseases are not suffered by dogs. Nevertheless, there are exceptions.

Most puppies are born with _Ascaris_ and can be a threat to children because of the possibility of the accidental ingestion of _Ascaris_ eggs; the eggs mature and subsequently migrate throughout the body. The incidence of this happening, however, is low.

A variety of bacteria, including _Salmonella, Campylobacter,_ and _Escherichia coli_ O157:H7 have been documented to be transmitted from pets to humans. To minimize the risk of pet-to-human transmission, practice good personal hygiene, particularly hand washing; deworm the dogs as necessary; clean up the area frequently where the dogs relieve themselves; and, most importantly, do not feed pets raw or undercooked fish or meats.

FIGURE B11.2 "Man's best friend." Dogs and other pets can be a source of bacterial, viral, worm, and fungal infections. This dog, Charlotte, is a Coton de Tulear, a relatively rare breed in the United States. Author's photo (RIK).

FIGURE 11.36 Moldy orange. Author's photo (TS).

Basically there are two distinct forms of fungi, namely, the unicellular microscopic yeasts and the branching filamentous molds characterized by their fuzzy appearance (**FIGURE 11.37**). Fungi are a difficult group to place in a taxonomical (classification) sense. There are over 50,000 species of fungi but surprisingly only about 25 cause disease in humans. Others are of serious nuisance and potentially may be harmful to humans and other life forms. Many fungi have complex life cycles exhibiting both asexual and sexual stages. Spore formation is the mode of multiplication (unlike bacterial spores). Spores are easily inhaled or land on the skin as manifested by the frequency of cutaneous and respiratory tract infections.

This chapter deals with fungi that can be a serious nuisance value and others that cause disease in humans. However, some fungi are beneficial in the production of certain foods. The soil mold _Penicillium notatum_ is the original source of penicillin; the mold _Streptomycin griseus_ is the original source of streptomycin, the first effective antibiotic against tuberculosis. Numerous other fungi are harvested as a source of antibiotics. _Penicillium roqueforti_ is used to produce blue cheese.

The authors have not attempted to cover fungi to the extent given to other microbes and worms; only a handful have been presented. Hence, the usual organization by mode of transmission does not apply.

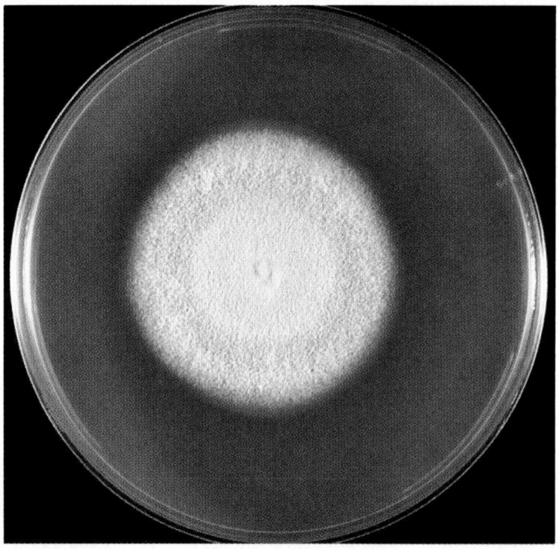

Fungi in the Environment

As pointed out, mushrooms and other fungi are common inhabitants in the environment and some mushrooms are quite beautiful, but don't be fooled. Others, like *Stachybotrys chartarum*, may take your house down. "Stachy" (as it is affectionately called, but I don't know by whom in their right mind) is truly a bad actor and is responsible for the sick house and the sick building syndrome. It is a black and rapidly producing mold capable of growing on the surface of many building materials including dry wall and surface tile. Once it gets into a building it is very difficult to get out and usually requires professional bioremediation; there are cases in which, as a last resort, homes were burned down to get rid of "Stachy."

FIGURE 11.37 Fungal colony. Courtesy of Dr. Libero Ajello/CDC.

Monster Category 4 Hurricane Charley hit land at Caya Costa, Florida, near Captiva Island in Southeastern Florida on August 14, 2004, with 140-mile-per-hour winds and water surges as much as 15 feet. Roofs went flying off, and in short order the interior of houses became soaked. The hurricane ended but the misery continued for months. After a few weeks, it was obvious that "Stachy" and other molds had invaded. As one homeowner stated, "After losing my roof, I wasn't prepared for mold taking over my house." Hurricane Katrina hit Louisiana and parts of neighboring states in Mississippi and Alabama in 2005. Again, once the hurricane was over, "Stachy" and other molds became a major problem causing entire blocks of houses to be torn down. Remember that these fungi grow well on dry wall, a major building material. However, it does not take a hurricane for "Stachy" and other molds to move into a house or into a building; leaky and poorly placed gutters, drains, and poor landscaping may allow seepage (FIGURE 11.38). In buildings

FIGURE 11.38 A room covered in mold after Hurricane Katrina and the flooding that occurred after the levees broke. © Julie Dermansky/Photo Researchers, Inc.

Yes, you may love your beautifully decorated blinking tree with its fresh pine scented aroma, but it may not love you (**FIGURE B11.3a**). The tree may well bear invisible spores of the mold *Phytophthora cinnamoni* or other species that can bring you misery in the form of itchy and teary eyes, stuffy nose, headache, sore throat, sneezing, and coughing, all of which are the usual symptoms of a mold allergy that occurs in about 15% of the population. A "Grinch," in the form of a mold, might steal your holiday causing you to,

instead of loving the tree, end up approaching it with a chain saw!

The Fraser fir is one of approximately ten species of fir trees; it has graced the blue room of the White House as a Christmas tree more than any other tree based on its fragrance, long limbs, and delayed loss of needles. In its natural environment it has its own problems as a result of infection by flagellated spores produced by *Phytophthora* that cause root rot (**FIGURE B11.3b**).

(a)

(b)

FIGURE B11.3 **(a)** Fraser fir, one of the most common Christmas trees, are especially vulnerable to fungal infections. Airborne mold counts increase after live Christmas trees are placed in a room. © casenbina/iStockphoto. **(b)** Root rot caused by *Phytophthora cinnamomi*. Courtesy of Stacy Clark, USDA Forest Service.

constructed in the last two decades there was great effort to build them "air tight" and unknowingly not allowed the building to breathe in the sense of outdoor-indoor air exchange; further, some of the materials used were found to exude noxious odors that induced the usual allergic-like symptoms. The new "Green Regulations" hopefully, will minimize construction-based future problems associated with sick homes and sick buildings. Despite these negative scenarios, fungi play very significant roles based on their fermentative capability in the food and alcoholic beverage industries. It's hard to beat a generous slice of Roquefort cheese on a slice of bread with a glass of beer to wash it down!

There are many examples of interactions between humans and other animals and plants and numerous species of fungi (**BOX 11.3**). In the wake of the tornado that hit Joplin, Missouri on May 22, 2011, at least 13 people developed a laboratory-confirmed rare skin infection called necrotizing soft-tissue infections caused by mucormycetes, resulting in 5 deaths. Mycormycetes is an old name for Zygomycetes, a class of fungi. Genera in this class include *Mucor, Rhizopus, Absidia, Rhizomucor,* and others. They are found in nature on vegetation and if the vegetation gets driven into the skin via wind or water, they can cause disease. The organism is an opportunist, causing problems for individuals with untreated diabetes and those who have compromised immune systems (as in AIDS, causing pneumonias) are particularly vulnerable. Mucormycetes have been reported following other natural disasters.

Fungal Diseases of Animal Species

"A Froggy Would A-Wooing Go" or "Where Have All The Froggies Gone?"

Walk the streets of Chinatown in San Francisco and you will note many street vendors selling bullfrogs from their tanks for use in laboratories, as pets, or as food. Perhaps you dissected a frog in an early course in biology. Care is taken by the vendor to prevent these amphibians from escaping from their tank or that during a sale drops of tank water do not fall to the pavement. Why the concern? It turns out that many of the frogs are infected with the fungus *Batrachochytrium dendrobatidis* (the worst tongue-twister yet cited) causing a thickening of their skin that even Botox wouldn't help. The thickening of their skin interferes with gas exchange and water absorption but does not usually cause death. Most of these frogs were imported from Taiwan. "Batra" can spread to native frog populations by an infected frog or by water from the holding tank. Efforts to ban the importation of frogs from Taiwan has met with strong resistance from the large Asian-American population in the area where many make their living selling the frogs. Further, the bullfrogs are a part of their cuisine. As stated by one Chinese resident of the community, "For over 5,000 years, it has been the practice of both the Chinese and Asian Americans to consume these particular animals. They are a part of our staple. They are a part of our culture. They are a part of our heritage."

The other side of the battle is presented by the nonprofit organization, Save the Frogs!, whose membership petitioned the governor to take action banning the importation and sale of bullfrogs in California fearing that a "Batra"-infected bullfrog would escape into the environment and threaten the native species. The solution may be that the importation of frogs will be prohibited unless there is appropriate documentation certifying that they are free of "Batra."

Bats

There may not be as many bats in the belfries around Washington, D.C. these days because millions of the "little brown bats," without even batting an eyelash, have been destroyed by the fungus *Geomyces destructans*. The bats develop white-nose syndrome characterized by a powdery white substance that develops on their muzzle, ears, and wings. The interaction between fungal species and plant and animal species is endless.

Fungal Diseases of Humans

Fungi are frequently secondary invaders in a number of viral and bacterial diseases in humans. The leading cause of death in AIDS is a pneumonia caused by the fungus *Pneumocystis jiroveci*; other **opportunistic** fungal infections are associated with AIDS because of the patients compromised immune system (TABLE 11.5). Fungi are the primary cause of a variety of human diseases known as **mycoses** (TABLE 11.6); a few are further detailed in this chapter. Other fungi exist "in the wild" (TABLE 11.7). Most of the diseases are not fatal, but once contracted, could be a constant source of irritation and can lead to permanent scarring. Fungal diseases don't occur as epidemics like we have found with bacterial disease outbreaks.

TABLE 11.5 Fungal Infections in Humans with Compromised Immune Systems

Fungal Pathogen	Disease
Candida albicans	Candidiasis (thrush, yeast infections)
Histoplasma capsulatum	Histoplasmosis
Penicillium marneffei	Penicilliosis
Coccidioides immitis	Coccidioidomycosis or "Valley Fever"
Aspergillus sp.	Aspergillosis
Pneumocystis jiroveci	Pneumonia

TABLE 11.6 Fungal Infections in Humans with Normal Immune Systems

Fungal Pathogen	Disease
Candida albicans	Candidiasis (yeast infections)
Coccidioides immitis[a]	Coccidioidomycosis or Valley Fever (lung infection)
Histoplasma capsulatum[a]	Histoplasmosis (lung infection)
Blastomyces dermatitidis[a]	Blastomycosis (lung infection)
Trichophyton[a]	Athlete's foot (*Tinea pedis* or ringworm of the foot)
Trichophyton[a]	Ringworm of trunk, arms, or legs (*Tinea corporis*)
Trichophyton[a]	Jock itch (*Tinea cruris* or ringworm of the groin)

[a] Also infects animals (e.g., dogs and cats).

TABLE 11.7 Fungal Infections in the Wild

Pathogen	Host	Disease
Batrachochytrium dendrobatidis	Amphibians (e.g., wild and pet frogs)	Chytridiomycosis
Phytophthora cinnamomi	Fraser fir (Christmas) trees; infects many plant and tree species (firs, oaks, avocado)	Severe root rot
Geomyces destructans	Hibernating bat populations in the U.S.	White-nose syndrome
Stachybotrys chartarum	Grows on building materials after water damage; can make humans sick.	"Sick building syndrome"
Apophysomyces trapeziformis	Humans wounded during tornado	Cutaneous mucormycosis or zygomycosis (rare infection following Joplin tornado, May 2011)

Histoplasmosis

Histoplasmosis is a worldwide fungal disease that affects the lungs; in the United States it is found primarily in the eastern and central United States and is endemic in the Mississippi and Ohio River valleys. The **etiological or causative agent** is *Histoplasma capsulatum*. Frequently, it causes no symptoms. However, the disseminated form is another story. It can be fatal unless treated and occurs primarily in those with AIDS or other forms of immunosuppression.

Histoplasma grows in soil and in materials contaminated with bird and bat droppings (guano). Whenever the ground is disturbed, be it by a shovel or bulldozer, spores of all sorts are released, become airborne, and may be breathed in. (Remember this every time you see a bulldozer preparing land for another useless mall when the one down the street is half empty.) Spores have been found in litter used in poultry houses, bird roosts, and areas harboring bats. The infection is also called "cave disease" or "Ohio Valley Disease."

Histoplasma has an interesting lifestyle. In the environment it grows as a brownish patch-like form (mycelium), but at body temperature it morphs into budding yeast-like stage. Antifungal medication is available and is a lifesaver for those with the disseminated form assuming treatment is started early in the top of the game.

Ringworm

Ringworm or the *Tinea* group of mycoses is caused by dermatophytic fungi (Table 11.6). The fungus thrives on keratin, the dead protein layer of the surface of skin. The condition once thought to be due to a worm, but, although this is not the case, the name persists. Heat and moisture favor the growth of this fungus resulting in it being found primarily in skin folds as in the groin and between the toes. The word *Tinea* (Latin) refers to the overall worm-like appearance, but the full name takes on the location of the outbreak on the skin. For example, *Tinea capitis* identifies ringworm of the scalp, *Tinea cruris* refers to ringworm of the groin area ("jock itch"), and *Tinea pedis* signifies ringworm between the toes ("athlete's foot"; FIGURE 11.39). A personal account of the disease is presented by one of the authors (TS; BOX 11.4). The symptoms are manifested by itchiness, scaly appearance, and, if on the scalp or in the beard, patches of baldness. Athlete's foot (and you don't have to be an athlete to get it) is quite common, and easily acquired by walking on wet surfaces as around a pool or on a shower-room floor. It won't kill you but it is a big nuisance with sometimes almost extreme burning and itchiness. It can be transmitted by person-to-person contact, and by sharing combs, brushes, clothing, and other personal articles. Dog and cat lovers beware—these pets can carry and transmit the fungus; you are well advised to stick to goldfish!

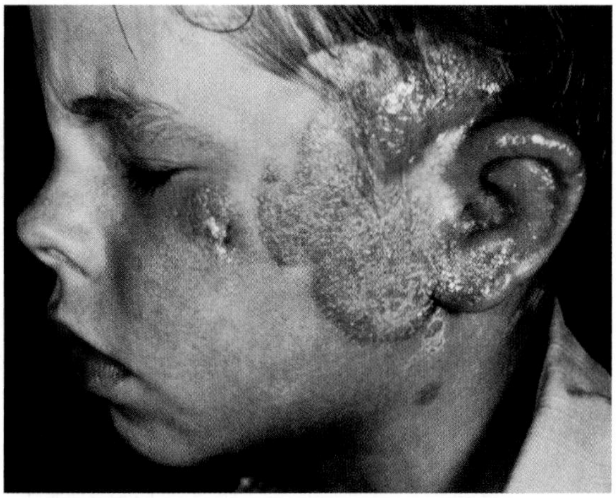

FIGURE 11.39 A child with a ringworm (*Tinea*) fungal infection on the left side of his face and ear. Courtesy of CDC.

Candidiasis

Candidiasis is caused by yeast cells; recall that yeasts are unicellular fungi. *Candida albicans* is the most common species, but numerous other species of *Candida* can cause mycoses (fungal infections). A common yeast infection is thrush, in which the yeast is in the oral cavity and the throat. It can affect both males and females. In females *Candida* grows in the vagina and is influenced by

Ringworm: An Author's Experience (TS)

I have poignant memories of a ringworm infection when I was about five years old growing up on a dairy farm in central Minnesota. There were many litters of kitties running around in the barn and I used to love holding them to my chest and playing with them. The barn kittens had very sharp claws and could crawl up the old brick walls of the barn; frequently they would scratch me. My memories are those of ringworm patches or lesions all over my tummy or chest. Every day, for about two weeks, my mother would put iodine on the ringworm patches. Eventually the ringworm went away. The iodine probably did nothing but this was in the 1960s, a time before there were antifungal over-the-counter medicines.

The culprit of ringworm is not a worm but instead common types of soil fungus known as **dermatophytes**. *Trichophyton* is the cause of ringworm in companion animals (dogs, cats, rabbits, rats, guinea pigs, horses, etc.). Infected animals may show no signs or symptoms. Ringworm or dermatophytosis affecting the trunk of the body is called *Tinea corporis*. Anyone at any age can get ringworm. It is diagnosed by the classic rash (**FIGURE B11.4**). Ringworm responds well to topical antifungal creams such as miconazole, clotrimazole, ketoconazole, and terbinafine, which must be applied twice a day to the lesions for at least three weeks.

Ringworm is highly contagious and is spread by direct contact, which is common between pets and kids, especially pets that often get smothered with attention like the barn kittens in my case. Good hygiene practices, particularly attention to hand washing, can certainly help, but some degree of risk will remain. *Trichophyton* spores can survive for more than a year. Therefore, once a pet has ringworm it is suggested that the pet's bedding be thrown away. Any grooming tools should be treated with bleach diluted with water at 1:10, or a 0.2% enilconazole solution. Carpets should be vacuumed frequently to remove hair and spores. Any areas that the pet frequents should be washed and sterilized. Hands should be thoroughly washed after touching an infected pet. Controlling ringworm outbreaks can take time and be frustrating but it's not a serious disease and is controllable.

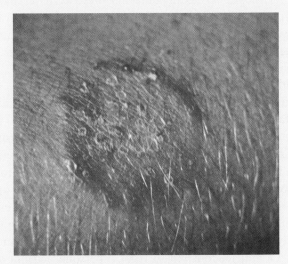

FIGURE B11.4 Ringworm or *Tinea corporis* present on the arm of a patient. Courtesy of Dr. Lucille K. Georg/CDC.

vaginal acidity and hormonal changes resulting in intense itching, burning, and discharge. Yeast infection is about the second most common cause of females seeing their physician.

Candida is a component of the normal body flora and is found on the skin and mucous membranes. In both oral thrush (**FIGURE 11.40**) and vaginitis, the infection is due to an imbalance of the normal microbiota of the area as caused by overgrowth of yeast. Consider, for example, that some individuals on an antibiotic may develop thrush due to the fact that the antibiotic kills off not only the pathogen for which the medication was given, but also members of the normal flora. The consequence is somewhat Darwinian-like in the sense that the medication does not kill the yeasts and, thereby reduces the competition giving the yeast a "growth kick." One of the authors (RIK) developed thrush after surgery and a course of antibiotics and can personally tell you it "isn't fun," but it is not at all life threatening. Your mouth and throat are painfully dry and covered with white patches of yeast and it feels like your tongue is sticking to the roof of your mouth. Antifungal mouthwashes solve the problem usually within a few days.

Invasive candidiasis, unlike thrush and vaginitis, which are considered superficial infections, is a serious and potentially life-threatening systemic condition. It is more likely to occur in those with an immunocompromised immune system as in AIDS, or those with compromised immune system due to another cause. The yeast enters the blood causing bloodstream infection and spreads throughout the body. Drug therapy is available.

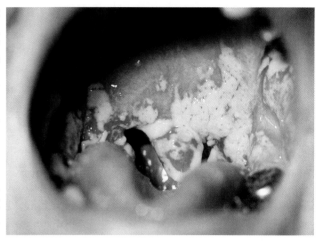

FIGURE 11.40 Oral thrush. Courtesy of CDC.

■ Overview

Protozoans are microscopic, eucaryotic, unicellular organisms; helminths are macroscopic, eucaryotic, multicellular organisms. Some fungi are microscopic, eucaryotic, unicellular organisms and other macroscopic, eucaryotic, and multicellular. All of these groups contain species that are pathogens of humans. Depending on the specific organism, transmission is by food and water, vectors, sexual contact, and direct (other than sexual) contact.

The diagnosis of protozoan and worm diseases is based largely on microscopic examination of body discharges, blood, and secretions for the presence of parasite eggs, larvae, or mature forms. In some cases examination of skin biopsies, bone marrow, and spinal fluid may be necessary. Symptoms and history, particularly travel history, are important parts of diagnostics.

Many of the protozoan and helminthic diseases are found in the rural sections of tropical and subtropical areas that suffer from poverty and poor sanitation. Strategies of prevention and control are targeted at vector abatement, minimizing contact between humans and parasites, and implementation of sanitary measures. Sanitation is extremely important and includes the use of latrines, provision of safe drinking water, avoidance of nightsoil (human feces as fertilizer), proper cooking and preparation of foods, and the wearing of shoes.

Drugs are available for the treatment and, in some cases, prevention of these diseases, but the treatment may be worse than the disease. There are no vaccines. Protozoans are unicellular eucaryotic organisms, as are humans, meaning that drugs toxic to them are also toxic to humans. The same is even more true for multicellular eucaryotic worms.

The fungi are an extremely diverse group and include organisms that are unicellular, multicellular, microscopic, and macroscopic; their life cycles may be extremely complex. Some fungi are beneficial, others are a nuisance value and others cause disease in humans. Hence, it is a juggling act to choose the drugs that are more toxic to protozoans, worms, and fungi that cause diseases in humans than they are to the human host.

■ Self-Evaluation

PART I: Choose the single best answer.

1. Mosquitoes are vectors for
  a. leishmaniasis **b.** filariasis **c.** malaria **d.** more than one of the above

2. Intestinal blockage is most likely to occur after infection with
  a. *Ascaris* **b.** tapeworms **c.** earthworms **d.** filarial worms

3. Which disease is one of the most common infectious diseases among soldiers on duty in Iraq and Afghanistan?
  a. malaria **b.** amebiasis **c.** leishmaniasis **d.** trypanosomiasis

4. Drinking what appears to be clean water from a mountain stream may put you at risk for infection with
  a. guinea worms **b.** *Giardia* **c.** *Cryptosporidium* **d.** tapeworms

5. Which parasite can be acquired from ingestion of eggs?
  a. tapeworm **b.** guinea worm **c.** *Ascaris* **d.** more than one of the above

6. Which disease can be acquired from stepping on larvae in the soil?
  a. hookworm **b.** tapeworm **c.** guinea worm **d.** elephantiasis

7. Ringworm is caused by a
  a. dermatophyte **b.** neophyte **c.** saprophyte **d.** parasite

8. Which of the following is not a fungal infection?
  a. *Tinea pedis* **b.** candidiasis **c.** jock itch **d.** schistosomiasis

9. Which of the following causes thrush?
  a. *Candida albicans* **b.** *Trichophyton* **c.** *Toxoplasma* **d.** *Plasmodium*

10. Which of the following causes "sick building syndrome"?
  a. *Stachybotrys* **b.** *Histoplasma* **c.** *Batrachochytrium* **d.** *Penicillium*

PART II: Match the statement on the left with the disease (or microbe) on the right by placing the correct letter in the blanks. A letter can be used more than once or not at all.

1. worm migrates out through skin _____
2. elephantiasis _____
3. dysentery _____
4. ingested cysts _____
5. pneumonia _____
6. chytridiomycosis
7. possible neurological damage _____
8. river blindness _____
9. ingestion of eggs _____
10. cats are carriers _____
11. heart condition _____
12. cryptosporidiosis _____
13. amebiasis _____
14. ring worm _____

a. onchocerciasis
b. *Pneumocystis jiroveci*
c. *Giardia*
d. tapeworm
e. malaria
f. *Histoplasma*
g. Chagas disease
h. *Cryptosporidium*
i. *Candida*
j. *Stachybotrys*
k. leishmaniasis
l. amebiasis
m. hookworm
n. dracunculiasis

15. chills and fever _____   o. blocked lymphatic vessels
16. schistosomiasis _____   p. intestinal blockage
17. soldiers in Iraq cases _____   q. Aswan Dam
18. thrush _____   r. *Naegleria*
19. dermatophyte _____   s. *Batrachochytrium*
20. Ohio Valley disease _____   t. *Toxoplasma*
21. sick building syndrome _____   u. *Enterobius*
22. pinworm disease _____   v. *Trichophyton*

Fill in the vector for each disease.

Disease	Vector (Transmission)
1. leishmaniasis	_____
2. Chagas disease	_____
3. sleeping sickness	_____
4. malaria	_____
5. onchocerciasis	_____
6. dracunculiasis	_____
7. filariasis	_____

PART III: Answer the following.

1. Choose a protozoan disease that is transmitted by food or water and describe its cycle. A diagram would be helpful.
2. You are a health worker involved in the eradication of guinea worm disease. Prepare an article to be handed out to the inhabitants of a village telling them about the disease and how to avoid becoming infected.
3. You are the director of public health services in an African country that is plagued with a variety of protozoan and helminthic diseases. Prepare a document outlining your plans to reduce the number of people who will become infected over the next years. (A budget will be developed based on your agenda.)
4. Compare and contrast the characteristics of parasites versus fungi.
5. Create a short list of fungi that produce antibiotics. Discuss why soil fungi produce antibiotics.
6. Explain why is it difficult to remediate buildings that are contaminated with *Stachybotrys*?
7. You will be traveling outside the United States to an area with malaria. What precautions will you take to prevent coming down with malaria?
8. Explain why a pregnant woman should not maintain a pet cat's litter box.
9. Why is trichinellosis relatively rare in the United States today?
10. Why are there more cases of helminthic diseases in the tropics compared to the United States?

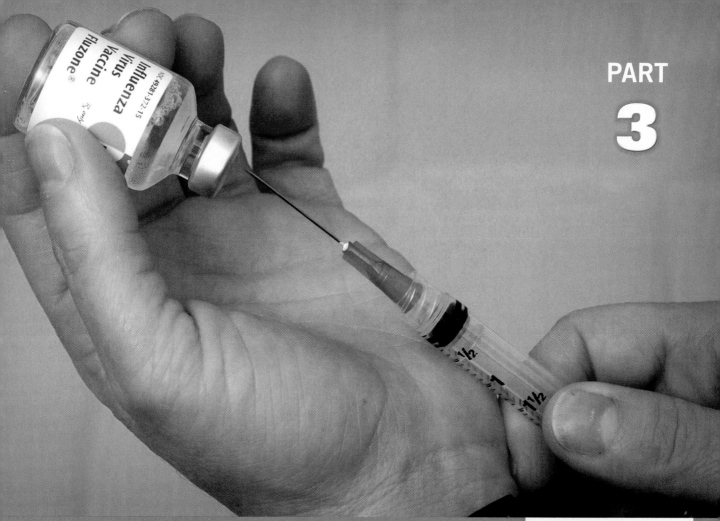

Meeting the Challenge

CHAPTER 12

The Immune Response

The immune system is complex; if it was simple, it would be practically useless.

—Unknown

■ Preview

The word "immunity," in its broadest sense, means freedom from a burden, be it legal action, taxes, or, in the present context, disease. In a legal connotation, the term means to be immune from civil liability that can be granted to a charitable or nonprofit organization: a hospital is a good example. Witnesses who may have been involved in criminal activity can be granted immunity from prosecution so they may testify against those involved in more serious crimes. In the early years of the colonization of the United States, the battle cry of the colonists was "immunity from taxes."

The ultimate outcome of infection is the result of the dynamic interplay between two opposing forces summarized by the expression $D = nV/R$. In the equation D refers to microbial disease, the numerator (n) refers to the microbe and its virulence (V) factors, and the denominator (R) is a function of the host immune system. The interplay determines whether no infection, mild infection, or severe and, possibly, fatal infection will result. The early hours of encounter between the microbe and the immune system are particularly crucial in determining the eventual outcome.

The immune system functions to recognize and destroy foreign molecules as embodied in invading microbes and their products and in mutant, damaged, and worn out cells. This is accomplished by both nonspecific (inherent) immunity and specific (acquired) immunity, functions accomplished by a variety of body systems working together. Nonspecific immunity is characterized by physiological defenses that act to prevent microbes from gaining access into the body and facilitate the elimination of those that have penetrated. Specific immunity targets microbes for elimination, based on the recognition of their foreign antigens (molecules), by antibody-mediated and cell-mediated immunity. Impairment of the immune system as occurs in AIDS, leukemia, use of immunosuppressive drugs, or congenital disease renders the immunocompromised individual subject to repeated life-threatening infections.

AUTHOR'S NOTE (RIK)
In writing this text I was initially in a dilemma. Should the microbes and their virulence mechanisms precede the body's immune defense, or the other way around? It is the "what came first, the chicken or the egg?" puzzle. Because a strong focus in this text is on disease prevention, it seemed more logical to first present what it is that the immune system is combating. I hope you agree!

Basic Concepts

The immune system functions to recognize and destroy molecules that are **foreign** to the body, including molecules on invading microbes and their secretions, toxins, and enzymes and foreignness on "non-normal" self cells. Functionally, the immune system consists of two components: **nonspecific** or innate (inherent) immunity and **specific** or adapted (acquired after birth) immunity (FIGURE 12.1). Both mechanisms are the strategies by which the body eliminates "foreignness."

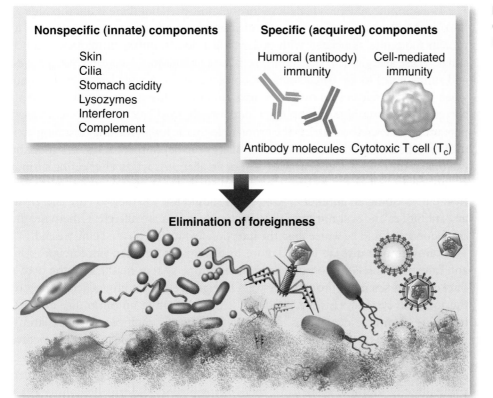

FIGURE 12.1 Nonspecific and specific components of the immune system protect against foreignness.

There is no one good way to measure the effectiveness of the immune system because of its many complex interactions. In fact, ability to cope with the constant barrage of microorganisms is not constant, nor is it measurable. Age, sex, race, nutrition, and state of health, along with physical and mental stress, play a significant role in immune status. Think about final exam time. Lights are on late in dormitories and in study halls around the campus, and fast foods are consumed at an even more rapid rate than usual as you review a semester's work for four or five courses in about a week's time. There is no doubt that your immune system is temporarily compromised, making you more vulnerable to infection.

Immune status varies within the individual and between individuals. Consider, for example, that a hundred people may be crowded into a lecture hall, movie theater, or restaurant and that all are exposed to circulating cold viruses. Some will "catch the cold," and others will not. Obviously, the numerator in the equation $D = nV/R$ is the same for all who are exposed, and, therefore, whether infection results is largely a function of R, the immune status at that time. A recent news article cited two deaths that occurred from babesiosis, a tickborne protozoan disease that is rarely fatal unless the individual has an underlying health problem. This was the case in both patients who died. Genetics, too, plays a key role; it is well established that there are ethnic and racial differences in susceptibility to microbial infection.

Although the primary role of the immune system is as a mechanism of defense against microbial infection, it is also a system of internal surveillance leading to elimination of tumor cells and destruction of old, worn out, and damaged red blood cells. These cells are abnormal and, like microbes, are recognized as foreign and trigger an immune response.

The immune system is beneficial for the most part, but there is another side to the coin. Many people suffer from **allergies**, which are adverse immune responses to protein molecules associated with pollens, dust, foods, mites, antibiotics, and bee stings. These allergies run the gamut from being a nuisance (sneezing, watery eyes, and runny nose) to life-threatening anaphylactic shock. Anaphylactic shock occurs when the body releases an overwhelming amount of histamine in response to an allergen. The histamine causes the blood vessels to dilate, which lowers blood pressure, and severely constricts the bronchioles in the lungs, making breathing very difficult. The truth of the matter is that an allergy can develop at any time. A dramatic example is the case of a student who died of an allergic reaction after eating shrimp in the dining hall at his college. Unfortunately, there are many similar examples. Before prescribing an antibiotic, a physician should ask whether you are allergic to any antibiotics and will not prescribe one to which you are allergic. Otherwise, the consequences could be worse than the infection for which you are being treated.

Autoimmune disease is the failure to distinguish between nonforeign (self) and foreign (nonself). In other words, the immune system is attacking the body's own cells and tissues as though they were foreign, a clearly nonbeneficial aspect of the immune system. This leads to a host of debilitating autoimmune diseases, including rheumatoid arthritis, lupus erythematosus, multiple sclerosis, and a variety of anemias. Paul Ehrlich, a pioneer immunologist, referred to this dysfunction as "horror autotoxicus," a fear of self-poisoning.

The major obstacle to transplantation of organs is rejection of transplanted organs by the immune systems of their recipients. Should this be considered a nonbeneficial aspect of the immune system? Most immunologists would not think so, because, after all, the recipient is responding to foreign molecules of the donor cells in a manner nature evolved to thwart microbial invasion and to conduct internal surveillance. Transplantation technology is not a natural phenomenon but is the result of advanced medical science.

Anatomy and Physiology of the Body's Defenses

Having now established a few basic concepts involving the immune system, attention needs to be focused on anatomical and physiological considerations. Perhaps in a precollege course in biology you dissected a frog; recall that the digestive system is a series of defined anatomical structures starting with the oral cavity, leading sequentially to the esophagus, the stomach, and intestinal tract. The same organization can be said for all the systems of the body. There is a hierarchy of cells, tissues, and organs in biological systems. The immune system is, however, unique in that the anatomical structures are, for the most part, shared with more defined systems. The spleen, in most anatomy manuals, is described with the circulatory system, and the lymph nodes are considered components of the lymphatic system, but these structures are also parts of the immune system. The major structures of the immune system are shown in FIGURE 12.2.

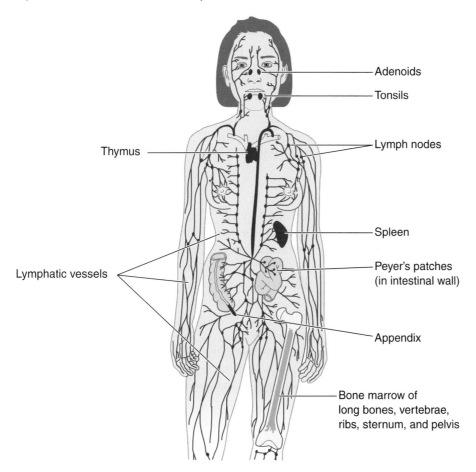

FIGURE 12.2 Anatomy of the immune system.

Adenoids

Tonsils

Lymph nodes

Thymus

Spleen

Peyer's patches (in intestinal wall)

Lymphatic vessels

Appendix

Bone marrow of long bones, vertebrae, ribs, sternum, and pelvis

Mechanical and Chemical Barriers

The skin is the first line of defense against microbial infection. The intact skin is a physical barrier that blocks the entry of microorganisms into the body based on its structural composition. Obviously, there are circumstances in which the skin is broken, allowing possible penetration by microbes. Cuts and abrasions, insect bites, and injections with hypodermic needles are familiar examples. Staphylococci, normal residents on the surface of the skin, are particularly prone to enter the body as opportunists through these breaks and establish infection. Many diseases, including malaria, yellow fever, babesiosis, and sleeping sickness (other than in the classroom), are transmitted by insects whose bites penetrate the skin.

The acidity of the stomach is detrimental to most microbes. Certainly, food and eating utensils are not sterile, and, hence, large numbers of microbes gain access to the stomach. In addition, mucous membranes line the internal body cavities and act as internal barriers. For example, the digestive tract running from the mouth to the anus is lined by a mucous membrane that protects the body from invasion by microbes that reside within the digestive tube. Penetration of this mucous membrane by intestinal microbes, even those that normally reside within the digestive tract, is extremely serious and leads to a life-threatening infection known as **peritonitis**. Surgeons are very much aware of this when performing procedures on the digestive system; nicking the lining of the intestinal tract allows microorganisms to spill out into body cavities where they do not belong.

The airways in certain areas of the respiratory system have cells with hairlike projections, called **cilia**, that propel the microbes into the pharynx where they are swallowed and expelled with fecal material. This system is called the mucosal-ciliary escalator system. Smokers are more prone to respiratory infections than nonsmokers, because smoke damages this system.

The enzyme **lysozyme** destroys the cell walls of some bacteria. Lysozymes are found in saliva, tears, and sweat (but don't sweat it!)

Blood

Several components of blood are vital defense mechanisms of the body and contribute to nonspecific immunity, specific immunity, or to both. A summary of the components of blood is presented in TABLE 12.1. The globulin fraction is particularly important to the function of the immune system; a subfraction, **gamma globulin**, contains most of the **antibodies**. Antibodies are molecules produced in response to antigens and act to target them for removal. Conversely, **antigens** are molecules that bring about antibody production. Antigens and antibodies are described more fully in a later section.

There are three categories of blood cells, namely **red blood cells** (erythrocytes), **white blood cells** (leukocytes, FIGURE 12.3), and **platelets** (thrombocytes). The red blood cells and platelets play no direct role in immunity and are not further described.

White blood cells are divided into five categories (TABLE 12.2). Neutrophils, basophils, and eosinophils are referred to as **granulocytes** because microscopic examination reveals the presence of granules in their cytoplasm. Lymphocytes

TABLE 12.1 Composition of Blood

Component	Function
Plasma	
Water (80–90%)	Solvent
Proteins	
Albumin	Blood volume
Globulins	Antibody molecules
Fibrinogen	Blood clotting
Complement	Immune amplification
Interferon	Antiviral properties
Blood cells[a]	
Red blood cells (erythrocytes)	Oxygen and carbon dioxide transport
White blood cells (leukocytes)	Antibody formation, CMI, phagocytosis
Thrombocytes (platelets)	Blood clotting

[a]In each microliter of blood there are about 4.5 to 5.5 million erythrocytes, 5,000 to 10,000 leukocytes, and 250,000 to 400,000 thrombocytes.

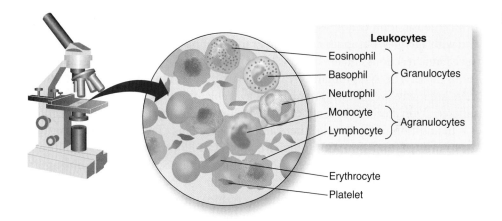

FIGURE 12.3 Several types of white blood cells.

and monocytes do not contain cytoplasmic granules, and these white blood cells are referred to as **agranulocytes**. The **lymphocytes**, of which there are two subsets, the **B lymphocytes** and the **T lymphocytes**, are the key cells of immunity. The B lymphocytes are the only producers of antibodies but are dependent in most cases on **T-helper** (T_H) cells, a subset of T cells for this function (discussed under Specific Immunity below).

Neutrophils and **monocytes** are phagocytic cells. Phagocytosis, or "cell eating," is an important defense mechanism by which microbes are engulfed and destroyed; phagocytosis is described in more detail below under Nonspecific Immunity.

A **differential count** reflects the ratio of the white blood cell components and is an important tool in the diagnosis of suspected infection. A drop of blood is smeared onto a microscope slide, covered with a specific stain, rinsed, and examined under the microscope to perform a differential count. Frequently, in bacterial (versus viral) infections an elevated neutrophil count is found, justifying antibiotic therapy while

TABLE 12.2 Categories of White Blood Cells (Leukocytes)

Cell Type	Total Leukocytes (%)	Function(s)
Agranulocytes		
Lymphocytes	25–35	Antibody formation, cell-mediated immunity
B lymphocytes		Produce antibodies
T lymphocytes		Necessary for specific immunity
Monocytes	3–8	Phagocytosis
Granulocytes		
Neutrophils	60–70	Phagocytosis
Eosinophils	2–4	Inflammatory response, limited phagocytosis
Basophils	0.5–1	Not clear; contain histamine

a more definitive laboratory diagnosis is awaited. By comparison, with the viral disease infectious mononucleosis, there is an increase in lymphocytes.

The role of **basophils** and **eosinophils** is not entirely clear. Eosinophilia (an elevated eosinophil count) is associated with the presence of helminth infections and with allergies. Basophils are rich in granules of histamine that are released during allergic reactions and are responsible for watery eyes, itchy throat, sneezing, and runny nose. The symptomatic treatment is antihistamine medication, of which there are many over-the-counter choices available. You may be one of the unlucky sufferers. Don't feel too bad; approximately 50% of the population has allergies.

In addition to blood cells, there are blood proteins (Table 12.4) that contribute to immune function. **Complement**, a series of blood proteins, is a significant defense mechanism against potential disease-causing microbes. Complement is a system of **biological amplification.** Consider the role of the amplifier in a music system as an analogy. It does not actually produce the music; if the amplifier were unplugged the music would continue, but the quality would suffer. In much the same way complement enhances phagocytosis, a nonspecific defense mechanism by which bacteria are engulfed and destroyed. Further, complement can bring about lysis (destruction by chemical attack) of the bacterial cell wall, causing it to "spill its guts" through the now-leaky membrane.

Interferon, another component of blood, was discovered in 1957 and was so named because of its ability to interfere with viral replication. It is not a single component but a group of related compounds. The product is released from virus-infected cells and acts to signal other cells of impending danger by triggering them to release virus-blocking enzymes. When interferon was discovered, the medical community had lofty expectations that it would prove to be a powerful antiviral drug that would be as effective against viruses as antibiotics are against bacteria. Unfortunately, interferon has not lived up to expectations. It has been only partially successful in the treatment of certain forms of hepatitis.

A group of proteins, called **septins**, are of increasing importance as a defense mechanism against infections caused by pathogenic bacteria. Septins are found in many organisms and have been particularly recognized for their role in building scaffolding-like supports for reproduction in yeast cells. More recently, they have been found to be associated with mammalian defense mechanisms. Studies indicate that septins build "cages" around bacterial pathogens trapping them and minimizing their spread. One study provides evidence that *Shigella*, a species that moves from host cell by "tails," are stopped in their tracks as they are caught in septin cages.

Lymphatic System

The **lymphatic system** is anatomically intertwined with the blood circulatory system. **Lymph**, a tissue fluid derived from blood but without blood cells, is circulated throughout the lymphatic capillaries and veins and opens into the blood circulatory system, allowing return of lymphatic fluid. In elephantiasis worms block the lymphatic vessels, leading to the grotesque elephant-like limbs. Hence, the two systems are interdependent: lymph is derived from blood and drains back into blood (FIGURE 12.4). Situated along the path of the lymphatic veins are the **lymph nodes**. Swollen lymph nodes in the armpits, groin, or neck indicate that these nodes have filtered out invading microorganisms; an infection is in progress. These nodes are highly significant because they contain both B and T cells, enabling antibody production. Lymph nodes also contain phagocytic cells, which destroy microbes and remove damaged or worn-out cells of the body. Thus lymph nodes contribute to both the nonspecific and specific immune systems.

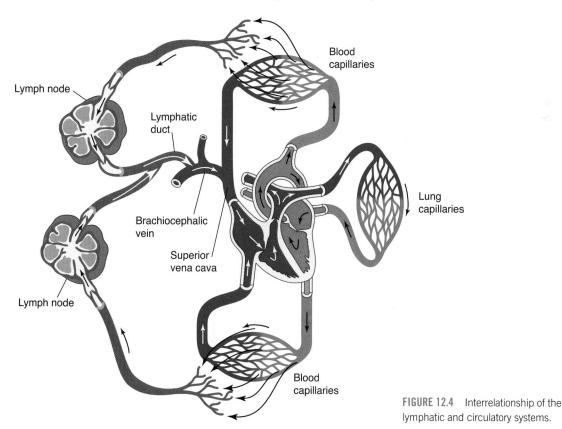

FIGURE 12.4 Interrelationship of the lymphatic and circulatory systems.

Primary Immune Structures

Bone Marrow

The **bone marrow**, a **primary immune structure**, plays a highly significant role in immune defense mechanisms; it is the source of all blood cells (red blood cells, white blood cells, and platelets) and produces them from stem cells located in the red bone marrow of the sternum (breastbone), vertebrae, and the upper ends of the long bones of the body. There are two pathways of blood cell maturation (FIGURE 12.5), the **myeloid path** and the **lymphoid path**. The myeloid lineage leads to platelets, red blood cells, and four of the five categories of leukocytes (monocytes, neutrophils, eosinophils, and basophils), whereas the lymphoid lineage leads to lymphocytes. B lymphocytes complete their maturation in the marrow and are released into the blood, from which they seed secondary structures of the immune system. Ultimately, all categories of blood cells are released into the general circulation as mature cells.

Thymus Gland

T lymphocytes are released from the marrow before their maturation, and some end up in the **thymus gland** (not thyroid), an organ located behind the sternum and just above the heart (Figure 12.2). In the thymus gland these cells mature into T cells that seed secondary structures of the immune system.

At birth, the thymus is a relatively large organ and attains its maximum size by puberty. The early years are the ones in which numerous childhood diseases and immunizations are experienced; the immune system is maturing, and the thymus plays a significant role. Some children are born without a thymus gland, a condition called **DiGeorge syndrome**. They suffer the consequences of deficient

FIGURE 12.5 Pathways of blood cell maturation.

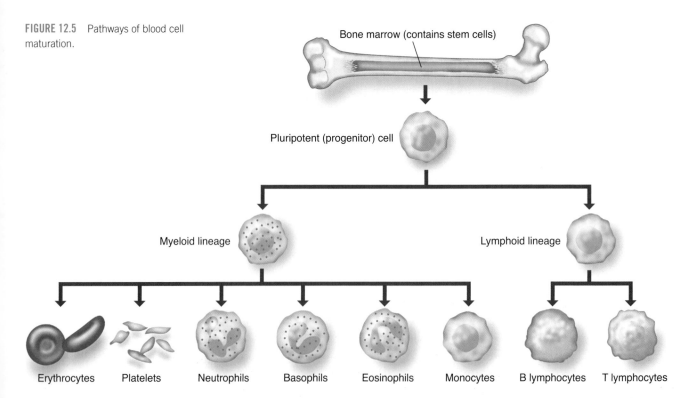

Bone marrow (contains stem cells)

Pluripotent (progenitor) cell

Myeloid lineage

Lymphoid lineage

Erythrocytes   Platelets   Neutrophils   Basophils   Eosinophils   Monocytes   B lymphocytes   T lymphocytes

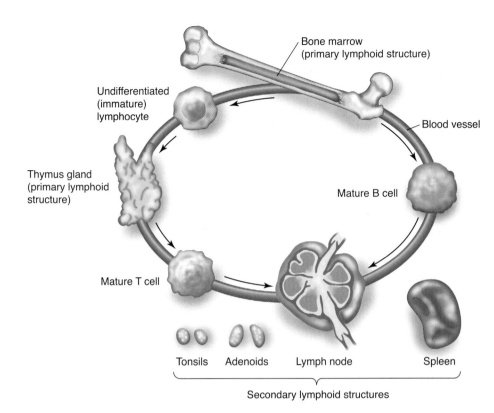

FIGURE 12.6 Maturation and "seeding" of B and T lymphocytes.

Bone marrow
(primary lymphoid structure)

Undifferentiated
(immature)
lymphocyte

Blood vessel

Thymus gland
(primary lymphoid
structure)

Mature B cell

Mature T cell

Tonsils   Adenoids   Lymph node   Spleen

Secondary lymphoid structures

immunity throughout their lifetimes, as manifested by one potentially life-threatening infection after another. After adolescence the thymus gradually degenerates and assumes a lesser significance; its function is taken over by the bone marrow.

To recapitulate, the thymus is the site of maturation of T cells; the bone marrow is the site of maturation of B cells. These are the primary structures of the immune system and serve as the beds for the maturation of T and B lymphocytes that are then seeded into the secondary structures of the immune system (FIGURE 12.6).

Secondary Immune Structures

The **spleen, tonsils, adenoids**, lymph nodes, and patches of tissue associated with the intestinal tract constitute the **secondary immune structures**. The spleen contains phagocytic cells and both mature T and B cells, endowing it with immunological functions. The spleen is a spongy, fist-sized organ located in the upper left portion of the abdominal cavity. Because of its location and its relatively thin connective tissue covering, it is subject to injury as might be sustained in an automobile accident. Severe trauma to this organ may result in its rupture, with severe hemorrhaging necessitating its removal (splenectomy).

The tonsils and adenoids are located at the back of the throat, just in front of the pharynx. Their lymphocytes play a role in protection against microbes entering through the nose and throat. A generation or so back it was routine for young children to undergo removal of their tonsils (tonsillectomy) because of the frequency of throat and ear infections resulting from infected tonsils; this is no longer a common procedure because of the availability of antibiotics. Doctors in the early 1900s often performed these operations in the home, perhaps on the kitchen table. Occasionally, tonsillectomy is performed on individuals with repeated infections that do not respond well to antibiotics. (As a benefit, you are indulged with a lot of ice cream and popsicles to help alleviate a severe sore throat!) Patches of tissue, called **Peyer's patches**, similar to the tonsils and adenoids, are distributed in the lining of the gastrointestinal, respiratory, and urinary tracts and contribute to immunity. Almost all antigens entering the body (usually as microbial components) end up in lymph or blood and are then carried to lymph nodes or to the spleen, where they encounter cells of the immune system.

■ Duality of Immune Function: Nonspecific and Specific Immunity

Recall that the immune system consists of two components: nonspecific or innate immunity and specific or acquired immunity (FIGURE 12.7). Specific immunity, in turn, has two arms: humoral, or antibody-mediated, immunity and cell-mediated immunity.

FIGURE 12.7 Nonspecific and specific immunity.

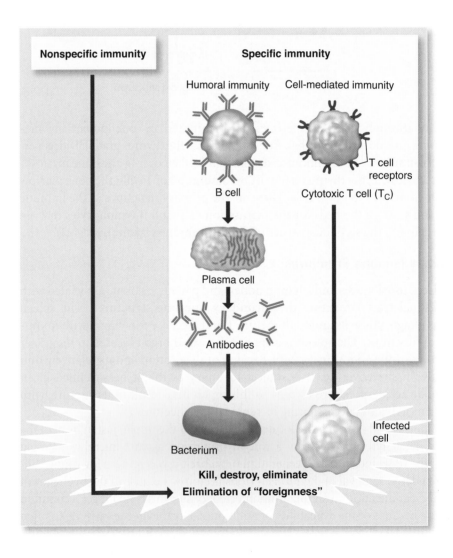

Nonspecific Immunity

The foregoing presentation identifies the blood, bone marrow, thymus, and the secondary immune structures as the key elements in the body's protection against foreignness. **Nonspecific immunity** is characterized by physiological defenses that operate either to prevent microbes in the external environment from gaining access into the body or to eliminate those that have penetrated into the body. TABLE 12.3 summarizes nonspecific immunity. These physiological defense mechanisms are present at the time of birth and do not involve specific targeting of any specific microbe: They act against all microbes in the same fashion. A number of nonspecific internal defense mechanisms, including complement and interferon, as previously described, are antibacterial to those microbes that breach these mechanisms.

TABLE 12.3 Nonspecific Immunity

Mechanism	Function
Skin and mucous membranes	Mechanical barriers
Cilia	Found along respiratory tract and have "upward" motion, pushing microbes up to pharynx where they are swallowed
Phagocytosis	A system of phagocytic cells in the blood and scattered throughout the body
Lysozyme	Found in tears; breaks down bacterial cell walls
Interferon	Found in blood; has antiviral properties
Acid pH of stomach	Many microbes are killed by strong acid environment

Phagocytosis

Phagocytosis is a highly significant nonspecific defense mechanism by which monocytes and neutrophils engulf and destroy foreign substances, including microbes (FIGURE 12.8). Consider the following familiar scenario: Several hours after getting a splinter in your finger, the wound displays the four cardinal signs of **inflammation**: redness (rubor), heat (calor), swelling (edema), and pain (dolor). Microbes from the skin, most likely staphylococci, have invaded through the site of injury. What is the significance of the inflammatory reaction? The injury causes small blood vessels in the area to become engorged with blood (vasodilation), accounting for the redness and the heat; some of the blood leaks from the vessels into the surrounding tissue spaces, causing swelling and pain as a result of increased pressure and the effect of products released by the injured tissue on nerve endings. It hardly seems like a defense mechanism. Within these blood-engorged capillary beds, however, phagocytic cells pile up and stick to the vessel walls surrounding the site of the injury. These cells, attracted in large numbers to the inflamed tissue (chemotaxis), migrate out of the capillaries and kill the bacteria, preventing or minimizing infection. The discovery of phagocytosis by Elie Metchnikoff

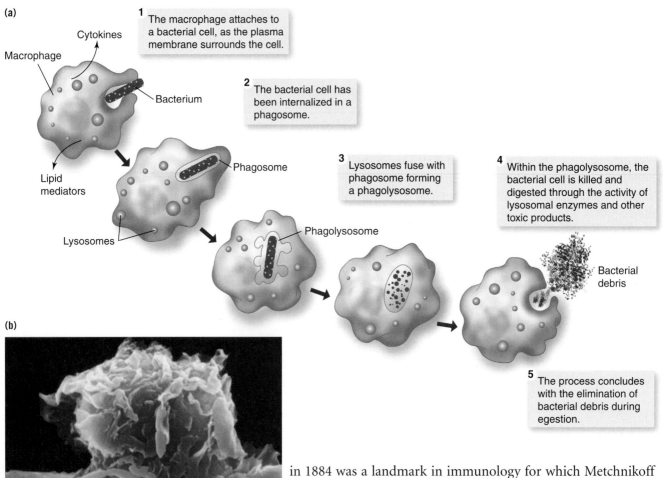

(a)

Cytokines

Macrophage

1 The macrophage attaches to a bacterial cell, as the plasma membrane surrounds the cell.

Bacterium

2 The bacterial cell has been internalized in a phagosome.

Lipid mediators

Phagosome

Lysosomes

3 Lysosomes fuse with phagosome forming a phagolysosome.

4 Within the phagolysosome, the bacterial cell is killed and digested through the activity of lysosomal enzymes and other toxic products.

Phagolysosome

Bacterial debris

5 The process concludes with the elimination of bacterial debris during egestion.

(b)

FIGURE 12.8 Phagocytosis. **(a)** Stages of phagocytosis. **(b)** A false-color scanning electron micrograph of a macrophage (blue) engulfing *E. coli* cells (green) on the surface of a blood vessel (red). © Phototake/Alamy Images.

in 1884 was a landmark in immunology for which Metchnikoff received a Nobel Prize in 1908 (FIGURE 12.9).

Phagocytosis is an extremely efficient host defense mechanism in its own right but is rendered even more efficient in the presence of antibodies and complement. Phagocytosis is not limited to cells of the bloodstream; a variety of other cells has phagocytic capabilities and is strategically located throughout the body, establishing an efficient system of surveillance. **Macrophages** are monocytes that have migrated out of the blood but retain their phagocytic capability. Some macrophages become fixed at particular sites; other macrophages are called **wandering macrophages** because they move freely about the tissues. Collectively, monocytes, neutrophils, and macrophages are referred to as "professional phagocytes"; they are located strategically throughout the body and play a key role in the destruction of microbes.

Specific Immunity

The specific immune system responds to the presence of foreign microbes that breached the external and internal nonspecific defense mechanisms. At this point an all-out war between microbes and host is in effect, as represented by $D = nV/R$. A key to understanding immunity is the concept of specificity. Think of antigen-antibody specificity as the complementarity that exists between a lock and a key or between two pieces of a puzzle that fit together.

As previously mentioned, the specific immune system is divided into two categories: **humoral (antibody-mediated) immunity** and **cell-mediated immunity (CMI)**. These two categories are the "big guns." Although these systems operate in very different ways, the mission is a common one: to eliminate foreign antigens—the hallmark of specific immunity. Most antigens are large protein molecules associated with microbes, tumor cells, damaged cells, pollens, dust, and foods. They (1) trigger the production of antibodies specific for that antigen (humoral immunity) or (2) bring about the production of T lymphocytes directed against that antigen (CMI). These lymphocytes are said to be "sensitized."

T Lymphocytes

T lymphocytes have a role in both humoral immunity and CMI, so it is important to first understand these cells. There is more than one category of T cells, collectively referred to as the **T-cell subset** (TABLE 12.4). Each category of T lymphocyte has a specific role and is identifiable by molecules referred to as **cluster of differentiation (CD)** molecules that are acquired in the thymus during the T-cell maturation process.

Some T cells, called **cytotoxic T (T_C) cells**, are the effectors of cell-mediated immunity. The **T_C cells** are capable of becoming sensitized to ("angry at") the foreign molecules (antigens) carried by the invading microbes. They are identified by **CD8** receptor molecules; their activity requires **T helper (T_H) cells**.

The T_H cells, identified by **CD4 receptor molecules**, are the regulatory or control cells and are another category of cells. The T_H cells are crucial in that the B and T_C lymphocytes are functionally crippled in the reduction or absence of T_H cells. During antigenic stimulation, T_H cells secrete molecules, known as **cytokines**, which activate B and T_C cells.

FIGURE 12.9 Elie Metchnikoff (1845–1916). © North Wind Pictures Archives/Alamy Images.

TABLE 12.4 T-Cell Subsets

Cell Type	Abbreviation	Clusters of Differentiation (Receptors)	Function(s)
T-helper cell	T_H	CD4	Activates B cells to produce antibodies; activates cytotoxic T cells
Cytotoxic T cell	T_C	CD8	Killer cell that works against cells with "foreign" intracellular antigens, including viruses and bacteria
T suppressor cell[a]	T_S	CD8	Regulatory cell works in concert with T_H cells
T delayed-type hypersensitivity cell[a]	T_{DTH}	CD4	Plays role in allergic responses; activates hypersensitivity macrophages

[a]Not discussed in text.

Antibody-Mediated (Humoral) Immunity

The concept of using an infectious agent to prevent and recover from certain diseases has been realized for centuries. Before the 1700s medical practitioners had some success preventing what later was termed smallpox by picking scabs from pox-infected individuals. The scabs were dried, pulverized, and introduced into the nose of those susceptible. Many of those treated contracted a mild case of smallpox, from which they recovered and were resistant to subsequent infections. In 1796 Edward Jenner introduced a vaccination against smallpox by inoculation of the cowpox virus, resulting in a benign infection in humans. Consider also the practice of intentionally injecting relatively avirulent forms of leishmania protozoan to immunize against the life-threatening species of the parasite. What these early pioneers in immunology did not know was the biological mechanisms behind the resistance—the production of antibodies.

Humoral immunity is mediated by antibodies, products of **B cells** (with the aid of T_H cells) in response to antigens. The antibody molecule is a four-chained structure (FIGURE 12.10) that binds with antigens on bacterial cells, viral particles, toxins, and internal cells, including tumor cells and dead cells.

The binding of antibodies to antigens has several outcomes (FIGURE 12.11), each of which facilitates destruction of the foreign antigen-bearing microbe. B lymphocytes that are "revved up" by contact with antigen and T_H lymphocytes release antibodies and at that stage are called **plasma cells**. Estimates are that a single plasma cell can produce thousands of specific antibodies per second. There are actually five categories of antibodies, or **immunoglobulins (Igs)**, termed **IgG, IgA, IgD, IgE**, and **IgM**, produced with identical specificity to each antigen. Each type has unique properties. IgG accounts for approximately 80% of the antibody molecules and is the best characterized. The antibody molecule binds with antigens on bacterial cells, viral particles, toxins, and internal cells, including tumor cells and dead cells, resulting in a "kill, destroy, or eliminate" outcome.

Although it is true that antigen-driven B cells are the only antibody producers, T_H cells are required to help these B cells produce antibodies; they do so by

FIGURE 12.10 Antibody molecules, a four-chain structure.

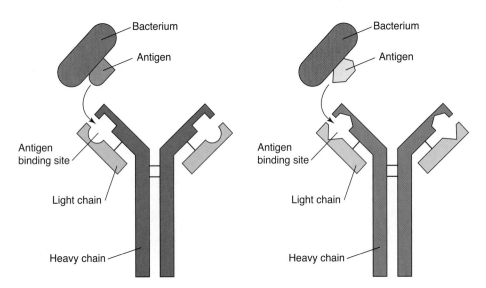

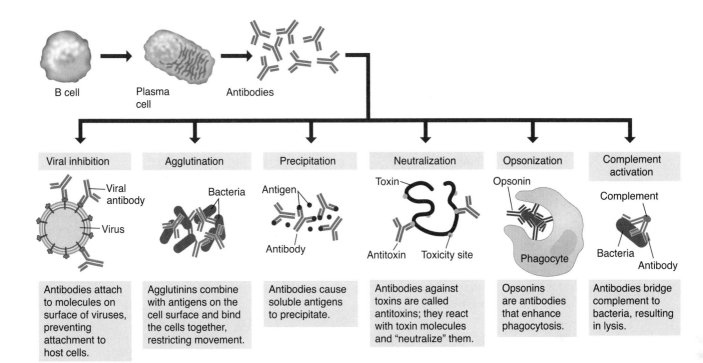

Viral inhibition	Agglutination	Precipitation	Neutralization	Opsonization	Complement activation
Antibodies attach to molecules on surface of viruses, preventing attachment to host cells.	Agglutinins combine with antigens on the cell surface and bind the cells together, restricting movement.	Antibodies cause soluble antigens to precipitate.	Antibodies against toxins are called antitoxins; they react with toxin molecules and "neutralize" them.	Opsonins are antibodies that enhance phagocytosis.	Antibodies bridge complement to bacteria, resulting in lysis.

FIGURE 12.11 Protective effects of antibodies binding to antigens (humoral immunity).

releasing molecules called **cytokines** (FIGURE 12.12). How do B cells know which antibody specificity to produce? The **clonal selection theory** (FIGURE 12.13) is the explanation. According to this theory, a population of B cells for every possible antigen exists at the time of birth. Each B cell displays many copies of a (single) specific antibody molecule on its surface. By random contact antigens "dock" with surface antibodies of corresponding specificity (lock and key), thus stimulating those B cells to undergo cellular reproduction. Hence, the population of B cells that was selected by the antigen is expanded. Antibodies that are copies of the surface antibody are produced and released. After a lag time of approximately ten to twelve days, antibodies are present in the blood in amounts large enough to be detected.

Some of these B cells may not progress to antibody-producing plasma cells but may remain for years as **memory cells**. This accounts for the fact that, except in rare circumstances, certain so-called childhood diseases (measles, German measles, mumps, and chicken pox) are acquired only once in a lifetime despite repeated exposure. Assume, for example, that you had measles as a child. In later years your child has measles and, despite close contact with the child, you do not acquire the disease. This is because a part of the B-cell population making antibodies against the measles virus was reserved as memory cells. If exposed to the measles virus again these preprogrammed memory cells respond in a matter of hours by pouring out measles antibodies at a rate and quantity sufficient to target the virus for destruction.

Cell-Mediated Immunity

Most bacterial pathogens are **extracellular**; that is, they take up residence on the surface of the cells. For example, streptococci, the causative agents of strep throat, colonize the surface of the throat and pharynx, and the organisms do not penetrate

FIGURE 12.12 Antibody-mediated immunity and CMI.

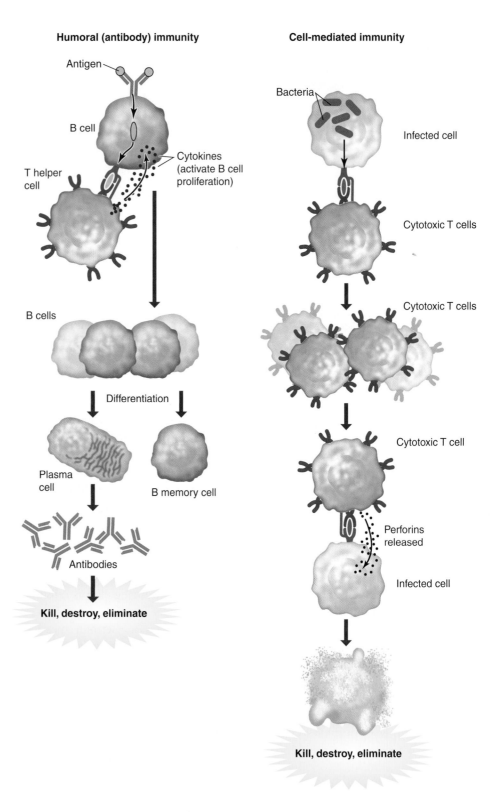

Humoral (antibody) immunity

Antigen

B cell

T helper cell

Cytokines (activate B cell proliferation)

B cells

Differentiation

Plasma cell

B memory cell

Antibodies

Kill, destroy, eliminate

Cell-mediated immunity

Bacteria

Infected cell

Cytotoxic T cells

Cytotoxic T cells

Cytotoxic T cell

Perforins released

Infected cell

Kill, destroy, eliminate

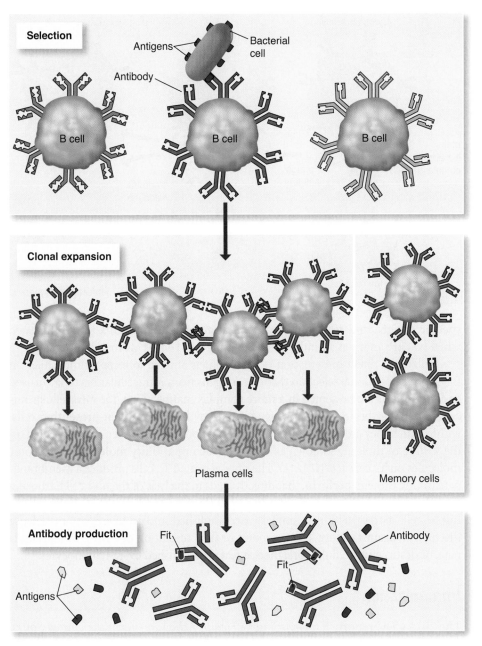

FIGURE 12.13 Clonal selection theory.

Selection

Antigens

Bacterial cell

Antibody

B cell

B cell

B cell

Clonal expansion

Plasma cells

Memory cells

Antibody production

Fit

Antibody

Fit

Antigens

into the cells. Antibody-mediated immunity is the body's defense strategy for coping with extracellular microbes, but what about **intracellular** bacteria and viruses (all of which are intracellular) once they penetrate cells, and what about mutant and damaged cells in the body? Antibodies play little role in protection against intracellular microbes. Viruses during their early, extracellular stage before cell invasion are an exception. It would be pointless to immunize against viral diseases if this were not the case. So what is the defense? Cell-mediated immunity (CMI) is the major defense strategy against intracellular bacteria, protozoans, viruses, and tumor cells. CMI is mediated by T_C cells that have become sensitized (angry) to a

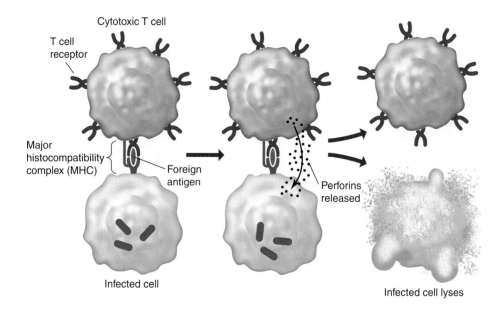

FIGURE 12.14 The protective effects of perforin.

Cytotoxic T cell

T cell receptor

Major histocompatibility complex (MHC)

Foreign antigen

Perforins released

Infected cell

Infected cell lyses

specific antigen and react only against that antigen. (Think of CMI in this way: when you are angry at an individual, your anger is vented against that person.)

Unlike the direct presentation of extracellular antigen to receptor molecules on B lymphocytes (clonal selection theory), antigens from intracellular bacteria, viruses, and tumor cells are presented in a more complex manner. First, the intracellular microbes bearing the foreign antigens are phagocytized by **antigen-presenting cells**, within which they are chewed up. Pieces of these processed antigens are presented on the surface of molecules known as **major histocompatibility molecules** to receptor molecules on T_C cells (**FIGURE 12.14**). The now-activated T_C cells produce a membrane-penetrating protein, **perforin**, that directly leads to the lysis of the host cell harboring the microbes or damaged or mutant body cell (Figures 12.12 and 12.14. If the cell is a tumor cell, its destruction is similarly accomplished and is an important defense mechanism against cancer. The end point or final resolution of CMI is to kill, eliminate, or destroy intracellular foreigners, a feat accomplished by T_C lymphocytes.

■ Immunization

The 2011 CDC report, *Ten Great Public Health Achievements in the United States, 2001–2010,* includes vaccination (immunization) as one of the achievements. No wonder, considering the millions of lives saved around the world as a result of vigorous vaccination campaigns targeted against bacterial and viral diseases. In the United States alone the lives of three million children are saved each year because of routine immunization. For some diseases the decline over the past century in the United States is 100%, or close to 100% (**TABLE 12.5**). As example, since 1988 polio has decreased by almost 100% and by 1997 eliminated from the United States: The Western hemisphere was free of polio by 1991.

Smallpox was the first disease to be eradicated thanks to Edward Jenner's pioneering work with vaccination and to the fact that the disease has no animal reservoirs. In 1881, almost a century after Jenner's work, Louis Pasteur developed

TABLE 12.5 Impact of Twentieth and Twenty-First Century Public Health Achievements on Selected Diseases in the United States (2009)

Disease	% Decline
Smallpox	100
Diphtheria	100
Hepatitis A	93
Hepatitis B, acute	86
Poliomyelitis (paralytic)	100
Measles	>99
Haemophilus influenzae type B	99
Mumps	99.6
Rotavirus, hospitalizations	55
Congenital rubella syndrome	99.4
Rubella	>99
Tetanus	96
Pertussis	86
Varicella	90

Adapted from Hinman, A. R. et al., 2011. Vaccine-Preventable Diseases, Immunizations, and MMWR—1961–2011. *MMWR* Supplements 60(04):49–57.

a vaccine against anthrax in animals and, only a few years later (1885), developed a vaccine against human rabies. Before 1900 vaccines against three additional diseases were available, and during the twentieth century twenty-one additional diseases were added to the list. Since 2006 nine more vaccines have been approved by the U.S. Food and Drug Administration (FDA) (TABLE 12.6). The FDA is the agency responsible for issuing a license to vaccine manufacturers in the United States, allowing the vaccine to be widely distributed. Polio is near eradication, and other diseases are on the "hot list" for eradication including guinea worm and measles. Vaccine development and improvement of existing vaccines is a top priority. Malaria, AIDS, and tuberculosis are responsible for suffering and deaths by the millions throughout the world; many children in developing countries do not grow up to become adults.

The availability of vaccination is a story of "good news and bad news." The good news is that much of the world is immunized against a variety of microbial diseases, whereas the bad news is that underdeveloped countries have not shared in this victory. Tragically, whooping cough, measles, and tetanus take the lives of over three million children each year, despite the fact that immunization is available. The Bill and Melinda Gates Foundation's Child Vaccination Program, The Global Fund for Children, The Clinton Foundation, and The Global Alliance for Vaccines and Immunization are but a few examples of organizations aimed at delivering vaccines to the world's poorest countries.

Active Immunization

TABLE 12.7 outlines the categories of immunization. Active immunization is the result of stimulating a person's immune system to produce antibodies and memory cells and

TABLE 12.6 Vaccine-Preventable Diseases

Disease	Year of Vaccine Development or U.S. Licensure
Smallpox (V)	1798
Rabies (V)	1885
Typhoid (B), cholera (B)	1896
Plague (B)	1897
Pertussis (B)	1926
Tetanus (B), tuberculosis (B)	1927
Influenza (V)	1945
Yellow fever (V)	1953
Poliomyelitis (V)	1955
Measles (V)	1963
Mumps (V)	1967
Rubella (German measles) (V)	1969
Anthrax (B)	1970
Meningitis (B)	1975
Pneumonia (B)	1977
Adenovirus (V)	1980
Hepatitis B (V)	1981
Haemophilus influenzae type b (B)	1985
Japanese encephalitis (V)	1992
Hepatitis A (V), chickenpox (V)	1995
Lyme disease (B)	1998
Pneumococcal conjugate (B)	2000
Meningococcal conjugate (B)	2005
Shingles (V), rotavirus (V), human papilloma virus (V)	2006
Avian influenza vaccine for humans (V)	2007
H1N1 vaccine (V)	2009
Human papillomavirus (V)	2009
High-dose inactivated influenza vaccine approved for elderly	2009
Pneumococcal 13-valent conjugate vaccine against *Streptococcus pneumoniae* (B)	2010

B, bacterial disease; V, viral disease.

Adapted and updated from CDC, *Morbidity and Mortality Weekly Report* 48 (1999):241–243 and the Immunization Action Coalition for Health Professionals. http://www.immunize.org/timeline/.

generally confers immunity over a relatively long time. There are two categories of active immunization—natural and artificial. Natural active immunity is achieved by the process of recovering from a particular disease. Analysis of an individual's blood serum frequently reveals the presence of antibodies against which there is no clinical history of disease, indicating that the disease at a subclinical level had occurred.

The use of vaccines, however, is artificial in the sense that vaccines are administered into the body to provoke an antibody immune response as a future

TABLE 12.7 Outline of Immunization

Active immunization (individual makes own antibodies)
Natural (subclinical or clinical disease and recovery)
Artificial (vaccines for immunization "shots")
Live, attenuated microbes
Killed microbes
Toxoids and other purified microbial components
New and experimental vaccines
Passive immunization (individual receives preformed antibodies)
Natural (in utero mother-to-infant passage; breast milk)
Artificial (use of immune globulin)

protective measure. As indicated in Table 12.7, artificial active immunization can be accomplished in six ways; the method chosen reflects the best potential for the particular disease. All strategies must meet the three basic requirements of safety, effectiveness, and stability (TABLE 12.8). Additionally, an ideal vaccine needs to be affordable to developing countries.

Safety issues are of prime importance and are further addressed under Vaccine Safety below. For a vaccine to be effective it must stimulate an immune response affording protection to vaccine recipients. Vaccine preparations need to be stable over time to make them cost effective. Some vaccines can be stored at room temperature, whereas others need to be refrigerated, presenting a problem in distribution of these vaccines to the developing world. On too many occasions vaccines need to be destroyed because refrigeration requirements are not observed. Strict attention needs to be paid to expiration dates and to conditions of storage.

Types of Active Artificial Vaccines

How are live, disease-producing microbes turned into non-disease producing, antibody-stimulating agents? As indicated in Table 12.7, artificial active vaccines can be produced in four ways.

TABLE 12.8 An Ideal Vaccine

General Vaccine Requirements	Ideal Vaccine Requirements
Safety	Safety
Effectiveness	Effectiveness
Stability	Stability
	Affordability
	Administration as a nasal spray or edible vaccine
	No need for refrigeration; stability at ordinary "tropical" temperatures
	One dose or one shot
	Long shelf life

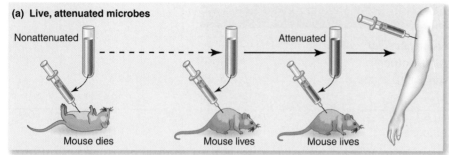

(a) Live, attenuated microbes

Nonattenuated — — — → Attenuated

Mouse dies | Mouse lives | Mouse lives

FIGURE 12.15 Preparation of a vaccine **(a–d)**. The dashed line in panel **a** indicates that the number of transfers continue until attenuation takes place as determined by loss of virulence as tested in mice.

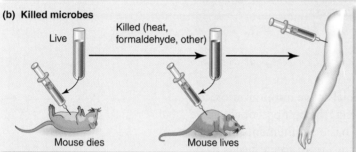

(b) Killed microbes

Live → Killed (heat, formaldehyde, other)

Mouse dies | Mouse lives

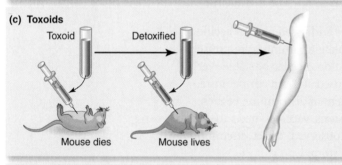

(c) Toxoids

Toxoid → Detoxified

Mouse dies | Mouse lives

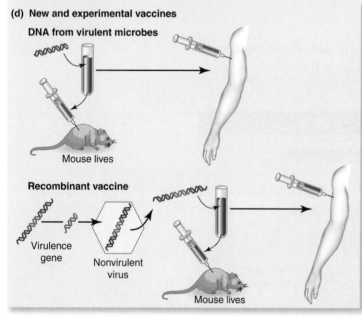

(d) New and experimental vaccines

DNA from virulent microbes

Mouse lives

Recombinant vaccine

Virulence gene → Nonvirulent virus

Mouse lives

Live Attenuated Microbes

Vaccines made from live attenuated microbes, most of which are viruses, confer long-lasting, frequently lifetime immunity. (Poetic license is taken with the term "live" to describe viruses.) The word "attenuated" means weakened virulence. These vaccines may produce mild and limited symptoms but not overt clinical disease as manifested by nausea, headache, fatigue, and soreness at the site of injection for about 24 hours; frequently, there are no symptoms. Some live attenuated viral vaccines have been achieved by serial (repeated) transfer in tissue culture (**FIGURE 12.15**), allowing the production of random and unpredictable mutants. These mutants are tested in laboratory animals, and those strains producing no symptoms are selected for further trial. The **bacillus Calmette-Guérin (BCG) vaccine** against tuberculosis continues to use a tuberculosis strain (*Mycobacterium bovis* BCG) attenuated by repeated subculturing between 1908 and 1918 on laboratory media with no reversion to virulence for over eighty years. Pasteur's vaccine against rabies was the result of gradually drying virally infected spinal cords of dogs and rabbits, a procedure that rendered the live virulent virus nonvirulent but continued to provoke protective antibodies against the virus. The major concern of the use of live attenuated vaccines is the possibility of reversion to virulent forms. Examples of vaccines using the live attenuated microbes are the Sabin polio, measles, and yellow fever vaccines.

Killed (Inactivated) Microbe Vaccines

Killed microbe vaccines are used when attenuation has not been accomplished or when reversion to the virulent type is considered to be too risky. Virulent microbes are heat killed or killed with particular chemical reagents. They present no risk but are not as effective at stimulating antibody production as those containing live microbes. Some require multiple doses to maintain protective antibody levels. The Salk polio vaccine and the vaccines against plague, influenza, hepatitis A, and cholera are examples.

Antitoxins

Some of the most serious bacterial diseases (diphtheria, tetanus, cholera, and botulism) result from the production of very potent microbial exotoxins. These toxins can be inactivated by heat or formaldehyde, resulting in a loss of toxicity, but they retain the property of stimulating specific antibody production; inactivated toxins are referred to as toxoids. The **diphtheria-tetanus-pertussis (DTaP)** vaccine, commonly administered to children at about the age of two months, contains diphtheria and tetanus **toxoids** (inactivated toxins). To make antitoxins, pharmaceutical companies inject toxoids into horses (or sheep or goats in some cases). After a brief time the animal's immune system produces antibodies to the toxin (antitoxins); the blood is taken and processed to recover the antitoxin component as an effective immunization strategy. Diphtheria and tetanus antitoxins are examples.

New and Experimental Vaccines

These vaccines, also called recombinant DNA vaccines, are based on DNA technology. They are substitutes for "whole-agent" vaccines. A "virulence gene" is inserted into a nonvirulent host microbe and that gene is then expressed and replicated in a new host. In this new environment the gene can now be safely used as a vaccine. The vaccine against hepatitis is an example of a subunit vaccine.

DNA vaccines are another new and promising approach to immunization. In this strategy microbial DNA is inserted into plasmids that are then injected directly into the host. Subsequently, these proteins are recognized as foreign by the host's immune system and stimulate an immune antibody response. DNA vaccine development for a variety of microbial diseases, including influenza, tuberculosis, malaria, Lyme disease, and hepatitis C, is under way but has not been shown to be effective.

Passive Immunization

Active immunization is based on stimulating a recipient's immune system to produce antibodies and memory cells (FIGURE 12.16). In passive immunization, by contrast, the recipient receives ready-made preformed antibodies (immune serum) from human or animal sources by injection. A big advantage to passive immunization is that antibodies are present immediately at the time of infection and can be of lifesaving value, as in cases where exposure or symptoms have already occurred. The immunity gained, however, is relatively short lived and limited to the duration of the administered antibodies in the recipients; there is no immunological memory. TABLE 12.9 summarizes the distinctions between active and passive immunization.

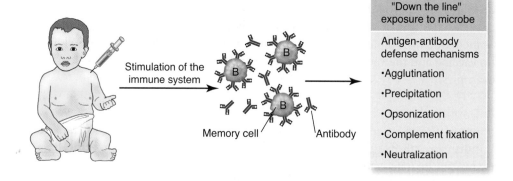

FIGURE 12.16 How immunization works.

TABLE 12.9 Properties of Active Versus Passive Immunization			
	Protection	**Duration**	**Adverse reactions**
Active	Waiting period	Extended memory	Possible
Passive	Immediate	Limited (no memory)	Possible

Before the 1940s and the advent of antibiotics, immunotherapy, the use of immune serum was a common practice, particularly for diphtheria, tetanus, and pneumococcal pneumonia. The first two diseases are toxemias, meaning that they are a manifestation of the production of a toxoid (a lethal toxin). As described earlier, antitoxins (antibodies against toxoids) are produced by the injection of toxoids into horses to produce immune serum. In the treatment of diphtheria, about all that could be done in the days of the "horse and buggy" doctors was the administration of the antitoxin-containing immune serum in a desperate attempt to save lives.

Immunotherapy can be hazardous because of possible complications arising from the fact that the antibodies, along with other (blood) serum components are "seen" as foreign protein by the recipient's immune system, resulting in antibody production against the foreign protein in the serum, causing a condition known as **serum sickness**. Serum sickness is characterized by the formation of antigen-antibody complexes that are deposited in the skin, kidney, and other body sites. Nevertheless, under certain circumstances immunotherapy is still used when immediate antibody protection is required.

The use of human immune serum, taken from individuals after vaccination or during their convalescence period from a specific disease, minimizes but does not eliminate the risk of serum sickness in the recipient. This human immune serum (also called immune globulins, referring to the globulin fraction of the blood) contains high levels of specific antibodies. For example, tetanus immune globulin is rich in antibodies against tetanus, and varicella-zoster immune globulin has a high concentration of antibody against the virus that causes chickenpox and shingles. Consider a case in which an individual reports to an emergency room having sustained a puncture and is at risk of tetanus. Should that person receive active immunization by a booster shot with tetanus toxoid or passive immunization with tetanus immune globulin, or both?

The answer depends on the person's immune history. If the individual has, within the last ten years, received tetanus toxoid as a booster, all that is necessary is another booster to effectively stimulate those memory cells preprogrammed to produce tetanus antibodies almost immediately. However, if the individual has not received (or is not certain of) past immunizations against tetanus, immediate protection against the tetanus toxin is necessary. There is not sufficient time to make antibodies from scratch, in which case tetanus immune globulin for immediate protection should be administered, along with tetanus toxoid for future protection.

Antitoxins against the deadly toxins injected into the body by certain species of snakes and arachnids (e.g., spiders and scorpions) are examples of lifesaving passive immunization. If bitten by a poisonous snake or scorpion, immediate antibody protection is vital.

AUTHOR'S NOTE (RIK)
While in Guatemala I was about to stick my bare foot into my slipper without first checking it out as I had been advised to do. Suddenly a large scorpion—at least I think it was a scorpion—crawled out. It pays to follow advice! While there, I did also see one of the world's most poisonous snakes and captured it on film. That was probably the fastest close-up I have ever taken!

Vaccine Safety

Vaccines have greatly reduced the burden of infectious diseases around the world, but as these diseases declined attention has focused on the risks associated with vaccines: How safe are the vaccines on the market? No vaccine (or other medication) is 100% safe and without risk. The better question to be asked is, "Do the benefits of the vaccine outweigh the risk?"

For example, oral polio vaccine, a live vaccine preparation, carries a risk of polio of about one in seven million people. Now that polio is nearly eradicated, it is unfortunate that unwarranted concern is sometimes paid to the one possible risk case. This is ironic in the sense that if polio were still prevalent, little attention would be paid to the rare and unfortunate mishap that might occur. The realization is lost that seven million people minus one individual were candidates for paralytic polio. The current strategy for polio immunization is that all four doses consist of the Salk killed virus vaccine to eliminate the slight risk of vaccine-associated paralytic polio that might occur with the Sabin live attenuated oral polio vaccine.

Vaccines are constantly monitored and modified or withdrawn as circumstances dictate. For example, in 1976 Fort Dix, New Jersey was threatened by the appearance of a new and deadly strain of swine flu viruses; in response a vaccine was quickly developed, and forty-five million people were vaccinated. Unexpectedly, in some cases the vaccine triggered a debilitating and potentially fatal neuromuscular disease called **Guillain-Barré syndrome**. Rotavirus infection is a potentially fatal disease in children, and the development of the RotaShield vaccine in 1998 was heralded as a preventive measure against this disease. But, one year later RotaShield was withdrawn because of an association with the occurrence of intussusception (twisting and obstruction of the bowel) in 23 infants 1 to 2 weeks after vaccination. As disappointing as this was, the rapid response remains as a tribute to the FDA because of their rapid response to this unexpected circumstance. Subsequently, **RotaTeq**®, a replacement for RotaShield®, was approved by the FDA in 2006.

The pertussis (whooping cough) component of the combined diphtheria, tetanus, and whooping cough vaccine, formerly made from whole cells, has, since 1991, been derived from a component of the microbe, resulting in fewer adverse effects. The new DPT vaccine is called **DTaP vaccine** because of the use of acellular pertussis (aP). The old DPT, no longer used in the United States, had been associated with brain damage, autism, and learning disabilities. Autism is further discussed in BOX 12.1.

Although it is true that some adverse reactions have occurred after vaccination, it is also true that vaccines may be falsely blamed because unrelated events may coincidentally occur shortly after vaccine administration. Although adverse reaction examples are worrisome, there is some comfort in the realization that the FDA attempts to stay on top of the situation.

In response to the vaccine safety concerns, Congress passed the National Childhood Vaccine Injury Act in 1986, mandating that healthcare providers furnish a vaccine information sheet to recipients describing the risks and benefits of the vaccine. Providers are also required to report certain side effects after vaccination to the FDA's Vaccine Adverse Event Reporting System. Further, "no fault" vaccine compensation is provided to those injured by vaccines.

The FDA, the licensing agent, does not approve a vaccine unless initial trials indicate the benefits clearly outweigh the risks. Licensure is a rigorous process involving three phases of clinical trials and may take 10 or more years. Vaccines are subject to particularly high safety standards, because, unlike other health treatments, they are given as preventives to healthy people.

Vaccines are manufactured by pharmaceutical companies; each batch of vaccine must be approved by the FDA before it can be released for use by health providers. Issues of safety, effectiveness, sterility, and purity are all evaluated by laboratory procedures, and postmarketing surveillance is conducted to identify undesirable side effects that might occur in large groups of people over long periods.

Oral vaccines and vaccines administered by nasal sprays are on the horizon and may make vaccines easier to implement in developing countries. FluMist®, introduced in 2006, is an example. Vaccines have come a long way since the pioneering work of Jenner and Pasteur.

Childhood Immunization

The burden of infectious disease has been reduced throughout the world, most notably in the United States and in other industrialized countries, through the routine practice of childhood immunization. Examples have been cited in this text. FIGURE 12.17 illustrates the 2012 recommended childhood immunization schedule (aged newborn to six years) in the United States, supported by the CDC Advisory Committee on Immunization Practices, the American Academy of Pediatrics, and the American Academy of Family Physicians. Included are routine immunizations against fourteen diseases and immunizations against one other (hepatitis A) for selected populations. Some of the immunizations are against bacterial diseases, but most are against viral diseases; attenuated, killed, subunit, and genetically engineered vaccines are all represented. All these immunizations are recommended to be started before the age of fifteen months.

Despite the low cost and effectiveness of immunization, thousands of children and adults have never had basic immunizations or are not up to date. Almost 100,000 adults die every year from influenza, pneumonia, and other vaccine preventable diseases.

BOX 12.1 Autism, Vaccines, and Television

A handful of parent anti-vaccine groups are convinced childhood vaccines containing thimerosal, a mercury-based substance used to prevent bacterial contamination in vaccines, particularly in MMR and DTaP vaccines, trigger autism. Autism is now recognized as a spectrum of neurological disorders from mild to severe characterized by social, behavioral, and communication problems. Even though thimerosal has not been used in childhood vaccines (except in some flu shots) since 2001, a number of studies show there has been no decline of autism. The CDC, the American Academy of Pediatrics, and other respected organizations conclude that there is no link between autism and vaccines and continue to assure parents that vaccines are safe and life-saving.

The popular media is not helping to spread the positive message of vaccine safety to the public, however. In early January 2008 the Immunization Action Coalition (IAC), a nonprofit organization dedicated to promoting immunization, became aware that the American Broadcasting Company (ABC) was going to televise an upcoming legal drama episode in which the lawyer sues a vaccine manufacturer on the grounds that the thimerosal in the vaccine caused a child's autism. The IAC sent the letter reproduced here to ABC expressing their disapproval. Unfortunately, the letter did not persuade ABC to pull the episode. Instead, the episode was aired with the following disclaimer, "The preceding story is fictional and does not portray any actual persons, companies, products or events" any directed viewers to the CDC autism website. The IAC letter follows:

January 25, 2008

ABC, Inc.
500 C. Buena Vista Street
Burbank, CA 91521-4551

Dear Sir or Madam:

I was dismayed to learn that ABC is planning to debut the legal drama "Eli Stone" with a script riddled with misinformation about the safety of life-saving vaccines routinely given to infants and children. My understanding, based in part on a *New York Times* article published on January 23, is that lawyer Eli Stone sues a vaccine manufacturer on behalf of the mother of an autistic child. Stone argues that the mercury-containing preservative in a vaccine the child received caused the child's autism. At the end, the jury awards the mother 5.2 million dollars.

Scientific research conducted in the past decades has decisively and repeatedly refuted the claim that childhood exposure to thimerosal, the mercury-containing preservative used in some vaccines, is a cause of autism. Most recently, the medical records of millions of Californian children who were vaccinated in the 12 years between 1995–2007 were studied. During those years, thimerosal (other than in trace amounts) was removed from all the routinely administered childhood vaccines except the influenza vaccine. The California study determined that the incidence of newly reported cases of autism *increased* during a period when the presence of thimerosal in vaccines was significantly *decreased*. The California study, along with five major studies conducted in the United States, Denmark, the United Kingdom, and Sweden, found no association between childhood vaccinations with thimerosal-containing vaccines and the developments of autism.

The misinformation the Eli Stone script communicates to viewers about the safety of childhood vaccines is irresponsible and dangerous. It will cause many parents to believe that vaccines are potentially hazardous to their children's health. In some instances, healthcare providers will not be able to overcome parents' fears, and the United States could see an increase in rates of life-threatening diseases, as rates of childhood immunizations decline. Such fears travel to the international community. Two years ago, Nigeria experienced outbreaks of polio because of misinformation communicated by leaders in the Islamic community about the supposed "dangers" of polio vaccines.

I urge you, as a leader in the production of "Eli Stone," to do the responsible thing. Work to block the airing of this episode. In doing so, you will protect and promote the health of children within the United States and worldwide.

Sincerely,

Deborah L. Wexler, MD
Executive Director
Immunization Action Coalition

Vaccine	Age											
	Birth	1 month	2 months	4 months	6 months	9 months	12 months	15 months	18 months	19–23 months	2–3 years	4–6 years
Hepatitis B	HepB	HepB			HepB							
Rotavirus			RV	RV	RV							
Diptheria, Tetanus, Pertussis			DTaP	DTaP	DTaP		DTaP					DTaP
Haemophilus influenzae type b			Hib	Hib	Hib		Hib					
Pneumococcal			PCV	PCV	PCV		PCV					PPSV
Inactivated Poliovirus			IPV	IPV		IPV						IPV
Influenza						Influenza (yearly)						
Measles, Mumps, Rubella							MMR					MMR
Varicella							Varicella					Varicella
Hepatitis A							HepA (2 doses)					HepA series
Meningococcal						MCV4						

☐ Range of recommended ages   ☐ Certain high-risk groups   ▨ Range of recommended ages and high-risk groups

FIGURE 12.17 Childhood immunization recommendation schedule in effect as of December 23, 2011. The CDC publishes a schedule for recommended immunizations that is approved by the Advisory Committee on Immunization Practices, the American Academy of Pediatrics, and the American Academy of Family Physicians. Data from CDC and American Academy of Pediatrics.

Immunocompromised individuals are particularly vulnerable to fungal diseases. Candidiasis, for example, ranks high as a cause of invasive infection into the bloodstream. Vaccine development for fungi has been delayed by the belief that patients would be too immunologically impaired to effectively respond to vaccination. Recent research, however, has shown that this is not necessarily the case, leading to increased activity to develop antifungal vaccines.

Clinical Correlates

Despite these elaborate mechanisms of defense against potential pathogens, people become ill and sometimes die because of microbial diseases. It may be that the immune system is defeated in the dynamic interplay between microbe and immunity in a virtual tug-of-war by highly infectious agents such as Ebola, HIV, and rabies viruses, and bacteria that causes meningococcemia, or HIV. The person's own immune system may be impaired as is the case with leukemia, or the person may have an immunodeficiency. Treatment with immunosuppressive drugs also increases a person's susceptibility to infection.

Human Immunodeficiency Virus

HIV, responsible for AIDS, destroys T_H cells, resulting in the failure of an infected person to mount an appropriate immune response either by antibody formation or by T_C. The HIV-infected individual becomes severely **immunocompromised** (i.e., has a weakened immune system), which results in one infection after another or several infections at one time, eventually causing death from infection. Blood from AIDS-infected individuals is routinely monitored to determine the level of T_H lymphocytes. The prognosis is grave when the number of T_H cells drops significantly, signaling that the individual no longer has the capacity to combat invading microbes via the specific immune system. When you hear that an individual with AIDS has a low CD4 level, this refers to a low level of T_H cells.

Leukemia

Leukemia, unfortunately, is an all too familiar term. It is a form of cancer characterized by uncontrolled reproduction of white blood cells. There are two major categories of leukemia, reflecting the two pathways of blood cell maturation (Figure 12.5). **Myeloid leukemia** is the result of overproduction of monocytes and granular leukocytes (neutrophils, basophils, and eosinophils). Consider that immature neutrophils may be produced in excess and are unable to carry out phagocytosis, resulting in an immunocompromised individual. **Lymphocytic leukemia** is the result of overproduction of lymphocytes, leading to abnormally large numbers of immature and nonfunctional lymphocytes and their spread by **metastasis** into tissues throughout the body, crowding out normally functioning cells. Leukemia is classified as acute or chronic. In **acute leukemia** the symptoms appear suddenly and progress rapidly; death occurs in a few months unless the condition is successfully treated. **Chronic leukemia** can remain undetected for many months, and life expectancy without treatment is somewhere around three years. Acute lymphocytic leukemia is the most common form of leukemia in children, but fortunately treatment of this condition has a high success rate. The Dana-Farber Cancer Institute (originally the Jimmy Fund Building) in Boston is world renowned for its success in treating children with leukemia. Fans of baseball history will be delighted to know that baseball legend Ted Williams of the Boston Red Sox was a frequent visitor and contributor to this institution.

Immunodeficiencies

A variety of immunodeficiencies may be present at birth as a result of inheritance, including deficiencies in complement production, phagocytosis, B cells, and T cells. A properly functioning immune system requires the interplay of both nonspecific and specific mechanisms; impairment results in an immunocompromised individual subject to repeated life-threatening infections. TABLE 12.10 gives a sampling of immunodeficiency disorders, three of which are described here.

Severe combined immunodeficiency (SCID) is a devastating congenital disease in which individuals lack functional T and B cells and therefore cannot mount either an antibody- or cell-mediated immune response. Death is a certainty, resulting from repeated infections. This disorder is commonly known as "bubble boy disease," because infants born with this condition must be kept in a germ-free environment, such as a plastic bubble. (One of John Travolta's early movies, *The Boy in the Plastic Bubble* (1976), dramatized the plight of a boy who spent his first twelve years in a germ-free bubble because he was born with SCID.) Those afflicted cannot even have contact with their parents, and all items introduced

TABLE 12.10 Sampling of Immunodeficiency Disorders

Deficiencies in complement
   C3 (complement factor 3) deficiency
   C5 (complement factor 5) deficiency
Deficiencies in phagocytosis
   Chronic granulomatous disease
   Chediak-Higashi syndrome
Deficiencies in B lymphocytes
   X-linked agammaglobulinemia
   Bruton's syndrome
Deficiencies in T lymphocytes
   DiGeorge syndrome
   Wiskott-Aldrich syndrome
   Ataxia telangiectasia
Deficiencies in both B and T lymphocytes
   Severe combined immunodeficiency disorder (SCID)

into the bubble, including air, food, and water, must be sterilized. SCID is the result of a genetically caused enzyme deficiency. In recent years gene therapy and early bone marrow transplants (within three months of birth) have proved to be effective in some cases.

DiGeorge syndrome is a disorder resulting from the absence or incomplete development of the thymus gland. T-cell maturation is abnormal, resulting in impairment of both humoral immunity and CMI.

Chronic granulomatous disease is an inherited disorder of phagocytes. Because of the inability of phagocytes to kill, serious, life-threatening, and persistent infections result.

■ Overview

The immune system is a defense mechanism against invading microbes and, additionally, an internal surveillance system. The ultimate outcome of exposure to pathogenic microbes is the result of the dynamic interplay between the disease-producing (virulence) mechanisms of the parasite and the immune system of the host ($D = nV/R$). The recognition of foreignness is the key element in triggering mechanisms of immune defense. The immune system may incorrectly target self as foreign and mount an immune response against self, resulting in a variety of autoimmune diseases.

Anatomically, the immune system is unique in that it shares cells, tissues, and organs with other functional systems of the body. Blood and blood cells, thymus gland, bone marrow, tonsils, lymph nodes, and spleen are key components of the immune system.

The nonspecific immune system includes the skin, phagocytic cells, components of blood, and ciliated cells that line part of the respiratory tract. Specific immunity responds to the presence of foreign and potentially harmful microbes by antibody-producing (B) cells and by cell-mediated cytotoxic (T$_C$) cells. Both specific immune mechanisms target microbial invaders for destruction.

Disorders of the immune system may be the result of infection, cancer, or immunosuppressive drugs, or may be congenitally inherited. Allergic reactions and autoimmune diseases are adverse reactions of the immune system. Allergic reactions are reactions of hypersensitivity (exaggerated responses). Autoimmune diseases are characterized by the misrecognition of self as nonself (foreign), resulting in the immune system mounting an attack against self. Immunization developed from its obscure and nonscientific beginnings to result in the eradication of smallpox and the elimination, or near elimination of infectious diseases. New and improved vaccines are sought and continue to become available.

■ Self-Evaluation

PART I: Choose the single best answer.

1. Which type of white cell is primarily involved in phagocytosis?
   a. neutrophils **b.** eosinophils **c.** basophils **d.** halophils

2. Which one of the following is a primary immune structure?
   a. tonsils **b.** lymph nodes **c.** thymus gland **d.** blood cells

3. SCID disease is characterized by a lack of
   a. B cells **b.** T cells **c.** both B and T cells **d.** phagocytic function

4. Which of the following represents innate, nonspecific immunity?
   a. Ig molecules **b.** acidity of stomach **c.** phagocytosis **d.** more than one of the above

5. The AIDS virus specifically attacks
   a. T$_H$ cells **b.** red blood cells **c.** T$_C$ cells **d.** all of the above

6. Myeloid stem cells give rise to all the following *except*
   a. lymphocytes **b.** neutrophils **c.** platelets **d.** basophils

7. Which of these statements is true regarding the antibody molecule?
   a. It is also called the immunoglobulin molecule. **b.** It is a four-chain molecule **c.** Antibodies "fit" into specific receptors. **d.** Items a, b, and c are all correct.

8. Which one of the following is not considered a secondary immune structure?
   a. spleen **b.** tonsils **c.** bone marrow **d.** lymph nodes

9. Serum sickness can result from
   a. attenuated vaccines **b.** live vaccines **c.** toxoid administration **d.** passive immunization

10. What group of individuals should get the most vaccines to prevent disease?
    a. children **b.** elderly **c.** immune compromised individuals **d.** individuals with AIDs

PART II: Fill in the blank.

1. B cells mature in the _____.
2. Antibody-producing plasma cells may remain for years as _____ cells.
3. _____ diseases fail to distinguish between nonforeign and foreign cells.
4. The immune system functions to recognize and destroy that which is _____.
5. _____ or leukocytes play a direct role in immunity.
6. CMI stands for _____.
7. Categories of antibodies are referred to as _____.
8. _____ are live or inactivated viruses or microbes administered to prevent infectious diseases.
9. _____ theory explains how B cells know which antibodies to specifically produce.
10. _____ is a form of cancer characterized by uncontrolled reproduction of white blood cells.

PART III: Answer the following.

1. What are autoimmune diseases? Discuss the immunological basis of these diseases.
2. Explain the differences between specific and nonspecific immunity.
3. An individual with AIDS has a low T_H count. What does this mean? What is the significance of this low count?
4. Explain why there aren't vaccines available for every infectious microbe or virus.
5. Why are immune compromised individuals more at risk for microbial infections?
6. List the different subsets of T-cells and describe how each subset functions within the immune system.
7. List the different protective effects of antibodies.
8. Why are memory cells important?
9. Distinguish between active and passive immunization.
10. List some requirements or properties of an ideal vaccine.

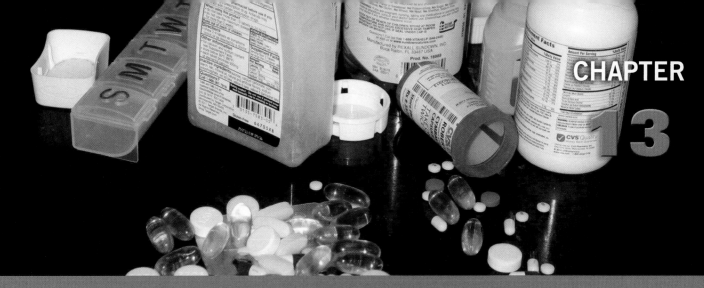

Control of Microbial Diseases

It is a disturbing fact that Western Civilization, which claims to have achieved the highest standards of health in history, finds itself compelled to spend ever increasing sums for the control of disease.

—Rene Dubos, 1987

■ **Preview**

Advances in public health during the twentieth century have decreased the burden of microbial disease on a worldwide basis, particularly in the United States and in other industrialized nations. Sanitation and clean water, food safety, immunization, and antibiotics are major factors in the control of microbial diseases. Each of these factors is discussed in this chapter.

Is the general health of your generation better than that of your parents' or grandparents' generations? Most decidedly, your response would be "yes," despite the current problem of new and reemerging infections. Society, particularly in developed countries, is no longer "plagued by plagues," but thousands of citizens in the United States and in other developed areas live in pockets of poverty and disease similar to the developing world. Even in developing countries the burden of disease has been reduced, although certainly not to the degree enjoyed by the richer nations of the world.

Author's photo (RIK).

Consider that a person born in the United States in the early 1900s could anticipate an average life span of forty-five years and that the death rate at birth was slightly higher than 10%. Back then tuberculosis was the leading cause of death. Today, life expectancy in the United States, as estimated by the Central Intelligence Agency *Worldfact Book*, is 78.37 years (75.92 for males and 80.93 for females), an increase of thirty-three years. The overall worldwide life expectancy is 78.37 years according to the 2011 CIA *Worldfact Book*. Life expectancy ranges from a high of 89.73 years in Monaco to a low of 38.76 in Angola. Surprisingly, the United States ranks fiftieth in a list of 221 countries despite the fact that it spends the most on health. Wealthy Americans are among the world's healthiest people, whereas Americans on the bottom rungs have a life expectancy characteristic of sub-Saharan Africa.

An eminent historian stated, "The retreat of the great lethal diseases was due more to urban improvements, superior nutrition, and public health than to curative medicine." In 2011 the Centers for Disease Control and Prevention (CDC) published a report entitled *Ten Great Public Health Achievements—United States, 2001–2010* (TABLE 13.1); three of those achievements focused directly on a reduction of infectious diseases and are further discussed (FIGURE 13.1).

TABLE 13.1 Ten Great Public Health Achievements in the United States, 2001–2010[a]

Health Achievement	Impact
Vaccine-preventable vaccines	Impact of the pneumococcal and rotavirus vaccines have been particularly striking.
Prevention and control of infectious diseases	TB rates decreasing, better diagnosis of HIV/AIDS, improved screening of blood donors (e.g., prevention of West Nile virus infections through blood transfusions).
Tobacco control	More states have enacted smoke-free laws. FDA requires more warning labels on cigarette packaging.
Maternal and infant health	Screening of newborns for genetic and endocrine disorders.
Motor vehicle safety	Seat belt and car seat policies reduce motor vehicle deaths.
Cardiovascular disease prevention	Stroke rates and coronary heart disease rates have decreased.
Occupational safety	Lower back injury rates in nursing homes, less farm youth injuries, lower crab fisher fatalities.
Childhood lead poisoning prevention	More states have lead poisoning prevention laws; steep decline in the percentage of children with lead blood levels, from 88.2% in 1976 to 0.9% in 2008.
Cancer prevention	Improved cancer screening rates, declining death rates.
Public health preparedness and response	Surveillance and response capacities improved (e.g., H1N1 pandemic).

[a]Not ranked in order of importance.

Adapted from CDC. *Morbidity and Mortality Weekly Report* 60(19);619–623.

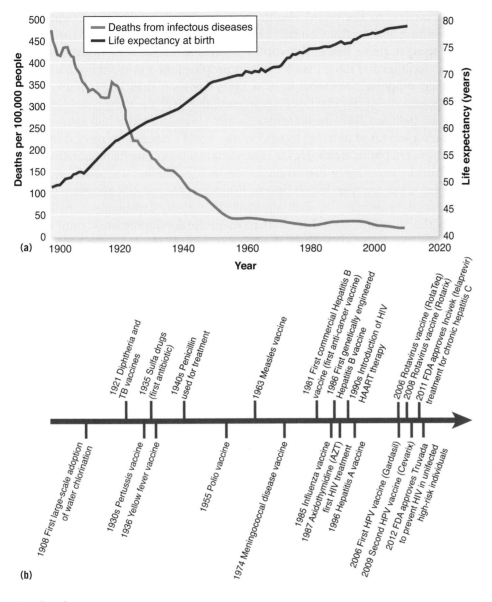

FIGURE 13.1 Major public health events and their influence on crude death rate, 1900–2012. (a) Infectious diseases were a major cause of morbidity and mortality in the early 1900s. (b) Timeline of public health intervention focused on improvements in sanitation and hygiene, implementation of antibiotics, and immunization programs. These strategies were effective over the generations. Modified from The Hamilton Project at Brookings Institution.

▨ Sanitation

Part of the daily routine each morning as you prepare for the day is attending to matters of sanitation and personal hygiene, including showering, brushing your teeth, and using a clean flush toilet. These activities are taken for granted, but they are luxuries. On the other side of the globe in a poverty-stricken and war-torn part of Indonesia, refugees occupy makeshift shelters in camps and live under miserable conditions with inadequate sanitation, food, and water. A newspaper account tells the plight of one family: "Mr. _____ watched cockroaches crawl over his wife and three children asleep on the dirt floor of a tent they share with three other families. He lamented the lack of food and medicine to treat the diarrhea, measles, eye infections, flu, and other ailments that spread quickly here. There is nothing but misery and sorrow in this place."

According to the World Health Organization (WHO)'s Water Supply and Sanitation Collaborative Council, 2.6 billion of the world's citizens (half of the developing world or 35% of the world's population) lacked basic hygiene and sanitation facilities and more than 1.5 billion people in the world did not have access to a daily supply of clean water in 2006. Every day, more than 4,000 children under the age of 5 die of diarrheal diseases because they lack the most basic sanitation. Between 1990 and 2008, the share of the world's population that had access to basic sanitation increased only 7% to 61% of the world's citizens. In most developing countries, cell phone technology is expanding at a faster rate than sanitation!

World leaders agree that hygienic means of sanitation and a safe supply of drinking water are basic human needs. Kofi Annan, a past UN Secretary General, said it all in his words: "We shall not finally defeat AIDS, tuberculosis, malaria, or any of the other infectious diseases that plague the developing world until we have also won the battle for safe drinking water, sanitation, and basic health care." Clean water, improved sanitation, and hygiene have the potential to prevent at least 9.1% of world diarrheal diseases and 6.3% of all deaths.

Development of Sanitation

Urbanization is not a new phenomenon but one that dates back millennia to the times when hunter–gatherers first saw a benefit in pooling their meager resources by living together in villages. These villages have evolved into today's cities. Population growth and the resulting urbanization in the mid-nineteenth century are major factors contributing to the challenge of infectious disease control. The germ theory and the idea of contagion had yet to be developed, and urban centers were struggling to establish infrastructure to keep pace with the burgeoning masses. Little regard was paid to public health measures. Industrial development and immigration led to an influx into the cities, which, in turn, fueled poor housing, overcrowding, lack of clean water, and lack of facilities for disposal of human waste. John Cairns, a British biologist, termed cities the "graveyards of mankind." The nineteenth century outbreaks of cholera in London illustrate the consequences of inadequate sanitation and hygiene. Arno Karlen, in his book, *Man and Microbes* (Simon & Schuster, 1996), wrote the following:

> "*The city's seven sewer systems were uncoordinated and relied on defective pipes. They received tons of human and animal feces, dead animals, waste from abattoirs [slaughter houses], effluvia from hospitals and tanneries, the occasional human corpse, and contaminated ground water from cemeteries.*"

All of London's refuse ended up in the Thames, which provided most of the city's water. Cholera arrived; there were eight separate water companies and just one experimental filtration system. Water not taken from the Thames came from wells, many as badly polluted as the river. The city drank, cooked, and washed in its own filth. With the crowding and dirt, once a waterborne disease was established, further person-to-person transmission was virtually assured.

London was not the only city to be so afflicted. Filth and squalor prevailed in Europe and around the world. Cities in the United States were hardly models of cleanliness and sanitation. Sewage disposal systems were few, and outhouses,

FIGURE 13.2 Factors in the spread and emergence of infectious disease. **(a)** People living in squalor without adequate shelter and basic sanitation are at increased risk for infectious disease. © Vishal Shah/ShutterStock, Inc. **(b)** Despite major advances over the past century, substandard levels of living such as those shown in this old drawing persist in underdeveloped countries and in pockets of developed countries. © National Library of Medicine.

(a)

overflowing cesspools, and garbage-littered streets flourished, as did tuberculosis, diphtheria, scarlet fever, and typhoid fever. Somewhere about the middle of the nineteenth century a sanitary reform movement gradually arose from the ashes of human corpses, debris, and human and animal wastes, perhaps fired by the third cholera epidemic to hit London (on the heels of the second epidemic). The combination of disease, filth, and lack of shelter (FIGURE 13.2) led to the enactment of laws relating to sanitation, including sewage and water treatment, garbage collection, and other public health measures. In the 1880s Louis Pasteur and Robert Koch triumphed with their discoveries. The germ theory of disease was established, and microbes were at last linked to sanitation and (microbial) disease.

The germ theory was embraced in Europe and in the United States. Sanitation was "in," and sanitary engineers and bacteriologists (a term that preceded "microbiologists") flourished (FIGURE 13.3). In 1887 the Marine Hospital Service was established and charged with monitoring cholera in immigrants on ships coming into New York. This facility was the forerunner of today's National Institutes of Health. Other public health laboratories and organizations were established in major cities around the world, and bacteriologists and sanitary engineers worked in concert. Public health statutes promoting sanitation and good hygiene were passed and implemented, and by 1900, 40 states had health departments.

Over the twentieth and twenty-first centuries the focus of health departments changed from meeting the urgent and immediate needs of basic sanitation to delivering health services (TABLE 13.2). The 1920s through the 1950s witnessed great strides in public health strategies to control infectious diseases (FIGURE 13.4). Great attention

FIGURE 13.3 At the beginning of the twentieth century, the age of sanitation began to emerge. © National Library of Medicine.

TABLE 13.2 Changing Role of Health Departments

Health department services in 1900 (driven by urgent needs)

Sewer construction

Water supply inspection

Sewage disposal

Nuisance and pest control

Privy inspection or removal

Milk supply inspection

Infectious disease control (tuberculosis, diphtheria and croup, scarlet fever, smallpox, and typhoid fever)

Health department services in 1999 (developing an organized approach)

Monitoring community health status to identify potential hazards

Investigating disease outbreaks and safety hazards in the community

Mobilizing community partnerships to solve health problems

Developing policies and plans that support individual and community health efforts

Enforcing laws and regulations that protect health and ensure safety

Linking populations with needed personal health services and ensuring the provision of health care when otherwise unavailable

Ensuring a competent public health and personal health care workforce

Evaluating effectiveness, accessibility, and quality of personal and population-based health services

Researching new ideas and innovative solutions to health problems

U.S. state public health agencies top programs and functions since 2005

Preparedness and disease monitoring

Data collection: vital statistics

Maintaining public health laboratories

Tobacco prevention and control

Environmental health

Food safety

Health facilities, drinking water, and environmental regulations

Source: CDC and Beitsch, L. M. et al., 2006. Public health at center stage: new roles, old props. *Health Affairs*, 25(4):911–922.

AUTHOR'S NOTE (RIK)

In New Delhi, India, there is the Museum of Toilets, which has a collection of artifacts, pictures, and objects illustrating the historical development of toilets since the year 2500 B.C. On display at the museum is a replica of the throne of King Louis XIII with its built-in commode he used while giving audience.

FIGURE 13.4 Public health statutes were passed and enforced by health officers. Quarantine signs were required by law to be placed on homes in which an individual was suffering from a "contagious" disease. Courtesy of San Antonio Metropolitan Health District. Used with permission.

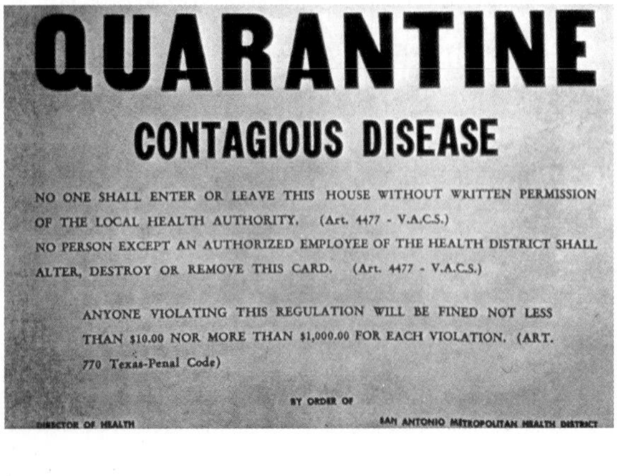

QUARANTINE
CONTAGIOUS DISEASE

NO ONE SHALL ENTER OR LEAVE THIS HOUSE WITHOUT WRITTEN PERMISSION OF THE LOCAL HEALTH AUTHORITY. (Art. 4477 - V.A.C.S.)

NO PERSON EXCEPT AN AUTHORIZED EMPLOYEE OF THE HEALTH DISTRICT SHALL ALTER, DESTROY OR REMOVE THIS CARD. (Art. 4477 - V.A.C.S.)

ANYONE VIOLATING THIS REGULATION WILL BE FINED NOT LESS THAN $10.00 NOR MORE THAN $1,000.00 FOR EACH VIOLATION. (ART. 770 Texas-Penal Code)

BY ORDER OF

DIRECTOR OF HEALTH   SAN ANTONIO METROPOLITAN HEALTH DISTRICT

was paid to the construction of water and sewage treatment facilities, chlorination, better housing, control of tuberculosis and venereal diseases (now called sexually transmitted diseases, or STDs), food production and distribution, animal and pest control measures, and garbage disposal (FIGURE 13.5). The public was bombarded with information regarding the evil of "germs" and their transmission from the sick to the healthy by various modes of transmission (FIGURE 13.6). The "gospel of germs" was accepted, and rub-a-dub-dub, scrub, dust, and clean-clean-clean were heralded. Somewhere along the line the expression "Cleanliness is next only to godliness" became a household dictate; some say this expression is attributed to Mahatma Gandhi of India as his "battle cry" during his efforts in the 1920s to 1930s to clean up the villages. These new efforts paid off. Malaria, plague, tuberculosis, and other diseases were markedly reduced; the last major outbreak of plague in the United States occurred during 1924 and 1925 in Los Angeles. The twentieth century could be termed the "Golden Age of Public Health" based on the gain of more than 60% life expectancy that is directly correlated with public health knowledge of infectious diseases and vaccinations along with improved sanitation, food and water safety practices, and regulations.

Today the breadth and scope of public health programs are vast, ranging from disease surveillance and data collection to environmental regulation and medical or mental health and education programs (Table 13.2).

FIGURE 13.5 Garbage disposal is an important public health measure. Garbage accumulation attracts rats and other rodents and animals that serve as reservoirs and vectors of infectious disease. Author's photo (RIK).

FIGURE 13.6 Departments of health fostered in the population an appreciation of good personal hygiene in an effort to control the dissemination of infectious disease. Reproduced from Robertson, J.D.A. *Report on an Epidemic of Influenza in the City of Chicago in the Fall of 1918*. Chicago, 1918.

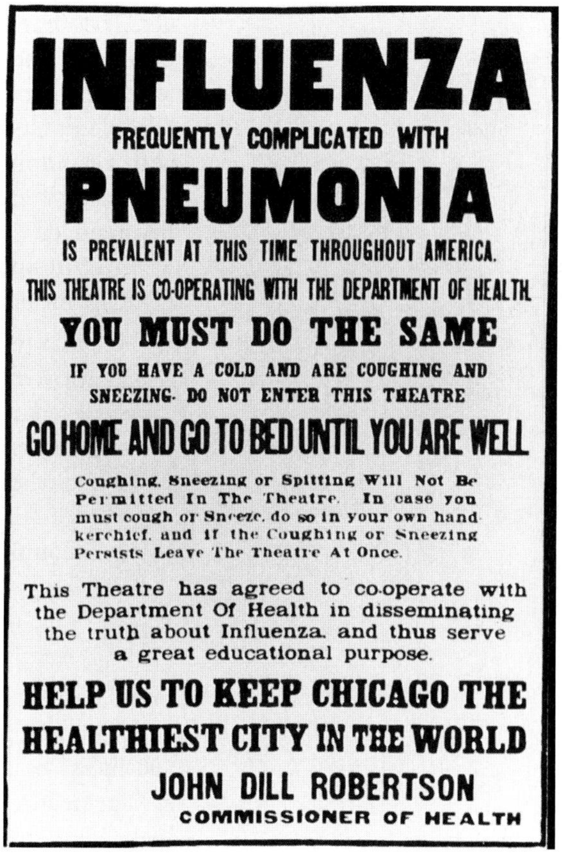

Human Waste Disposal

Your first impulse may be to laugh at learning there is actually a World Toilet Organization, founded in 2001 to promote sanitation. Almost three billion people across the globe lack access to appropriate toilet facilities. Further, 200 million tons of human waste goes uncollected and untreated around the world because of the lack of toilets (**FIGURE 13.7**). World Toilet Day has been celebrated every November 19th since the inception of the World Toilet Organization.

The safe disposal of human excreta is central to sanitation, and its significance cannot be overemphasized. The General Assembly of the United Nations (UN) declared 2008 as an International Year of Sanitation in an effort to accelerate progress on worldwide improvements in sanitation in recognition of the fact that 2.6 billion people in the world lack proper sanitation facilities, particularly in developing countries and in poverty pockets in developed countries. Estimates are that only 40% of the population has no choice but to squat and defecate in the open directly onto the ground.

In India, as an example, it is estimated that over 100 million households have no toilets and 10 million households use buckets for waste disposal. Further, 900 million liters of urine and 135 million kilograms of fecal material need to be disposed of each day. About half a million children die every year in India because of dehydration resulting from diarrheal diseases that are frequently traceable to open defecation. The subject of toilets and defecation is hardly dinnertime conversation, but it is a fact of life and an integral part of the history of human hygiene.

Organizations such as the United Nations and the World Bank are working on the development of sewage disposal systems in developing countries. Bathrooms and flush toilets, such as our society is accustomed to, are not necessary goals; certainly,

FIGURE 13.7 Human waste disposal. **(a)** Contamination of food and water by fecal material is a major cause of many infectious diseases. Author's photo (RIK). **(b)** Primitive and simple toilets are relatively inexpensive and efficient if properly constructed. Outer barrel is for privacy; inner barrel is for urination only. Author's photo (RIK).

(a)

(b)

less luxurious and primitive facilities, be they outdoor or indoor, are affordable and effective (Figure 13.7). Programs to improve poor sanitation are in progress in slums and squatter settlements of the world's poorest countries. The Kampung Improvement Program in Indonesia, a highly successful program, has focused on covering open sanitation drains and on bringing reasonably clean water to families. The Orangi Pilot Project reached 650 thousand people in a poor neighborhood in Karachi, Pakistan. Orangi is the largest squatter settlement in Karachi, with approximately one million inhabitants. According to a public health official, "People don't need a flush toilet in every home or a faucet in every room. But with a standpipe for every three units, with adequate pit latrines, and other forms of [waste] treatment, the services can be there and the health of the children maintained."

A 2010 update about the progress on sanitation published by the WHO and UNICEF includes the following statistics:

- On a typical day, more than half the hospital beds in sub-Saharan Africa are occupied by patients suffering from fecal-related diseases.
- Seven out of 10 people without improved sanitation live in rural areas.
- Worldwide 1.1 billion people still defecate in the open (FIGURE 13.8a).
- Diarrhea is the second highest cause of child mortality after pneumonia.
- More than 5,000 children under the age of 5 die every day from diarrhea accounting for 17% of this age group.

The improved sanitation facilities worldwide are depicted in the map in Figure 13.8b. Virtually the entire population in developed regions uses improved facilities. This is not true for developing regions in which only around half the population use improved sanitation.

A human produces roughly 132 gallons (500 liters) of urine and 13.2 gallons (50 liters) of feces every year. The United States has about 16,000 municipal sewage treatment plants that convert human waste into **sludge** (also referred to as biosolids). Thirty years ago, thousands of cities dumped their raw sewage directly into the nation's rivers, lakes, and bays. Since the 1990s, sludge has been sold or given away free to farmers for use as fertilizer to improve soils and stimulate plant growth. The Environmental Protection Agency (EPA) has promoted the use of sludge as an environmentally friendly way to recycle sewage. Local governments decide whether or not to recycle the biosolids as fertilizer, or to incinerate or bury it in a landfill. About 50% of all sludge is recycled to land and takes place in all 50 states. The biosolids are used on less than 1% of the nation's agricultural land.

Given the fact that human waste contains microbes, including a small percentage that may be pathogenic (**TABLE 13.3**), rigorous testing and stringent guidelines were created to address the risk of infectious disease and its consequence to public health. The National Academy of Sciences reviewed the current practices, public health concerns, and regulator standards and have concluded that the use of sludge in the production of crops for human consumption, in conformance with federal and state regulations set, presents negligible risk to the consumer, to crop production, and to the environment. There are two classes of biosolids. Class A biosolids are treated or sanitized and contain no detectable levels of pathogens. Class B biosolids are treated but still contain detectable levels of pathogens. Class B biosolids are restricted regarding public access and crop harvesting.

Clean Water

"Water, water everywhere, nor any drop to drink." This famous line from the classic 1798 poem, *The Rime of the Ancient Mariner*, by Samuel Taylor Coleridge, depicts the desperate plight of an old mariner surrounded by a sea of undrinkable water. It can also serve to depict the desperation of one-fourth of the world's

FIGURE 13.8 **(a)** The 81% of people that defecate in the open live in these ten countries. **(b)** World map showing the use of improved sanitation facilities in sub-Saharan Africa and Asia is low (2008). Modified from *Progress on Sanitation and Drinking Water—2010 Update*. WHO and UNICEF.

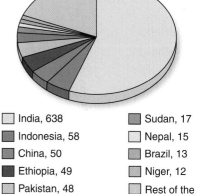

India, 638
Indonesia, 58
China, 50
Ethiopia, 49
Pakistan, 48
Nigeria, 33
Sudan, 17
Nepal, 15
Brazil, 13
Niger, 12
Rest of the world, 215

(a)

(b)

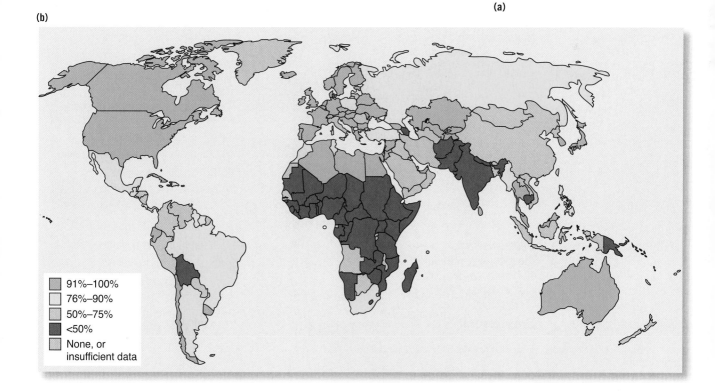

91%–100%
76%–90%
50%–75%
<50%
None, or insufficient data

TABLE 13.3 Water, Sanitation, and Hygiene-Related Microbial Diseases in Humans[a]

Type of Pathogen	Pathogen	Disease	Association
Virus	Norovirus	Gastroenteritis	Contaminated water
Virus	Rotavirus	Gastroenteritis	Contaminated water and poor sanitation
Bacterium	*Campylobacter jejuni*	Campylobacteriosis	Contaminated water
Bacterium	*Chlamydia trachomatis*	Trachoma	Sanitation and hygiene
Bacterium	*Escherichia coli*	Gastroenteritis	Contaminated water and poor sanitation
Bacterium	*Leptospira* sp.	Leptospirosis	Contaminated water
Bacterium	*Mycobacterium ulcerans*	Buruli ulcer	Contaminated water related to environmental changes
Bacterium	*Salmonella* sp.	Salmonellosis	Contaminated water
Bacterium	*Salmonella typhi*	Typhoid fever	Contaminated water
Bacterium	*Shigella sonnei*	Shigellosis	Contaminated water and poor hygiene
Bacterium	*Vibrio cholerae*	Cholera	Contaminated water, inadequate sanitation and hygiene
Protozoan	*Cryptosporidium* sp.	Cryptosporidiosis	Contaminated water, inadequate sanitation and hygiene
Protozoan	*Cyclospora cayetanensis*	Cyclosporiasis	Contaminated water
Protozoan	*Entamoeba histolytica*	Amebiasis	Contaminated water and poor sanitation
Protozoan	*Giardia intestinalis, Giardia lamblia,* or *Giardia duodenalis*	Giardiasis	Contaminated water and poor sanitation
Helminth	*Ascaris* sp., *Trichuris* sp., *Anclostoma* sp., *Necator* sp.	Soil-transmitted helminthiasis	Sanitation and hygiene
Helminth	*Dracunculus medinensis*	Dracunculiasis (guinea-worm disease)	Contaminated water
Helminth (common liver fluke)	*Fasciola hepatica*	Fascioliasis	Contaminated water
Helminth	*Schistosoma mansoni, S. haematobium,* or *S. japonicum*	Schistosomiasis	Contaminated water
Helminth	*Wuchereria bancrofti*	Lymphatic filariasis	Sanitation and hygiene
Fungi	*Tinea* sp.	Ringworm	Sanitation and hygiene
Arthropod (lice)	*Pediculus humanus capitis* (head louse); *Pediculus humanus corporis* (body louse, clothes louse); *Pthirus pubis* ("crab" louse, pubic louse)	Lice	Sanitation and hygiene
Arthropod (mites)	*Sarcoptes scabiei var. hominis*	Scabies	Sanitation and hygiene

[a]List does not include all arthropod-associated diseases associated with water.
Source: CDC

(a)

(b)

(c)

population that have only limited access to water that may or may not be safe (FIGURE 13.9a). Less than 2.5% of the Earth's water can be used and reused as freshwater. Less than half of that is readily available or accessible with some effort. FIGURE 13.10a is a world map showing that sub-Saharan Africa faces the greatest challenge in increasing the use of improved drinking water. There is a sharp contrast in the U.S. and in other developed countries where water is plentiful and wasted by the gallons. Row after row of jugs, bottles, flavored and carbonated waters in a variety of containers are displayed (Figures 13.9b and 13.9c) in markets despite the availability of clean water in just about every household. There may be exceptions but there is generally no shortage of clean water for bathing, showering, washing the dishes, and washing the dog.

The WHO estimates that over one and half billion people worldwide lack access to clean water; in some villages people, primarily women and children, spend a major part of their day carrying buckets to a source of clean water. The UN Millennium Development Goals aim to reduce the number of people lacking access to water by 50%, proclaiming the years 2005 to 2015 as the International Decade for Action "Water for Life." The proportion of households in major cities connected to piped water (house or yard connection) is shown in Figure 13.10b. Not surprisingly, there is a strong correlation between access to safe drinking water and child health (Figure 13.10c). Table 13.3 lists specific diseases that occur when water, sanitation, and hygiene are compromised.

A 2010 update about the progress on drinking water published by the WHO and UNICEF includes the following statistics:

- In developing regions, 84% of the population uses an improved source of drinking water.
- The number of people living in rural areas who do not use an improved source of drinking water is over five times the number living in urban areas.
- Worldwide, 37% of people not using an improved source of drinking water are living in sub-Saharan Africa.

FIGURE 13.9 (a) Clean water: a luxury. For much of the world's population, water is not piped into their homes and must be transported in containers. Bodies of water may serve for clothes washing, bathing, water for animals, and drinking water. © Marcus Brown/ShutterStock, Inc. (b) Rows of water jugs. Author's photo (RIK). (c) Rows of regular and carbonated drinking water at a superstore. Author's photo (RIK).

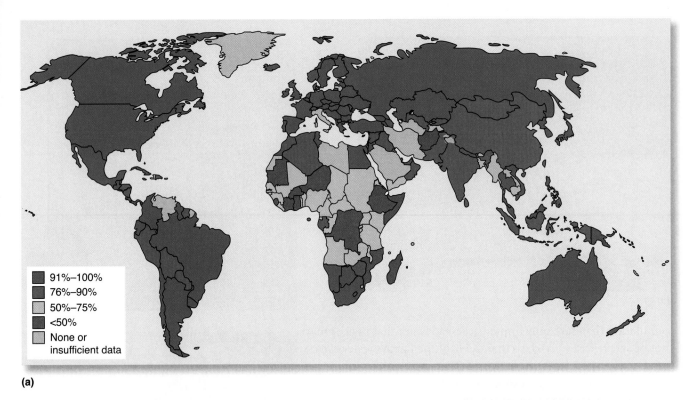

(a)

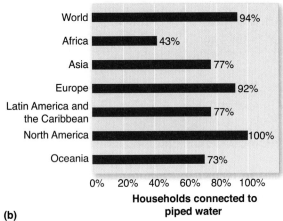

(b)

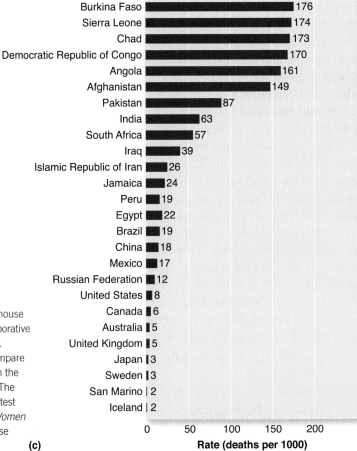

FIGURE 13.10 **(a)** World map showing the improved drinking-water sources (2008). Sub-Saharan Africa continues to face the greatest challenges. Adapted from *Progress on Sanitation and Drinking Water—2010 Update*. WHO and UNICEF. **(b)** Graph showing the proportion of households in major cities connected to piped water (house or yard connection). Data from the Water Supply & Sanitation Collaborative Council (http://www.wsscc.org/). **(c)** Water quality and child survival. Access to safe drinking water correlates with low child mortality. Compare the data for Chad with those for the United States; the differences in the availability of safe drinking water and in child mortality are striking. The graph is intended to illustrate the concept rather than provide the latest figures. Adapted from UNICEF Childinfo: *Monitoring the Status of Women and Children: Under-Five Mortality* 2010. UNICEF statistical database available from http://www.childinfo.org/mortality_underfive.php. **(c)**

The WHO estimates that at any given time, perhaps one-half of all people in the developing world are suffering from one or more of the six main diseases associated with drinking contaminated water (Table 13.3). The poorest people in the world are paying many times more than their richer compatriots for the water they need to live and are getting more than their share of deadly diseases because supplies are dangerously contaminated. Control of waterborne diseases is a particularly difficult problem when excreta are disposed of in a way that allows fecal material to gain access to water sources and food supplies.

Improvement of water quality was, and continues to be, a major public health priority aimed at the decline of waterborne and water-associated diseases. Anywhere from a 20% to 80% decline in morbidity and mortality is possible with improved water sanitation. In some countries people are infected with guinea worms, which bite their way through their victim's flesh, or with cholera so devastating that in several hours their life is threatened by severe dehydration resulting from massive loss of water through diarrhea. In the village of Chiladi in rural Brazil some end up with brown spots on their hands and serious symptoms due to drinking the arsenic-contaminated water that was supposed to be clean (BOX 13.1).

However, even in developed countries vigilance needs to be maintained. A safe water supply can never be taken for granted, as the citizens of Milwaukee, Wisconsin painfully discovered in 1993 when *Cryptosporidium parvum* caused the largest outbreak of waterborne disease in U.S. history. When developed countries experience fires, flood, hurricanes, wars, and other catastrophic events, they suffer the same misery of land and water becoming fouled with human and animal wastes and carcasses as the developing countries. Hurricane Floyd hit North Carolina with a vengeance in September 1999 and the rivers filled with human waste, animal waste, and other pollution. Hurricane Katrina, one of the most monstrous storms over the past century, hit the U.S. Gulf Coast on August 29, 2005, and caused widespread damage over the area, particularly in New Orleans because of the resulting flooding. In the aftermath the population was exposed to contaminated drinking water and other public health hazards.

The CDC and the U.S. Environmental Protection Agency have maintained a collaborative surveillance system since 1971 and continue to report on the occurrence of waterborne diseases with the goal of characterizing and identifying the causative agents. Over the past century legislation has been implemented in an effort to regulate the nation's water supply. In 1972 the Clean Water Act came into effect as a response to the pollution of water in the United States with industrial and human wastes; bodies of water were becoming the nation's dumping ground. The goal was to reduce waterborne diseases and other adverse outcomes. Two years later, in December 1974, the Safe Drinking Water Act became effective and established measures to ensure the safety of drinking water at the tap. The Act was updated in 1986 and in 1996. As shown in Figure 13.10, the initiatives have paid off.

AUTHORS' NOTE (RIK AND TS)
Over the last several years we have watched students trudge into class with water bottles of all sizes, shapes, colors, and labels to get through the upcoming class of less than an hour. It almost appears like an army of soldiers readying for an extended march through Sinai or some other desert area.

Food Safety

Safer foods are considered one of the 10 great public health achievements of the United States during the twentieth century. President Obama signed the U.S. Food and Drug Administration (FDA) Food Safety Modernization Act on January 11, 2011.

BOX 13.1 Arsenic in the Well and in the Wood

Sometimes, in an effort to alleviate a problem, well-meaning public health officials initiate interventions that backfire. Consider the following example.

Until about 30 years ago millions of people in Bangladesh and West Bengal, poor and densely populated regions, drank surface water contaminated with disease-producing microbes from shallow hand-dug wells, streams, and ponds resulting in a high burden of disease. To combat the high incidence of death and disease resulting from contaminated drinking surface water, international aid agencies such as UNICEF and local health officials installed tube wells to tap groundwater.

A tube well is a simple device constructed of steel pipes sunk deep into the ground fitting with a pump handle. You might find one at a roadside picnic area where piped water is not available. The pump is sealed topside to prevent water leaking back down the pipe. Microbes are filtered out as groundwater trickles through the aquifer, resulting in microbiologically safe water.

An estimated 3.5 million wells gave millions of people in the area access to the groundwater and was expected to be the answer to the epidemics associated with the use of contaminated unsafe surface water. Although the number of waterborne diseases was markedly reduced, the price was too high; the groundwater was contaminated with naturally occurring arsenic in concentrations well above the accepted levels. As it turns out, farmers were fertilizing their soil two to three times per year with a fertilizer that contained 20 mg of arsenic per kilogram resulting in contaminated groundwater in the tube wells. By the mid-1990s

thousands of people had been diagnosed with arsenic poisoning, and a new crisis existed.

Chemically, arsenic is categorized as a heavy metal, as is mercury and lead. It is usually excreted from the body, but if excess amounts are ingested it accumulates. Arsenic is very toxic and interferes with essential enzyme systems, resulting in death due to multiorgan failure. Historically, the use of arsenic as a poison for political assassinations dates back several centuries. Some historians believe Napoleon was killed by food and beverage tainted with arsenic. ("Arsenic and Old Lace" was a hilarious and highly popular play that opened on Broadway in 1941. It was a comedy about two elderly sisters who poisoned lonely old men with elderberry tea containing arsenic, strychnine, and a "touch" of cyanide.) Interestingly, salvarsan, developed about 1909, one of the first drugs to treat syphilis, was an arsenic-containing compound.

It seems that the people of Bangladesh and West Bengal have unwittingly gone "from the frying pan into the fire." The choice may be between drinking arsenic-free, microbiologically contaminated water and drinking arsenic-contaminated, microbiologically safe water. This problem continues but there are immediate alternatives such as using arsenic-free safe water traps at shallow depths and rainwater harvesting. Since the discovery of arsenic-contaminated groundwater in Bangladesh, recent studies have found arsenic contamination to be a worldwide occurrence. Arsenic can be found in the groundwater in the United States, contaminating aquifers and wells. For example, arsenic in groundwater is a known public health issue in Maine.

It is the largest reform of food safety in more than seventy years. The law will be fully implemented over time. It calls for rules and regulations related to the design, production, labeling, promotion, manufacturing, and testing of regulated food products. The WHO has expanded its food safety initiatives in response to new challenges, including health implications of genetically engineered foods. In the United States the Food Safety Inspection Service of the U.S. Department of Agriculture (USDA) is charged with the safety, labeling, and packaging of the nation's commercial supply of meat, poultry, and eggs. Nevertheless, outbreaks of foodborne disease continue to occur.

Leaders of the sanitation movement, initiated in the early 1900s, recognized that typhoid fever, tuberculosis, scarlet fever, botulism, and other diseases were transmitted by food (including milk) and water and advocated safer food-handling procedures, including pasteurization, refrigeration, hand washing, safer food processing, and pesticide application. *The Jungle*, the powerful 1906 novel by Upton Sinclair, portrayed the unsanitary practices in the Chicago meatpacking industry. An excerpt from the book follows:

"There were some [cattle] . . . that had died, from what cause no one could say; and they were all to be disposed of here in darkness and silence. "Downers," the men called them; and the packing houses had a special elevator upon which they were raised to the killing beds, where the gang proceeded to handle them, with an air of business-like nonchalance which said plainer than any words that it was a matter of everyday routine . . . . and in the end Jurgis saw them go into the chilling rooms with the rest of the meat, being carefully scattered here and there so that they could not be identified."

The public reacted so strongly that within a year after publication of Sinclair's book the Pure Food and Drug Act was passed. Even today, news shows such as "Dateline" and "20/20" periodically air alarming stories about food processing. Despite the advances in the safety of the food supply, foodborne diseases remain a major cause of morbidity and mortality. Well over twenty-five microbes, including bacteria, viruses, protozoans, and worms, are causative agents of foodborne diseases. Further, changes in food-processing technologies, personal eating habits, and food distribution have contributed to the presence of new emergent pathogens. In 2011, the CDC estimated that 1 in 6 Americans (or 48 million people) suffers from a foodborne illness in the United States each year, with 128,000 hospitalizations and 3,000 deaths. There are 31 known pathogens that cause foodborne disease. Many of these are tracked by public health systems that monitor diseases and outbreaks.

How does food become contaminated along the pathway from farm to plate? What are the sources of contamination? In 2009 the American Society for Microbiology issued a report called *Global Food Safety: Keeping Food Safe from Farm to Table* that cited urgently needed technologies to improve food safety at the pre-processing stage; these technologies are listed in TABLE 13.4.

Ultimately, foods are prepared and consumed in the home—the last stage in the journey from farm to plate. Cleanliness in the kitchen has received considerable

---

TABLE 13.4 Cost-Effective Technologies Used to Significantly Prevent Food Contamination

Means to detect water quality breaches

Practical water treatment devices

Effective vaccines for livestock and inoculants to prevent pathogen colonization

Pathogen-resistant varieties of plants and animals

Effective decontamination techniques for dried products and fresh produce to prevent uptake of pathogens by produce during postharvest processing

Improved produce-harvesting equipment to reduce risk of contamination that can be easily sanitized on a regular basis

Better storage structures in developing countries to prevent contamination by pests

Better means for managing potential mycotoxin contamination

Adapted from American Society for Microbiology.
Global Food Safety: Keeping Food Safe From Farm to Table, 2009.

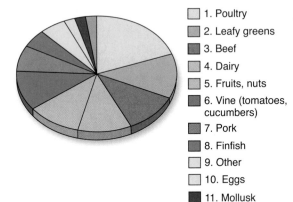

1. Poultry
2. Leafy greens
3. Beef
4. Dairy
5. Fruits, nuts
6. Vine (tomatoes, cucumbers)
7. Pork
8. Finfish
9. Other
10. Eggs
11. Mollusk
12. Grains, beans

attention in recent years. A biologist who calls himself the "Sultan of Slime" claims that "our bathrooms may be cleaner than our kitchens"; perhaps a course in "kitchen microbiology" is not a bad idea. Environmental microbiologist Dr. Charles Gerba, also known as Dr. Germ, became famous for his study on toilet splatter, tests offices, households, and public places for bacterial pathogens. Dish towels, sponges, counter surfaces, and hands are all culprits involved in the spread of foodborne pathogens. Using common sense can minimize the problem.

New measures to increase food safety, some of which are "gimmicky," are in vogue. The (noninspiring) debate regarding the use of wooden versus plastic cutting boards still exists. One entrepreneurial company now advertises "cut and toss" disposable cutting boards. Toxin Alert, a Canadian company, has developed an "intelligent" food wrap, covered with antibody sensors, that changes color when it is contaminated with *Salmonella, Campylobacter, Escherichia coli,* and other bacteria. Pasteurized eggs are available and marketed under the brand name Davidson's Pasteurized Eggs®; they look and taste like unpasteurized eggs. This new technique, allowing in-shell pasteurization, may "egg on" other producers to make pasteurized eggs more available.

The most common causes of single food product foodborne outbreaks from 2003 to 2008 is shown in FIGURE 13.11.

There is some comfort in knowing that city and state health departments, as well as agencies at the national and international levels, including the CDC, the United States Department of Agriculture (USDA), and the WHO, are all working to develop more efficient strategies of surveillance and of standardization of sampling procedures. In the United States the largest meat and poultry processing plants have been required by the USDA since 1998 to implement the Hazard Analysis and Critical Control Point program targeted at reducing contamination during food processing. In 1997 PulseNet, a network designed to track foodborne illnesses at a national level, was established, allowing comparison of DNA fingerprint patterns through electronic communication. About a year later PulseNet revealed that isolates of *Listeria* responsible for outbreaks of infection involving an estimated 100 cases and 22 deaths were linked to contaminated hot dogs and deli meats from a single processing plant. *Food Safety News* is a web-based newspaper dedicated to reporting illnesses associated with food safety. The CDC also has a Food Safety web page with information on food safety tips, laws and regulations, prevention, education, current safety challenges, and outbreaks.

It is convenient to rely on the food industry and governmental agencies as guardians to protect consumers from the perils of foodborne diseases. On the other hand, the specter of mad cow disease, a prion-caused disease, and continued outbreaks of foodborne disease sound a word of caution the world over. Consumers need to share in the responsibility of minimizing the consumption of contaminated foods by exercising common sense without becoming food safety fanatics.

Disinfection and Disease Control

The connection between germs and disease was validated in the late nineteenth century and has led to important innovations such as antibiotics, new antibacterial products, and hygiene practices (FIGURE 13.12). Disinfection is one of the cornerstones of microbial infection prevention and control. Our world is becoming sterilized, sanitized, and "germ-free." How many times have you washed your hands today? Global Handwashing Day (October 15th) was created for children and schools in 2008 to raise awareness about the benefits of hand washing and to foster hand washing with soap in every country. Hand sanitizer dispensers have become commonplace. There is a growing sense of germ phobia.

Disinfection is defined as killing or inactivating microbes that cause disease. Disinfection applications date to the eighteenth century, a time when humans were ignorant of pathogenic microbes yet employed methods that controlled diseases of animals and humans. Methods of disinfection used were classified into three categories: chemical (by derivatives of sulfur, mercury, copper, alkalis, and acids), physical (fumigation, heating, and filtration), and biological (burial). Today, physical and chemical methods remain the important means to control and prevent infectious diseases.

FIGURE 13.12 Cleaning products on shelves in a store. Author's photo (RIK).

Disinfection Methods

TABLE 13.5 is a list of physical and chemical methods used to control microbes. Heat is one of the most common physical methods to kill microbes. For example,

TABLE 13.5 Disinfection Methods

Physical Methods	Chemical Methods
Heat: pasteurization, autoclaving (steam heat), dry heat (oven), boiling, incineration, dehydration, refrigeration, and freezing (retards growth)	Heavy metals: mercury (e.g., topical mercurochrome), silver (e.g., silver sulfadiazine dressings on wounds), copper (e.g., copper bedrails in hospitals)
Radiation: X rays, gamma rays, ultraviolet radiation, microwaves	70% alcohol (e.g., hand sanitizer or skin antiseptic, instrument disinfectant)
Filtration of air and liquids	Aldehydes (used to preserve dead animals)
	Halogens: chlorine (e.g., swimming pools), bromine, iodine (e.g., topical on wounds), fluorine
	Quaternary ammonium compounds (QUATS): used as a disinfectant in hospital and healthcare settings)
	Salt (for preservation of foods)
	Hydrogen peroxide (wound treatment)
	Ethylene oxide gas (used to sterilize plasticware)

TABLE 13.6 List of Microbes and Their Resistance to Disinfection

Infectious Agent	Examples (disease or specific infectious agents)	
Prions	Creutzfeldt-Jakob disease (CJD), chronic wasting disease (CWD)	Most Resistant
Bacterial spores	*Bacillus, Clostridium*	
Helminth eggs	*Ascaris*	
Mycobacteria	*Mycobacterium tuberculosis, M. leprae*	
Small, nonenveloped viruses	Poliovirus, norovirus, papilloma viruses	
Protozoan cysts	*Giardia*	
Fungal spores	*Aspergillus, Penicillium*	
Gram-negative bacteria	*Escherichia, Salmonella, Pseudomonas*	
Yeast	*Candida*	
Gram-positive bacteria	*Staphylococcus, Streptococcus*	
Enveloped viruses	HIV, influenza, herpes simplex virus	Least Resistant

Adapted from: G. McDonnell, Burke, P. 2011. Disinfection: Is It Time To Reconsider Spaulding? *Journal of Hospital Infection* 78:163–170.

autoclaves (steam under pressure) are used to inactivate biohazardous materials collected in a hospital or to sterilize instruments used in surgical procedures or in a dentist office. **Sterilization** involves destroying all live microbes, spores, and viruses present on an object. Some microbes are harder to inactivate than others (TABLE 13.6). For example, prions are the most difficult infectious agent to inactivate. Prion-contaminated samples must be placed in sodium hydroxide or undiluted fresh household bleach and autoclaved at 130°C for 4.5 hours whereas materials contaminated with all other infectious agents can be sterilized for 15 minutes at 121°C.

Pasteurization, named after Louis Pasteur, was originally invented to keep beer from going bad! Yes, beer! In 1856, Pasteur, a chemistry professor researching fermentation, studied spoiled and fresh beer from local breweries for over a year. He observed with a microscope that good beer was full of round yeast cells and that spoiled beer was swimming with long microbes. After heating the beer for a brief period, the long microbes died but the yeast cells were still thriving. Today, heating liquids to reduce the number of pathogenic microbes is known as pasteurization. At the turn of the century, scientists discovered that cows could spread disease through milk (e.g., tuberculosis). By 1907, pasteurization of milk was implemented and the number of disease outbreaks associated with milk plummeted.

Heavy metals are very reactive and have an **oligodynamic effect** on microbes. In other words, it takes very few molecules of a heavy metal to kill or be toxic to microbes. For example, a few drops of copper sulfate kill algae in a fish aquarium. Recent studies have shown that copper room surfaces in intensive care units (ICUs) kill 97% of hospital bacteria, reducing infections by 40%. Don't be surprised if hospital bed rails will soon be refitted with copper rails, sinks, and toilets (FIGURE 13.13)! The mechanisms of the killing action of chemicals used in disinfection methods varies from binding, denaturing, or oxidizing proteins; dissolving membranes causing them to leak; or reacting with nucleic acids. Essentially they are all "dirt cheap" inorganic molecules but we pay "top dollar" for their packaging!

FIGURE 13.13 Water running into a copper sink. The copper will inhibit bacteria from forming biofilms on sink surfaces. © iStockphoto/nkbimages.

Cleaning Products, Soap, and Handwashing

There is a vast array of disinfectants and antiseptics for hospital and general use to choose from (Figure 13.12) that differ by odor, color, and mechanism of dispersal (foam, spray, direct from bottle, diluted, not diluted). Disinfectant and antiseptics are defined by their properties. A **disinfectant** can be used on inanimate objects and surfaces but are too toxic to use on body tissues. **Antiseptics** can be used on body tissues (e.g., a wound). Both are able to kill or slow the growth of microbes. We pay high prices for these products but only a few different chemistries are involved. The labels of the products will advertise their effectiveness and scents. For example, a product containing n-alkyl and dimethyl benzyl ammonium chlorides contains a label that emphasizes it "kills 99.9% of germs in 60 seconds and powers through tough grease and grime!" Some products will advertise they have an original fresh lemon or citrus scent. The newest trend are natural cleaning products that are "chlorine free" but contain ingredients that are at least 95% naturally derived, e.g., citric acid, enzymes, boric acid, lactic acid, ethanol, fragrance with essential oils, calcium chloride, alkyl polyglucosides from plants, sodium lauryl sulfate, sodium gluconate, and others.

FIGURE 13.14 Employees are reminded to wash their hands after going to the bathroom at work. Author's photo (RIK).

Humans have been living with bacteria for centuries. In fact, there are ten times more bacterial cells than the number of human cells. Bacteria live inside us, on us, and all around us. A small percentage of bacteria can make us sick. Microbes strengthen our immune systems. If we are not adequately exposed to germs, our immune systems will atrophy such that the next time a pathogen is encountered, we will have an extreme reaction. Yet, extensive manufacturing and advertising has convinced consumers to buy soaps containing antibacterial ingredients such as triclosan or tetrasodium EDTA. Triclosan is found in toothpaste, mouthwashes, deodorants, hand soaps, and lotions. Experts believe that we have gone too far and continue to remind that good old-fashioned hand soap is sufficiently effective after using the bathroom (**FIGURE 13.14**). All that is needed is to get the microbes off your hands and down the drain. Killing them using antibacterial chemical smart-bombs is unnecessary. To date, *Escherichia coli*, *Mycobacterium*, and *Staphylococcus* triclosan-resistant strains have been reported, raising concerns that these pathogens may develop cross-resistance to other antimicrobial agents.

FIGURE 13.15 Technology gadgets may harbor pathogenic bacteria, including bacteria from human feces. Author's photo (TS).

Tech Gadgets and Germs

This is the age of technology gadgets. Most of us own at least a cell phone (**FIGURE 13.15**). "Mashable," an online news site that covers digital culture, social media, and technology, ran a story about "tech germs" on November 9th, 2011. It stated that keyboards are five times dirtier and have sixty times more microbes on them than toilet seats. The article includes an infographic created by "Keeping It Kleen," a website

formed by a group of individuals that provides resources and information to keep families safe by reducing the risk of foodborne illness. The infographic includes more enlightening statistics such as:

- 16% of cell phones were found to have fecal material on them.
- Light switches contain 217 bacteria per square inch.
- The TV remote is the single most hotbed of germs in a typical home, hotel room, or hospital.
- The Wii Fit Balance board attracts germs because many use it barefooted, a potential way to spread athlete's foot fungus.

Surely, a newsy note like this will inspire anyone to decontaminate his or her gadgets!

▨ Antibiotics

Antibiotics can rightly be considered the single most important discovery for the treatment of microbial-caused diseases in the history of medicine (BOX 13.2). They serve as testimony that "nature knows best." The secretion of metabolic products

---

BOX 13.2 __Discovery of Penicillin__

The story of the discovery of penicillin centers on the observations of Alexander Fleming. Others had previously described antibacterial properties of the *Penicillium* mold, but it was Fleming who followed through on a serendipitous (chance) observation. (Serendipity is a significant factor in several important scientific discoveries. Louis Pasteur wrote, "In the field of observation, chance favors only the prepared mind." In other words, it is not just sheer luck but rather the ability to recognize the significance of unexpected.) Fleming was studying staphylococci and left a Petri dish streaked with this organism on his lab bench while he went away on a two-week vacation. On returning, he noted that the plate was contaminated with a common (*Penicillium*) mold and that the staphylococci failed to grow only in the vicinity of the mold; the mold had produced an inhibitory substance—penicillin (FIGURE B13.2). Fleming was a humble person and later stated "Nature created penicillin. I only found it."

Penicillin's therapeutic potential was not fully investigated until several years after its discovery, when Ernst Chain, Howard Florey, Edward Abraham, and Norman Heatley purified penicillin and successfully cured mice that had been injected with fatal doses of bacteria. Human trials were initiated and proved to be very successful. World War II

triggered the large-scale production of penicillin, saving thousands of soldiers' lives. By 1944 supplies of penicillin were abundant and were released for civilian population. In 1945 Fleming was awarded the Nobel Prize in Physiology or Medicine, along with Chain and Florey, who helped develop penicillin into a widely available medical product.

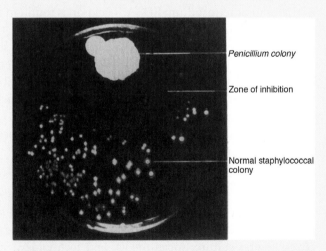

FIGURE B13.2   Alexander Fleming's Petri dish showing the inhibition of staphylococcal colonies in the immediate vicinity of a mold contaminated, which produced penicillin. © National Library of Medicine.

by soil bacteria and fungi that inhibit the growth of other microbes is an example of ecological antagonism at the microbial level.

History of Antibiotics

The first "wonder" drugs were the sulfonamide (sulfa) drugs, introduced in the preantibiotic era. These drugs, although they are antimicrobials, are not antibiotics, because they are synthetic compounds and not products of microbes. The sulfa drugs saved millions of lives in World War II; when sulfa drugs were not available, medics were taught to sprinkle sulfur on wounds sustained on the battlefields. An article written by Lewis Thomas, an esteemed physician, scientist, and author, appeared in the *Annals of Medicine* in 2003:

> *"Then came the explosive news of sulfanilamide, and the start of the real revolution in medicine. I remember the astonishment when the first cases of pneumococcal and streptococcal septicemia were treated in Boston in 1937. The phenomenon was almost beyond belief. There were moribund patients, who would surely have died without treatment, improving in their appearance within a matter of hours of being given the medicine and feeling entirely well within the next day or so."*

The antibiotic era was ushered in with the first use of penicillin. On February 12, 1941 Police Constable Robert Alexander of Oxford, England was the first person in the world to receive penicillin. Alexander was seriously ill with a staphylococcal infection that started with a small sore at the corner of his mouth. Despite treatment with sulfonamides, the staphylococci spread uncontrollably into his bloodstream, resulting in numerous abscesses over his body and the spread of infection to the rest of his face, eyes, and scalp, necessitating removal of his left eye. Death seemed imminent. Miraculously, 24 hours after receiving penicillin he was much improved: His lesions showed signs of healing, his elevated body temperature dropped toward normal, and within several days his right eye was almost normal. Unfortunately, the small amount of penicillin that was available was insufficient for continued treatment. In a heroic effort to save the patient's life, doctors extracted penicillin from his urine and injected it back into his bloodstream. But the microbes gained the upper hand, and Alexander's condition deteriorated. He died on March 15, 1941.

After this dramatic event, antibiotics—along with improvements in sanitation and hygiene, safer foods, cleaner water, and implementation of vaccines—began contributing to the decline in infectious diseases.

From about the mid-1940s through the 1960s several antibiotic-producing microbes were isolated from samples of soil. Streptomycin is a product of the soil bacteria *Streptomyces* discovered by Selman A. Waksman in 1943 and was the first antibiotic effective against tuberculosis. Waksman received the Nobel Prize in Physiology or Medicine in 1952 for his work. The "Waksman story" was and remains a controversial issue based on a lawsuit by Dr. Albert Schatz, at the time a graduate student in Waksman's laboratory, that he should be cited as co-founder; eventually, the case was settled out of court. There are numerous other stories involving graduate students and their professional mentors. The search for new antifungal and antibacterial compounds produced by soil microbes remains an active area of research by pharmaceutical companies continuing to dig and test soils from around the world. Many of the antibiotics are losing their punch as antibiotic resistance increases.

Penicillin became a prescription drug in the mid-1950s. Chemists learned how to manipulate and modify the penicillin molecule, giving rise to semisynthetic penicillin derivatives, including methicillin, ampicillin, and penicillin V, each with distinctive and beneficial properties. In the post-World War II period many other antibiotics were discovered, and their use brought about a rapid decline in deaths due to diseases caused by bacteria.

Types of Antibiotics

There is no such thing as a universal antibiotic any more than there is a universal disinfectant. Bacteria vary in their antibiotic susceptibility, and each antibiotic has a spectrum of activity against certain bacteria. Some antibiotics are more effective against gram-positive organisms, whereas others exhibit greater activity against gram-negative bacteria. A broad-spectrum antibiotic is inhibitory to a large variety of gram-positive and gram-negative bacteria, whereas a narrow-spectrum antibiotic is inhibitory to a limited range of bacteria. Some antibiotics are extremely effective but, unfortunately, exhibit marked toxicity, rendering them not useful. Some antibiotics are very expensive, whereas others are not, and some are more prone to result in antibiotic-resistant strains than others. In prescribing an antibiotic from among the many that are available, cost and antibiotic resistance are considered, but effectiveness and lack of toxicity are the central factors. Broad-spectrum antibiotics are generally prescribed when an individual is seriously ill and the causative bacteria have not been identified, so as to target a broad range of suspects. The downside of a broad-spectrum antibiotic is that a large number of bacterial species of the normal flora are killed, causing ecological disruption and allowing non–antibiotic-susceptible organisms to flourish. Some people receiving antibiotics develop oral thrush, a painful yeast infection, resulting from disruption of the normal bacterial flora and allowing yeasts to overgrow the normal flora. The tongue has a whitish appearance as a result of the colonies of yeast that have colonized it; fortunately, thrush responds well to mouth rinses with anti-yeast drugs. In females receiving a course of antibiotics, vaginal thrush may develop and cause pain, burning, and itching, treatable by vaginal rinses. Narrow-spectrum antibiotics cause less disruption in the ecological balance of microbes and also minimize the likelihood of antibiotic resistance.

Mechanisms of Antimicrobial Activity

Antibiotics act by interfering with or disrupting vital structures and metabolic pathways of the bacterial cell (FIGURE 13.16). Each antibiotic has a specific mechanism of action, although there is some overlap. For example, penicillin interferes with the ability of bacterial cells to synthesize cell walls, rendering these disabled cells subject to lysis, whereas erythromycin inhibits protein synthesis. Some antibiotics are **bactericidal** (they kill directly), whereas others are **bacteriostatic** (they keep the population from growing, thus allowing the body's defense mechanisms to get rid of the invaders). **Selective toxicity** is exhibited by penicillin because this drug interferes with cell wall production, and human cells do not have cell walls.

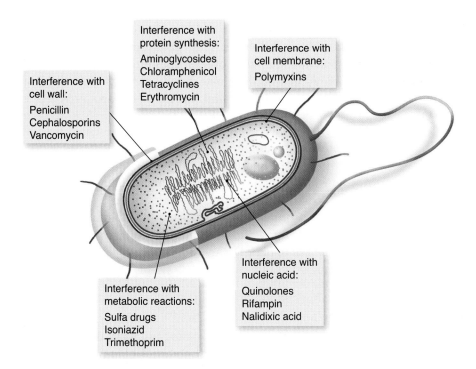

FIGURE 13.16 Mechanisms of antimicrobial activity.

Interference with protein synthesis:
Aminoglycosides
Chloramphenicol
Tetracyclines
Erythromycin

Interference with cell membrane:
Polymyxins

Interference with cell wall:
Penicillin
Cephalosporins
Vancomycin

Interference with nucleic acid:
Quinolones
Rifampin
Nalidixic acid

Interference with metabolic reactions:
Sulfa drugs
Isoniazid
Trimethoprim

On the other hand, amphotericin B is quite toxic because it acts on the bacterial cell membrane, a structure also present in human cells. The five mechanisms of action of antibiotics are described in the following sections.

Interference With Cell Wall Synthesis

Bacterial cells have rigid cell walls (peptidoglycan) that afford protection against lysis when bacteria are exposed to the low osmotic pressure of body fluids. The antibiotics penicillin and cephalosporins contain structures (beta-lactam rings) that interfere with enzymes responsible for cell wall synthesis. Vancomycin, sometimes considered the "last antibiotic stronghold," blocks a crucial reaction necessary for cell wall synthesis.

Interference With Protein Synthesis

Protein synthesis is an integral part of a cell's activity and is the culmination of expression of DNA. Bacterial ribosomes, cytoplasmic structures on which protein synthesis takes place, are targets for some antibiotics because they differ in size and structure from human ribosomes. (Bacterial cells are procaryotic and their ribosomes are 70S, whereas human cells are eucaryotic and their ribosomes are 80S. "S" represents Svedberg units, a measurement of sedimentation rates.) Streptomycin is a powerful antibiotic that was discovered in 1944; its use is now usually reserved for treatment of tuberculosis. The tetracyclines, chloramphenicol, and erythromycin are commonly used antibiotics that, like streptomycin, interfere with protein synthesis by binding with procaryotic ribosomes. Chloramphenicol is used for the treatment of typhoid fever and for other serious infections despite the fact that it can cause aplastic anemia, a potentially fatal condition in which the bone marrow ceases to produce red blood cells.

Interference With Cell Membrane Function

Cell membranes function in a vital capacity as "gatekeepers." Based on their chemical and physical structure, they control what goes into and out of the cell. Polymyxin B is an antibiotic that binds to and distorts the bacterial cell membrane, resulting in increased permeability and leakage of important molecules out of the cell.

Interference With Nucleic Acid Synthesis

The replication and synthesis of the nucleic acids, RNA and DNA, are steps in the expression of DNA, a long and complicated series of chemical reactions that can be targeted for antibiotic activity. Rifampin and nalidixic acid block RNA synthesis. Quinolones are a large family of synthetic drugs that act by inhibiting the action of an enzyme called DNA gyrase that is responsible for the supercoiling of bacterial DNA, enabling the cell to pack DNA. Mammalian cells use different enzymes for this activity and, hence, are not affected by this antibiotic—another example of selective toxicity.

Interference With Metabolic Activity

Metabolism, the ability to carry out energy-generating reactions, is a key characteristic in the distinction between life and nonlife. Antimetabolites are drugs that are structurally similar to natural compounds involved in metabolism and that competitively bind with these enzymes, rendering them inactive; this is called **molecular mimicry**. The sulfa drugs work in this fashion. They mimic folic acid, a component of the microbial cell, resulting in interference with cell multiplication. Mammalian cells do not make folic acid but obtain this compound from their diet; hence, sulfa drugs can be used in human therapy.

Acquisition of Antibiotic Resistance

The experience of the past fifty years has revealed that bacteria could be thought of as "smart" because of their development of mechanisms of resistance to the antibiotics designed to bring about their death. (You really can't blame them!) But they are not really smart; the development of antibiotic resistance is a manifestation of the Darwinian process of natural selection—"survival of the fittest"—resulting from the widespread misuse of antibiotics. The development of antibiotic resistance is a major international public health problem contributing to the threat of emerging infections and demands attention.

How do cells develop resistance to antibiotics? What are the biological factors involved? The development of antibiotic resistance is based on genetic changes. The bacterial cell, like all cells, has DNA, and some cells possess plasmids, extra bits of DNA independent from the DNA in the chromosome. Some plasmids are R (resistance) plasmids, which can be transferred from one cell to another; in this context they are infectious agents. Antibiotic resistance can result from mutations in the chromosomal DNA or plasmid DNA. Mutations in DNA occur spontaneously and randomly in populations of growing cells at a rate higher than 1 in 10 million; they are not caused by selective pressure but rather are selected for survival by selective pressure. Only the survivors multiply and, in so doing, pass on their new "survival genes" along with all the other genes. The outcome is that the next generation carries these new survival genes. This is the basis for Charles Darwin's concepts of survival of the fittest and of evolution applied at the microbial level.

As an analogy, consider a hypothetical population of trees in a geographical area that has suffered a drought for several years. Gradually, most of the trees die, except for a few survivors that are "lucky" enough to have random preexisting mutations in their genes that allow them to survive with less water. In the years to come the forest will be populated by drought-resistant trees. The drought did not cause the mutations but selected those trees with spontaneous, random, and pre-existing mutations that allowed the trees to survive—hence, survival of the fittest. In the same fashion, antibiotics do not cause mutations but select those preexisting mutations that confer antibiotic resistance; the genes with those mutations are passed on to successive generations during cell division.

The genes for antibiotic resistance result either from chromosomal mutations or from transfer of R plasmids from antibiotic-resistant strains to antibiotic-sensitive ones. Chromosomal mutations usually confer resistance to only a single antibiotic, whereas R plasmids can confer resistance to several antibiotics at one time, a phenomenon that was first reported in Japan in 1959 when lab personnel noted the emergence of *Shigella* bacteria that were resistant to several antibiotics. The origin of these R plasmids is not known.

Transposons, or "jumping genes," are another strategy of antibiotic resistance. They may carry genes for antibiotic resistance and can integrate into chromosomes or plasmids allowing rapid dissemination of antibiotic resistance.

AUTHOR'S NOTE (RIK)
You frequently hear the expression, "I'm resistant to antibiotics"; it is not you but your bacteria that display antibiotic resistance. You may be allergic, but that is a whole different story.

Mechanisms of Antibiotic Resistance

Antibiotic-resistant microbes counter the effects of the antibiotic (FIGURE 13.17) by different strategies. In some cases antibiotic resistance is the result of the production of enzymes that bring about inactivation of the antibiotic (e.g., penicillin or

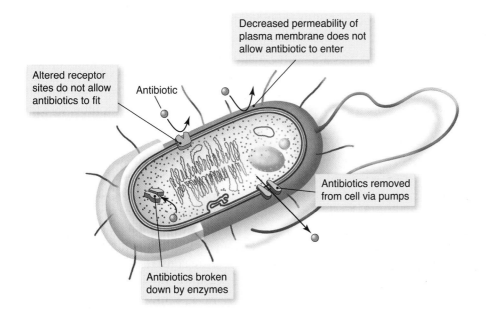

FIGURE 13.17 Microbes fight back.

Decreased permeability of plasma membrane does not allow antibiotic to enter

Altered receptor sites do not allow antibiotics to fit

Antibiotic

Antibiotics removed from cell via pumps

Antibiotics broken down by enzymes

a cephalosporin), rendering it ineffective. Some bacteria alter the uptake of the antibiotic; they possess pumps that actively transport antibiotics out of the cell. A variety of gram-positive and gram-negative bacteria are resistant to the antibiotic tetracycline by this mechanism. Drug uptake may be altered by a decrease in the permeability of the cell membrane to certain antibiotics. Another mechanism of countering the effects of antibiotic activity is by modification of the drug receptor site or of an essential metabolic pathway, preventing binding of the antibiotic. Penicillin resistance in streptococci and methicillin resistance in staphylococci are the result of alteration in the antibiotic receptor sites. Finally, as pointed out above, sulfonamides and some antibiotics act by interfering with essential metabolic pathways, but some bacteria develop an alternative pathway, rendering the antibiotic ineffective.

Antibiotic resistance was not caused by the misuse of antibiotics; the genes for antibiotic resistance were present long before. Pathogenic (and other) bacteria acquired these preexisting genes from other bacteria through **horizontal gene transfer**, a mechanism by which genes are passed from one mature bacterial cell to another. (Vertical gene transfer is the passage of genes from parent to offspring.) The widespread misuse of antibiotics has fostered the emergence of antibiotic-resistant strains.

It appears that evolutionary forces are constantly at play in humans' attempts to rein in bacteria; we fight bacteria with new drugs and they fight back by adaptation. Darwin was right. Nobel Laureate Joshua Lederberg stated, "Pitted against microbial genes we have mainly our wits."

Antibiotic Misuse

Antibiotic misuse is a global problem, but the meaning of antibiotic "misuse" reveals a paradox between developed and developing countries. In developing countries antibiotics are either unavailable or affordable to the majority of the population. For those fortunate enough to procure antibiotics, it is common for people to take only a few pills and to save the rest (FIGURE 13.18) for a later time. In developed countries, where antibiotics are readily available, "misuse" is due to antibiotics being too readily available. Further, in some countries antibiotics can be purchased over the counter in markets and pharmacies without a prescription. Millions of prescriptions for antibiotics are written around the world each year, of which many are unnecessary. Prescriptions are written for colds and flu, despite the fact that antibiotics are not effective against these viral illnesses. The justification is "just in case." Too

FIGURE 13.18 The paradox of antibiotic misuse.

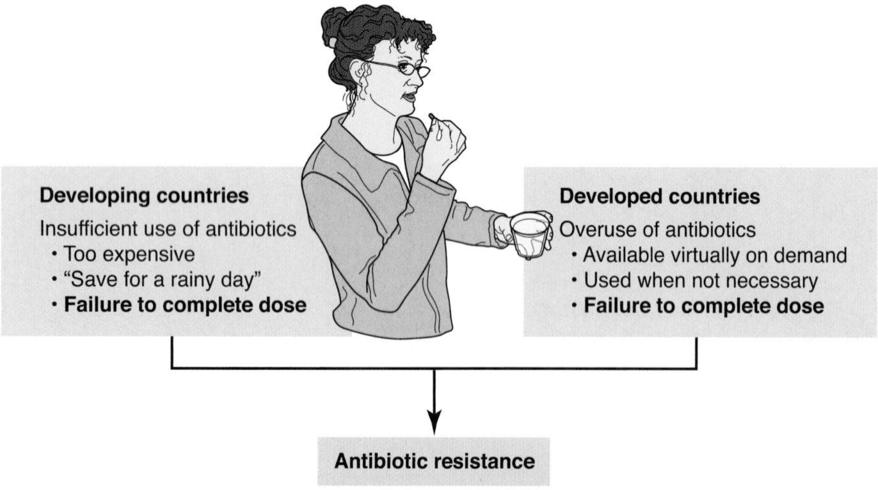

Developing countries
Insufficient use of antibiotics
- Too expensive
- "Save for a rainy day"
- **Failure to complete dose**

Developed countries
Overuse of antibiotics
- Available virtually on demand
- Used when not necessary
- **Failure to complete dose**

Antibiotic resistance

frequently, patients demand and receive antibiotics from their physicians even though antibiotics are not indicated; the particular disease is running its normal course, or the infection is caused by a virus. Studies have indicated that patients who walk out of their physician's office without a prescription for an antibiotic will often complain that their physician "billed them for nothing." A patient's lack of knowledge about antibiotics adds to the problem. The patient begins to feel better after a few days and, because he or she is not well informed, does not take the remaining doses. These practices favor the selection of antibiotic resistance, because only the more susceptible strains will be wiped out by only a few doses. The consequence of this misuse of antibiotics, many biologists and health professionals warn, is that we may be forced back to the pre-antibiotic era.

Here are a few alarming facts resulting from the misuse of antibiotics:

- Gonorrhea is increasingly more difficult to treat because of antibiotic resistance. In Hawaii, resistance to the quinolone antibiotics jumped from 1.4% in 1997 to 9.5% in 2000. The resistance of gonorrhea to a newer antibiotic, azithromycin, is on the increase.
- More than 90% of the strains of *Staphylococcus aureus* are resistant to penicillin and other antibiotics (BOX 13.3).

BOX 13.3 An Urgent Public Health Problem: Methicillin-Resistant *Staphylococcus aureus*

Have you heard about methicillin-resistant *Staphylococcus aureus*? Perhaps you are more familiar with its acronym, MRSA. According to the CDC MRSA infections accounted for nearly 10,000 life-threatening illnesses and close to 19,000 deaths in 2005. It kills more Americans each year than AIDS. The appearance of MRSA infections in schools prompted these schools to close and allow cleaning personnel armed with mops and buckets of disinfectant to march in and sanitize buses, classrooms, cafeterias, and gymnasiums.

MRSA is not an infection but refers to a property of the *Staphylococcus*, namely its resistance to the anti-staphylococcal drug methicillin and other antimicrobials, including oxacillin, penicillin and amoxicillin, and cephalosporins. Staphylococci are part of the normal flora of the skin and, for the most part, are not disease producers. *Staphylococcus aureus* is the species commonly involved in contact disease. About 50% of the population carries *S. aureus* on their skin, hair, and in their throat, and about 25% harbor this organism in their nose.

Natural selection, however, has enabled some strains of *S. aureus* to adapt to antibiotics. These strains can become dangerous when they penetrate the skin or mucous membranes and cause infections. *S. aureus* causes disease ranging from pimples and localized skin infections to life-threatening infections when it invades the blood and colonizes the internal organisms. It produces a variety of toxins that lead to serious damage.

In the past most cases were associated with hospitals and other care facilities. Disturbingly, many new cases are found in people who have no known exposure to these facilities; these cases are referred to as "community associated." Person-to-person exposure is the usual mode of transmittal through contact with infected skin lesions, nasal discharges, and contaminated hands or by contact with recently contaminated objects. The signs and symptoms of a MRSA infection include a pimple or cut that turns red, swollen, and purulent (pus-filled) with presence of yellow or white center or "head," draining pus, and fever.

- Ear infections (otitis media) are the second leading cause of office visits to physicians and account for over 40% of all outpatient antimicrobial use in children. An alarming number of bacterial strains that cause this condition are now antibiotic resistant.
- Resistance to vancomycin, once considered the "last stronghold," has been reported to occur in staphylococci and enterococci.
- Drug-resistant strains of the tuberculosis bacterium are increasing worldwide.
- Approximately half of all the antibiotics produced are used for disease control and promotion of growth in animals destined for the table, a practice that fosters the selection of antibiotic-resistant pathogens in animals and a potential risk of possible transmission to humans.
- The marked increase in domestic and international travel allows exposure to antibiotic-resistant pathogens that can be spread within a country and between countries. The antibiotic-resistant strains of the gonorrhea bacterium that originated in Africa and in Asia are now prevalent throughout the world.

Working Toward the Solution

The seriousness of antibiotic resistance as an impending global crisis has been established. What is the solution? The answer lies in the hands of physicians and patients, both of whom share the responsibility for the misuse and overuse of antibiotics resulting in the emergence of "superbugs," and therefore both are obligated to work toward the solution. Perhaps you unintentionally misuse antibiotics and demand that your physician prescribe an antibiotic when you are ill with what seems to be a cold. You may be guilty of not following instructions to take the full dose, because after a few days you feel better. In so doing you contribute to the emergence of antibiotic-resistant bacteria. Another too common scenario is that you do not feel well and pull some leftover antibiotic out of the medicine cabinet or ask your roommate if he or she has any antibiotics on hand without knowing the identity of the microbe. Physicians share in the responsibility for the antibiotic crisis and in its control; too often, they fail to spend the time explaining to patients why they do not need an antibiotic and succumb to patient pressure. Also, physicians fear being sued by a patient, claiming that his or her illness is a result of not being "put on an antibiotic."

Supermarket shelves are loaded with a tremendous variety of sprays, mists, and bubble-producing solutions, all designed to kill bacteria, viruses, molds, and mildew. Mattresses, cribs, playpens, and toys now boast that they contain antibacterial materials. Have we gone too far? The overuse of products containing antibacterials could eventually enable the evolution of antibiotic-resistant strains, akin to the situation that exists as a result of overexposure of microbes to antibiotics. You can fight back by frequently washing your hands with regular soaps rather than hand sanitizers. The judicious use of antibiotics can stave off and minimize an already impending antibiotic-resistance crisis, allowing society to continue to enjoy the benefit of Fleming's serendipitous observation of the antagonism between a mold and a bacterium.

Antiviral Agents

There are few effective nontoxic antiviral agents. A virus infiltrates the host cell, so trying to destroy the virus while keeping damage to the cell at a minimum poses a problem. To be effective, antiviral drugs must penetrate a cell and target a stage in the viral replication cycle to block the release of new viruses. Viral replication involves the steps of adsorption, penetration, replication, assembly, and release, offering a variety of targets for effective antiviral therapy (FIGURE 13.19). As pointed out, antibiotics are not effective against viruses because they lack the target components against which antibiotics are directed. In certain circumstances, the use of antibiotics for individuals with viral infections is justified when secondary bacterial infection is a potential threat. For example, senior citizens with influenza are at risk for bacterial pneumonia, a potentially fatal disease, and are frequently put on antibiotics as a preventive measure.

A number of antiviral agents are available, and research is ongoing to develop new ones (TABLE 13.7). At some point antiviral chemotherapy may parallel antibiotic chemotherapy. The AIDS pandemic and, more recently, influenza and the

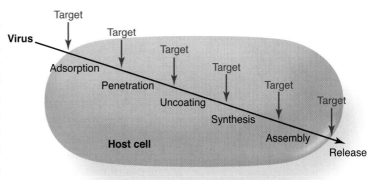

FIGURE 13.19 Mechanism of antiviral activity.

TABLE 13.7 Sampling of Antiviral Agents and Their Activity

Activity	Mechanism of Activity
Viral entry (to synthesis)	
Amantadine	Interferes with entrance of influenza virus by blocking fusion of virus with host cell
Fuzeon	Interferes with binding of HIV to host cell receptors
Tamiflu, Relenza	Interferes with influenza neuraminidase spikes necessary for entry of virus into host cell
Viral synthesis and replication	
Acyclovir	Interferes with viral DNA replication in herpes virus
Cidofovir	Interferes with RNA, DNA synthesis; used against cytomegalovirus
Interferons	Interferes with viral replication
Nevirapine	Interferes with binding site of reverse transcriptase
Zidovudine (AZT)	Interferes with DNA replication in HIV by targeting reverse transcriptase
Viral assembly, release	
Saquinavir	Interferes with action of HIV protease, resulting in noninfectious viruses

potential of an avian flu outbreak drive the search for new and improved antiviral agents.

In 1999 two new antiflu drugs, zanamivir (Relenza®) and oseltamivir (Tamiflu®), effective against influenza A and B viruses, were approved by the FDA and are currently recommended by the CDC to treat the flu. They do not prevent or cure the flu, but if taken early they decrease duration of the illness by a few days. This may not sound like much, but if you have ever had the misfortune of having the flu, a few days is a lot!

The drug zidovudine (also called azidothymidine, or AZT), an inhibitor of reverse transcriptase, and a group of **protease inhibitors** have achieved some success in AIDS therapy, but, not surprisingly, drug resistance is an emerging problem. A new drug and the first integrase inhibitor, raltegravir (Isentress®), was approved by the FDA in October 2007 to be used with other anti-HIV agents.

■ Overview

The twentieth century witnessed an increase in life expectancy in many nations of the world. U.S. residents live 33 years longer, on average, than they did in 1900. At least 25 of those gained years are attributable to public health achievements in sanitation, clean water, food safety, immunization, and antibiotics—the topics of this chapter.

During the 1900s departments of public health designed to promote health and longevity flourished in industrialized countries of the world. Implementation of the 1972 Clean Water Act and the 1974 Safe Drinking Water Act and advances in food safety along the complex path from farm to table were major steps forward. The availability of penicillin, followed by other antibiotics, dramatically improved the treatment of microbial diseases.

Industrialized countries of the world share in these successes. The sad part is the tremendous life span and quality of life disparity that exists between the "haves" and the "have-nots"—a highly significant world public health problem. The disparity is so great that the average life span in Macau is 84, whereas in Swaziland the life span is only 32 years, a 52-year difference. Poverty is at the root and leads to lack of clean water, inadequate sanitation, and a host of other public health deficiencies. Programs like the Kampung Improvement Program in Indonesia and the Orangi Pilot Program in Pakistan are underway, but progress is slow and disappointing.

More than half of Americans perceive foodborne illness from bacteria as the most important food safety issue today. President Obama signed the FDA Food Safety Modernization Act on January 11, 2011. It is the largest reform of food safety in more than 70 years. The law will be fully implemented over time. It calls for rules and regulations related to the design, production, labeling, promotion, manufacturing, and testing of regulated food products.

The connection between germs and disease was validated in the late nineteenth century and has led to important innovations such as antibiotics, new antibacterial products, and hygiene practices. Society is becoming germ phobic. Our world is becoming sterilized, sanitized, and "germ-free."

Self-Evaluation

PART I: Choose the single best answer.

1. The relationship between safe drinking water and child survival is
   a. inverse **b.** direct **c.** country related **d.** without correlation

2. The Kampung Improvement Program in Indonesia focused on
   a. immunization programs **b.** providing antibiotics **c.** improving sanitation **d.** food safety

3. All the following are antibiotics with the exception of
   a. chloramphenicol **b.** penicillin **c.** Relenza® **d.** erythromycin

4. Less than _____% of the Earth's water can be used and reused as freshwater.
   a. 2.5 **b.** 10 **c.** 25 **d.** 50

5. On a typical day, more than half of the hospital beds in sub-Saharan Africa are occupied by patients suffering from _____ diseases.
   a. respiratory **b.** fecal-related **c.** liver **d.** kidney

6. Disinfectants cannot be used on which of the following?
   a. countertops in the kitchen **b.** human skin **c.** toilet **d.** sinks

7. Which of the following have oligodynamic action?
   a. Alcohols **b.** Heavy metals **c.** Halogens **d.** QUATS

8. Which of the following technology gadgets likely has the most germs on it?
   a. cell phone **b.** TV remote **c.** laptop keyboard **d.** light switch

9. Which of these methods is used to reduce the number of pathogen microbes in milk?
   a. Sterilization **b.** Disinfection **c.** Fumigation **d.** Pasteurization

10. The most common single food product associated with foodborne outbreaks according to CDC data (2003 to 2008) is
    a. dairy **b.** fruits and nuts **c.** leafy greens **d.** poultry

PART II: Fill in the blank.

1. The mechanism of activity of penicillin is _____.

2. _____ is commonly used as an indicator of clean water.

3. Penicillin is not toxic for humans because it interrupts bacterial _____.

4. _____ can be used on skin before drawing blood.

5. _____ uses steam heat to penetrate and kill microbes, spores, and viruses.

6. _____ are the hardest infectious agents to destroy or inactivate.

7. _____ has become a serious contaminant of well water in India and groundwater in the United States.

8. _____ discovered that *Penicillium* produces an antibiotic compound.

9. _____ is a cleaning disinfectant used in hospitals.

10. _____ is/are an example of a drug that has a molecular mimicry mechanism of action.

PART III: Answer the following.

1. Sanitation was "in" during the twentieth century. Explain (give examples of) this statement.

2. Describe the work of local health departments as partners in microbial disease control.

3. From a Darwinian point of view, describe the emergence of antibiotic-resistant strains.

4. Discuss why experts believe that old-fashioned soap and water is a wiser choice than using antibacterial soaps to wash your hands?

5. List six cleaning products and their active ingredients.

6. Explain why the twentieth century witnessed an increase in life expectancy in many nations of the world.

7. List three or four microbes or infectious agents that are difficult to inactivate with disinfectants. What characteristics make them resistant to disinfection?

8. Explain at least two different mechanisms of antibiotics used to treat bacterial infections.

9. Why are hygienic means of sanitation and a safe supply of drinking water basic human needs?

10. Over one billion people still defecate in the open worldwide. Discuss why this is happening and what steps would you take to remedy this?

Partnerships in the Control of Infectious Diseases

I am my brother, and my brother is me.

—Ralph Waldo Emerson

Preview

Enormous strides have been made in the twentieth century in decreasing the burden of infectious diseases worldwide, particularly in developed countries. These successes are the result of collaborative partnerships in pooling funds, talents, and resources toward common goals. Continued progress depends on partnerships that range from local to national and international levels and require cooperation between the public and private sectors. The disparity in health care between more developed and less-developed countries requires the implementation of strategies to ensure the poorer nations of the world receive their fair share of the benefits of progress.

Background

"The attainment for all people of the world by the year 2000 of a level of health that will permit them to lead a socially and economically productive life" was the stated goal that emerged from the International Conference on Primary Health

Photo courtesy of Dr. Edwin P. Ewing, Jr./CDC.

411

Care held in Alma-Ata, USSR (now Almaty, Kazakhstan) from September 6 to 12, 1978. Further it was declared:

> *"The Conference strongly reaffirms that health, which is a state of complete physical, mental, and social wellbeing, and not merely the absence of disease and infirmity, is a fundamental human right and that the attainment of the highest possible level of health is a most important world-wide social goal whose realization requires the action of many other social and economic sectors in addition to the health sector."*

Three decades have passed since the Alma-Ata conference. Where do we stand? Has there been progress toward the attainment of these ambitious goals? Life expectancy has dramatically increased since the beginning of the twentieth century, and Alma-Ata and other international alliances have contributed to these gains. The World Health Assembly adopted the slogan "Health for all by the year 2000." That year has come and gone, and the goal has yet to be realized. Nevertheless, progress has been made, partially as a result of Alma-Ata and its emphasis on primary health care. Health for all is an elusive and moving target and may not be realistic, but striving toward it can only have positive consequences. The achievements of public health and advances in medical science have not been equally distributed across the board, presenting an ongoing challenge.

The attainment of health in all its dimensions, including reducing the burden of infectious diseases, is a matter of public health concern at several levels ranging from the individual to the community, to state departments of health, to national agencies, and to international organizations. Barry Bloom, former dean of the Harvard University School of Public Health, stated, "One of the myths of the modern world is that health is determined largely by individual choice and is therefore a matter of individual responsibility." Realistically, individuals are limited in their efforts to achieve and maintain good health, necessitating public health strategies aimed at populations to reduce the burden of infectious and other diseases.

The eradication of infectious diseases has been a goal since the establishment of Koch's postulates and the germ theory of disease. Smallpox was the first microbial disease in humans to be eradicated. Rinderpest, also known as "cattle plague," is the second disease to be wiped from the world as declared by the United Nations (UN) in June 2011. Thomas Jefferson, a few years after the introduction of smallpox vaccination in 1796, commented, "One evil more [smallpox] is withdrawn from the condition of man." In 1892 a contagious pleuropneumonia of cattle (which was imported into the United States in 1847) was declared eradicated from the country as the result of a five-year, $2 million campaign to identify and slaughter infected animals. The Rockefeller Foundation ambitiously campaigned to eradicate yellow fever and hookworm disease in the early years of the twentieth century but failed because of the complexity of eradication programs. Malaria eradication seemed plausible in the period from 1955 to 1965 but was unsuccessful, primarily because of the emergence of drug-resistant malaria parasites and mosquitoes.

The International Task Force for Disease Eradication (ITFDE) convened for the eighteenth time since 1989 at the Carter Center in April 2011. The task force focused their assessment on the elimination strategies for two worm diseases—onchocerciasis and lymphatic filariasis in Africa. The ITFDE defines elimination strategy as the "reduction of infection and transmission to the extent that interventions can be stopped but post-surveillance is still necessary." Further, elimination strategies generally refer to a limited geographical area (a single country or continent), whereas eradication is used in a global sense.

The eradication of smallpox in 1980 stands as a public health triumph of the twentieth century. This generation, and those who follow, will never know the horrors of this disease, other than through photographs (FIGURE 14.1). Several characteristics of smallpox and the smallpox virus (variola) were unique leading to eradication and its establishment as the criterion by which to evaluate other diseases as targets for eradication:

- It is a disease only of humans; there are no natural reservoirs or biological vectors.
- The infection is easily diagnosed because of a characteristic rash.
- The duration and intensity of infectiousness is limited.
- Recovery establishes permanent immunity.
- A safe, effective, inexpensive, easily administered, stable (even in tropical climates), one-dose vaccine is available.
- Vaccination confers long-lasting immunity.
- Vaccination usually results in a permanent and recognizable scar, allowing for detection of immune versus nonimmune individuals in a population.

The degree to which other diseases mimic smallpox reflects their potential for eradication, but these are not absolute criteria (BOX 14.1). For example, a biological vector is a part of the guinea worm life cycle, polio immunization requires four doses, and neither disease produces visible early manifestations. Despite these considerations, both diseases are on the "hot list" for eradication, and considerable

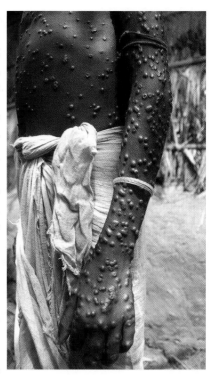

FIGURE 14.1 Smallpox: a disease of the past. The disease is characterized by the appearance of pustules on the body. The use of smallpox by terrorists is a potential worldwide threat.
Courtesy of Jean Roy/CDC.

BOX 14.1 ___Criteria for Assessing Eradicability of Diseases and Conditions___

Scientific Feasibility

- Epidemiological vulnerability (e.g., existence of a nonhuman reservoir, ease of spread, natural cyclical decline in prevalence, naturally induced immunity, ease of diagnosis, and duration of any relapse potential)
- Effective, safe, long-lasting, easily deployed intervention available (e.g., a vaccine or other primary preventive, a curative treatment, and a means of eliminating the vector)
- Demonstrated feasibility of elimination (e.g., documented elimination from an island or other geographical unit)

Political Will and Popular Support

- Perceived burden of the disease (e.g., extent, deaths, or other effects; true burden may not be perceived; the reverse of benefits expected to accrue from eradication; relevance to rich and poor countries)
- Expected cost of eradication (especially in relation to perceived burden from the disease)

Source: CDC *Morbidity and Mortality Weekly Report* 1993; 42(RR-16):1–25.

TABLE 14.1 Mortality of Selected Infectious Diseases	
Cause of Death	**Annual Deaths**
Lower respiratory infections	4.1 million
Diarrheal diseases	2.1 million
HIV/AIDS	2 million
Tuberculosis	1.4 million
Malaria	889,185
Measles	423,710
Hepatitis C	350,000
Sexually transmitted diseases	128,472
Rabies	55,000
Polio	1,195

Adapted from Schlipkoter, U., Flahault, A. 2010. Communicable diseases: achievements and challenges for public health. *Public Health Reviews* 32:90–119.

progress has been made toward that achievement. The last case of wild polio in the Americas occurred in Peru in 1991. Not all diseases reviewed by the ITFDE are considered candidates for eradication highlighting the complexity of eradication programs. Infectious diseases are the leading causes of death in many developing nations (TABLE 14.1).

Partnerships in Infectious Disease Control

Reducing the incidence of infectious diseases is a tremendous challenge. The successes to date are largely the result of collaborative partnerships involving the sharing of funds, talents, and resources toward common goals. Continued successes in public health will depend on partnerships within and between the public and private sectors. Numerous agencies have been cited for their leadership; this chapter describes some of them.

Today's crowded societies and sharing of resources are far removed from hunter–gatherer societies, where family units were relatively isolated and depended only on their own efforts and ingenuity to stay alive. As populations grew and urbanization developed, a sharing of community responsibilities emerged in all aspects of life, including health. A negative aspect of all this "togetherness" is that the sharing of pathogens also increased. Individuals were limited in measures that could be taken to minimize exposure to pathogens. In a collective effort, the early 1900s saw the establishment of community and state departments of health that evolved into a complex network from the state level to the national and international levels and to partnerships in the public and private sectors.

At the Local Level

Community, City, and State Health Departments
Every state has a department of health, together with subordinate health departments at the community and city levels. Although the organizational charts vary from state to state, they are to a large extent a reflection of the size of the population covered. Health departments focus on the prevention of disease and the

promotion of health and safety of the people within their jurisdiction. They watch for the hazards of foodborne and waterborne diseases, including bacterial, viral, protozoan, and helminthic (worm) diseases. A major responsibility of health departments is to establish and implement safety regulations pertaining to food and water sanitation. For example, there is rigorous control of the milk industry in each state involving the health of dairy cows and conditions "down" at the dairy farm. Every dairy must submit milk and milk products on a strict schedule to state departments of public health laboratories to ensure the safety of the product before delivery to markets.

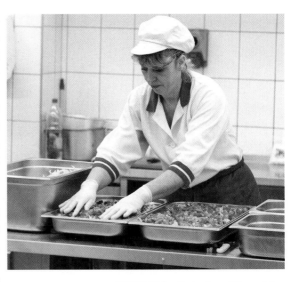

FIGURE 14.2 Protection of the public. Local departments of health require food handlers to wear disposable gloves. © Sebastian Czapnik/Dreamstime.com.

Public health restaurant inspectors make spot visits to eating establishments to ensure adherence to proper temperatures for the cooking and storing of foods; the absence of mice, rats, roaches, and other vermin; and the practice of appropriate measures of sanitation and hygiene by food workers. Food workers are also required to submit stool specimens to avoid a repeat of "Typhoid Mary" (**FIGURE 14.2**). Salad bars must have shields to protect the food from coughs and sneezes. Increasingly, food workers must use disposable gloves.

Some communities, in addition to imposing fines and closing noncompliant establishments, make restaurant inspection reports available to the public on a website and have implemented other methods to alert the public. Los Angeles requires restaurants to post inspection grades in the windows, and the local newspapers in central Florida and other areas publish the detailed results of restaurant inspections and the fines imposed on those failing to achieve a clean bill of health. In some cases the results are hard to swallow (pun intended). Violations include potentially hazardous, uncooked food held at unsafe temperatures; food handlers preparing foods with their bare hands; raw chicken stored over other raw meat in the cooler; foods like feta cheese, ranch dressing, and cheddar cheese kept at improper temperatures; and roaches crawling over counters and food. Additionally, some communities require that food workers, in those restaurants failing to pass inspection, attend a hospitality education program.

FIGURE 14.3 A reporting form that is filled out in the event of a foodborne illness and sent to a state health department. Courtesy of Massachusetts Department of Public Health.

Massachusetts Department of Public Health
Foodborne Illness Complaint Worksheet

Date: ___/___/___
#: ______

Please Complete and Send or Fax to:
MDPH Food Protection Program
305 South Street, Jamaica Plain, MA 02130
Fax: (617) 983-6770

Questions? Call:
Food Protection Program: (617) 983-6712
Division of Epidemiology: (617) 983-6800
Enterics Laboratory: (617) 983-6609

Person Completing Information
Name: ______ ☎: ( ) ___-___
Affiliation: ☐ Local BOH *(town)*: ______ ☐ State DPH *(division)*: ______ ☐ Other: ______

Reporter/Complainant
Name: ______ ☎: ( ) ___-___
Affiliation: ☐ Consumer ☐ Laboratory ☐ Local BOH ☐ Medical Provider ☐ State DPH ☐ Other
specify: division, *facility*, address, *town, etc.*

Illness Information
Persons ill: ☐ Symptoms: *(mark if reported for anyone)*:
☐ Diarrhea ☐ Vomiting ☐ Nausea ☐ Abdominal cramps
☐ Fever ☐ Bloody stool ☐ Headache ☐ Muscle aches
☐ Chills ☐ Loss of appetite ☐ Fatigue ☐ Dizziness
☐ Burning in mouth ☐ Other symptoms: ______
Onset: Earliest Date: ___/___/___ Time: ___:___ ☐AM ☐PM
Latest (if > 2 ill) Date: ___/___/___ Time: ___:___ ☐AM ☐PM
Duration: ☐ Less than 24 Hours ☐ 24-48 Hours ☐ More than 48 Hours ☐ Ongoing ☐ Unknown
Ill Persons:

	Name	Address/Town	☎	Age (yrs)	Occupation	Med. Provider/ ☎
1	☐ same as reporter (above)					
2						
3						
4						

Medical attention received *(by anyone)*? ☐ Yes ☐ No ☐ Unknown *If Yes, specify above:*
Stool specimens submitted *(by anyone)*? ☐ Yes ☐ No ☐ Unknown To SLI [1]? ☐ Yes ☐ No ☐ Unknown
Medical diagnosis reported?

[1] State Laboratory Institute, 305 South St., Jamaica Plain, MA, 02130: 617-522-3700 Sept 2005
[2] Always record Time if possible; otherwise, choose B=breakfast, L=lunch, D=dinner [3] Total # persons (both ill and well) who consumed indicated food(s)

It is not uncommon for foodborne infections to hit college campuses. Students suffer from the usual and unpleasant symptoms of gastroenteritis. Rotaviruses, noroviruses, and *Escherichia coli* are frequently identified as the causative pathogens. Each incident and its containment illustrate the epidemiological detective work necessary for tracking down the source and implementing preventive measures for the future. Investigation of such outbreaks requires partnerships at the college, community, state, and, in some cases, the national and international levels. State health departments are required to notify the Centers for Disease Control (CDC) of specific diseases so that a national network of surveillance and communication can be maintained (**FIGURE 14.3**).

Surveillance and control of infectious diseases are major functions of local health departments to which they respond with control strategies, including vaccination. Vaccine campaigns and implementation of

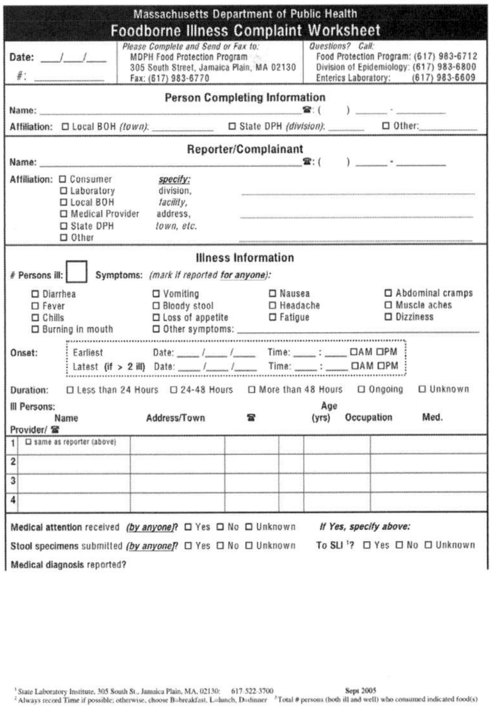

FIGURE 14.4 The CDC headquarters in Atlanta, Georgia. The CDC conducts infectious disease surveillance and works in conjunction with state and local health departments and with the WHO and other agencies. Courtesy of James Gathany/CDC.

regulations requiring immunization against infectious diseases are another function of state health departments.

Local departments of health are involved in numerous other endeavors to foster the health and welfare of citizens, including toxicology, vital statistics, public awareness health programs, prevention and treatment of drug abuse, cancer surveillance, and lead paint screening.

■ At the National Level

Centers for Disease Control and Prevention (CDC)

The CDC based in Atlanta, Georgia, is the nation's premier public health facility, and its impact is global (FIGURE 14.4). Its functions are to:

- Detect and investigate health problems.
- Conduct research to enhance prevention.
- Develop and advocate sound public health policies.
- Implement prevention strategies.
- Promote healthy behaviors.
- Foster safe and healthful environments.
- Provide leadership and training.

The agency was founded in 1946 and employs nearly 15,000 people including approximately 840 commissioned corps officers) in 170 occupations. The CDC is a key member of partnerships with local health departments and with national and international organizations (BOX 14.2).

The CDC is the nation's main line of defense against threatening epidemics and plague. The CDC's research labs are in some ways like a large microbial zoo; within its locked freezers are every known microbe on Earth (including smallpox

---

BOX 14.2 Partners in Infectious Disease Control

Local Level
  Community Departments of Health
  City Departments of Health
  State Departments of Health
National Level
  Centers for Disease Control and Prevention
  Department of Homeland Security
  U.S. Public Health Service
  Federal Emergency Management Agency
  National Institutes of Health
  American Red Cross
International Level
  World Health Organization
  Pan American Health Organization

United Nations Foundation
United Nations Children's Foundation
Rotary International
Rockefeller Foundation for the 21st Century
Doctors Without Borders
World Bank
The Private Sector
  Pharmaceutical Companies
  The Bill and Melinda Gates Foundation
  The William J. Clinton Foundation

virus) caged in small vials under the watchful eye of a microbe keeper or in the live bodies of rabbits, mice, rats, and monkeys. Locked corridors house the "deadliest of the deadly" viruses, including those that cause HIV, rabies, Ebola hemorrhagic fever, and hantavirus pulmonary syndrome.

The CDC has come to the rescue on numerous occasions around the world helping to control or investigate incidents including the *Legionella* outbreak in Philadelphia in 1976, the Ebola hemorrhagic fever outbreaks in the Democratic Republic of the Congo in 1995 and in 2003, the SARS epidemic in China in 2002, the Rift Valley fever outbreak in Kenya in 2006, and the aftermath of the earthquake in Haiti in January 2010. The CDC fields about 1,000 calls for help each year. As deemed necessary, its "SWAT" teams of epidemiologists head into the fields, frequently in collaboration with partnership agencies, equipped with ready-to-go containers stocked with syringes, needles, vaccines, intravenous fluids, examination gloves, refrigerators to store samples of blood and other tissues, generators, stacks of questionnaires, and other items to conduct guerrilla warfare against the microbes. Intriguing and heroic tales of their battles against an invisible enemy have been recounted in the movies *Contagion* and *Outbreak,* in the books *The Coming Plague* and *The Hot Zone,* and in the television series *The Walking Dead.*

Frequently, these "disease detectives" must ship tissue samples from sick and dead victims to the CDC labs in Atlanta for identification. The labs are designed to work with deadly microbes, and their locked doors bear large biohazard signs (**FIGURE 14.5**). The CDC's Building Fifteen is the "hot zone"—a biosafety level four facility—prepared to handle the deadliest of microbes, including Ebola and hantaviruses, for which there is neither cure nor vaccine (**FIGURE 14.6**). Air leaving Building Fifteen is passed through a series of filters, water is boiled before entering sewer lines, and the fortress-like building has a camera trained on its one entrance. Researchers strip naked, shower, and don biohazard suits, known as orange suits, and as many as three pairs of gloves before entering the lab; air is supplied to them through a tube attached to the orange suit. Their protection against the deadliest of pathogens is limited to a layer of fabric, which can be penetrated by the jagged glass of a broken

FIGURE 14.6 Level four biocontainment. Ebola virus, hantaviruses, and other microbes are potentially deadly and must be handled in specially designed environments to protect laboratory personnel and the environment from the risks of contamination. Here, a lab worker in a protective suit showers inside a decontamination booth after handling pathogenic viruses. Courtesy of James Gathany/CDC.

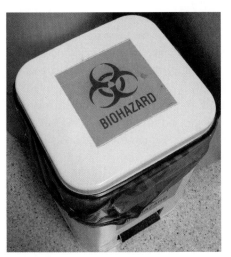

FIGURE 14.5 The biohazard symbol. Potentially infectious materials from hospitals, laboratories, and other facilities are disposed of in clearly marked biohazard containers. © joe outland/Alamy Images.

test tube or by a syringe needle. As much as possible, researchers work in pairs and monitor each other for fatigue and for tears in their gloves and suits.

In its role as the nation's watchdog, the CDC has developed a four-point strategy for the twenty-first century to counter the threat of new, emerging, and reemerging infections. These goals are elucidated in BOX 14.3.

Do you want to go on a cruise to some far off destination? If so, you would be well advised to consult the "Green Sheet," CDC's *Summary of Sanitation Inspections of International Cruise Ships*. The Vessel Sanitation Program was established in the early 1970s to minimize the risk of diarrheal diseases among passengers. All vessels with a foreign itinerary that carry more than thirteen passengers and call on U.S. ports are subject to unannounced twice-yearly inspections by Vessel Sanitation Program staff and to reinspections when necessary. To pass, the ship must score a minimum of 86 points on a 100-point scale. Further, the general cleanliness, personal hygiene, and physical condition of the crew, along with training programs in environmental and public health practices, are evaluated.

BOX 14.3   CDC Framework for Preventing Infectious Diseases: Sustaining the Essentials and Innovating for the Future

The CDC's roadmap for improving our ability to prevent known infection diseases and to recognize and control rare, highly dangerous, and newly emerging threats has three elements:

Element I: Strengthen public health fundamentals, including infectious disease surveillance, laboratory detection, and epidemiological investigation.

- Modernize infectious disease surveillance to drive public health action.
- Expand the role of public health and clinical laboratories in disease control and prevention.
- Advance workforce development and training to sustain and strengthen public health practice.

Element II: Identify and implement high-impact public interventions to reduce infectious diseases.

- Identify and validate high-impact tools for disease reduction, including new vaccines, strategies and tools for infection control and treatment, and interventions to reduce disease transmitted by animals or insects.
- Use proven tools and interventions to reduce high-burden infectious diseases, including vaccine-

preventable diseases, healthcare associated infections, HIV/AIDS, foodborne infections, and chronic viral hepatitis.

Element III: Develop and advance policies to prevent, detect, and control infectious diseases.

- Ensure the availability of sound scientific data to support the development of evidence-based and cost-effective policies.
- Advance policies to improve prevention, detection, and control of infectious diseases to help integrate clinical infectious disease practices into U.S. health care: (1) increase community and individual engagement in disease prevention efforts; (2) strengthen global capacity to detect and respond to outbreaks with potential to cross borders; (3) address microbial drug resistance; and (4) promote "One Health" approaches to prevent emergence and spread of zoonotic diseases.

Adapted from *A CDC Framework for Preventing Infectious Diseases: Sustaining the Essentials and Innovating for the Future*. CDC, October, 2011. http://www.cdc.gov/oid/docs/ID-Framework.pdf.

The cruise ship industry has been subject to waves (pun intended) of criticism because of outbreaks caused by noroviruses among passengers and staff. Some of the cruises and their ships have been dubbed "ships from hell" by passengers as they experienced their dream vacations explode into nightmares. The midnight buffets (eating orgies) and advertised "eat all you want" snacks and meals were vomited over the ship's side. In June 2010, guests aboard *Sapphire*, a jewel in the Princess Cruise lines fleet, suffered just such an experience as did those on deck on Britain's Grand Princess ship on holiday in 2007; in both cases many were tempted to "walk the plank."

Which shots do you need before traveling to the Amazon, Mozambique, or Tahiti? These exotic foreign destinations are potential sources of disease that can make you very ill and even kill you. Before embarking, you can consult the CDC's website (http://www.cdc.gov/travel) for up-to-date traveler's health information, including vaccinations, availability of safe food and water, and disease outbreaks in the land of your dreams.

Department of Homeland Security

The Department of Homeland Security was proposed by President George W. Bush in June 2002 in response to the September 11, 2001, terrorist attacks on the United States. The Department replaced the earlier office of Homeland Security and entailed a major reorganization of government agencies. The Department of Homeland Security's mission, as applicable to the use of biological weapons (public health), is to:

- Reduce the vulnerability of the United States to terrorism.
- Minimize the damage and assist in the recovery from terrorist attacks that do occur within the United States.

U.S. Public Health Service

The Public Health Service (PHS), headed by the U.S. Surgeon General, consists of a 6,000-member Commissioned Corps and support staff and is a component of the U.S. Department of Health and Human Services (Box 14.2). It originated as the Marine Hospital Service in 1889; in 1912 the name was changed to the PHS. Initially, the PHS focused on sailors and their medical care in an attempt to alleviate the burden on public hospitals in caring for merchant seamen. In 1891 the service was charged with being the nation's medical gatekeeper by providing medical inspection of arriving immigrants to weed out "idiots, insane persons, persons likely to become a public charge, and persons suffering from a loathsome or a dangerous contagious disease." The commissioner general of immunization clearly stated in 1902 that America should not become "the hospital of the nations of the earth." Immigrants were screened by PHS officers at Ellis Island for "germ diseases," including cholera, typhus, plague, smallpox, yellow fever, and trachoma (blindness). Ellis Island was dubbed "The Island of Hope" and also "The Island of Tears"—"hope" for the new and promising way of life for those who made it through the inspection line, but "tears" for those who were separated from their families and returned to their place of origin.

The PHS has taken a leading role, particularly during wars, in campaigns against sexually transmitted diseases (formerly called venereal diseases), in immunization campaigns, in vector control programs, and in public health awareness programs (FIGURE 14.7).

The mission of the PHS is to improve the health of every individual by conducting research, engineering systems for safe delivery of water and disposal of waste, overseeing food and drugs, studying and developing means to contain or eliminate disease, and promoting a safe and healthful environment at work or home. As America's uniformed service of public health professionals, the Commissioned Corps achieves this mission through:

- Rapid and effective response to public health needs
- Leadership and excellence in public health practices
- Advancement of public health science

Federal Emergency Management Agency

The Federal Emergency Management Agency (FEMA) was created in 1979 by President Carter as an independent agency of the federal government that reports directly to the president. Its slogan is "Helping People, before, during, and after disaster." Natural disasters such as hurricanes, floods, and tornadoes can destroy the public health infrastructure, leading to polluted waters and compromised sanitation followed by disease. It is common to hear in such instances that the president has declared a community to be eligible for FEMA funds and assistance in coping with the disaster. FEMA works in partnership with local, national, and international agencies in performing its role (Box 14.2).

Hurricane Katrina presented a major challenge to FEMA in August 2005; Katrina is considered to be the largest natural disaster in the United States. The agency came under severe criticism because of a delayed response time, but, to its credit, has since implemented major policy changes and has led the way in the nation's emergency response strategies. FEMA responded to a crippling snowstorm in Buffalo, New York in the winter of 2006, to a devastating tornado that hit Dumas, Arkansas in February 2007, to wildfires in California in October 2007, and to a catastrophic EF5 multiple-vortex tornado that struck Joplin, Missouri in May 2011.

U.S Food and Drug Administration

The U.S. Food and Drug Administration (FDA) is part of the PHS. It focuses on the safety of foods, vaccines, antibiotics, other medicinal (biological) products, and medical devices, all of which play a role in the prevention and control of infectious disease. This agency has the last word before these products are approved for release. The safety of the nation's blood supply system is under the umbrella of the FDA; their inspectors routinely test blood and blood products for contamination. The agency has the authority to direct withdrawal of products, either voluntarily or legally, as in the withdrawal of the RotaShield® vaccine against gastroenteritis shortly after its approval. Products not directly related to infectious diseases also fall under the scrutiny of the FDA. The agency is charged with protecting the consumer by enforcing the Federal Food, Drug, and Cosmetic Act and related public health laws, including the truth in labeling laws.

DON'T VISIT HOUSES OF ILL-FAME

If in doubt call at the nearest Hospital for free and confidential advice

SAVE YOURSELF & YOUR FAMILY FROM V.D.

(a)

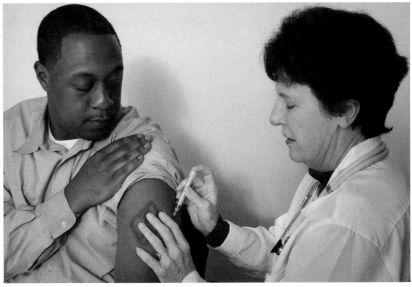

(b)

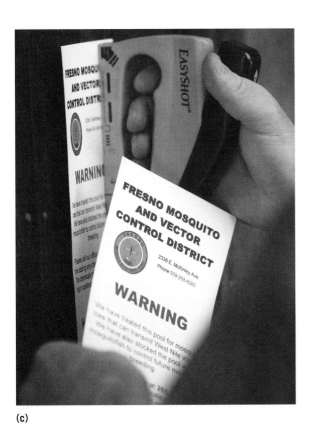

(c)

Understanding AIDS

A Message From The Surgeon General

This brochure has been sent to you by the Government of the United States. In preparing it, we have consulted with the top health experts in the country.

I feel it is important that you have the best information now available for fighting the AIDS virus, a health problem that the President has called "Public Enemy Number One."

Stopping AIDS is up to you, your family and your loved ones.

Some of the issues involved in this brochure may not be things you are used to discussing openly. I can easily understand that. But now you must discuss them. We all must know about AIDS. Read this brochure and talk about it with those you love. Get involved. Many schools, churches, synagogues, and community groups offer AIDS education activities.

I encourage you to practice responsible behavior based on understanding and strong personal values. This is what you can do to stop AIDS.

C. Everett Koop, M.D., Sc.D.
Surgeon General

Este folleto sobre el SIDA se publica en Español. Para solicitar una copia, llame al 1-800-344-SIDA.

U.S. Department of Health
& Human Services
Public Health Service
Centers for Disease Control
P.O. Box 6003
Rockville, MD 20850

BULK RATE
CARRIER ROUTE PRESORT
POSTAGE & FEES PAID
PHS/CDC
Permit No. G-284

Official Business

POSTAL CUSTOMER

HHS Publication No. (CDC) HHS-88-8404. Reproduction of the contents of this brochure is encouraged.

(d)

FIGURE 14.7 **(a)** The wartime fight against sexually transmitted diseases. These diseases remain a very significant public health problem in the United States and around the world. Today, AIDS overshadows the presence of other sexually transmitted diseases. Courtesy of CDC. **(b)** A person being immunized in a PHS immunization campaign. The PHS plays a major role in immunization campaigns around the world to halt epidemics by immunizing the population. Courtesy of James Gathany/CDC. **(c)** In April, 2009, in Fresno, California, a notice of inspection and treatment for mosquito larvae is stapled to a foreclosed house. © Fresno Bee/MCT/Landov. **(d)** Back cover of the brochure sent to every household in the United States in the early 1990s to help Americans understand AIDS. Courtesy of CDC/U.S. Department of Health and Human Services.

National Institutes of Health

The National Institutes of Health (NIH) is a component of the U.S. Department of Health and Human Services. The NIH is one of the most distinguished medical research centers in the world. It has come a long way since its founding in 1887 as the one-room Laboratory of Hygiene to seventy-five buildings spread over a 300-acre campus in Bethesda, Maryland (FIGURE 14.8). Composed of 27 institutes and centers, one of which is the National Institute of Allergy and Infectious Diseases, a primary function of the NIH is to administer and support biomedical research at over 3,000 sites in the United States and abroad. Many of the research findings have direct applicability to public health measures.

FIGURE 14.8 Building 1 at the NIH. This grant-awarding and research-oriented organization is located in Bethesda, Maryland. Courtesy of National Institutes of Health.

At the International Level

World Health Organization

The days are long gone when a continent's, or a country's for that matter, public health problems were unique and limited to that geographical area. The planet's microbes know no boundaries and travel freely in or on their hosts without passport, from hemisphere to hemisphere in a matter of hours. The World Health Organization (WHO) has its headquarters in Geneva, Switzerland, functions as a command post in its extensive partnerships, and uses sophisticated systems of surveillance and communication to keep track of microbial diseases on a global level. WHO's activities are multifold and are not limited to microbial disease surveillance and control.

The WHO partnerships date back to the early years of the organization; its influenza surveillance network is responsible for identifying each year the strains of influenza to be used in the next year's vaccine, thereby serving as a global watchdog for surveillance of influenza. The WHO has entered into many partnerships in an effort to combat diseases including AIDS, rabies, malaria, leprosy, sleeping sickness, filariasis, and guinea worm disease.

The growth of information technology provides increased opportunities for disease surveillance and response, requiring rapid assessment to initiate control efforts with minimal delay and to screen out unsubstantiated reports. The WHO has an innovative approach to global disease surveillance aimed at improving epidemic disease control by rapid verification and response of potentially significant outbreaks to health professionals around the globe (FIGURE 14.9). The WHO, in its alliance with 193 member countries, is in an ideal position to monitor infectious disease surveillance and control.

Pan American Health Organization

The Pan American Health Organization (PAHO) has "100 years of experience in working to improve the health and living standards of the countries of the Americas." It is a component of the United Nations and serves as the Regional Office for the Americas of the WHO. Its member states include all thirty-five

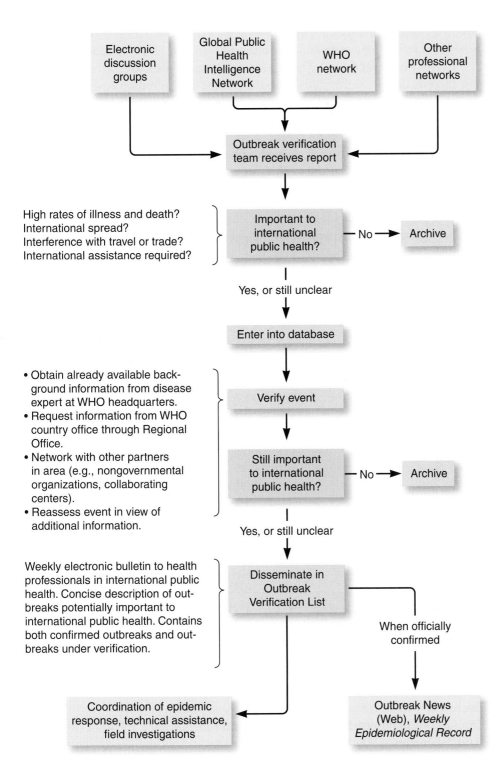

FIGURE 14.9 Outbreak verification program. This organization has teams ready to go in the event of verification of an impending outbreak of disease. Adapted from Grein, T.W., et al., *Emerging Infectious Diseases* 6 (2000):97–102.

countries in the Americas. Its mission is achieved in association with other governmental and nongovernmental agencies, universities, and community groups.

PAHO promotes primary health strategies and assists countries in combating cholera, dengue, tuberculosis, and AIDS and is committed to ensuring that blood for transfusion is safe and not a vehicle of disease. Further, the organization aims to eliminate all vaccine-preventable diseases. The agencies polio eradication efforts started in 1985 paid off, resulting in a declaration of polio-free Americas in 1994; the last case was identified in August 1991. Improvement of drinking water, sanitation, and health care remains a top priority for PAHO, with a focus on equity.

United Nations Foundation

The United Nations (UN) Foundation was created in 1988 with entrepreneur and philanthropist Ted Turner's historic U.S.$1 billion gift to support UN causes and activities. The UN Foundation builds and implements public–private partnerships to address the world's most pressing problems and also works to broaden support for the UN through advocacy and public outreach; the UN Foundation is a public charity. The UN Foundation and its partners, including the American Red Cross and the CDC, have raised large sums of money for the Measles Initiative.

World Bank

The World Bank is an international financial institution founded in 1946 that provides loans to developing countries; it has a history of project support to low- and middle-income countries to assist them in meeting their developmental needs (FIGURE 14.10). The Bank has evolved to become the world's largest external funder of health providing more than $1 billion annually; it has supported public health projects associated with onchocerciasis, tuberculosis, HIV/AIDS, malaria, cholera, meningitis, and other diseases associated with microbes and fostering immunization. A part of its function is "to overcome poverty, enhance growth with care of the environment, and to create individual opportunities and hope."

FIGURE 14.10 World Bank headquarters in Washington DC. © iStockphoto/qingwa.

Doctors Without Borders

Doctors Without Borders is a volunteer organization founded in France (Médecins Sans Frontières—MSF) in 1971 by physicians and journalists to assist those "whose survival is threatened by violence, neglect, or catastrophe primarily due to armed conflict, epidemics, malnutrition, exclusion from health care, or natural disasters (FIGURE 14.11)." The organization co-founded the project "Drugs for Neglected Diseases" in 1999 and was awarded the Nobel Peace Prize the same year for its humanitarian work.

The Private Sector

The role of government agencies as partners in the control of infectious diseases has been outlined, but the success achieved over the past century is, in no small measure, also attributable to the role of the private sector and nongovernmental organizations as equal partners in the ongoing battle to lessen the world's burden of infectious diseases. Rotary International has been a key player in the polio eradication program since 1985, when it launched Polio Plus, a commitment that was to cost over $500 million by the projected polio eradication date of 2010. The Carter Center has commited to a variety of projects, including partnerships with both governmental and nongovernmental organizations, for the eradication of guinea worm disease, onchocerciasis, and filariasis. In January 2001 Microsoft billionaire Bill Gates pledged over $100 million in pursuit of an AIDS vaccine and challenged others to pitch in. Pharmaceutical companies, too, have been significant partners. Ivermectin, an important drug in the onchocerciasis control program, has been donated free of charge by Merck & Co. to countries where this disease is prevalent. SmithKline Beecham (now part of GlaxoSmithKline) has supplied the drug albendazole, an orally administered broad-spectrum anthelmintic drug, to countries as needed. Other drug companies have also made generous contributions.

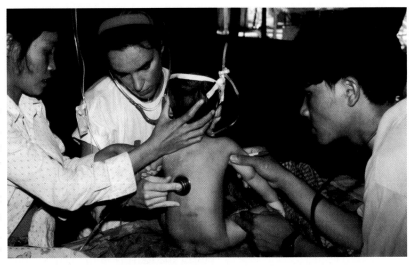

FIGURE 14.11 Doctors Without Borders in a refugee camp in Thailand. © BSIP/Photo Researchers, Inc.

The William J. Clinton Foundation was established in 1997 by former President Bill Clinton. The Clinton HIV/AIDS Initiative, created in 2002 and extended to malaria in 2007, aims to bring affordable, high-quality treatment to infected persons in developing areas.

Partnerships: The Way to Go

The enormous public health strides of the last century have resulted in an increased life span for people in many countries and to decrease the burden of microbial diseases around the globe. These achievements are largely the result of partnerships from the community level to the international level, as described in this chapter. The potential threat of microbes is too enormous to be handled without teamwork, particularly given the ease of transmission resulting from globalization and the increased threat of new, emerging, and reemerging infections.

Partnerships are the key to preventing and coping with epidemics and pandemics and in responding with minimal delay to populations endangered by the ravages of infectious disease as a consequence of floods, earthquakes, and other natural disasters. History and current events provide many examples of the misery and deaths that occur in the aftermath of catastrophic events because of the collapse of the public health infrastructure. The lack of clean drinking water and safe foods, sanitation, and personal hygiene, coupled with overcrowded and makeshift living quarters, create an environment that is conducive to outbreaks of diarrheal, respiratory, and a multitude of other diseases. Whereas the primary motive of these partnerships is humanitarianism, there is also a selfish motivation of self-protection.

Examples of partnerships in action in response to catastrophic events are abundant. A massive earthquake, the worst in fifty years, rumbled through western India on January 28, 2001, causing thousands of houses to collapse in Bhuj, a city of 150,000 close to the quake's epicenter, and spreading damage in its path. The earthquake was quickly followed by the arrival of international disaster teams and pledges of financial aid from Britain's International Rescue Corps, the Swiss Red Cross, the International Federation of the Red Cross, the United Nations, the European Union, Germany, Turkey, Norway, and the Netherlands. Pakistan, India's archrival, was among the first to offer condolences and help and demonstrated that human tragedy transcends politics.

The earthquake in India (2001), the SARS epidemic in China (2003), the tsunami in Indonesia (2004), Ebola outbreak in central Congo (2005) and in Uganda (2012), Hurricane Katrina in the United States (2006), and cholera outbreaks in Zimbabwe (2008) and Haiti (2010) are excellent examples of partnerships in action at the local, national, and international levels, involving both public and private sectors, working together to prevent, monitor, and control outbreaks. The world is threatened by possible pandemics of avian and swine flu, and if it were to happen, partnerships at all levels will be called on to develop and implement containment strategies. There is no doubt that everything works better when everyone works together. Partnerships are the way to go.

◼ Overview

The twentieth century witnessed tremendous accomplishments in public health that led to a worldwide decrease in the burden of infectious diseases, particularly in developed countries. The attainment of "health for all" is a goal yet to be realized. Further successes require teamwork at the local, national, and international levels and between the public and private health sectors. The players on the team are described in this chapter.

Surveillance and control of microbial disease at the community level are the responsibility of local and state departments of health. They are responsible for alerting the CDC to the occurrence of specified communicable diseases in order that a national network of surveillance and communication be maintained. At the national level a number of agencies, most notably the CDC, cooperate to bring about the containment of infectious diseases. Microbes spread readily and quickly from continent to continent, creating the need for international agencies like the WHO, that have the capacity to respond to outbreaks of disease at the global level.

◼ Self-Evaluation

PART I: Choose the single best answer.

1. Early in its history, the Rockefeller Foundation was involved in attempts to eradicate
   a. smallpox **b.** hookworm **c.** AIDS **d.** malaria

2. Which characteristic of smallpox was significant in contributing to its eradication?

   a. treatable with antiviral drugs **b.** disease only of humans **c.** requires only two doses of vaccine **d.** can be grown on Petri dishes in the lab

3. The expression "microbial zoo" applies to

   a. WHO **b.** Bronx Zoo **c.** NIH **d.** CDC

4. Which agency is the primary support of medical research in the United States?

   a. PAHO **b.** FDR **c.** NIH **d.** FDA

5. Which agency focuses on the safety of foods, vaccines, antibiotics, and other medicinal and medical products?

   a. CDC **b.** FDA **c.** MSF **d.** Ronald McDonald

6. Which entity was created in 1979 by President Carter as an independent agency whose slogan is "Helping People"?

   a. FEMA **b.** WHO **c.** ZOMBIE **d.** MSF

7. Which entity is charged with reducing vulnerability in the U.S. to terrorism?

   a. FDA **b.** Department of Homeland Security **c.** Keeping it Kleen **d.** Secret Police

8. Which bank loans money to developing countries?

   a. U.S. Bank **b.** Citizensfirst Credit Union **c.** World Bank **d.** Wells Fargo

9. All states have

   a. CDC laboratories **b.** vaccine manufacturers **c.** departments of health **d.** nuclear reactors

10. Key player in the polio eradication program since 1985 is

    a. Promega Corporation **b.** Bill and Melinda Gates Foundation **c.** Rotary International **d.** FBI

PART II: Fill in the blank.

1. ITFDE stands for _____ .

2. "Hot" microbes, such as Ebola virus, are dangerous to work with and require special facilities at biosafety level _____ .

3. The William J. Clinton Foundation was established to _____ .

4. Public health achievements are the result of _____ .

5. The Public Health Service is headed by the _____ .

6. The cruise ship industry has been subject to waves (pun intended) of criticism because of outbreaks caused by _____ .

7. _____ improves the health and living standards of the countries of the Americas.

8. _____ was eradicated from the world in 1980.

9. _____ are used as effective, safe, preparations that induce long-lasting immunity against specific microbes or viruses.

10. _____ is the city that is the headquarters of the CDC.

PART III: Answer the following.

1. Smallpox is a model chosen by ITFDE for evaluating eradication of other microbial diseases. Discuss four or five characteristics of smallpox.

2. Describe the work of local health departments as partners in microbial disease control. Cite a few specific examples.

3. You, as director of the WHO, are well aware of the disparity between the developing and developed countries in matters of public health. Outline your approach to address this issue.

4. Research and create a list of vaccines you would need to travel to the Republic of the Marshall Islands? Jekyll Island? Peru?

5. Create a list of microbes that must be handled in a level four biocontainment facility.

6. Explain why partnerships are needed to make enormous health strides in many countries.

7. Create a list of public or private entities that respond to natural disasters.

8. If there is an avian influenza pandemic, what partnerships will be called on to develop and implement containment strategies?

9. Why are there so few level four biocontainment laboratories in the world?

10. Why haven't diseases like malaria or filariasis been eradicated from the world?

Current Challenges

Author's photo (RIK).

Biological Weapons

They shall beat their swords into plowshares and their spears into pruning hooks. Nation will not lift sword against nation; there will be no more training for war.

—*Isaiah 2:4*

■ Preview*

"I Ain't Gonna Study War No More" was the anthem and rally cry of the 1960s antiwar protesters at a time when the United States was engaged in the Vietnam War; society has still not been able to achieve that lofty goal. Wars continue and the conventional weapons of warfare have evolved from the days of hand-to-hand combat with spears and knives to highly sophisticated weaponry for killing from carrying warriors and supplies on horses and mules to using aircraft carriers, nuclear powered submarines, and jet aircraft that can deploy thousands of military personnel and support equipment from one corner of the world to another in less than twenty-four hours.

Current events bear witness to this all too well. Ironically, the progress in microbiology, geared to alleviate humankind's suffering by minimizing and controlling infectious diseases, has been subverted to biological weapons in the form of

*Credit is given to *Textbook of Military Medicine, Part I, Medical Aspects of Chemical and Biological Warfare* (Office of the Surgeon General, United States Army, 1997) as a major reference for the history of biological warfare in this chapter.

Author's photo (RIK).

microbes designed to bring about mass destruction. These weapons are deadly and can be deployed inside intercontinental missiles, in aerosol cans, or by crop dusters. Those who make light of the threat of biological warfare and terrorism are, like the proverbial ostrich, hiding their heads in the sand.

Biological warfare has a long history. Armies learned that the outbreak of disease is deadly to the opposition. Biological weapons used for warfare or for acts of bioterrorism are described as the "poor man's weapons of mass destruction" because of the relative ease and low cost of their deployment. Despite international treaties prohibiting biological warfare, this threat has escalated during the early years of the twenty-first century and persists because of political unrest in the Middle East. In all wars, armies had to defend against microbes within their own troops as a natural consequence of battlefield conditions, as well as against their opposition. In stark contrast, however, the deliberate use of microbes to kill is a heinous crime and challenges the rules of warfare.

The use of biological weapons is a constant threat. The U.S. Department of Health and Human Services (DHHS), of which the Centers for Disease Control and Prevention (CDC) is an agency, is charged by presidential directive with the responsibility of preparing a national response to emergencies resulting from the use of biological weapons.

■ Biological Warfare and Terrorism

Biological weapons are the tools of **biological warfare** and **bioterrorism**. They are microbes deployed to grow in (or on) their target host or microbial products that are deliberately used to produce a state of clinical disease with the intent of incapacitating or killing individuals or producing mass casualties. Biological warfare is the sanctioned use of biological weapons by nations in the conduct of war. Biological terrorism is the use of biological weapons by nonstate government groups, including religious cults, militants, and crazed individuals.

No country and no community are safe from warmongers or from terrorists. In 1998 then Secretary of State Madeleine Albright remarked that "biological weapons know no boundaries." In 1999 President Bill Clinton stated that the possibility of germ attack kept him awake at night and, further, "a chemical attack would be horrible, but it would be finite . . . but a biological attack could spread . . . kind of like the gift that keeps giving."

■ War and Disease

Historically, microbes have always played a role in battles and wars; in many cases microbes were perhaps even more influential than the conventional weaponry forces brought to bear by the opposition. Battlefield conditions are conducive to the spread of infectious diseases because of unsanitary conditions, particularly resulting from the problem of disposal of large amounts of human fecal material, the unavailability of clean drinking water, poor personal hygiene, crowding, inadequate supplies of safe foods, and overwhelmed and inadequate medical facilities. A variety of diarrheal diseases can quickly bring soldiers to their knees, as happened in the Battle of Crecy in 1346. An army with its pants down is not ready to fight!

Napoleon's army was defeated by typhus fever in the early 1800s, transmitted from person to person by the feces of body lice. During the Civil War bacterial, viral, and protozoan diseases, including typhoid fever, staphylococcal and streptococcal infections, battlefield gangrene, smallpox, chickenpox, measles, malaria, and amebic dysentery, killed and incapacitated more soldiers than did rifles, cannons, and other weapons. During World War I and previous wars, horses and mules played a significant role in the transport of soldiers and supplies and would seriously cripple an army anytime they became ill or died.

The circumstances of war herd together large groups of people living in proximity under unsanitary and unhygienic conditions (e.g., refugees, prisoners of war, and inhabitants of concentration camps) that promote outbreaks of microbial diseases. International relief agencies, including the International Red Cross and the United Nations (UN), play a vital role in attempting to minimize human misery.

History of Biological Weaponry

Documentation of the use of biological weapons is difficult because of several factors, including confirmation of allegations of use, the lack of reliable microbiological and epidemiological data, propaganda, the secrecy inherent in the use of biological weapons, scare tactics, and the occurrence of naturally caused diseases, particularly during war. Nevertheless, there is ample evidence that biological weapons have been used since ancient times. The history of the use of biological weapons is considered over three time frames: early history to World War II, World War II to the 1972 Biological Weapons Convention, and 1972 to September 11, 2001.

Early History to World War II

Ancient warriors recognized the devastating impact of infection and poor sanitation on armies. Human excrement and human and animal carcasses were used to produce epidemics of infection and to pollute wells and other water sources.

In preparation for naval battle against King Eumenes II of Pergamum in 184 B.C., Hannibal, the famous Carthaginian leader, instructed his army to fill earthen pots with "serpents of every kind," which were then hurled onto enemy ships to terrorize warriors. Hannibal's forces were victorious as the opposition, crazed with fear, now had to battle two forces. During the siege of Kaffa (now Feodossia, Ukraine) in 1346, Tatar military leaders recognized that diseased corpses could be used as weapons and catapulted their own dead soldiers into the midst of the enemy. According to one account:

> "The Tatars, fatigued by such a plague and pestiferous disease, stupefied and amazed, observing themselves dying without hope of health, ordered cadavers placed on their hurling machines and thrown into the city of Kaffa, so that by means of these intolerable passengers, the defenders died widely."

During a plague outbreak bodies of dead soldiers along with large quantities of excrement were hurled into the ranks of the enemy at Karlstein, Bohemia in 1422.

In 1710 Russian troops battling Swedish forces in Reval (now Tallinn, Estonia) threw their plague victims over the walls into the midst of the opposition. The agent of smallpox was a potent weapon long before its viral nature was known. In the sixteenth century Spanish conquistadors were reported to have given smallpox-contaminated clothing to South American natives. General W. T. Sherman in his *Memoirs* complained that the Confederate troops were deliberately shooting farm animals in ponds so that their "stinking carcasses" (Sherman had a way with words!) would contaminate the water supplies of the Union forces, resulting in troops weakened and demoralized by the onset of gastrointestinal disease.

The end of the nineteenth century and the beginning of the twentieth century marked the emergence of modern microbiology, a period known as the "Golden Era of Microbiology." The pioneering work of Louis Pasteur, Robert Koch, and other microbe hunters established that infectious diseases are caused by microbes and not by miasmas—"bad air" or "swamp gas." Hence, the capability for the production of selected pathogens that could be used for biological warfare was born as the result of, and on the heels of, advances designed to control microbial infections—a terrible irony. During World War I, Germany was accused of conducting biological warfare by shipping infected horses and cattle to the United States and to other countries. Further, it was alleged, but never proved, that in 1915 German forces attempted to spread cholera into Italy and Russia and air-dropped bacterially contaminated fruit, chocolate, and children's toys into Romania. Germany was exonerated by the League of Nations because of the lack of hard evidence.

After the war the Protocol for the Prohibition of the Use in War of Asphyxiating, Poisonous or other Gases and of Bacteriological Methods of Warfare, more frequently referred to as the 1925 Geneva Protocol, has been signed by 137 nations, including Iraq (in 1931), as of November 2010. This was a landmark event; it represented the first multilateral agreement prohibiting the use of biological weapons. The term "bacteriological" was subsequently accepted to be synonymous with the term "biological" to include viruses and fungi. There were no provisions prohibiting the use of microbes for basic research or mandating inspection for purposes of monitoring compliance. These limitations were at the heart of Iraq's defiance of UN inspections for the alleged production of bioweapons, prompting escalation of the Persian Gulf War in 2002. The Geneva Protocol allows the right of retaliatory use of chemical or biological weapons. Any treaty, not just those restricted to matters of warfare, is only as good as the intent of those who sign it.

World War II to 1972

The occasions on which biological weapons were or may have been used during the next period of approximately fifty years (to the 1972 Biological Weapons Convention) are numerous, and the evidence is confounded with allegations, secrecy, intrigue, charges, and countercharges. Japan used biological warfare agents in Manchuria, China from 1932 until the end of World War II. The infamous Japanese Army Units 731 and 100 were biological warfare units and conducted unforgivable experiments on prisoners in China. Unit 731 carried out experimental studies in Ping Fan, Hancheng, Changchun, and other sites on prisoners from many countries, including Manchuria, the Soviet Union, the United States, Great Britain, and Australia. The prisoners were injected with pathogens, including *Bacillus anthracis, Neisseria meningitidis, Vibrio cholerae, Shigella* species, and *Yersinia pestis*. At least 3,000 prisoners died at Ping Fan. Further allegations against Japan, in at least eleven Chinese cities, included contaminating water and food supplies with a variety of pathogens, throwing cultures into homes, or spraying infective aerosols from aircraft over cities. Japan's attack on Changteh in 1941 using *Vibrio cholerae* as a weapon backfired, resulting in 1,700 deaths among Japanese troops. This illustrates a potential pitfall to the aggressor using biological weapons. Although some of these incidents have not been conclusively substantiated, the Japanese government later stated that its conduct was "most regrettable from the viewpoint of humanity."

In the events leading up to World War II and during the war, there is no documentation that Adolf Hitler's Nazi Germany conducted biological warfare. Winston Churchill, Prime Minister of the United Kingdom from 1940 to 1945 and again from 1951 to 1955, is said to have considered anthrax as a retaliatory measure if Germany used biological agents against Britain. During the years after World War II, allegations regarding the use of biological agents were carelessly thrown about. It is true that epidemics of disease occurred, but were they the result of biological weapons? Certainly, the devastation of war infringes strongly on a country's infrastructure, including a subsequent decline in public health measures and a resulting increase in disease. To some extent, allegations of the use of biological weapons were purposely used as blatant propaganda and scare techniques.

In response to the international threat of the use of biological weapons, the United States embarked on an offensive biological weapons program in 1942. The Soviet Union also began such a program during World War II. The U.S. initiative was under the direction of the War Reserve Service, a civilian agency headquartered at Camp Detrick (later renamed Fort Detrick) in western Maryland. At its peak in 1945 almost two thousand scientific and military personnel worked at this maximum-security campus, which had the dubious distinction of being the world's most advanced biological warfare unit. After the Korean War the U.S. program focused on retaliatory capability and was centered at Fort Detrick (FIGURE 15.1) under the jurisdiction of the U.S. Army. Weapons systems disseminating a variety of pathogens were tested at Fort Detrick, major cities in the United States, and remote desert and Pacific sites. New York and San Francisco were unknowingly used as models to test aerosolization and dispersal methods using "harmless bacteria." In 1966 *Bacillus globigii,* a nonpathogenic sporeformer related to *B. anthracis,* was released into the subway system of New York City to determine the effectiveness of aerosolization and the city's vulnerability. The results indicated that the entire tunnel system can be infected by release of spores into a single station because of the air currents generated by the trains. It was also determined, however, that the bacteria could be carried on the trains and then emitted onto station platforms when the doors open, which was and still is a bigger threat.

During the 1960s international concern peaked regarding the risk, unpredictability, lack of control measures, and the failure of the 1925 Geneva Protocol for

(a)

(b)

(c)

FIGURE 15.1 Fort Detrick, the location of the U.S. Army's main facility for research on biological warfare defense. **(a)** An old photo of the main gate, probably taken in the 1950s. **(b)** Technicians working with highly virulent microbes in biological containment hoods. **(c)** Huge vat used to culture microbes on a large scale. All photos courtesy of Fort Detrick Public Affairs Office/U.S. Army.

preventing biological weapons proliferation. Nevertheless, in a surprise move in November 1969, President Richard Nixon terminated the U.S. offensive biological weapons program, stating, "We'll never use the damn germs. . . . If someone uses germs on us we'll nuke 'em." The 1972 Convention on the Prohibition of the Development, Production, and Stockpiling of Bacteriological (Biological) and Toxin Weapons and on their Destruction, commonly known as the Biological Weapons Convention, was convened, and a treaty was signed by 103 nations. One point of the treaty prohibited the development of delivery systems intended to dispense biological agents and, further, required signatory nations to destroy their stocks of biological agents, delivery systems, and equipment within nine months of signing. The treaty was ratified in April 1972 and went into effect in March 1975.

1972 to September 11, 2001

As a result of the Biological Weapons Convention of 1972 and the termination of the offensive biological warfare program, the biological arsenal at Fort Detrick was destroyed, including the causative agents of the potentially lethal diseases anthrax, tularemia, and Venezuelan equine encephalitis and of botulinum toxin, staphylococcal enterotoxin B, and several anticrop biological agents. The focus at

AUTHOR'S NOTE (RIK)

I spent my first sabbatical leave from Providence College at Fort Detrick in 1965; security was tight and enforced. Before my appointment at Fort Detrick, I was extensively investigated to get a security clearance. On occasion, reports of hospitalization of Fort Detrick laboratory workers surfaced. For approximately ten years after leaving, I periodically received a questionnaire inquiring as to my health and whether I had suffered from any unusual illnesses. Fortunately, all I got was athlete's foot—a fungal disease.

Fort Detrick shifted from offensive to defensive strategies and was placed under the auspices of the U.S. Army Medical Research Institute of Infectious Diseases. The availability of high-level containment facilities at the Fort Detrick facility still allows for the study of diseases caused by highly virulent pathogens.

As was the case following the 1925 Geneva Protocol, the 1972 Biological Weapons Convention has proved to be ineffective. The Soviet Union continued an offensive biological weapons program through the 1980s under the control of the Ministry of Defense. As an example, on April 2, 1979, inhabitants of the community of Sverdlovsk (now Yekaterinburg), a city 4 kilometers downwind of a Soviet military microbiology facility, reported to hospitals with symptoms of anthrax. At least 77 cases and 66 autopsy-confirmed deaths resulted, the largest epidemic of inhalational anthrax in history. The outbreak also caused the death of livestock. Western intelligence suspected the Sverdlovsk facility was a biological warfare research center and that the epidemic was due to the accidental airborne release of *Bacillus anthracis* endospores. The Soviets denied the allegation and insisted that the incident was due to ingestion of contaminated meat. However, in 1992, almost 15 years later, Russian President Boris Yeltsin admitted that the microbiology facility was part of an offensive biological weapons program and that failure to activate air filters resulted in the accidental release of *Bacillus anthracis* endospores. Yeltsin proclaimed he would terminate biological warfare initiatives. Some in the government believe that a large number of workers are still employed in the biological weapons business.

Iraq was considered to be a major threat in terms of biological warfare. Between 1986 and the end of the first Gulf War in April 1991, Iraqi scientists investigated the potential of a large number of biological weapons, including bacteria, fungi, and viruses, as well as botulinum toxin. As preposterous as it may sound, the *Clostridium botulinum* culture used for the mass production of botulinum toxin was obtained from the United States. In 1990 the United Nations imposed sanctions on Iraq because of its invasion of Kuwait and the threatened use of biological weapons. Iraq's failure to allow inspection of its biological weapons by the UN Special Commission led to air strikes by U.S. and British forces in December 1998. The alleged possession of weapons of mass destruction in Iraq was a major factor in that country being invaded by a multinational force in March 2003, which began the second Gulf War.

■ Emergence of Biological Terrorism

In the past several years the threat of biological weaponry has taken on a new face: that of biological terrorism. Nongovernmental agents, including religious cults, terrorists, and individuals, can use biological weapons to further their own personal or political agendas. In some respects biological terrorism is an even greater threat than biological warfare; it is more difficult to detect and control and can strike without warning. In 1981 members of a religious commune, followers of Bhagwan Shree Rajneesh, intentionally contaminated ten restaurant salad bars in an Oregon community with *Salmonella enterica* serovar *typhimurium*. Over 750 people fell ill with *Salmonella* gastroenteritis in a community in which fewer than five cases per year are usually reported. This was intended as a "rehearsal" for a plan to cause an epidemic on election day to influence the outcome of the county elections. The terrorist threat posed by radical groups was again demonstrated in 1995 by the Aum Shinrikyo ("Supreme Truth") cult, possibly the world's most infamous apocalyptic sect, when they released sarin gas into a Tokyo subway, killing at least eleven people and making thousands of others ill. Perhaps even scarier is the fact that the cult previously attempted at least nine biological attacks with the intent of causing mass murder. Their lack of sophistication, not their lack of determination, resulted in the failure of these activities. The cult possessed pathogenic microbes as well as sarin. In still another incident, on May 5, 1995, Larry Harris, a septic tank and well water inspector in Ohio, received three vials of *Y. pestis*, the causative agent of plague, from the American Type Culture Collection (ATCC). His intent was not clear, but the fact that he needed only a credit card and used a false letterhead is frightening. This is only one example of numerous others indicating the necessity of establishing tighter controls on microorganisms to minimize their getting into the hands of terrorists. To this accord, in 1997 the CDC was assigned the responsibility of monitoring the movement of microbes and toxins (TABLE 15.1).

The United States and other countries can no longer focus on biological warfare alone but also must heed the threat of bioterrorism from nonstate actors. In this regard, three major statutes and two directives (Table 15.1) were passed by the U.S. Congress: (1) the Biological Weapons Act of 1989, (2) the Chemical and Biological Weapons Control and Warfare Elimination Act of 1991, and (3) the Anti-Terrorism and Effective Death Penalty Act of 1996.

TABLE 15.1 U.S. Statutes and Directives Against Biological Terrorism

Law	Description
Biological Weapons Act, 1989	The knowing development, manufacture, transfer, or possession of any biological agents, toxin, or delivery for use as a weapon is to be considered a federal crime
Chemical and Biological Weapons Control and Warfare Elimination Act, 1991	Stipulates a system of economic sanctions and export controls by Congress to curb the proliferation of biological weapons; to be imposed on international companies in countries designated by the United States as terrorist states
Presidential Decision Directive 39, 1995	In response to the threat of biological weapons, President Clinton ordered the DHHS to develop countermeasures
Anti-Terrorism and Effective Death Penalty Act, 1996	Extended the Biological Weapons Act of 1989 by broadening criminal provisions to include threats or attempts to develop or to use biological weapons, including DNA, or other technological advances to create new and more virulent pathogens
CDC charged with regulation and transfer of biological agents, 1997	24 infectious agents and 12 toxins identified as biological hazards to be regulated by CDC

TABLE 15.2 CDC-Regulated Biological Agents/Diseases[a]

Viruses	Plague
Eastern equine encephalitis	Typhus fever
Venezuelan equine encephalitis	Glanders
Ebola	Melloidosis
Marburg	Brucellosis
Smallpox	Psittacosis
Hantavirus	Q fever
Yellow fever	**Toxins**
Bacteria	Staphylococcal enterotoxin B
Anthrax	Ricin toxin
Botulism	*Clostridium perfringens* epsilon toxin

[a]This is a partial list.

In 1997 the CDC was charged with the regulation and transfer of twenty-four infectious agents and twelve toxins that pose a significant public health risk (**TABLE 15.2**).

■ Assessment of the Threat of Biological Weaponry

Nations of the world have not yet been able to meet the aspirations of the quote on the opening page of this chapter. War and civil strife continue to pose a threat. National security is one of the utmost concerns to the population of all countries, large or small, developed or developing. Epidemics and pandemics of infectious diseases negatively affect national security because they subject populations to the threat of disease and create the potential for transmission beyond a country's borders. The nature of outbreaks of infectious disease is devastating, not only to human populations but also to the economic development of a country, as history has revealed from the ancient "plagues" to current events. The ongoing pandemic of HIV/AIDS on the African continent is witness to this point. The use of biological weapons, whether by nations or by terrorists, is focused on promoting outbreaks of disease. This focus adds to the public health burden, which must protect not only against naturally occurring outbreaks but also those intentionally caused. An article, "Security and Public Health: How and Why Do Public Health Emergencies Affect the Security of a Country?" from the Monterey Institute for International Studies Center for Nonproliferation Studies (2007) cites four ways in which a public health emergency, including those caused by biological weapons, threatens a nation's security: (1) pressure on a country's economy at both a micro and macro level, (2) social disruption, (3) political destabilization, and (4) affecting national defense.

The numerous events of using biological weaponry throughout history make the point all too obvious: It is within the capability and resources of many countries to use microbes as tools of mass murder. Biological agents are accessible because of their use in legitimate research activities throughout the world. The equipment and facilities for their mass production are found in the pharmaceutical, agricultural, and food industries, unlike both nuclear and chemical warfare, which require special facilities that would make their detection relatively easy. Nevertheless, conversion to bioweapons on a large scale is not at all impossible, and because of its covert nature an exact list cannot be known. The weapons used to deliver biological agents are proliferating and are cost-effective. A group of technical experts estimated that "for a large-scale operation against a civilian population, casualties might cost about $2,000 per square kilometer with conventional weapons, $800 with nuclear weapons, $600 with nerve gas weapons, and $1.00 with biological weapons."

The use of biological weapons, whether by governments, groups, or individuals, has inherent advantages and disadvantages (TABLE 15.3). The advantages of biological weapons include deadly or incapacitating effects on the target population, low cost, continued microbial proliferation, difficulty of immediate detection,

---

TABLE 15.3 Advantages and Disadvantages of Biological Weapons

Advantages

The potential deadly or incapacitating effects on a susceptible population

The self-replicating capacity of some biological agents to continue proliferating in the affected individual and, potentially, in the local population and surroundings

The relatively low cost of producing many biological weapons

The insidious symptoms that can mimic endemic diseases

The difficulty of immediately detecting the use of a biological agent, due to the current limitations in fielding a multiagent sensor system on the battlefield as well as to the prolonged incubation period preceding onset of illness (or the slow onset of symptoms) with some biological agents

The sparing of property and physical surroundings (compared with conventional or nuclear weapons)

Disadvantages

The danger that biological agents can also affect the health of the aggressor forces

The dependence of effective dispersion on prevailing winds and other weather conditions

The effects of temperature, sunlight, and desiccation on the survivability of some infectious organisms

The environmental persistence of some agents, such as spore-forming anthrax bacteria, which can make an area uninhabitable for long periods

The possibility that secondary aerosols of the agent will be generated as the aggressor moves through an area already attacked

The unpredictability of morbidity secondary to a biological attack, since casualties (including civilians) will be related to the quantity and the manner of exposure

The relatively long incubation period for many agents, a factor that may limit their tactical usefulness

The public's aversion to the use of biological warfare agents

Reproduced from the *Textbook of Military Medicine, Part I, Medical Aspects of Chemical and Biological Warfare.* Office of the Surgeon General, United States Army, 1997.

---

and lack of physical damage to the area. The disadvantages include danger to the health of the aggressors, the effects of physical factors such as weather conditions on the success of an attack, public backlash, and the environmental persistence of some agents.

A large number of microbes and biological toxins are potential agents of biological weaponry directed against humans; the list consists of bacteria (including rickettsiae), viruses, fungi, toxins, and genetically modified forms of these agents (Table 15.2). Further, some of these agents can be used against animals and crops. Considering the numerous factors necessary for effective deployment, relatively few biological weapons meet the criteria; anthrax, smallpox, plague, and botulinum toxin are the frontrunners.

The psychosocial responses after even the threat of a biological attack add another dimension to the horror of biological weaponry. The fear of "bugs" and infection is terrifying; the resulting panic would produce staggering numbers of psychiatric casualties that could overwhelm healthcare facilities. The threat, in itself, of the release of biological agents in a particular community, transportation system, school, or shopping mall causes a crippling effect. On December 26, 1998, 750 people were quarantined after police received a call claiming that the anthrax-causing bacteria had been released into a popular nightclub in Pomona, California. This incident was reported to be the sixth anthrax hoax in the area in two weeks.

■ Category A Biological Threats

Biological agents and the diseases they cause have been categorized into three groups based on their risk to the national security (BOX 15.1). Anthrax, botulism, plague, smallpox, tularemia, and viral hemorrhagic fevers make up Category A. They are summarized in TABLE 15.4. More needs to be said about anthrax and smallpox here.

▨ Anthrax

Anthrax is the most likely biological weapon to be used. The anthrax microbe is thought to be in the arsenal or in the process of being developed in at least ten countries. There are also fears that genetically engineered strains of anthrax capable of producing new toxins may exist.

What is anthrax, and why is it at the top of the list of biological weaponry? Anthrax is a rapidly progressing acute infectious disease caused by the organism *B. anthracis* (FIGURE 15.2), a rod-shaped, spore-forming bacterium. It is thought to have caused the fifth Egyptian plague around 500 B.C.; during the Middle Ages it was known as the **Black Bain** and was responsible for nearly destroying the cattle herds of Europe. The ability of *B. anthracis* to form spores makes anthrax a choice weapon of mass destruction, because these spores can be disseminated by bombs, artillery shells, aerial sprayers, and other methods of dispersal. Spores are condensed bits of the organism's protoplasm, including the genetic material, surrounded by a complex layer of coats rendering them resistant to the usual lethal effects of sterilization by heat and chemical disinfection; in other words,

BOX 15.1 _Categories of Biological Diseases and Agents_

Category A Agents

The U.S. public health system and primary healthcare providers must be prepared to address varied biological agents, including pathogens that are rarely seen in the United States. High-priority (Category A) agents include organisms that pose a risk to national security because they can be easily disseminated or transmitted from person to person, cause high rates of mortality with the potential for major public health impact, might cause public panic and social disruption, and require special action for public health preparedness.

Category A agents include variola virus (smallpox), *B. anthracis* (anthrax), *Y. pestis* (plague), *C. botulinum* toxin (botulism), *Francisella tularensis* (tularemia), filoviruses (Ebola and Marburg viruses), and arenaviruses (Lassa fever).

Category B Agents

The second highest priority agents include those that are moderately easy to disseminate, cause moderate morbidity and low mortality, and require specific enhancements of the CDC's diagnostic capacity and enhanced disease surveillance. Category B agents include *Coxiella burnetii* (Q fever), *Brucella* species (brucellosis), *Burkholderia mallei* (glanders), ricin toxin (from *Ricinus communis*—castor beans), epsilon toxin (from *Clostridium perfringens),* and *Staphylococcus* enterotoxin B.

Category C Agents

The third highest priority agents include emerging pathogens that could be engineered for mass distribution in the future because of availability, ease of production and dissemination, and potential for high morbidity and mortality and major health impact. Category C agents include Nipah virus, hantaviruses, tickborne hemorrhagic fever viruses, tickborne encephalitis viruses, yellow fever virus, and multidrug-resistant *Mycobacterium tuberculosis.*

From CDC. *Morbidity and Mortality Weekly Report* 2000, 49(RR-4):1–26.

TABLE 15.4 Category A Agents and Diseases They Cause

Disease	Agent(s)	Method of Transmission	Incubation Time[a]	Symptoms	Treatment
Anthrax	*Bacillus anthracis*	Inhalation	1–7 days	Flulike symptoms, fatigue, breathing difficulty, possible death	Antibiotics (early administration), prevention by vaccination
Botulism	*Clostridium botulinum* toxoid	Foodborne	12–36 hours	Double vision, blurred vision, slurred speech, difficulty swallowing, dry mouth, muscle weakness, paralysis of breathing muscles leading to death	Antitoxin
Plague (pneumonic)	*Yersinia pestis*	Inhalation	1–6 days	Fever, headache, cough, weakness	Antibiotics
Smallpox	Variola virus	Inhalation	7–17 days	Characteristic rash, fever, fatigue	Prevention by vaccination (within 4 days after exposure)
Tularemia	*Francisella tularensis*	Inhalation	1–14 days	Fever, ulcerated skin lesions, chills and shaking, headache, weakness; can progress to respiratory fever and shock	Antibiotics
Viral hemorrhagic fevers	Ebola virus, Marburg virus, Lassa virus	Close contact with infected person or his or her body fluids	7–14 days (Ebola fever), 5–10 days (Marburg fever), 6–21 days (Lassa fever)	Fever, fatigue, dizziness, weakness, internal bleeding, bleeding from body orifices, shock, delirium	Supportive therapy, ribavirin (depending on circumstances)

[a]Considerable variation.

Data from CDC Emergency Preparedness & Response Information.

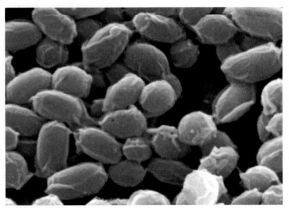

FIGURE 15.2 Colorized scanning electron micrograph of anthrax bacteria. Courtesy of Janice Haney Carr/CDC.

they are hard to kill and reside in soils worldwide for many years. A case in point is that in 1942 the British used Gruinard Island, a site three miles (about 5 kilometers) off the northwest coast of Scotland, as a site for anthrax experiments. The soil remained contaminated with *Bacillus anthracis* endospores for over 50 years, making the area unsafe for human entry. Spores are like seeds in that under appropriate conditions, as are met on and in the body, they germinate into actively multiplying and toxin-producing bacteria. The reasons responsible for anthrax as a "top contender biological weapon" are summarized in **TABLE 15.5**.

TABLE 15.5 **Anthrax as a Biological Weapon**
High fatality rate if untreated
Treatment only partially successful and only if administered within 24–48 hours
Microbe (*Bacillus anthracis*) easy to grow
Microbe produces spores that remain viable for years
Can be easily dispersed by missiles, rockets, aerial bombs, crop dusters, and sprayers
No indication of exposure

There are three varieties of anthrax disease based on the route of transmission and symptoms. First, cutaneous (skin) anthrax is a hazard to those employed in occupations requiring frequent close contact with the hides, hair, wool, bones, or bone products of infected animals. The bacterial spores enter the body through a break in the skin. About 20% of untreated cases result in death, but death is rare with appropriate antibiotic therapy. Second, gastrointestinal anthrax is rare but can be acquired by the ingestion of inadequately cooked contaminated meat. Death occurs in 25% to 60% of untreated cases. Third, inhalation anthrax poses the greatest threat with the most deadly consequences. Initially, symptoms resemble a common cold, which progress to severe breathing problems. The incubation period is 1 to 7 days; death usually results in 1 to 2 days after the occurrence of acute symptoms. All it takes is a millionth of a gram (8,000 to 10,000 spores) of *B. anthracis* to constitute a lethal dose; 1 kilogram could kill billions of people. Human-to-human transmission does not occur. The death rate is 99% in nonimmunized and untreated individuals. Antibiotic therapy can suppress the disease but only if administered within the first 24 to 48 hours after exposure. If antibiotic treatment is not begun within that period, the death rate is about 95%.

During the early 1990s, UN-sanctioned inspection teams reported that Iraq had produced 8,000 liters of anthrax spores, an amount believed capable of killing every man, woman, and child on Earth. In 1995 Iraq admitted to the United Nations that it had loaded *Bacillus anthracis* spores into warheads during the Gulf War. A few years later a series of bioterrorist threats of exposure to spores were

BOX 15.2 A Sample of Anthrax Threats, October 30 to December 23, 1998

Indiana

A threatening letter was sent to a health clinic. Thirty-one adults considered to be possibly exposed to spores were detained for three hours and decontaminated. They placed their clothing and personal effects in plastic bags and showered using soap and water plus a dilute bleach solution. The desktop where the letter was found was washed with household bleach (50% hypochlorite solution). All 31 persons were transported to local emergency departments to receive treatment. Some underwent additional decontamination.

The letter was found to be negative for *B. anthracis* by the Indiana State Department of Health and by the U.S. Army Medical Research Institute of Infectious Diseases in Fort Detrick, Maryland.

Kentucky and Tennessee

A health clinic in each state received a threatening letter. The circumstances and decontamination procedure were similar to those in Indiana. The letters were found to be negative for *B. anthracis* by the University of Louisville Hospital Clinical Microbiology Laboratory and by the U.S. Army Medical Research Institute of Infectious Diseases.

California

Four threats of *B. anthracis* contamination were received by businesses or government offices from December 17 to 23, 1998; all premises were found to be negative for *B. anthracis.* The first threat was made in a letter mailed to a private business; all 28 "at-risk" adults were decontaminated and given antibiotics. The second threat was made by a telephone caller to a government building claiming to have contaminated the building's air-handling system. The decision was made not to decontaminate approximately 95 adults, but all received chemoprophylaxis.

The third threat was made by a telephone caller to 911 claiming to have contaminated the air-handling system of a federal building; 1,200 to 1,500 persons (including at least one pregnant woman and two children) were at risk. Decontamination and chemoprophylaxis were not practiced, but all were advised to wash their vehicles with bleach, place clothing in plastic bags pending laboratory results, and shower. The fourth threat was made by a caller to 911 claiming to have contaminated an air-handling system; approximately 200 persons were at risk. Decontamination and chemoprophylaxis were not practiced.

---

reported by the CDC (BOX 15.2). In those incidents letters alleged to be impregnated with *B. anthracis* spores were sent to health clinics in Indiana, Kentucky, and Tennessee and to a private business in California. Additionally, three telephone threats warning of anthrax contamination of ventilation systems in private and public buildings were made. All threats turned out to be hoaxes, but they indicate the need for a state of readiness for the psychological stress that accompanies threats of biological weapons.

After a three-year study, on May 18, 1998, then Secretary of Defense William Cohen approved the Pentagon's plan to vaccinate all U.S. military service members for anthrax, a decision causing continuing controversy. As a show of confidence Secretary Cohen was immunized. Full immunization required six injections administered over an eighteen-month period. Implementation of the immunization program in the military was designed as a three-phase strategy with a completion date of 2003; as further described below the plan did not achieve its target date. At the time it had a price tag of about $60 to $80 per person for the full six-dose regimen and a total cost in excess of $200 million for the estimated 2.4 million personnel. Current cost reports are not available, but despite a decrease to five (instead of six) doses, the program remains at a very high cost.

The anthrax vaccine is a cell-free filtrate, meaning that it has no whole bacteria, alive or dead, and thus is incapable of causing infection. Hence, one of the arguments against use of the vaccine, namely, that it might cause anthrax, has no basis. The vaccine was first approved by the U.S. Food and Drug Administration (FDA) in 1970 as a measure to protect at-risk workers with high exposure to anthrax against the cutaneous form of the disease.

Criticism of the vaccine is based on the following major points:

1. It has a difficult immunization schedule (five or six shots are required over a period of 18 months).
2. Full immunity is not conferred until after 18 months.
3. There is a need for an annual booster.
4. There is a lack of proven effectiveness against a large dose of aerosolized spores.
5. The vaccine was approved by the FDA in 1970 for prevention of cutaneous anthrax; opponents claim that the vaccine has not been proven to be effective against inhalation anthrax.

In December 2003 the U.S. District Court ruled that the Department of Defense could not make anthrax vaccination in the military mandatory (unless through presidential order). Three years later, in October 2006, the Department of Defense announced its intention to resume mandatory anthrax vaccination for all service personnel and civilian employees deployed for fifteen or more consecutive days to Afghanistan, Iraq, and the Korean Peninsula (South and North Korea), areas determined to be "high threat." A recent directive from the Military Vaccine Agency, under the Department of Defense, advises of a change in the number of doses (six to five) and the route of vaccine administration (subcutaneous to intramuscular).

Smallpox

Smallpox, the first disease to be eradicated, is now a top biological candidate weapon. Smallpox has unique properties that made its eradication possible. Routine immunization against smallpox ended in 1972 in the United States, and within a few years ended around the world. Ironically, the science that led to the eradication of smallpox resulted in a world population under the age of thirty-five that is susceptible to smallpox (including about half the population of the United States). Those who were immunized have a waning immunity. If smallpox is introduced into the world population, it has the potential to spread like wildfire through the (nonimmunized) population because of a lack of herd immunity.

If smallpox has been eradicated from the face of the Earth, why is it a threat? The answer is that the disease has been eradicated but not the virus; it exists in freezers under lock and key at a designated facility in Russia and at the CDC in Atlanta, Georgia. The problem is that covert stocks of the virus are thought to be in the hands of potential terrorists and unreliable foreign governments.

The early signs of smallpox include high fever, fatigue, headaches, and backaches, followed in a few days by an extensive rash over the body. The rash is characterized by flat red lesions that become filled with pus and crust over in the second week; the lesions break out at the same time. (The rash of chickenpox appears in waves.) The crusts dry up and fall off, leaving deeply pitted scars, particularly on the face. The case fatality rate is about 30%.

Smallpox is transmitted directly from person to person by virus-infected droplets of saliva expelled from an infected individual onto a mucosal surface of another. Smallpox virus-contaminated clothing and bedding can also spread the virus. The first week of illness is the most infectious time because of the high numbers of viruses or viral particles in the saliva.

There is concern that the amount of smallpox vaccine necessary in the United States to immunize and treat the public is deficient, but that is not the case. The government continues to stockpile smallpox vaccine. Thus the country is in good shape with 300 million doses affording 100% coverage. Other countries, including Germany, the United Kingdom, France, and Israel, are also at 100% coverage, whereas still other countries are by no means fully covered, with coverage down to only 1%. Tommy Thompson, Secretary of DHHS from January 2001 to January 2005, stated, "We hope our smallpox vaccine stockpile will serve as a deterrent to those who may consider using smallpox as a weapon. We will have the necessary medicine to save and protect every American should there be an outbreak."

The CDC's plan advocates that the vaccine should not be used until after an attack occurs and then should be rapidly deployed. (Recall that the basic function of the immune system is to protect from foreignness as inherent in microbes. Smallpox is caused by a virus and can be prevented and, to some limited extent, treated by active immunization using live vaccinia, a virus closely related to smallpox.) Others call for mass inoculations, fearing that an epidemic may develop too rapidly to be contained, particularly if an attack were to occur simultaneously in several cities. But Anthony Fauci, director of the National Institute of Allergy and Infectious Diseases, warns against this. According to Fauci, "If we could vaccinate people with virtually no incidence of any serious toxicity . . . we tomorrow could eliminate the threat of a smallpox bioterrorist attack. Unfortunately, that is not the case." Estimates are that if all Americans were vaccinated, 180 to 400 people could die just from the vaccine's side effects.

Countermeasures to Biological Weaponry

Perhaps you found the foregoing account of biological warfare and terrorist activities unsettling and fearsome. If you did, it is with good reason. Compared with the use of conventional weapons or even chemical or nuclear warfare, the use of biological weapons presents a unique set of challenges to any country's public health infrastructure:

- An attack might not be immediately discernible and would remain unnoticed until individuals became seriously ill; incubation periods vary as a function of the biological agent involved and also from person to person.
- If an attack were to occur in a subway station, a train terminal, or, even worse, in an airport, victims would be widely dispersed and, within twenty-four hours, could spread the disease worldwide before even being aware of their own illness.

FIGURE 15.3 Public flyers created by the EPA to educate and alert communities about water security. Courtesy of U.S. Environmental Protection Agency.

- If the agent is communicable from person to person, each infected individual would seed the agent into an ever-increasing circle of disease in successive waves.
- Studies have shown that bioterrorist events trigger chaos, panic, and civil disorder to an extent greater than that resulting from other forms of terrorism.
- Bioterrorist attacks are directly targeted at human populations.
- Dispersal by aerosolization or drinking water can be inexpensively accomplished with crop dusters, spray cans, and other simple devices (FIGURE 15.3).
- The first responders would most likely be healthcare personnel, who would succumb to infection, thwarting further efforts of response.
- The pathogen selected would be one that is not routinely seen in a population. This would mean that the population would be particularly susceptible due to a lack of immunity (e.g., in the case of smallpox) because vaccination is no longer carried out.
- Diagnosis may be delayed because the disease would most likely be one with which medical personnel have little or no familiarity or one in which the symptoms are vague and may resemble the symptoms of a variety of microbial infections. For example, smallpox has not been seen for over 35 years, and anthrax initially has coldlike symptoms.

Given these factors, what is the nation's response, particularly to the threat posed by terrorists? How prepared are we, and what precautions are being taken? Presidential Decision Directive 39, issued by President Clinton in 1995, designated the DHHS as "the lead federal agency to plan and prepare for a national response to medical emergencies arising from terrorist use of weapons of mass destruction." The CDC is an agency within DHHS.

Initiatives and strategic plans for coping with potential bioterrorism are as follows:

- Deterrence by controlling access to and handling of dangerous pathogens, including monitoring of the facilities and procedures currently in use with these agents. The CDC is charged with this responsibility.
- Surveillance and rapid detection, based on the number of individuals becoming ill and their symptoms, of whether a biological agent has been released.
- Medical and public health response aimed at strengthening the response at the local level, including the capability of mass immunization or prophylactic management, safe disposal of the deceased, infection control, and assessment of extent of the problem.
- Creation of a stockpile of pharmaceuticals as a national resource, because it would be beyond the resource of local governments to develop and maintain such a facility. Appropriate pharmaceuticals will be deployed to reach victims within twenty-four hours. CDC has the responsibility for developing the stockpile. President George W. Bush passed legislation to create Project BioShield in July 2004 to develop and stockpile vaccines and drugs.

- Expansion of support for research and development related to pathogens that might be used as biological weapons, particularly the development of rapid diagnostic methods because "time is of the essence." New vaccines and new or improved antiviral and antibiological agents are critical.

More recently the United States has placed emphasis on the coordination of clinical, public health, and law enforcement activities to deal with suspected bioterrorism. Hospitals, fire and police departments, and other agencies charged with protecting the public have developed strategic plans to cope with disaster (FIGURE 15.4). President George W. Bush created a new government agency, the Office of Homeland Security, in 2003 to develop, coordinate, and implement measures to protect the nation. A plan of readiness is vital at all levels from local to national and is continually updated by the CDC and other agencies as the nation recognizes the use of biological weapons as a real threat. Being prepared to intelligently meet a threat is the best rational approach. The use of biological weapons is a clear and dramatic example of the interfaces of biology, public health, international law, and ethics.

FIGURE 15.4 Emergency workers in biosafety suits. Courtesy of Photographer's Mate 1st Class William R. Goodwin/U.S. Navy.

Acts of biological terrorism emerged in the mid-1920s and pose a threat greater than that of biological warfare. Terrorism can be accomplished by a small group of individuals and even by a single person making it difficult to detect, whereas warfare, by its magnitude, can hardly escape detection.

Anthrax is a major threat for several reasons, but primarily because *B. anthracis,* the causative microbe, is a spore-forming bacterium. Microbes and their toxins are considered "the poor man's weapons of mass destruction," based on their cost-effectiveness, putting them within reach of many individuals, groups, and nations. The CDC, an agency within the DHHS, is the lead agency responsible for a plan of readiness against the potential use of biological weaponry in warfare and in terrorism.

In the Aftermath of September 11, 2001

In the days following Al Qaeda's September 11, 2001 attack on the World Trade Center and the Pentagon, Americans were gravely concerned and anxiety stricken about what terror might follow. In a *Time*/CNN poll, 53% of those surveyed feared a chemical or biological attack, whereas 23% feared a nuclear attack would follow.

The reality of using *B. anthracis* spores as biological weapons surfaced only two weeks later when, on September 25, a 38-year-old assistant to NBC's Tom Brokaw developed the first case of cutaneous anthrax (BOX 15.3), presumably as the result of opening a letter, postmarked September 18, containing a powder. Over the next few weeks more cases of inhalational anthrax and cutaneous anthrax

BOX 15.3 ___Timeline of Anthrax Attacks in September 2001_____

September 18, 2001

A 38-year-old assistant to NBC News anchorman Tom Brokaw handles a letter containing powder. The letter was postmarked with the same date. On September 25 she notices a raised skin lesion on her chest, and over the next three days she experiences progressive erythema and edema. On September 29 she develops malaise and a headache. By October 1 the lesion has developed into a 5-cm oval with raised borders and satellite vesicles. There is left cervical lymphadenopathy (swollen and enlarged lymph nodes), and a black eschar (a scab-like lesion) soon develops. This turns out to be the first case of cutaneous anthrax. The patient recovers with antibiotics.

September 28, 2001

A seven-month-old son of an ABC producer is taken to his mother's workplace at ABC in New York. On September 29 a large weeping lesion is noted on his left arm. Over the course of days the lesion develops into an ulcer with a black eschar (scar). The child is hospitalized and develops hemolytic anemia and thrombocytopenia—a low blood platelet count. He is diagnosed as having cutaneous anthrax and recovers uneventfully with antibiotics. Subsequently, cases of cutaneous anthrax turn up at CBS, *The New York Times*, and the *New York Post*. All patients recover.

October 2, 2001

Robert Stevens, 63 years old, is admitted to a Lake Worth, Florida hospital gravely ill with the presumptive diagnosis of meningitis. The diagnosis of inhalation anthrax is made after testing. He died on October 5, the first bioterrorism casualty of this millennium. Anthrax bacteria are found at his workplace at American Media Inc. A coworker, Ernest Bianco, is admitted to a Miami hospital with the diagnosis of pneumonia, which is later changed to inhalation anthrax. He eventually recovers.

October 14, 2001

An aide to Senate Majority Leader Tom Daschle opens a letter to the senator postmarked from Trenton, N.J. The letter is loaded with high-grade, light, fine-textured *B. anthracis* spores. Three days later, twenty-eight persons test positive for exposure.

October 16, 2001

Joseph Curseen, Jr., 47 years old, a Washington, D.C. postal worker, falls ill with flulike symptoms but is not hospitalized. He returns to the hospital on October 22 and dies of inhalation anthrax. More than 2,200 Washington postal workers are placed on a 10-day supply of ciprofloxacin.

October 19, 2001

Thomas Morris, Jr., fifty-three years old, an employee of the Brentwood post office, which handles all mail delivered into Washington, D.C. (including the mail sent to Senator Daschle's office), is admitted from the emergency room at Inova Fairfax Hospital in Fairfax, Virginia. He is the third person to be diagnosed as having contracted inhalation anthrax. Morris dies on October 21.

October 23, 2001

Attorney General John Ashcroft releases the text from the anthrax letters sent to Daschle, Brokaw, and the *New York*

targeting primarily those in the media emerged; New York City was particularly hard hit, with seven cases of cutaneous anthrax (two at NBC, one at CBS, three at the *New York Post*, and one at ABC) and one case of inhalational anthrax. The buildings housing these offices tested positive for anthrax spores. The offices of Senate Majority Leader Tom Daschle and Senator Patrick Leahy and dozens of personnel were contaminated on October 14 by a finely milled version of "anthrax powder" sent in letters postmarked from Trenton, N.J. mail facilities to the Washington, D.C. area became centers for cross-contamination of mail delivered into the city by mail trucks serving as vectors of disease. Two postal workers died of inhalation anthrax, and over 2,000 postal workers in the nation's capital were

Post. The letters dated September 11 contain the phrases "Take the penicillin now," "Death to America," "Death to Israel," and "Allah is Great." Meanwhile, the CDC faces public criticism at a bioterrorism hearing at the Capitol. Defenders cite the CDC's lack of adequate resources. Two cases in point: the CDC operates in World War II-era buildings and bad wiring caused a power outage that delayed by 15 hours the CDC's ability to identify the anthrax case at NBC News.

October 24, 2001

A fifty-nine-year-old Washington, D.C. postal worker is admitted to an emergency room with a fever of 100.8°F, sweats, myalgia (muscle pain), chest pain, nausea, vomiting, diarrhea, and abdominal pain. Blood cultures lead to a diagnosis of anthrax.

October 25, 2001

Homeland Security Director Tom Ridge reports that the anthrax powder sent to Daschle's office was highly concentrated and designed to be disseminated and inhaled easily. The Postal Service decides to begin environmental testing at 200 postal facilities along the East Coast. The number of confirmed anthrax cases rises to 13, of which 7 are inhalation and cutaneous anthrax. Most of the cases are linked to mail passing through New Jersey, New York City, or Washington, D.C. An estimated 10,000 people are placed on prophylactic antibiotics.

The same day, Kathy Nguyen, a 61-year-old woman who works in the stockroom at Manhattan Eye, Ear, and Throat Hospital, falls ill with myalgia and malaise. She is hospitalized on October 28, and over the course of the next few days she develops shortness of breath, chest discomfort, and a cough with blood-tinged sputum. Cultures of her blood grow *B. anthracis*, the anthrax bacillus. On October 31 Nguyen dies of inhalational anthrax.

October 28, 2001

Officials confirm a new case of inhalation anthrax in a New Jersey postal worker. He works at the facility that processed the three anthrax-laden letters going to New York City and Washington, D.C.

October 29, 2001

The CDC reports two new cases of anthrax in New Jersey. One is inhalation anthrax in a postal worker, and the other is cutaneous anthrax in a private citizen who may have contracted the disease from mail.

November 16, 2001

A 94-year-old woman in Connecticut is diagnosed with inhalation anthrax. On November 21 she dies of this disease. How she contracted anthrax remains a mystery.

January 9, 2002

The CDC reports that a total of 23 cases of anthrax have occurred: 11 cases of confirmed inhalation anthrax and 12 cases of cutaneous anthrax (7 confirmed and 5 suspected).

March 4, 2002

Cutaneous anthrax is diagnosed in a laboratory worker.

From Rega P. *Biological Terrorism Response Manual*. Maumee, OH, 2001 and Mina B, Dym J P, and Kuepper F. *JAMA* 2002;287:858–862, 863–868.

placed on a ten-day course of the antibiotic ciprofloxacin (Cipro®). Since then, there have been many cases of anthrax hoaxes (BOX 15.4).

The microbes were not of the common or "garden variety" but had been in the hands of a trained microbiologist. The investigation by the Federal Bureau of Investigation (FBI) to identify the perpetrator was not completed until 2008, and its conclusion is hotly contested by some members of the scientific and the political community (BOX 15.5).

In the aftermath government bodies, including the U.S. Senate, the Central Intelligence Agency, and the U.S. Supreme Court, closed for various times; Americans were asked not to hoard ciprofloxacin following a buying frenzy of the

BOX 15.4 _Anthrax Hoaxes (2006–2008)_

- September 26, 2006: Letters are sent to Keith Olbermann, Jon Stewart, Sumner Redstone, David Letterman, Nancy Pelosi, and Charles Schumer containing a white powder identified as soap powder. Perpetrator is convicted on fourteen felony counts.
- September 27, 2006: Suspicious liquid is found, along with threatening signs, near the Lincoln Memorial, causing it to be temporarily closed.
- February 27, 2007: At the University of Missouri, Rolla Campus, a student makes a bomb threat and has a white powder that he claims to be *B. anthracis* spores, resulting in campus closing. The bomb threat is a hoax and the powder is ordinary sugar.
- October 22, 2008: More than two dozen J. P. Morgan Chase banks and other banks receive letters containing a white powder. No health threats were reported.
- November 11, 2008: Mormon temples in Los Angeles and Salt Lake City receive packages of a white powder in protest to the church's stance on marriage. Powder poses no threat.

BOX 15.5 _Search for the "Anthrax Killer"_

Who was responsible for the anthrax mail attacks that killed five people, sickened seventeen others, and crippled government agencies in 2001 in the aftermath of 9/11? The Federal Bureau of Investigation concluded, after a seven-year investigation, that Dr. Bruce Ivins, a microbiologist and anthrax expert at Fort Detrick, was the sole perpetrator. The truth will never be known because Ivins committed suicide on July 29, 2008, two days before prosecutors prepared to indict him.

Ivins had been employed at Fort Detrick for eighteen years. In 2003 he (and two others) had been awarded the Decoration for Exceptional Service, the highest honor a Defense Department civilian employee can receive. After the 2001 anthrax incidents, he was instrumental in helping the FBI analyze samples of suspicious powder that had been mailed.

But it seems that there was another side to Dr. Bruce Ivins. At various times he has been described as paranoid, depressed, homicidal, sociopathic, and "god-like," all indicative of a troubled mental history. There was a question as to why in 2001 Ivins committed a breach of protocol by not reporting an episode of *B. anthracis* contamination, raising questions about his veracity. As a student at the University of Cincinnati, he was obsessed with a particular sorority and sought to defame its reputation.

In late 2006 Ivins became the target of the FBI's anthrax investigation and was the subject of aggressive tactics by the agency. His son had been offered $2.5 million dollars to "rat" on his father, and his daughter had been shown photographs of anthrax victims for which, allegedly, her father was responsible. Central to the FBI's case was the genetic linkage between the 2001 anthrax strain used as the murder weapon and the strains found in Ivins' laboratory.

Not all are convinced of Ivins' guilt and don't agree with the FBI's conclusion that the case is closed. Only five years earlier the FBI's investigation resulted in Dr. Stephen Hatfill, an expert in bioweapons with access to *B. anthracis*, being labeled as a "person of interest" and subsequently relentlessly harassed and pursued. The case was later dropped, and Hatfill received a settlement of 5.82 million dollars. The FBI later apologized to Hatfill. A number of scientists at Fort Detrick, along with members of Congress, raised doubts and lingering questions as to Ivins' guilt and called for an independent investigation.

On July 29, 2008, Ivins died of an overdose of Tylenol after learning that the FBI was likely to file charges against him for his criminal connection to the 2001 anthrax attacks. The Justice Department officially ended its eight-year investigation in 2010. They released documents supporting that Ivins single-handedly committed the worst act of bioterrorism in U.S. history. The FBI's handling of the case was criticized by Ivins' colleagues and independent analysts who have pointed out the lack of physical evidence linking him to the anthrax letters.

antibiotic. The U.S. Postal Service instituted new security measures and distributed information about how to recognize and report suspicious mail (FIGURE 15.5). The nation's airlines were threatened with bankruptcy, and security measures in and around airports were heightened to protect the flying public. Northwest Airlines even went as far as to ban the sweetener Sweet'N Low because of its resemblance to the powdery substance associated with *B. anthracis* spores. Fire and police departments and ambulance services were bombarded with calls to check out "suspicious powders," and emergency room facilities were overwhelmed with people fearful that they had been exposed to anthrax spores.

Incidents, threats and near-panics continue in a society suffering from the events of 9/11. On December 25, 2009, a young Nigerian man walked on to a Northwest Airlines flight with 289 passengers bound from Amsterdam to Detroit. Apparently scanning had not been effective because sewn into his underwear was a plastic explosive later referred to as a "blessed weapon." Shortly before reaching Detroit he attempted to detonate the explosive but ended up burning his pants off and with burns on his thighs and genitals. He was subdued by others on board.

The country continues to be on alert in the post-9/11 world and worries, perhaps excessively, about the threat of bioterrorism. A magnitude 8.5 earthquake hit Virginia ten years later on August 23, 2011, causing tremors to be felt as far away as New York City. Buildings swayed, church spikes were damaged, two nuclear reactors close to the epicenter were automatically taken offline by safety systems, ceiling tiles fell, offices closed, and people scrambled for safety. Persons in the Empire State Building raced down dozens of flights of stairs. As one worker stated, "I thought we had been hit by a plane."

Reports of suspicious bioterrorist-type activity that turn out to be of no consequence continue to close buildings, streets, parks, schools, and other gathering places at the cost of anxiety-stricken people, lost wages, and millions of dollars spent in response. Suspicious-package reports and letters, all which need to be investigated, continue to stretch law-enforcement agencies around the country. New York and Washington, D.C., are particularly affected. New York City police have investigated over 83,700 calls since the terrorists attack in 2001. Washington, too, has been besieged by calls, partially due to the recently launched Department of Homeland Security program "If You See Something, Say Something."

Enacted on the heels of the terrorists attack, the Patriot Act of October 26, 2001, was designed to detect and prevent bioterrorism, some argue at the loss of personal privacy; as a result security has been beefed-up, particularly at airports. On November 13th, 2010, a 31-year-old male passenger at a San Diego airport refused to have a full body scan, sometimes referred to as a "naked scanner," that some claim is a strip search beneath the clothes. He was offered as an alternative an "enhanced" pat down allowing the inspector to use the front of his hands and fingers to touch passengers in the groin area. The passenger threatened suit on the grounds of sexual harassment and warned, "Don't touch my junk (a male term for genitals)" and then apparently changed his mind and left the airport. An investigation team of the incident is ongoing.

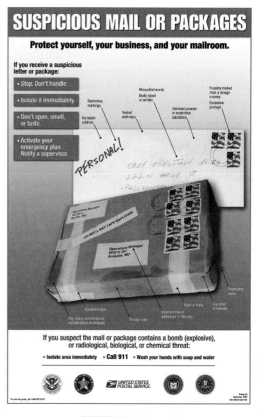

FIGURE 15.5 Postal Service instructions for dealing with suspicious mail. Poster #84 © 2006, United States Postal Service. All rights reserved. Used with permission.

FIGURE 15.6 Is avian influenza the next bioterrorist threat? © Victor Habbick Visions/Photo Researchers, Inc.

Influenza remains a cause of concern, particularly now that a team of virologists at Erasmus Medical Center in Rotterdam and an independent team at the University of Wisconsin-Madison have created in their laboratories a strain of airborne H5N1 virus, some details of which were presented at a conference in Europe in September 2011. The animal model was the ferret, the usual model for influenza studies, and the experiment resulted in influenza spreading by air within the colony. The original H5N1 is not airborne and is not easily transmitted from human to human; it is spread only by direct contact between persons and sick or deceased birds. The fear is that the modified "superstrain" could easily spread from human to human by a cough, sneeze, talking, and any other means of sending droplets into the air, potentially causing a pandemic rivaling the 1918 Spanish flu pandemic that killed over 40 million people in only 2 years. The fear is justified and focuses on the use of this information by terrorists groups to produce biological weapons as well as the possibility of accidental release of this altered virus. The genetic modification of microbes is dangerous territory as is the closely related topic of synthetic biology. In fact, the prospects are so frightening that the federal advisory board has delayed publishing details (FIGURE 15.6).

So what do we do? What is the take-home lesson? If only there was an easy answer. As individuals we must avoid panic, but we must always be on the alert. We must avoid an epidemic of fear. Perhaps we can look forward to better times and heed the admonition of the song title cited in the opening of this chapter, "I Ain't Gonna Study War No More."

■ Overview

Biological warfare has been deployed for centuries as armies realized the devastating effect of causing disease among their adversaries. Human excrement, dead human and animal bodies, and contaminated drinking water are crude methods of biological warfare that continue as threats into the twenty-first century.

Biological weapons are weapons of mass destruction used by rogue nations, religious cults and other groups, terrorists, and fanatic individuals to intentionally cause illness and death in populations. Ironically, the "Golden Era" of microbiology (the end of the nineteenth and the early years of the twentieth century), a time of increasing knowledge of microbes, made possible the development of microbes and their toxins as weapons. Biological warfare has been deployed since ancient times as armies realized the devastating effects of causing disease among the opposition. Acts of biological terrorism emerged in the twentieth century and pose a threat more difficult to detect than outright warfare. The horrors of biological warfare sparked the signing of the 1925 Geneva Protocol followed soon afterward by other treaties to prohibit their use.

Microbes and their toxins are the "poor man's weapons of mass destruction" based on their cost effectiveness in putting them into the reach of many nations. Anthrax, primarily because it is caused by a spore-forming organism, is considered to be the number-one threat. The U.S. military mandates anthrax immunization for personnel serving fifteen days or more of continuous service in certain areas of the world. The Centers for Disease Control is the lead agency responsible for a plan of readiness and implementation against biological weaponry.

Self-Evaluation

PART I: Choose the single best answer.

1. Napoleon's army was defeated by

    a. typhoid fever **b.** typhus fever **c.** diarrhea **d.** battlefield gangrene

2. When was the Geneva Protocol signed?

    a. 1920 **b.** 1925 **c.** 1948 **d.** 1972

3. In 184 B.C. Hannibal used a crude form of biological warfare. His act involved

    a. poisoning wells **b.** poisoning horses **c.** serpents **d.** catapulting human cadavers

4. In 1981 a religious commune purposely contaminated salad bars in an Oregon community. Which microbe was used?

    a. *Salmonella* **b.** *B. anthracis* **c.** influenza virus **d.** *Clostridium botulinum*

5. Which form of anthrax poses the greatest threat?

    a. cutaneous anthrax **b.** inhalation anthrax **c.** gastrointestinal anthrax **d.** waterborne anthrax

6. The anthrax vaccine consists of

    a. dead bacteria **b.** live bacteria **c.** attenuated bacteria **d.** cell-free filtrate

7. Which army facility has been involved in biological warfare defense?

    a. CDC **b.** Plum Island **c.** Fort Detrick **d.** Walter Reed Naval Hospital

8. Which of the following antibiotics was approved to treat anthrax in adults and children?

    a. ciprofloxacin **b.** erythromycin **c.** tetracycline **d.** sulfa drugs

9. Which of the following is a Category A biological agent?

    a. *Brucella* sp. **b.** variola virus **c.** *Mycobacterium tuberculosis* **d.** yellow fever virus

10. Which agency was charged with regulating and transferring biological agents in 1997?

    a. FDA **b.** NIH **c.** CDC **d.** FBI

PART II: Fill in the blank.

1. An epidemic of a _____ occurred in Sverdlovsk, Soviet Union in 1979.

2. In 1966 _____, a nonpathogenic sporeformer related to *B. anthracis* was released into the subway system of New York City to determine the effectiveness of aerosolization and the city's vulnerability.

3. No country and no community are safe from warmongers or from _____.

4. During a _____ outbreak, bodies of dead soldiers with large quantities of excrement were hurled into the ranks of the enemy at Karlstein, Bohemia in 1422.

5. In 1995, Larry Harris used his personal credit card to order bubonic plague (*Yersinia pestis*) bacteria from _____.

6. A relatively long _____ for many infectious agents plays a factor in their tactical usefulness.

7. Biological agents and the diseases they cause have been categorized into three groups based on their risk to the _____.

8. _____ is a Category A bacterium that causes double vision, blurred vision, slurred speech, difficulty swallowing, dry mouth, muscle weakness, and paralysis of breathing muscles, leading to death.

9. _____ is a Category A bacterium that causes fever, ulcerated skin lesions, chills and shaking, headache, weakness; can progress to respiratory fever and shock.

10. A pathogen selected as a bioterrorist agent would be one that is not routinely seen in the population. This would mean that the population would be particularly susceptible due to a lack of _____.

PART III: Answer the following.

1. Cite three landmark treaties associated with biological warfare. Describe each, and give an approximate date for each.

2. Briefly describe the three types of anthrax and how they are acquired. Why is anthrax on the list of biological weapons?

3. How do you feel about the Pentagon's decision to make the anthrax vaccine mandatory for those in the military? Defend your position.

4. Biological weapons are the least expensive method of mass destruction to use. What are the most expensive weapons?

5. Anthrax is a top choice as a biological weapon. Name two properties that make anthrax an ideal weapon.

6. What can you do to prepare for bioterrorism?

7. Biological agents can be spread through air, water, or in food. Provide examples of agents that represent each mode of transmission.

8. List at least three advantages and three disadvantages of biological weapons.

9. Explain how biological weapons threaten a nation's security.

10. What are the warning signs of a bioterrorism attack?

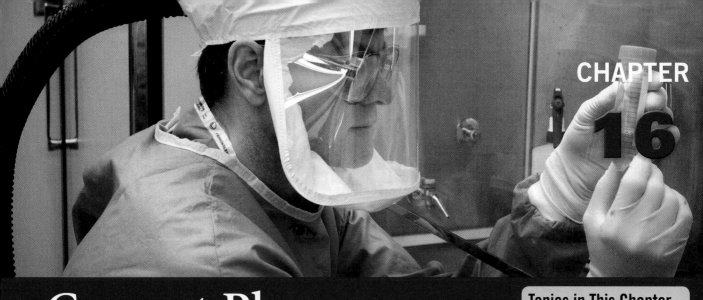

CHAPTER

16

Current Plagues

Even if you take every year the problems of malaria, tuberculosis and HIV/AIDS, which are horrible, pandemic influenza has the potential in just one fell swoop to kill so many more people than those diseases kill in decades.

—Michael Osterholm

■ Preview

The occurrence of large-scale outbreaks of disease, commonly referred to as plagues, is the result of the lack of adaptation between humans and microbes. A variety of microbe-caused plagues has decimated populations and influenced the course of civilizations since ancient times. The Black Death and plagues of small-pox are infamous. Plagues are not relegated to history; new diseases and new epidemics continue to emerge as humans and microbes match wits in the ongoing process of adaptation. Influenza, AIDS, and tuberculosis (TB) are modern-day plagues discussed in this chapter.

Photo courtesy of James Gathany/CDC.

Topics in This Chapter

- ■ Nature of Plagues
- ■ Influenza
  - Biology of Influenza
  - Transmission of Influenza
  - Influenza Strains and Vaccines
  - The Coming Flu Pandemic?
  - The First Flu Pandemic of the Twenty-First Century
  - Bird Flu Scare(s)
- ■ AIDS
  - Biology of HIV
  - Origin of AIDS
  - Transmission of AIDS
  - AIDS: The Disease
  - Cause of AIDS
  - AIDS Treatment and Prevention
  - Consequences of AIDS
  - AIDS in Africa, Caribbean, Eastern Europe, and Central Asia
  - AIDS in the United States
  - The Future of AIDS
- ■ Tuberculosis
  - Current Status
  - TB: The Disease
  - Diagnosis
  - Antibiotic Therapy
  - DOTS
  - Factors Contributing to Reemergence
  - Prevention

"A pox on you" and "a dose of the pox" were once expressions a person might use to wish evil on an enemy. The "pox" probably refers to the lesions associated with the smallpox virus, which had reached epidemic proportions in eighteenth-century Europe. Gradually, the term "pox" came to mean all forms of grotesque diseases manifested by horrendous sores over the body. It serves as a reminder that epidemics date back to the evolution and coexistence of humans and microbes. Disease is the result of a failure of adaptation; the more serious and fatal diseases are those with the least adaptation. In fact, today's normal flora may well have been yesterday's pathogens, and today's current pathogens may be tomorrow's normal flora. It is a matter of adaptation as each species battles to survive and to reproduce to guarantee the next generation; the battle of genetic wits leads to survival of the fittest.

On many occasions throughout history circumstances have caused humans and microbes to lock horns, resulting in a battle of epidemic or pandemic proportions. The term "plague," strictly speaking, refers to the bubonic and pneumonic plague of the Black Death, but the word is used in a popular and somewhat loose sense to describe outbreaks of disease. Epidemics and pandemics of the Black Death, smallpox, malaria, typhus fever, and yellow fever have sporadically wiped out populations and influenced the course of civilization; examples of the plagues are presented in this chapter. In time, plagues run their course and die out as the population of susceptible (nonimmune) individuals decreases. Implementation of public health measures and treatment of disease during the twentieth century were highly significant in interrupting the cycle of disease. Smallpox has been eradicated, and other diseases have been controlled, but the threat of other epidemics and pandemics remains. A case in point is that of the outbreak of H1N1 (swine) flu in Mexico that surfaced in April 2009 as the first pandemic of the twenty-first century. A highly pathogenic bird influenza, H5N1, which occasionally is able to infect people who have contact with infected birds, continues to receive high attention as a potential cause of a pandemic. Recently, in 2011, a highly contagious airborne strain of H5N1 that is transmitted from ferret to ferret was engineered by virologists. Some scientists feel this is Dr. Frankenstein's worst nightmare in the making. (An 1818 novel by Mary Shelley depicts Dr. Frankenstein as a scientist who created a monster using an unorthodox scientific experiment.) There are fears that this is the next bioterrorist weapon, labeling it the "Armageddon virus."

Microbiologists have described the outbreak of an epidemic as being like a forest fire. The human–microbe conflict is the result of the combination of environmental changes, population increase and urbanization, technological advances, and human behavior. These factors serve as kindling to an outbreak. Once ignited, a disease, like a fire, can spread to pandemic proportions as long as the susceptible population is large enough to sustain and foster its transmission, a phenomenon described as the R-nought factor. Again, like a fire, when all the trees have been consumed, the fire will die out.

The outbreak of diphtheria in the newly independent states of the former Soviet Union in the 1990s was kindled by the decline in the public health infrastructure resulting from internal wars and a devastated economy, accompanied by a decline in hygiene and in immunization of the population. Hence, soldiers returning from distant battlefields may have been the spark to fuel diphtheria in the nonimmune population.

Accounts in the Bible describe pestilence, and many were epidemics. Medical historians trace the natural history of infectious disease and epidemics to the days when nomadic hunter–gatherer societies began to form community partnerships to share their meager resources. But, in so doing, they also shared their microbes from one individual to another, a process that magnified as villages merged into cities and, ultimately, into urban sprawl. Wars through the ages left epidemics in their wake, resulting in devastation greater than that wrought by the war itself.

It seems that conflict between humans and microbes is a natural part of the biosphere. Adaptation between humans and microbes is like being on a merry-go-round and trying to capture the elusive brass ring. Along the ride, new diseases and new epidemics are spawned as progress is made against older diseases. The proof of this is evidenced by daily news headlines and news broadcasts heralding new disease outbreaks.

AUTHOR'S NOTE (RIK)
"Plague" is a commonly used word to describe circumstances denoting misery and disease outbreak. For example, today's society speaks of the plagues of violence, drug abuse, poverty, and alcoholism, to mention only a few. My exams have often been described as a plague and are, undoubtedly, thought of as a misery, particularly to those who do not study. On the other hand, correcting exams is a plague!

Influenza

The year was 1918 and the world was engaged in the horrors of World War I. As if that were not tragic enough, along came the **influenza** virus, a biological agent knowing no political or national loyalties. Influenza took a toll on human life greater than the ravages of the war itself; the pandemic that it created is regarded as one of the most devastating of all time (FIGURE 16.1). Little was known about viruses; it was not until about twenty years later that viruses were studied in earnest. Several million people died in the short span of only 120 days, including an estimated 500,000 to 675,000 in the United States. Estimates are that the virus killed 50 to 100 million people worldwide in a single year, including 17 million in India. The virus killed rapidly; a person could be fine in the morning, feel sick in the afternoon, and be dead by nightfall! Bodies piled up, and burying of the dead became a major public health problem. The overworked, underpaid, and at-risk gravediggers, perhaps in an effort to keep their sanity during those terrible times, developed their own jargon, stories, and poems. One poem that surfaced was the following:

> *I lost a little bird*
> *And its name was Enza*
> *I opened the window*
> *And in-flu-enza!*

FIGURE 16.1   A ward of patients with influenza. The 1918 influenza was a catastrophe and killed millions of people all over the world. Courtesy of the National Library of Medicine.

Biology of Influenza

The term "flu" is certainly familiar to you and is used somewhat loosely to describe coldlike respiratory symptoms accompanied by muscle pain, but in most cases this is probably not the "true" flu. "Stomach flu" is another vague term used to describe symptoms associated with gastroenteritis; the flu virus does not cause gastroenteritis. Influenza is caused by an RNA virus that is as distinct from the cold virus and the

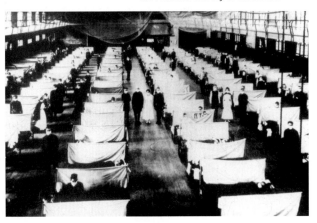

gastroenteritis-causing virus as dogs are from cats. So the term "flu" is frequently used inappropriately (particularly by students on exam days!), as are the "grippe" and the "24-hour" viruses. The real flu is characterized by exaggerated coldlike symptoms, including headache, high fever, muscle pain (particularly in the back and legs), severe cough, and congestion. The disease usually disappears in a week or two, and treatment, including bed rest and fluids, is aimed at relieving the symptoms. Some influenza viruses infect not only humans but other species, including seals, dogs, pigs, and birds (particularly ducks).

There are three categories of flu viruses, all RNA viruses, designated A, B, and C. **Influenza A virus** causes epidemics, and occasionally pandemics, and is associated with animal reservoirs, particularly birds. **Influenza B virus** is less severe, causing only epidemics, and there is no animal reservoir, whereas **Influenza C virus** does not cause epidemics and produces only mild respiratory illness. Influenza control is directed against types A and B.

Transmission of Influenza

Influenza is acquired from droplets and aerosols; fomites play a secondary role. The incubation time is twenty-four to forty-eight hours. Young children are particularly significant sources of the virus because of their relatively nonhygienic practices. Conditions of colder temperatures (5°C or 41°F), low humidity (20%), crowding, and close intermingling favor transmission, as in theaters, classrooms, nursing homes, and barracks. The disease usually peaks from about the middle of December through early March, the months in temperate climates when "fresh air" indoors is not possible. There is no seasonal variation in tropical areas.

Influenza Strains and Vaccines

Each year, in early October, preparation for the flu season begins; those over 65 years of age and other high-risk groups, including people with immunodeficiency diseases and, in some cases, pregnant women, are particularly urged to get a flu shot (FIGURE 16.2). Why is this necessary each year? Lifetime, or at least long-term, prevention is usually the case in immunization against diseases such as polio, measles, and mumps—but not so with influenza. The explanation lies in the biology of the influenza virus. Protruding through the viral membrane are **hemagglutinin (H)** spikes, of which there can be as many as 500 per virus, and **neuraminidase (N)** spikes, of which there are about 100 per viral particle (FIGURE 16.3). Both of these serve as virulence factors. The H spikes are essential in attaching the virus to the epithelial (lining) cells of the respiratory mucosa and in aiding their penetration into these cells. The N spikes play a role in the last stage of the viral replication cycle, namely, release of new virions, allowing for spread to other cells. Both of these spikes are antigens, which means they provoke the production of specific antibodies as part of the body's defense mechanism. Now, here is the reason why influenza vaccination must be given on a yearly basis. These spikes undergo variation in their antigenic structure. The different forms are assigned numbers, for example, H1, H2, H3, N1, N2,

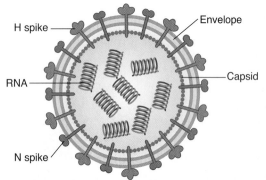

FIGURE 16.3 Influenza virus H and N spikes.

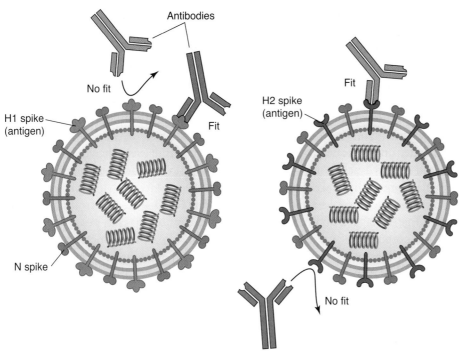

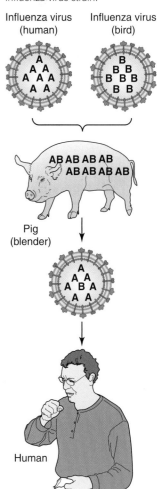

FIGURE 16.4 Lock-and-key arrangement of antigen and specific antibodies.

FIGURE 16.5 Origin of a new influenza virus strain.

and N3. The full nomenclature of a virus reflects the virus group (A, B, or C), where and when the virus was isolated, and the H and N types. An example is the A/Philippines/82/H3N2 influenza virus. There are 17 types of H and nine types of N, allowing for 153 possible combinations. Hence, antibodies produced against H1 react against H1 spikes but not against H2 spikes. Think of it as a lock-and-key arrangement—a key made against lock H1 does not fit into the modified H2 lock but only into the original H1 lock (FIGURE 16.4).

Influenza antigens continually undergo two types of changes, known as **antigenic shift** and **antigenic drift**. Antigenic shift occurs only with influenza type A. The shift is a major, abrupt antigenic change in the H or N spikes that results from the recombination of genetic material from different viral strains, creating a new influenza virus strain (FIGURE 16.5). Consider that if a pig were to be coincidentally infected by both a human strain and a bird strain, a new hybrid strain might develop that could then infect a human; the pig served as a **blender**, somewhat like a mixer creating a new hybrid virus. Flu pandemics commonly originate in China, where millions of pigs, birds, and people live in close quarters, allowing for new combinations of strains and enormous opportunity for species leaps. Possibly, the deadly 1918 influenza arose as a result of antigenic shift. Antigenic drift, on the other hand, is a minor change in the H and N spikes occurring over a period of years.

The upshot of both antigenic shift and antigenic drift is that they enable the virus to evade the antibody defense mechanisms of the host by "changing their coats." Because these changes continually occur, the viral strains that make up the vaccine year to year need to be adjusted. Antibody against last year's viral strains is effective only to the extent that some of these strains may, coincidentally, be the same. If the changes in the virus did not occur, a flu shot would provide long-term, maybe lifetime, immunity.

How do health officials determine almost a year in advance of the flu season which viral strains are to be included in the vaccine cocktail for the coming year? The World Health Organization (WHO), in cooperation with other agencies, has over 100 surveillance sites around the world that gather information regarding the circulating flu types. These data are assembled in the fall of each year, allowing vaccine-manufacturing pharmaceutical companies to begin production in February for the following flu season. Considering this "year in advance" process, it is not surprising that, depending on the year, there may be some mismatch between the viral strains anticipated and those that are included in the vaccine, but even when the match is not perfect, there is usually some cross-protection. Equally significant is the serious problem that some of the viral strains can mutate and become resistant to the antiflu drugs.

Influenza outbreaks are costly. The Centers for Disease Control (CDC) estimates that a staggering number of lost work days occur each year. The cost of vaccine preparation, vaccine administration, and loss of productivity runs into the billions of dollars in the United States alone. Following are some additional statistical and epidemiological facts:

- 5% to 20% of the population get the flu each year.
- Approximately 36,000 people die from the flu each year.
- More than 200,000 people are hospitalized each year.
- A person is contagious one day before symptoms appear and up to five days after symptoms disappear.
- Rate of infection is highest in children.
- Annual vaccination is best prevention:
  - Flu shot: inactivated ("killed") virus used in those six months of age or older
  - Flu mist: live attenuated (weakened) virus used in those aged two to forty-nine years

There is reluctance on the part of many individuals, including those aged sixty-five and older that are the most susceptible, to be vaccinated against influenza. Their refusal of the vaccine is risky, because secondary bacterial infections, possibly leading to a fatal pneumonia, are more likely to occur in the older population. This is why many physicians prescribe antibiotics for their older patients who have been diagnosed with the flu. Historically, it was thought that influenza was caused by a bacterium, *Haemophilus influenzae,* but it turned out not to be the case. In fact, *H. influenzae* is a less frequent cause of secondary bacterial pneumonia in influenza patients than are staphylococci and streptococci.

The FDA approved two antiflu drugs in 1999, **zanamivir** (Relenza) and **oseltamivir** (Tamiflu). Zanamivir and oseltamivir are effective against types A and B influenza viruses, and these two drugs continue as the only recommended anti-influenza drugs. The drugs do not prevent infection, nor are they cure-alls. If taken within about 48 hours after the appearance of symptoms, however, they can shave off a few days of illness and reduce the severity of a bout with the flu. They are not intended as substitutes for vaccines, and, as with all medicines, they have side effects. Furthermore, individuals need to recognize very early flu signs (that mimic many other microbial diseases) and visit their physician almost immediately to get a prescription for these drugs. **Peramivir**, an unapproved drug in phase 3 clinical trials received an Emergency Use Authorization in October 2009 to treat hospitalized patients with known or suspected H1N1 influenza who did not respond to zanamivir (Relenza®) or oseltamivir

(Tamiflu®). Peramivir was administered intravenously, provided rapid drug delivery at high levels. Peramivir treatment for emergency use was terminated by the FDA on June 23, 2010.

The Coming Flu Pandemic?

Historically, type A was responsible for the occurrence of pandemics. Antigenic shifts in type A occur periodically, and people the world over had no antibody protection against each new strain, leaving them vulnerable. A study of influenza epidemics and pandemics that have occurred over the centuries indicates a cyclical pattern of influenza outbreaks.

The 1918 flu pandemic remains a mystery in terms of understanding why this virus was so virulent. Many victims of the disease died because their lungs filled with fluid causing them to drown. Generally, the senior members of the population are usually more prone to die from influenza, but an unusual aspect of the 1918 pandemic is that the disease struck young adults. Because flu pandemics appear to be cyclical, for the sake of future readiness it is important to understand why the 1918 flu pandemic was so devastating. To this end, Johan Hultin, a 73-year-old pathologist from San Francisco, traveled to the Alaskan tundra to exhume the bodies of Inuits who died during the 1918 global influenza outbreak. Hultin brought back tissue specimens for study to determine the genes responsible for the unique virulence of the virus to provide valuable information about thwarting another potential pandemic (FIGURE 16.6). Hultin's samples were analyzed at the Armed Forces Institute of Pathology, and the evidence indicates it to be an H1N1 influenza A virus.

FIGURE 16.6 Johan Hultin collecting tissue samples in 1997 from 1918 flu victims buried in an Alaskan mass grave. Courtesy of Dr. John Hultin.

The First Flu Pandemic of the Twenty-First Century

During the twentieth century major pandemics occurred in 1918, 1957, and 1968. In fact, public health experts feared that we were overdue for an epidemic, possibly a pandemic. Contrary to what many virologists expected, the first pandemic of the twenty-first century started in North America and not in Southeast Asia or China. Mexico became the epicenter of the first pandemic of the twenty-first century (FIGURE 16.7). From the beginning of the outbreak in April 2009, through December 5, 2009, 208 countries reported cases. The influenza virus was identified as a new swine H1N1 influenza strain.

FIGURE 16.7 Newspaper headlines in April 2009, announcing that influenza deaths had risen to 236 in Mexico. © Dreamshot/Dreamstime.com

A striking observation of the 2009 pandemic was that more than 75% of infected people were younger than 30 years of age. The most affected group were 10 to 19-year-olds. Less than 3% of cases were persons over the age of 65. A plausible explanation was that older people born before 1957 still had some cross-reacting neutralizing antibodies that protected them from infection by the 2009 H1N1 strain; it was the same strain that dominated the other circulating strains during the pandemic in Mexico. Based on conservative estimates, the number of 2009 pandemic H1N1 deaths in the U.S. ranged from 7,500 to 12,000, less than half the number caused during a typical flu season. The first doses against the

FIGURE 16.8 Poster on handwashing etiquette on a wall in the Halsey Science Center at the University of Wisconsin Oshkosh. Author's photo (TS).

2009 H1N1 pandemic influenza strain were available in October 2009. Healthcare workers were given first priority to early vaccinations. Other high-risk groups were next on the list: pregnant women and children. This pandemic strain was included in the 2010 and 2011 seasonal influenza vaccines.

Response to a pandemic in the twenty-first century has technological advantages over responses to previous pandemics. The media responded with public service announcements on cough etiquette, hand washing reminders, school closings, and reminding sick individuals to stay home to minimize spreading the flu. Flyers and posters were hung on walls in schools and other public places (FIGURE 16.8). Website healthmaps and cell phone applications tracking influenza cases were created. Users could track outbreaks in their region and alert others by e-mail when new cases were reported in their area. The WHO and the CDC's Emergency Response Team played a vital role in assessing, guiding, and optimizing resources to reduce the impact of the pandemic on businesses, hospitals, schools, and the community. The "wildfire" was manageable this time around.

A swine H3N2 influenza strain was circulating in the United States during the summer of 2012. Over 200 human cases in more than 10 states were reported. Most cases were children showing hogs for 4H or FFA projects at county fairs. Fairgoers were advised to wash their hands and to avoid touching the pigs. The variant strain spread directly from pigs to humans but was not spreading from human to human. Dr. Michael Osterholm, Director of the University of Minnesota's Center for Infectious Diseases Research and Policy commented, "It's time to take what likely would be a very unpopular step—tell organizations this year, pigs should stay home from the fair. These pigs shouldn't be at fairs." Per his warning followed by human flu cases at other county fairs in Ohio, the Cuyahoga County Fair Board in Ohio banned the pig exhibits. Several other county fairs closed their pig barns following human influenza cases. Perhaps the nursery rhyme should be changed to

This little piggy went to the county fair,
This little piggy stayed home
This little piggy had roast beef,
This little piggy had none,
And this little piggy went wee wee wee wee all the way home.

Bird Flu Scare(s)

A near panic occurred in December 1997 when health officials throughout the world warned us of the so-called bird flu (also known as the "Hong Kong flu" or "chicken flu") outbreak in Hong Kong that caused 18 cases of influenza in humans, 6 of whom died. The Hong Kong government slaughtered all of the territory's 1.4 million chickens as a public health measure to break the transmission chain.

After a six-week ban health authorities allowed chickens to be brought back into the city, but, as a precaution, health workers conducted random tests among the 35,000 chickens before allowing them to be transported into the city to a wholesale poultry market. All the chickens tested negative for the bird flu virus. Some celebrated the return of the chickens with dinner parties featuring (you guessed it) chicken! The fear on the part of the public health authorities the world over was warranted; Hong Kong is one of Asia's most crowded cities, with over seven million people living in a small area, and is a major tourist destination.

Although the time is ripe for a major flu outbreak, today's health infrastructure is better equipped in terms of surveillance, diagnostics, and treatment to handle an onslaught of influenza and to avoid the horrors of the 1918 influenza pandemic. For a massive outbreak to occur a large segment of the population would have to be nonimmunized. How can this happen? As pointed out, if a new strain of the influenza virus were to arise by antigenic shift, the world population would be nonimmunized and, hence, susceptible.

The world currently faces the threat of a potential pandemic of avian (bird) flu caused by H5N1. Most epidemiologists and public health personnel would agree that it is not "if" but "when" an outbreak of avian flu will happen.

What is bird flu and what is H5N1? All species of animals (and plants) are hosts for a variety of microbes. Wild birds harbor the influenza virus, along with many other microbes, in their intestinal tract as normal flora usually without displaying signs of illness. The virus is shed into their feces, nasal secretions, and saliva and can be transmitted to other birds that might have contact with these materials or contact with contaminated surfaces. Similarly, domestic birds (poultry) can also pick up microbes. The avian flu virus is highly pathogenic for poultry and has a mortality rate approaching 100%; it can destroy a flock of chickens in only a few days.

There are many strains of type A influenza viruses, based on their combination of H and N spikes that circulate in birds. Rarely do these viruses infect humans, but human infection did happen in Hong Kong in 1997 when H5N1-infected poultry caused influenza in humans, marking the spread of H5N1 into the human population. The species leap—chickens to people—had occurred.

Toward the end of 2003 H5N1 was on the march and by the middle of 2005 it had progressed geographically into wild and domestic fowl in Europe, Asia, Africa, and the Near East. The WHO monitors the appearance of laboratory-confirmed cases of human H5N1 (FIGURE 16.9); their data indicate that, from 2003 through

FIGURE 16.9 Laboratory-confirmed human cases of avian influenza A (H5N1) 2003 to 2012. Data as of March 26, 2012 from the WHO/ Global Influenza Programme.

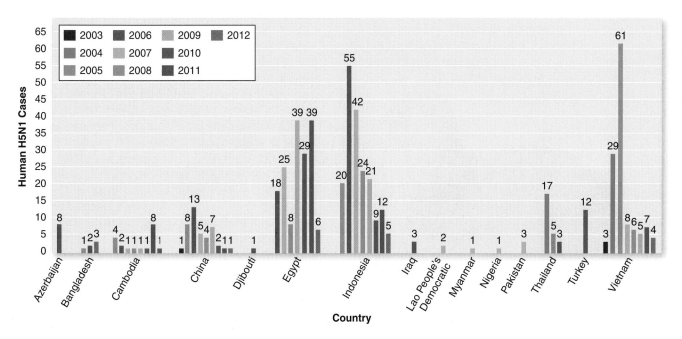

2012, the virus had been reported in 15 countries causing 587 cases and 346 deaths—a 59% mortality rate. These figures are considered to be underreported. Indonesia has been the hardest hit with 139 cases and 113 deaths, followed by Vietnam with 185 cases and 153 deaths. Cases in 2012 were reported in Cambodia, China, Egypt, Indonesia, and Vietnam. The World Organization for Animal Health (formerly the Office International des Epizooties, OIE) maintains up-to-date statistics on animal avian influenza cases.

The virus has never been detected in the United States in wild birds, poultry, seals, or humans. The worst-case scenario is that H5N1 will further adapt to a strain that will allow human-to-human transmission on a sustained basis—"sustained" is the key word—triggering a pandemic. Human-to-human transmission has already

---

BOX 16.1 ___The Armageddon Virus?___

Virologists made an alarming announcement to scientists attending the European Scientific Working Group on Influenza Conference held September 11th to 14th, 2011. They genetically manipulated an H5N1 influenza virus into a version of virus that could easily pass from ferret to ferret by airborne transmission (the best animal model for influenza research, **FIGURE B16.1a**). The collaborative research was led by Dr. Ron A. M. Fouchier from the Erasmus Medical Center in the Netherlands and Dr. Yoshihiro Kawaoka from the University of Wisconsin-Madison and the University of Tokyo.

While experts at the meeting acknowledged that this research helps influenza experts "prepare for the unpredictable," the National Science Advisory Board for Biosecurity (NSABB) began reviewing their findings and placed an embargo on the manuscripts for publication describing their research. The lab-made H5N1 viruses are kept under high security but there were concerns that publishing the recipe for these viruses in scientific journals could fuel the efforts of terrorists who would try to create a biological weapon or "Armageddon virus." Or worse yet, the Fouchier

FIGURE B16.1a The ferret model has been used to evaluate H5N1 viruses that might infect and cause illness in humans. © Jagodka/ShutterStock, Inc.

and Kawaoka viruses could escape the laboratory, triggering an influenza pandemic with millions of deaths! Their findings ignited intense public debates and criticism in the media on the benefits and potential harm of this type of research. Concern was so great that on December 7th, 2011, U.S. Secretary of State Hillary Clinton attended a summit on biological weapons held in Geneva. No American official of her ranking had attended the summit in decades. Just before Christmas, the NSABB advised that the manuscripts be published as long as the methods to create the viruses would be excised or too vague for would-be-terrorists. This decision put the burden of ethics on the shoulders of the editors of *Nature* and *Science*, two leading scientific journals.

Fouchier and Kawaoka braced themselves for the media storm. On January 20th, a letter signed by over 30 senior influenza researchers was published in *Nature* and *Science* declaring a two-month moratorium on H5N1 research that focused on what makes the H5N1 bird flu strain more transmissible between mammals. The intention of the researchers was to obtain knowledge that would benefit public health, allowing scientists to prepare aggressive measures if a mutated H5N1 virus showed up in the wild. The lab-made H5N1 virus allowed them to perform studies testing whether H5N1 vaccines and antivirals would be effective against a new H5N1 influenza strain.

In February 2012 at a meeting assembled by the United Nations agency, experts agreed that full disclosure of the information needed to create these lethal mutant H5N1 lab strains was preferable over a redacted version of the research of mutant flu viruses. Editors of both journals have been open to more discussions by influenza researchers, policy experts, and scientists outside of the immediate field before the manuscripts are published. Details of the secret experiments will inevitably leak out. Electronic information could be leaked or hacked.

happened in family clusters but rarely (probably underreported) and unsustained, but, nevertheless, establishing the potential for a pandemic. Two examples are presented. In Thailand in 2004 human-to-human transmission occurred as a result of prolonged and close contact between a mother and her sick daughter. The WHO reported human-to-human spread in Indonesia in 2006 in a family cluster in which eight people were infected, seven of whom died. There are other examples pointing to human-to-human transmission. The most alarming potential example is the creation of a laboratory mutant H5N1 strain that is highly contagious in ferrets by airborne transmission (BOX 16.1).

The likelihood of a pandemic of H5N1 is impossible to predict, but if sustained transmission within the human population occurs, the rates of illness and

Journal editors have faced biosecurity issues before. Another serious example of research affected by publishing considerations occurred as early as 2001 when scientists genetically modified a mousepox virus to be 100% fatal in mice by the addition of a single gene introduced into the mousepox virus. Mousepox is not known to hurt or infect humans. The work was published purely as a scientific paper at a time when there was no form of biosecurity review.

In 2005, a paper submitted by Stanford University researchers titled *"Analyzing a Bioterror Attack on the Food Supply: The Case of Botulinum Toxin in Milk"* was published. The researchers developed a mathematical model to predict human deaths if botulinum toxin contaminated thousands of gallons of milk in the U.S. In October 2011, the entire reconstructed genetic code of the ancient *Yersinia pestis* bacterium was published, comparing its sequence to modern *Yersinia pestis* genomes. *Y. pestis* caused Europe's fourteenth century scourge and is considered the deadliest pandemic in human history. The ancient bacterial DNA was isolated from the teeth and skeletons of Black Death victims buried in cemeteries in East Smithfield of London.

We are at a crossroads in which well-intentioned scientific research has the potential to be misused for nefarious purposes, a **dual-use dilemma**. When physicists observed atomic fission, they thought their research may have beneficial applications in medicine and energy production but on the flip side, it could also lead to the creation of a devastating atomic bomb. Some of the same discoveries that lead to advancements in medicine or public health can be adapted to the development of weapons of mass destruction (FIGURE B16.1b).

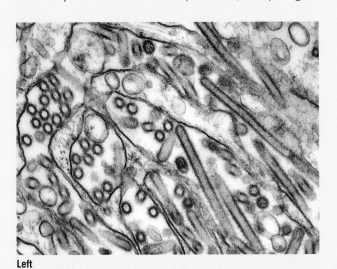
Left

Right

FIGURE B16.1b Examples of research with a dual-use dilemma. (left) H5N1 influenza viruses. Courtesy of Cynthia Goldsmith/CDC. (Right) Atomic bomb testing in 1946. © Scott Camazine/Photo Researchers, Inc.

death would be very high because of the lack of herd immunity. The H5N1 avian flu virus is a new assortment and, hence, the human population has no herd immunity, rendering it extremely vulnerable. Regarding treatment, H5N1 is resistant to two of the four antiviral agents commonly used to treat influenza, and resistance is developing to one of the other two drugs. A vaccine was approved in April 2007 by the FDA against H5N1 and is a part of the U.S. Strategic National Stockpile to be distributed if and as needed. The threat is worldwide, triggering other countries to take a similar measure. The CDC continues to provide state departments of health with updates regarding surveillance, diagnosis, and prevention of avian influenza. It does not restrict travel to countries with known outbreaks.

■ AIDS

Today's college students are living in the shadow of a pandemic of **AIDS**, a dreaded disease caused by the **human immunodeficiency virus** (**HIV**), which surfaced in 1981 and has since spread havoc and desperation throughout the world, particularly in Africa. A distinction needs to be made: HIV infection is not AIDS but rather signifies the presence of the virus. AIDS is a syndrome of many opportunistic diseases.

The stories of Mark Gardner Hoyle and Ryan White remain symbolic. Mark, a student of Case Junior High School in Swansea, Massachusetts, died of AIDS on October 26, 1986 at the age of fourteen. His name is inscribed on home plate at the school's playing field. Ryan was an Indiana teenager who died of AIDS on April 8, 1989 at the age of nineteen. The Ryan White CARE (Comprehensive AIDS Resources Emergency) Act, designed to assist the growing number of Americans living with AIDS but without adequate healthcare insurance, was founded in his remembrance. At the time both boys and their families struggled with an unsympathetic, hostile, and uninformed public intent on ostracizing them. Mark and Ryan contracted and died from this so-called gay disease when, in fact, both suffered from a bleeding disorder known as hemophilia and required blood products in their treatment. These products were contaminated with HIV, the virus that causes AIDS.

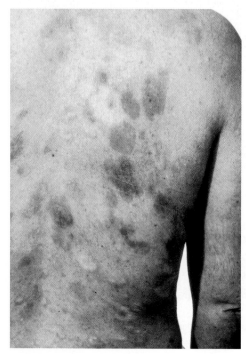

FIGURE 16.10 Kaposi's sarcoma in a young man with HIV. Courtesy of the National Cancer Institute.

In the United States the AIDS story began in the late 1970s when physicians in Los Angeles, San Francisco, and New York City observed clusters of patients with symptoms that included severe fungal pneumonia, swollen lymph nodes, sudden weight loss, a history of recent and new infections, and blue-purplish spots on their skin. These spots were later diagnosed as a rare vascular cancer called **Kaposi's sarcoma** (FIGURE 16.10) first seen in the late 1800s by Moritz Kaposi, a Viennese dermatologist. All patients were young, male, and homosexual.

The acronym "AIDS" first appeared in 1982 in the CDC's *Morbidity and Mortality Weekly Report* and was described as "a disease, at least moderately predictive of a defect in cell-mediated immunity, occurring with no known cause with diminished resistance to that disease." The 2008 definition is based on the CD4+ cell count and the presence of bacterial, viral, protozoan, and fungal opportunistic infections (BOX 16.2). Note that the number of CD4$^+$ T cells is critical in the revised definition of AIDS. In developing countries where diagnostic facilities do not have the capability to do CD4$^+$ T-cell counts, epidemiologists use a case definition based on the presentation of clinical symptoms and the exclusion of other known causes of immunosuppression, such as cancer and malnutrition. This less precise definition results in underreporting of the incidence of AIDS.

Biology of HIV

As in all viruses, the replication of HIV consists of the five stages of adsorption, penetration, replication, assembly, and release (Table 5.3). FIGURE 16.11 illustrates the virus "budding" out of host cells. The biology of this virus and other retroviruses is unique in that the flow of genetic information starts with RNA, not with DNA as the usual starting point.

There are at least two types of HIV: types 1 (HIV-1) and 2 (HIV-2). The best known is HIV-1, the most common cause of AIDS worldwide. In West Africa HIV-2 is most common. Within each group are a number of subcategories. This chapter focuses on HIV-1.

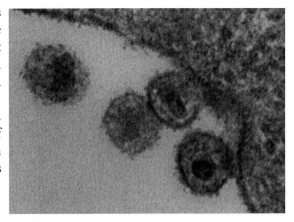

FIGURE 16.11 HIV "budding" out of host cells. Courtesy of Louisa Howard, Dartmouth College, Electron Microscope Facility.

Origin of AIDS

Where did AIDS come from, and why did it explode in the population during the last two decades of the twentieth century? There is considerable speculation regarding this question, but there is no definitive answer. Genetic studies link HIV to an African primate, specifically the chimpanzee *Pan troglodytes*. It is postulated that somewhere along the line cross-species transmission occurred between humans and the chimpanzees. This likely occurred when humans hunted and killed the chimpanzees.

(a)   **(b)**

FIGURE 16.12   **(a)** Robert Gallo, codiscoverer of HIV. Image courtesy of the Institute of Human Virology. **(b)** Luc Montagnier, codiscoverer of HIV. © Jacques Brinon/AP Photos.

The first well-documented case of AIDS occurred in an African man in 1959, although the diagnosis of AIDS was not made until decades later. Before the identification of HIV as the cause of AIDS, sperm antibodies, fungi, and recreational drugs were, at various times, named as the cause of AIDS. The discovery of HIV in 1982 was claimed by two research teams, one headed by Luc Montagnier of the Pasteur Institute in Paris and the other by Robert Gallo of the National Institutes of Health (FIGURE 16.12). A battle ensued as to which of these two investigators should be regarded as the discoverer of the virus. The awarding of the 2008 Nobel Prize in Physiology or Medicine to Montagnier and his colleague Françoise Barré-Sinoussi settled the dispute in the minds of some in the scientific community. Montagnier expressed his disappointment that Gallo had not shared in the award and stated to the Associated Press, "It is certain that he (Gallo) deserved this as much as us two."

Transmission of AIDS

In the early 1980s it seemed that everyone had an opinion on the mechanisms of transmission, and myths abounded. Insects, food, terrorist attacks, casual contact, and "God's curse" were all considered possible vehicles of transmission. The severity and prognosis of the disease, and its association with anal sex, led to fear, stigmatization, and harassment of HIV-infected individuals.

It is important to understand how AIDS is transmitted to break the epidemic cycle and to dispel myths, particularly the "I can't get it" myth that persists in many persons (BOX 16.3). HIV can be transmitted in five ways (FIGURE 16.13):

1. Sexual contact with an infected partner, whether it be male to male, male to female, female to male, or female to female. The virus can penetrate the lining of the vagina, vulva, penis, rectum, or mouth during sexual activity. Some sexual behaviors are considered more risky and unsafe than others. Anal sex is the most dangerous because the lining of the anus is more subject to tears and injury than is the lining of the vagina, allowing the AIDS virus (and other microbes) easier passage into the blood.

2. Contact with infected blood or blood products. Before identification of the cause of AIDS and the development of an HIV screening test in April 1985, persons needing transfusions of blood or blood products, including those with hemophilia, were hard hit. Many recipients of blood or blood products died as a result of AIDS, including tennis star Arthur Ashe, who died in 1995 as a result of having received contaminated blood ten years earlier during heart surgery.

Sexual contact

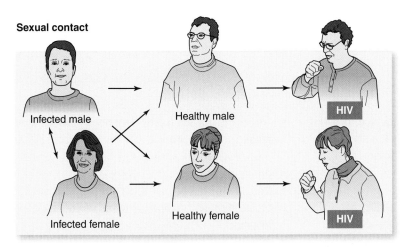

Infected blood or blood products

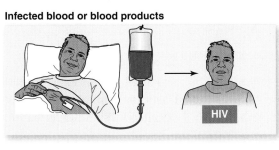

Contaminated needles or syringes

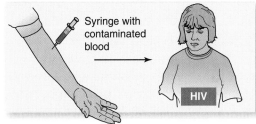

Mother to fetus

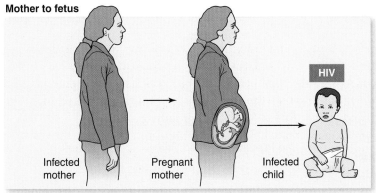

FIGURE 16.13 Transmission of HIV.

3. Sharing blood-contaminated needles and syringes, as is sometimes practiced by drug users when shooting up, runs a high risk of HIV transmission. All it takes is a minute amount of contaminated blood on a needle. Accidental needlesticks and glove tears are a nightmare to all healthcare workers, although the estimated risk of acquiring AIDS from such an incident is less than 0.5%.

4. Transmission from mother to unborn child through the passage of HIV across the placenta carries about a 25% risk if the mother does not receive antiretroviral therapy or breastfeeds. The risk is reduced to 2% if the mother receives antiretroviral treatment. The virus can also be transmitted at the time of birth by infected vaginal secretions or, after birth, through infected breast milk.

5. **Premastication:** prechewing food for infants can transmit HIV.

AUTHOR'S NOTE (RIK)
How comforting is the 0.5% statistic to someone who has accidentally stuck a finger? Not very, as my experience tells me; it happened to me 30 years ago when drawing blood from a patient at Hadassah Hospital in Jerusalem. Fortunately, I did not pick up any infections from the needle. In 1990 a dentist with AIDS was responsible for infecting six patients. Improperly sterilized needles used to deliver local anesthetic were found to be contaminated with the dentist's blood.

BOX 16.3 How You Can and Cannot Become Infected with HIV

How do you get HIV from sexual intercourse?

- HIV can be spread through unprotected sexual intercourse from male to female, female to male, or male to male. Female-to-female sexual transmission is possible but rare. Unprotected sexual intercourse means sexual intercourse without correct and consistent condom use.
- HIV may be in an infected person's blood, semen, or vaginal secretions. It is thought that it can enter the body through cuts or sores—some so small they are not detectable—on tissue in the vagina, penis, or rectum, and possibly the mouth.
- HIV is transmitted by anal, vaginal, or oral intercourse with a person who is infected with HIV.
- Because many infected people have no apparent symptoms of the condition, it is hard to be sure who is or is not infected with HIV. So, the more sex partners you have, the greater are your chances of encountering one who is infected and becoming infected yourself.

How do you get HIV from using needles?

- Sharing needles or syringes, even once, is an easy way to be infected with HIV and other germs. Sharing needles to inject drugs is the most dangerous form of needle sharing. Blood from an infected person can remain in or on a needle or syringe and then be transferred directly into the next person who uses it.
- Sharing other types of needles also may transmit HIV and other germs. These types of needles include those used to inject steroids and those used for tattooing or piercing.
- If you plan to get a piercing or a tattoo, make sure you go to a qualified tattoo artist.

If somebody in my class at school has AIDS, am I likely to get it too?

- No. HIV is transmitted by unprotected sexual intercourse, needle sharing, or infected blood. It can also be given by an infected mother to her baby during pregnancy, birth, or breastfeeding.
- People infected with HIV cannot pass the virus to others through ordinary activities of young people in school.
- You will not become infected with HIV just by attending school with someone who is infected or who has AIDS.

Can I become infected with HIV from "French" kissing?

- Not likely. HIV occasionally can be found in saliva, but in very low concentrations—so low that scientists believe it is virtually impossible to transmit infection by deep kissing.
- The possibility exists that cuts or sores in the mouth provide direct access for HIV to enter the bloodstream during prolonged deep kissing.
- There has never been a single case documented in which HIV was transmitted by kissing.
- Scientists cannot absolutely rule out the possibility of transmission during prolonged, deep kissing because of possible blood contact.

Can I become infected with HIV from oral sex?

- It is possible.
- Oral sex often involves semen, vaginal secretions, or blood—fluids that contain HIV.
- HIV is transmitted by the introduction of infected semen, vaginal secretions, or blood into another person's body.
- During oral intercourse, the virus could enter the body through tiny cuts or sores in the mouth.

AIDS: The Disease

A diagnosis of AIDS is terrifying because the best that an infected person can hope for is a life without too much misery. HIV does not directly cause death, as other microbes do, by their secretions of toxins or by tissue damage. Rather, HIV depletes the number of T-helper cells, resulting in the individual's becoming

As long as I use a latex condom during sexual intercourse, I won't get HIV infection, right?

- Latex condoms have been shown to prevent HIV infection and other sexually transmitted diseases.
- You have to use them properly. And you have to use them every time you have sex—vaginal, anal, and oral.
- The only sure way to avoid infection through sex is to abstain from sexual intercourse or to engage in sexual intercourse only with someone who is not infected.

My friend has anal intercourse with her boyfriend so that she won't get pregnant. She can't get AIDS from doing that, right?

- Wrong. Anal intercourse with an infected partner is one of the ways in which HIV has been transmitted.
- Whether you are male or female, anal intercourse with an infected person is very risky.

If I have never injected drugs and have had sexual intercourse only with a person of the opposite sex, could I become infected with HIV?

- Yes. HIV does not discriminate. You do not have to be homosexual or use drugs to become infected.
- Both males and females can become infected and transmit the infection to another person through intercourse.
- If a previous sex partner was infected, you may be infected as well.

Is it possible to become infected with HIV by donating blood?

- No. There is absolutely no risk of HIV infection from donating blood.
- Blood donation centers use a new, sterile needle for each donation.

I had a blood transfusion. Is it likely that I am infected with HIV?

- It is highly unlikely. All donated blood has been tested for HIV since 1985.
- Donors are asked if they have practiced behaviors that place them at increased risk for HIV. If they have, they are not allowed to donate blood.
- Today the American blood supply is extremely safe.
- If you are still concerned about the remote possibility of HIV infection from a transfusion, you should see your doctor or seek counseling about getting an HIV antibody test. Call the CDC National AIDS Hotline, 1-800-CDC-INFO (1-800-232-4636) which is available 24 hours a day, 7 days a week, or your local health department to find out about counseling and testing facilities in your area.

Can I become infected with HIV from a toilet seat or other objects I routinely use?

- No. HIV does not live on toilet seats or other everyday objects, even those on which body fluids may sometimes be found. Other examples of everyday objects are doorknobs, phones, money, and drinking fountains.

Can I become infected with HIV from a mosquito or other insects?

- You cannot get HIV from a mosquito bite. The AIDS virus does not live in a mosquito, and it is not transmitted through a mosquito's salivary glands like other diseases such as malaria or yellow fever. You cannot get it from bed bugs, lice, flies, or other insects either.

Source: Thirty years of HIV 1981–2011. *Morbidity and Mortality Weekly Report* 60(21):689–699.

immunocompromised and vulnerable to opportunistic diseases caused by an array of microbes (TABLE 16.1).

Infection with HIV does not constitute AIDS but does, with rare exceptions, progress to clinical AIDS over an incubation period that can vary from a few years to as many as 15 years—maybe even more. Those few HIV-infected individuals

TABLE 16.1 Opportunistic Infections Associated With AIDS

Microbe	Disease	Characteristics
Bacteria		
Mycobacterium tuberculosis	Tuberculosis	Primarily lungs, but dissemination possible
Mycobacterium avium,	Extrapulmonary tuberculosis,	Lungs and other organs
Mycobacterium intracellulare	Disseminated infection in immunocompromised persons	
Salmonella sp.	Salmonellosis	Gastrointestinal symptoms
Legionella pneumophila	Legionnaires disease	Pneumonia-like respiratory symptoms
Viruses		
Herpes simplex virus	Herpes	Lesions on skin and mucous membranes
Cytomegalovirus	Cytomegalovirus	Encephalitis, blindness
Varicella-zoster virus	Shingles	Painful, itchy lesions on skin
Epstein-Barr virus	Hairy leukoplakia	Lesions in mouth
Protozoans		
Pneumocystis jirovecii	Pneumonia	Pneumonia-like respiratory symptoms
Cryptosporidium parvum	Cryptosporidiosis	Diarrhea
Toxoplasma gondii	Toxoplasmosis	Brain and nervous system
Isospora belli	Isosporiasis	Diarrhea
Fungi		
Candida albicans	Candidiasis	Lesions on throat, pharynx, lungs
Cryptococcus neoformans	Cryptococcosis	Meningitis and other disseminated symptoms
Histoplasma capsulatum	Histoplasmosis	Pneumonia and other disseminated symptoms

who continue to survive after 10 years live in a state of uncertainty, not knowing when full-blown AIDS will develop. The CDC classifies HIV infection into four stages. It should be emphasized that no two cases are alike, and there might be considerable variation in the time and course of infection. Further, these stages progress from one to another without sharp lines of demarcation (FIGURE 16.14).

Stage 1, the **prodromal stage**, is characterized by fever, diarrhea, rash, aches, headaches, lymphadenopathy (enlarged lymph glands), and fatigue. The symptoms resemble mononucleosis and usually last only a few weeks to a few months. Laboratory evaluation indicates an initial drop in the number of CD4$^+$ T helper (TH) cells from its normal value range of 800 to 1,000 per microliter of blood, but the count remains sufficiently high for the immune system to function. The viral load—the number of HIV particles in the blood—can also be tracked in the laboratory.

Stage 2 is the **latency period**, which can last from 2 to 15 years, with an average of about 10 years. During this time most infected persons are free of symptoms because the virus is in hiding, and the number of CD4$^+$ T cells keeps ahead of the viral load, enabling destruction of virus-infected cells. An AIDS blood test, based on the detection of antibodies against HIV, is the only diagnostic measure and may result in a false-negative report, because the number of antibodies may be insufficient to be detected. Hence, persons may or may not be aware that they are infected, but they can infect others.

Stage 3, persistent generalized **lymphadenopathy**, is characterized by the appearance of swollen lymph nodes, recurrent fevers, night sweats, persistent diarrhea, persistent cough, extreme fatigue, and, in some cases, neurological impairment, including memory loss, confusion, and depression. Opportunistic infections, frequently caused by yeasts, result in the production of painful sores in the mouth and on the tongue (oral thrush) or in the vagina (vaginal thrush); herpesvirus infection resulting in painful eruptions on the skin around the genital area and around the mouth may also occur. This stage usually signals onset of stage 4 within a short time.

Stage 4 is manifested as full-blown AIDS, and death usually occurs within a few years. This stage is one of misery, because the patient is tormented by one debilitating and painful opportunistic infection after another. About the best that can be done medically is to treat the symptoms of each infection as it develops; frequently, this means hospitalization. Since the advent of AIDS, much has been learned about management of opportunistic infections, affording some measure of relief and extended life to those with AIDS. Opportunistic infections occur because the immune system is severely compromised (weakened) as a result of the CD4+ T-cell count dropping below 200 per microliter of blood and an accompanying degeneration of the lymph glands.

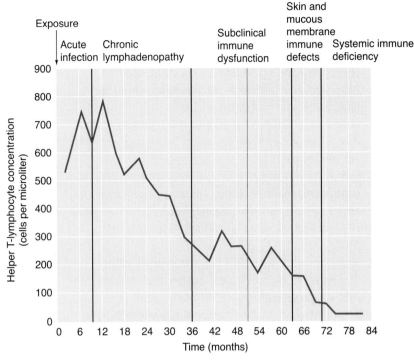

FIGURE 16.14   Stages of HIV infection.

Cause of AIDS

Virtually all AIDS researchers and other scientists agree that AIDS is caused by HIV, yet there is a dissident point of view advanced by a handful of scientists. In the interest of scientific objectivity, the accepted and the dissident points of view are presented in BOX 16.4.

Evidence That HIV Causes AIDS

The fulfillment of Koch's postulates are a series of cause and effect principles, developed in the late nineteenth century, are generally accepted as the criterion establishing the link between a pathogen and a disease. These postulates have been somewhat modified over the years, particularly with regard to viruses, but the basic principles have remained the same for more than a century and continue to serve as the litmus test of causality. The National Institute of Allergy and Infectious Diseases presents the following criteria based on Koch's postulates:

1. Epidemiological association: The suspected cause must be strongly associated with the disease.

BOX 16.4 HIV-AIDS Controversy

Evidence that HIV causes AIDS:

- HIV fulfills Koch's postulates as the cause of AIDS.
- AIDS and HIV infections are invariably linked in time, place, and population group.
- Severe immunosuppression and AIDS-defining illnesses occur almost exclusively in individuals who are HIV infected.
- Before the appearance of HIV, AIDS-related diseases were rare in developed countries; today, they are common in HIV-infected individuals.
- HIV can be detected in virtually everyone with AIDS.
- Nearly everyone with AIDS has antibodies to HIV.
- The specific immunological profile that typifies AIDS (a persistently low CD4+ T-cell count) is extraordinarily rare in the absence of HIV infection or other known causes of immunosuppression.
- Among HIV-infected patients who receive anti-HIV therapy, those whose viral loads are driven to low levels are much less likely to develop AIDS or die than patients who do not respond to therapy. Such an effect would not be seen if HIV did not have a central role in causing AIDS.

An argument that HIV does not cause AIDS:

- HIV antibody testing is unreliable.
- HIV cannot be the cause of AIDS because researchers are unable to explain precisely how HIV destroys the immune system.
- AZT and other antiretroviral drugs, not HIV, cause AIDS.
- Behavioral factors such as recreational drug use and multiple sexual partners account for AIDS.
- AIDS among transfusion recipients is due to underlying diseases that necessitated the transfusion rather than to HIV.
- High usage of clotting factor concentrate, not HIV, leads to CD4+ T-cell depletion and AIDS in hemophiliacs.
- The spectrum of AIDS-related infections seen in different populations proves that AIDS is actually many diseases not caused by HIV.
- HIV cannot be the cause of AIDS because the body develops a vigorous antibody response to the virus.

Source: Adapted from *NIAID Topics,* (updated 2010), *The Evidence That HIV Causes AIDS* (National Institute of Allergy and Infectious Diseases, Bethesda, MD).

2. Isolation: The pathogen can be isolated—and propagated—outside the host.

3. Transmission pathogenesis: Transfer of the suspected pathogen to an uninfected host, human, or animal results in disease in the host.

4. Reisolation: The pathogen must be reisolated from the infected host and be identical to the original pathogen.

Numerous studies around the world since 1981 indicate that virtually all individuals with AIDS are **HIV seropositive** (i.e., have antibodies against HIV), satisfying postulate one. In fulfillment of postulate two, the virus has been isolated in virtually all AIDS patients and in almost all HIV-seropositive individuals. Further, the presence of HIV genes in virtually all persons with AIDS and in those in the earlier stages of HIV disease has been determined by using techniques of molecular biology. Postulate three has been satisfied in tragic incidents, including the cases of three laboratory technicians that developed HIV disease after accidental exposure to the virus and the case of transmission of HIV from an infected dentist to six patients. The development of AIDS has also been documented in healthcare workers after accidental exposure to the virus, in blood transfusion cases, in mother-to-child transmission, and in injection by drug users. Koch's postulates have also been met in experiments with animals, including chimpanzees, monkeys, and certain strains of mice.

A Dissenting View About the Cause of AIDS
It should be clear from the preceding discussion that HIV causes AIDS. However, the dissenting point of view is championed by Peter Duesberg at the University of California, Berkeley, the most outspoken opponent of the HIV-AIDS hypothesis. He is no crackpot. He has top scientific credentials and is acknowledged as an authority on retroviruses. Duesberg and his associates, including people with PhDs and MDs, three of whom are Nobel Prize winners, are convinced that AIDS is not an infectious disease but is caused by lifestyle, environmental drugs, recreational drugs, and anti-HIV drugs. They believe these factors depress the immune system (Box 16.4). Duesberg and four of his colleagues published "HIV-AIDS hypothesis out of touch with South African AIDS—A new perspective" in *Medical Hypotheses* (a scientific journal) in 2009, but the paper was later withdrawn. Subsequently, a similar paper was published in the peer-reviewed *Italian Journal of Anatomy and Embryology* (IJAE) in 2011. The highly controversial paper challenged the mortality estimates of the HIV-AIDS epidemic in Africa and the effectiveness of antiretroviral drugs. Two members of *IJAE's* editorial board resigned in response to the scientific community's outcry of criticism.

The former South African president Thabo Mbeki has also expressed skepticism about HIV causing AIDS and about the treatment of AIDS. He expressed his dissenting opinion at the 13th International AIDS Conference on July 10, 2000, causing many delegates to walk out in protest during his speech. Mbeki's critics claim he is promoting confusion among his own people, who fail to protect themselves because the president (i.e., Mbeki) says that HIV does not cause AIDS. The policies based on Mbeki's oppositional views on the cause of AIDS resulted in the denial of antiviral medicines to AIDS patients and the premature death of 365,000 South African people from 2000 to 2005.

AIDS Treatment and Prevention

Treatment
The bottom line is that there is no cure for AIDS, and there is no preventive vaccine. From time to time, over the past twenty years, "breakthroughs" have been announced, but they always fall short of meeting expectations. Nevertheless, considerable progress has been made: The virus has been isolated, diagnostic laboratory tests have been developed, the dynamics of HIV are better understood, treatment of opportunistic infections has improved, and mother-to-child transmission has been drastically reduced thanks to the antiviral drug zidovudine (also called azidothymidine [AZT]). Further, society is more tolerant toward individuals with AIDS, be they homosexual or heterosexual. The world awaits a cure to deal with the millions of infected individuals and, even more important, a vaccine that will prevent AIDS. Perhaps due to the millions of dollars dedicated to AIDS research, the ultimate goals of prevention and cure will be met in the not too distant future.

The life cycles of all viruses are complex; this is particularly true of HIV, which can remain in a latent state (in hiding) for more than ten years before returning to a replicative cycle. Like all retroviruses, HIV defies the central dogma of biology by storing its genetic information in RNA and using the reverse transcriptase enzyme to get it back on the track to DNA. To prevent or cure AIDS, a vaccine and drugs must somehow interfere with the virus's life cycle. Potential targets are adsorption, penetration, inhibition of reverse transcriptase, and assembly of new viral particles (FIGURE 16.15). Drugs currently in use are directed

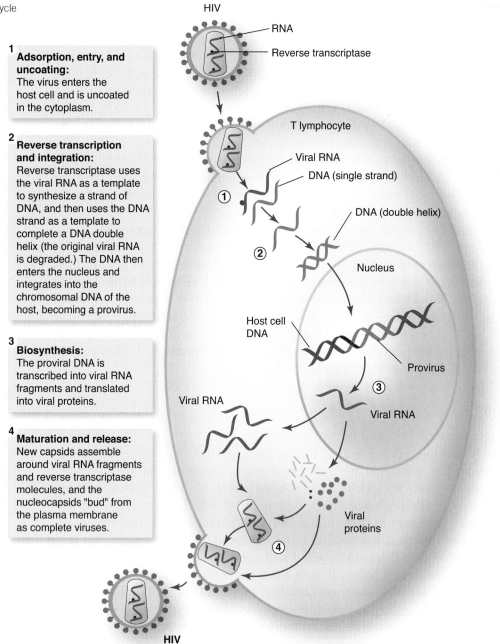

FIGURE 16.15 HIV replication cycle and potential drug targets.

1
Adsorption, entry, and uncoating:
The virus enters the host cell and is uncoated in the cytoplasm.

2
Reverse transcription and integration:
Reverse transcriptase uses the viral RNA as a template to synthesize a strand of DNA, and then uses the DNA strand as a template to complete a DNA double helix (the original viral RNA is degraded.) The DNA then enters the nucleus and integrates into the chromosomal DNA of the host, becoming a provirus.

3
Biosynthesis:
The proviral DNA is transcribed into viral RNA fragments and translated into viral proteins.

4
Maturation and release:
New capsids assemble around viral RNA fragments and reverse transcriptase molecules, and the nucleocapsids "bud" from the plasma membrane as complete viruses.

HIV
RNA
Reverse transcriptase
T lymphocyte
Viral RNA
DNA (single strand)
DNA (double helix)
Nucleus
Host cell DNA
Provirus
Viral RNA
Viral RNA
Viral RNA
Viral proteins
HIV

against some of these targets and, to varying degrees, inhibit viral replication, but only on a temporary basis.

An early treatment was to flood the body with soluble, free-floating CD4 molecules, which would act as decoys and tie up the "docking" **gp120** molecules on the surface of HIV (FIGURE 16.16). Recall that CD4 molecules are on T-helper and other cells that "fit" the gp120 molecules on HIV. However, flooding the body with decoy gp120 molecules to tie up the CD4 sites was unsuccessful.

AZT was the first clinically safe and effective drug used to treat HIV infection. AZT and other anti-HIV drugs are inhibitors of the reverse transcriptase enzyme and act at an early stage by preventing synthesis of DNA from RNA. The AZT molecule resembles an important building block in viral DNA synthesis, and when it is mistakenly used by the virus, DNA synthesis is brought to a halt. AZT and related drugs slow the spread of HIV in the body and temporarily forestall opportunistic infections. They are not cures.

A second class of drugs, called **protease inhibitors,** has been developed and approved for treating HIV infection. These drugs act by interfering with the final assembly and maturation of the virus; whereas AZT slows viral reproduction, protease inhibitors act to shut down the assembly line. Because HIV develops resistance to AZT and to protease inhibitors, combination treatment is now considered to be the best approach. The so-called **AIDS cocktail,** or HAART (highly active antiretroviral therapy), includes two reverse transcriptase inhibitors and one protease inhibitor, a "triple whammy," in an effort to block viral replication at two target sites. HIV-infected patients undergoing this combination treatment often become ecstatic because their T-cell counts rise, their viral loads decline, and their health improves remarkably. However, the improvement is not permanent.

In 1996 researcher David Ho announced a new strategy of AIDS treatment, namely, to administer the cocktail to patients during the first few weeks of infection. "Hit early, hit hard" was the strategy. Studies of a small group of patients indicated that this new strategy might actually eliminate—not just decrease or slow down—HIV from the body. The use of this strategy resulted in a decline of AIDS in the United States and elsewhere. David Ho was named *Time*'s 1996 "Man of the Year" for his work. AIDS began to be cautiously viewed for the first time as curable. Only four or five years later, however, the dark side of the "hit early, hit hard" approach emerged. The appearance of drug-resistant HIV strains increased, as did the toxic side effects, necessitating reconsideration about the wisdom of early intensive therapy. Some believe that the more toxic drugs should be saved

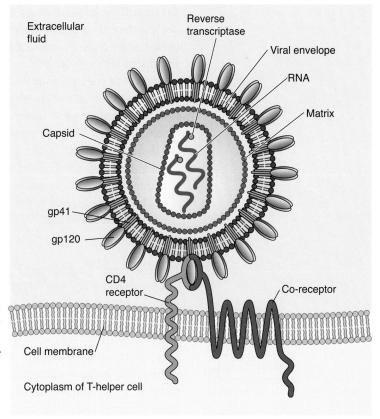

FIGURE 16.16 "Docking" between the HIV and the T-helper cell: CD4-gp120 complementarity.

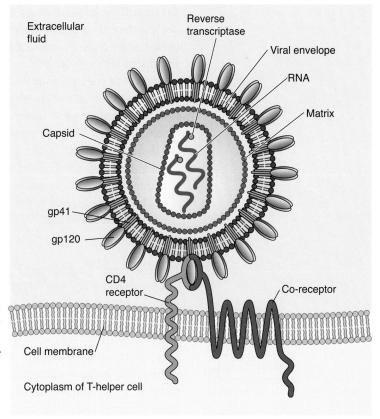

for later in the battle. The situation is complex and is one of those "damned if you do, damned if you don't" situations, forcing physicians and patients to choose between the consequences of early treatment and the consequences of HIV when treatment is postponed.

None of the currently available drugs cures HIV infection or AIDS; they all have side effects, some of which are very serious; they conflict with other drugs; they require strong motivation and self-discipline because of the large number of pills to be taken daily on a strict time schedule; and they cost over $20,000 a year. All these factors contribute to the frustration of less-developed countries as they witness the onslaught of AIDS. One major breakthrough is that the risk of HIV transmission from a pregnant woman to her baby is markedly reduced if she takes AZT during pregnancy, labor, and delivery; AZT is also given to the infant during the first six weeks of life.

Vaccines

Development of a safe and effective vaccine for AIDS is a priority; most scientists believe that the current AIDS pandemic will be controlled and future ones prevented only when a vaccine is available. A sense of urgency exists among AIDS researchers involved in vaccine development, but AIDS vaccines and drugs—"breakthroughs"—appear to come and go. The drug Viracept, produced by Roche Pharmaceuticals of Switzerland, which acts as a protein inhibitor, was withdrawn in July 2007 because of contamination with a dangerous chemical. Another major setback in the search for an AIDS vaccine occurred in September 2007 when Merck announced it was ending enrollment and vaccination of volunteers; production of the vaccine was partially funded by the National Institutes of Health. The vaccine was regarded as the most promising candidate for an AIDS vaccine. The prospects of finding a vaccine are diminished. The ideal vaccine would be a one-dose, safe, effective, oral vaccine that would establish lifelong protection against all subtypes of HIV.

So, what is the conclusion regarding prevention? The conclusion is that the best protection against AIDS is a matter of personal conduct. Theoretically, AIDS should be easy to control, if not eradicated, because there are no vectors or animal reservoirs. On the other hand, matters of behavior, particularly sexual behavior, are sensitive and difficult to deal with. Mechanisms of transmission of HIV were discussed previously; they point the way for preventive measures centered on abstinence or safe sex and avoidance of drug use. The AIDS cocktail and the implementation of early treatment dramatically reduced AIDS deaths, but recent CDC reports indicate that the decline is slowing. A false sense of complacency has emerged from the myth that the cocktail and early intervention cures AIDS, resulting in an increase in high-risk sexual behavior. The CDC, along with other agencies, plays a vital role in educating the public about AIDS and AIDS transmission (FIGURE 16.17). AIDS conferences are frequently held (FIGURE 16.18) to evaluate, improve, and develop strategies to counter AIDS.

The sex industry flourishes in many countries, and many females are forced into a life of prostitution to earn a living. Prostitution is an important factor in considering the spread of AIDS throughout Africa and other continents. The use of condoms is, to a large extent, not acceptable in many societies, and for the

AUTHOR'S NOTE (RIK)

Thirty years ago I lectured to my microbiology class that within the next several years, there would be a vaccine for AIDS. Several years later, I all but promised that by the end of the twentieth century AIDS would be defeated. So, what do I say now?

FIGURE 16.17 AIDS education posters. Left: Courtesy of NPIN/ National Center for US Department of Health and Human Services/CDC. Middle and right: Courtesy of CDC/NLM.

female sex partner to even suggest that her male partner use one is an insult to the man's character, dignity, and manhood. Females are frequently considered second-class citizens, and such suggestions are likely to be dealt with by vicious beatings and ostracism. Hopefully, time and education will lead to gentler, kinder, and wiser societies.

Consequences of AIDS

You are witnessing the AIDS pandemic and are constantly reminded of its consequences. Perhaps you even know someone with HIV infection or AIDS, or perhaps you know someone who has died of the disease. The virus knows no discrimination in terms of race, gender, or age. It attacks the rich and the famous, the young and the old. All countries, particularly developing countries, have been hit, and all countries, rich or poor, bear the scars of AIDS and its repercussions in politics, in the economy, and in the well-being of citizens. AIDS and its consequences have been called "crimes against humanity," and all countries are its victims. The positive news is that we are on the verge of a significant breakthrough in the AIDS response according to a recent UNAIDS report (2011) (TABLE 16.2). More people than ever are living with HIV, largely due to greater access to treatment. The annual number of new HIV infections fell 21% between 1997 and 2010 (FIGURE 16.19).

FIGURE 16.18 Bill Gates, left, co-chairman of the Bill and Melinda Gates Foundation, speaks as former U.S. President Bill Clinton listens at the 16th International AIDS conference in Toronto, Ontario, Canada on Monday, August 14, 2006. Clinton and Gates, whose charitable foundations have spent billions of dollars on AIDS, said inadequate prevention, testing, and healthcare services may be bigger obstacles than a lack of money in ending the epidemic. © J.P. Moczulski/REUTERS.

TABLE 16.2 Global Summary of the AIDS Epidemic, 2011

Number of People Living with HIV at the end of 2010

34 million (31.6–35.2 million)

Number of People Dying of AIDS-Related Causes Fell in 2010

1.8 million (1.6–1.9 million), which is down from a peak of 2.2 million in the mid 2000s

Proportion of Women Living with HIV in 2010

Remains stable at 50% globally

New HIV infections in 2010

2.7 million (2.4 million–2.9 million)

390,000 among children (340,000–450,000)

The number of people becoming infected with HIV continues to fall in 33 countries, 22 of them in sub-Saharan Africa, the region most affected by the AIDS epidemic.

Eastern Europe and Central Asia

250% increase in the number of people living with HIV from 2001–2010 in Eastern Europe and Central Asia.

The Russian Federation and Ukraine account for almost 90% of the increase.

Injecting drug use remains the leading cause of HIV infection in this region.

Data from: *How to Get to Zero: Faster. Smarter. Better.* UNAIDS World AIDS Day Report 2011, Joint United Nations Programme on HIV/AIDS.

AIDS in Africa, Caribbean, Eastern Europe, and Central Asia

The news out of Africa is particularly catastrophic, where AIDS adds a heavy weight to the existing burden of other microbial and nonmicrobial diseases. It is the top killer in some parts of the African continent and has left millions of children orphaned, accounting for a significant portion of the world's orphans. Since the late 1970s over twenty million Africans died from AIDS. Further, life expectancy in some African countries is less than 51 years. Hospital wards are overburdened, and the beds are filled with bodies wasting away as flesh disappears from the bones of AIDS victims. Corpses are piled high in morgues to the point where individual identity is obliterated, and the landscape is littered with unmarked graves. Shame, stigma, poverty, prostitution, sexual violence, and ignorance enter into the cycle, contributing to the paralysis of the already weak public health infrastructure. The AIDS pandemic has also caused grave political and economic problems in Africa. Hospital victims with AIDS may be told to "go home and die," because there are no drugs to help. Further, many choose not to be tested because they know that treatment is expensive and only temporary, so why bother? The positive news is that the number of new HIV infections in sub-Saharan Africa has dropped by more than 26% since the height of the epidemic in 1997. The number of AIDS-related deaths has steadily decreased as free antiretroviral therapy has become more readily available in the region.

Brazil might have at least a partial answer and may serve as a role model for other developing countries. In 1992 Brazil embarked on a controversial program of producing its own AIDS drugs and distributing them free of charge; its government

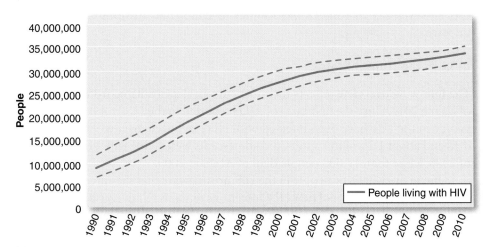

(a)

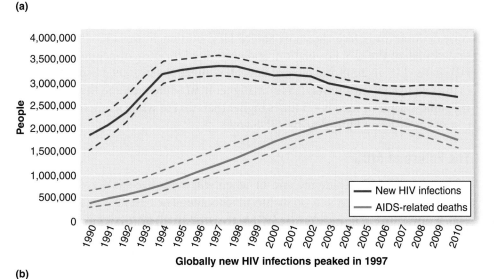

Globally new HIV infections peaked in 1997

(b)

FIGURE 16.19 Positive trends in AIDS response. **(a)** People living with HIV is slowing down. **(b)** New HIV infections and AIDS-related deaths fell 21% between 1997 and 2010. Adapted from *How to Get to Zero: Faster. Smarter. Better.* UNAIDS World AIDS Day Report 2011, Joint United Nations Programme on HIV/AIDS.

laboratories now produce at least five AIDS drugs. The number of AIDS-related deaths in Rio de Janeiro and São Paulo, the country's hot spots of AIDS, has plummeted. Developing countries in Africa and elsewhere are studying Brazil's example. The strategy flies in the face of politics, patent and copyright laws, and big (pharmaceutical) business, but it is working.

The Caribbean has the second highest HIV prevalence after sub-Saharan Africa. The number of new HIV infections in the Caribbean region has decreased by 25% since 2001 in the Dominican Republic and Jamaica and by 12% in Haiti. The decline has been attributed to access to HIV prevention services for pregnant women resulting in a steep decline in the number children newly infected with HIV and in AIDS-related deaths among children.

Several regions and countries do not fit the overall trend of HIV epidemic decline or stabilization. The number of people living with HIV in Eastern Europe and Central Asia rose 250% from 2001 to 2010! The Russian Federation and Ukraine are among the Eastern European countries with the fastest growing numbers of cases of HIV. The primary route of transmission is intravenous drug use; however, sexual contact and mother-to-child transmissions have begun to increase.

AIDS in the United States

The following overview reflects the most current data available from the CDC as of March 2012. Since the first cases of AIDS reported in the U.S. in 1981, 1.7 million people in the U.S. have been infected with HIV, including over 619,000 that have died and about 1.2 million adults and adolescents were living with HIV by the end of 2008. More than 17,000 people died of AIDS in 2009.

One in five (20%) of people in the U.S. are unaware that they are infected with HIV. The annual number of new HIV infections remains relatively stable. However, new infections continue at too high of a level, with approximately 50,000 Americans becoming infected with HIV each year. Men having sex with men accounted for 61% of all new HIV infections in the U.S. in 2009. Women accounted for 23% of estimated new HIV infections in 2009. Among racial/ethnic groups, African Americans face the most severe HIV burden in the U.S. Blacks represented 14% of the U.S. population, and accounted for 46% of people living with HIV in the U.S. in 2008 and 44% of new HIV infections in 2009. Hispanic/Latino men's rate of HIV infection is 2.5 times higher than white men and the rate among Hispanic/Latino women is 4.5 times higher than that of white women (data courtesy of the CDC and http://www.AIDS.gov.)

The Future of AIDS

When will the AIDS epidemic end or when will definitive AIDS treatment and prevention be found? Will it be within this decade? The best answer to this question is "hopefully so." Meanwhile, active research in all phases of HIV infection and AIDS is under way sponsored by the private and public sectors—another example of partnerships.

A major stumbling block is the fact that HIV is silently hiding in certain pools or reservoirs of cells that antiviral drugs cannot reach. The virus is in a latent state as a provirus that is integrated into the DNA of a host cell where it remains latent or eventually reactivates. A team of HIV researchers at Johns Hopkins described a 2012 study that offers hope that latent HIV can be "turned on," making it visible to the immune system's killer T-cells. Subsequently, antiretroviral drugs could aid in eliminating the newly activated viruses from the body. A different group of researchers led by Dr. David Margolis at the University of North Carolina, Chapel Hill have shown that the chemotherapy drug **vorinostat** (Zolinza®) was able to purge HIV that was dormant in T-cells. Their study was small (six HIV patients); however, their results demonstrate a baby step toward eradication of HIV from the body and a possible cure for HIV.

Tuberculosis

"Consumption," a nineteenth-century term for tuberculosis (TB), was so called because its victims, excessively thin, pale, and weak, appeared to be consumed by their illness. The disease is caused by *Mycobacterium tuberculosis,* also referred to as the tubercle bacillus. TB is included in this chapter because it exists at epidemic levels. Globally, TB is a leading cause of death from infectious disease. TB is not, by any means, a disease of the past. In the United States and other industrialized countries it is a reemerging disease, yet only forty years ago it was on the brink of extinction. The disease may be better considered an "emerged" disease that has returned despite the years of work toward its control. March 24th is World TB Day, which commemorates the date in 1882 when Dr. Robert Koch announced the discovery of *Mycobacterium tuberculosis*, the bacterium that causes tuberculosis. World TB provides awareness about TB-related problems. The U.S. slogan for the 2012 observance was "Stop TB in My Lifetime."

Paintings from tombs in Egypt and examination of mummies dating back to 4,000 B.C. indicate the antiquity of TB. The disease emerged in Neolithic times as human populations increased, settled down, and domesticated cattle. Presumably, human TB may have arisen from bovine TB. TB is as much a social disease as it is a microbial disease; in the eighteenth and nineteenth centuries its development as an urban plague was associated with poverty, poor housing, crowding, inadequate nutrition, and unemployment, all spawned in the wake of the Industrial Revolution. It is ironic that progress fueled the social conditions that allowed TB and other diseases to flourish. Before the discovery of the tubercle bacillus by Robert Koch in 1882, the environment of crowded tenements was associated with the cause of the disease.

The impact of TB was so profound in the nineteenth century that some feared it would bring about the end of European civilization. In many cities of America and Europe TB was the leading cause of death, accounting for as many as 15% to 20% of fatalities.

The treatment of TB centered on fresh air, bed rest, and good nutrition with plenty of fresh eggs, milk, and cream, all of which were provided in the TB sanatoria (as hospitals for the care of people with TB were called) around the country. Porches and decks were prominent in these facilities to allow the patients to be in fresh-air environments, even during the winter. In some cases surgical intervention to collapse a lung was practiced. The rationale was that in a "rested" state an infected lung would heal more quickly. (A tube was inserted into the chest cavity to allow air to enter and collapse the lung.) TB declined as social conditions improved during the first few decades of the twentieth century, accompanied by the development of immunization in the 1920s and antibiotics in the 1940s. Streptomycin, an antibiotic effective against the tubercle bacillus, was discovered in the 1940s, followed by the introduction in 1952 of isoniazid, a drug still widely used to treat TB.

Current Status

In 1993 the WHO declared TB a global emergency and estimated that between 2000 and 2020, nearly one billion people would be newly infected, 200 million people would get sick, and 35 million people would die from TB. So where does the world stand now? The absolute number of TB cases has been falling since 2006 (rather than rising slowly as indicated in previous WHO reports). According to the 2011 WHO 16th Report on Global Tuberculosis Control, as of 2010 there were 8.8 million incident cases of TB, 1.1 million deaths from TB among HIV-negative people, and 0.35 million deaths from HIV-associated TB. TB mortality rates have

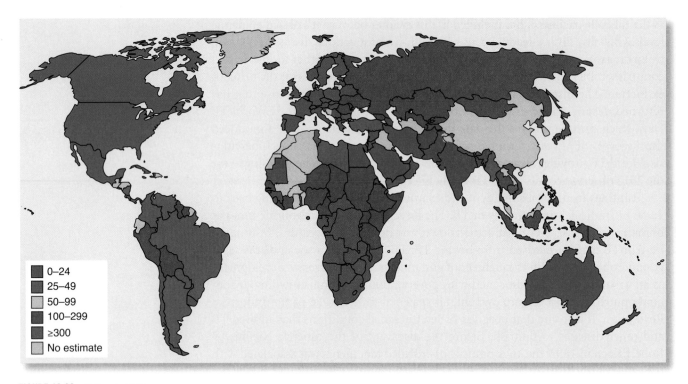

FIGURE 16.20 Estimated TB incidence rates, 2010. Modified from WHO Report: *Global Tuberculosis Control 2011.*

0–24
25–49
50–99
100–299
≥300
No estimate

fallen by just over one-third since 1990, and the world as well as five of six WHO regions (except the African region) are on track to achieve the Stop TB Partnership target of halving 1990 mortality rates by 2015.

Most of the estimated number of cases in 2010 occurred in Asia (59%) and Africa (26%), the Eastern Mediterranean region (7%), the European region (5%), and the region of the Americas (3%). The five countries with the largest number of incident cases in 2010 were India (2.0 million to 2.5 million), China (0.9 million to 1.2 million), South Africa (0.4 million to 0.59 million), Indonesia (0.37 million to 0.54 million), and Pakistan (0.33 million to 0.48 million). India accounted for 26% of all TB cases worldwide, and China and India combined for 38%. The proportion of TB cases coinfected with HIV is highest in countries in the African region. Overall, the African region accounted for 82% of TB cases among people living with HIV. The world estimates of TB incidence rates are shown in **FIGURE 16.20**.

TB: The Disease

The causative agent of TB is *M. tuberculosis.* TB is an infectious disease of the lower respiratory tract. The bacilli are acquired by the inhalation of infected droplets sprayed into the air by TB-infected individuals during coughing, sneezing, singing, talking, or laughing. Large numbers of TB bacilli are coughed up by infected individuals, and, because the waxy outer coat protects the bacilli from drying, transmission is efficient. Hence, people at the greatest risk are those who spend relatively long periods with an infected person, including family members, friends, and coworkers. Before pasteurization, milk was significant in the transmission of TB. Less commonly, the bacilli can be acquired through the skin, and cases have been reported in laboratory personnel who handled specimens containing the bacilli and in people who have received tattoos. An embalmer contracted TB from an infected corpse while preparing the body for burial. In most cases the primary infection is in the lungs, but infection can occur in the brain, spinal cord, kidney, bone, or cutaneous (skin) tissue (FIGURE 16.21).

The first exposure to tubercle bacilli results in primary infection; in most cases there are no symptoms, and the individual is not even aware that infection has occurred. Cell-mediated immunity walls off the bacilli in lesions known as **granulomas**. This is the usual response in about 90% of individuals with primary infection. In the other 10% the immune response is not adequate, resulting in the escape of bacilli from the granulomas. The course of TB infection is illustrated in FIGURE 16.22. The bacilli cause symptomatic primary TB manifested by fevers, night sweats, weight loss, fatigue, and the coughing up of blood-tinged sputum. A classical sign of active TB is a blood-stained handkerchief. Symptomatic

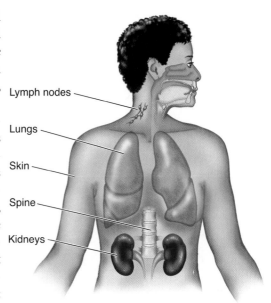

FIGURE 16.21 Common sites of TB infection.

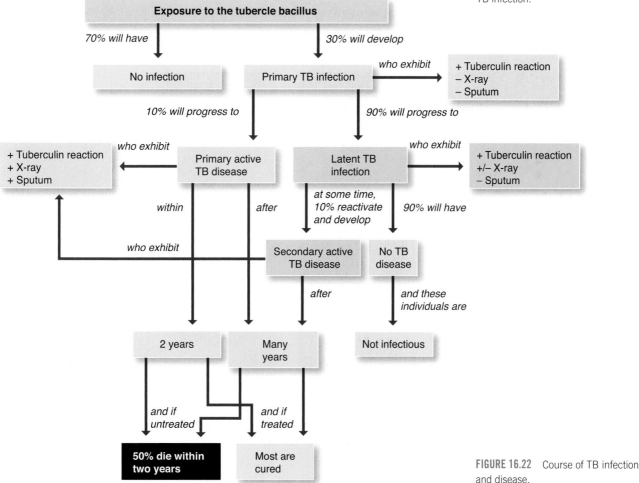

FIGURE 16.22 Course of TB infection and disease.

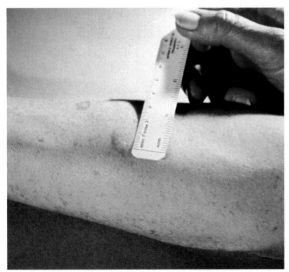

primary infection is more likely to occur in children, in the elderly, and in the immunocompromised, especially those with HIV.

Although most cases of TB infection do not progress to acute disease, reactivation TB, or secondary TB, can result because the bacilli can remain in a dormant state for years and be reactivated decades later. If this happens any of the anatomical sites seeded during the primary infection may be affected, including primarily the lungs but also the bones and joints. Perhaps Quasimodo, the hunchback in Victor Hugo's classic tale, *The Hunchback of Notre Dame*, was a victim of spinal TB.

■ Diagnosis

The early symptoms of pulmonary TB resemble those found in a plethora of microbial diseases. The presumptive diagnosis is based on the appearance of clinical signs and symptoms supported by a positive tuberculin skin test (FIGURE 16.23), x-ray examination (FIGURE 16.24), microscopic examination of sputum for the presence of tubercle bacilli (FIGURE 16.25), and culture of sputum specimens for the demonstration of *M. tuberculosis* (TABLE 16.3).

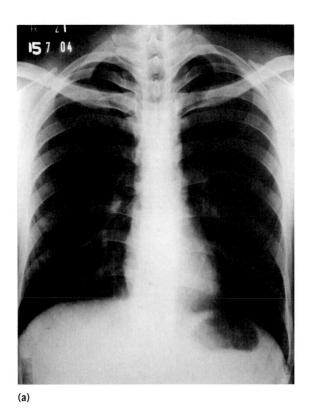

(a)

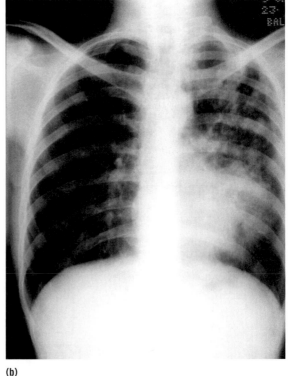

(b)

FIGURE 16.24 (a) A normal chest x-ray. © hald3r/ShutterStock, Inc. (b) Chest x-ray showing TB infection in the upper and lower lobes of one lung. © SPL/Photo Researchers, Inc.

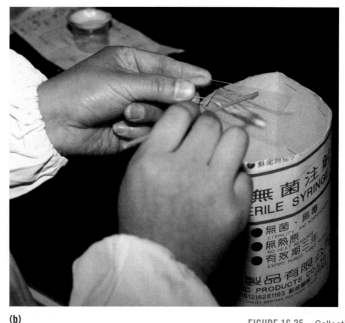

(a)   (b)

FIGURE 16.25 Collecting **(a)** and preparing **(b)** sputum for examination. Courtesy of WHO/TDR/Andy Crump. Used with permission.

The Mantoux **tuberculin skin test** is performed by the intradermal injection of a minute amount of **purified protein derivative (PPD)** from *M. tuberculosis;* the test is harmless because it does not use whole organisms. After forty-eight to seventy-two hours the injection site is visually examined for the presence of an induration (a red, raised lesion), which, if measuring, five mm or more in diameter, constitutes a positive skin test. Those testing positive may falsely jump to the conclusion they have TB, but a positive skin test only indicates previous exposure to the organism—a condition referred to as TB infection, not TB disease. In infection, the bacilli are walled off in the granulomas and there are no symptoms, whereas active symptomatic disease is the result of bacilli escaping from granulomas. Many people in the United States have come from countries with a high rate of TB and contracted TB. Although they may have completely recovered, they continue to manifest a positive skin test. Further, they may have received a TB vaccine (described below) as a preventive measure and will test positive, probably for their lifetime.

TABLE 16.3 Basis for Diagnosis of TB

1. Clinical signs and symptoms: weight loss, cough, fatigue, night sweats, low-grade fever
2. X-ray examination
3. Positive skin test (indurated lesion 5 mm or greater)
4. Demonstration of bacilli in sputum (presence of acid-fast bacilli on microscopic examination of sputum)
5. Culture of *M. tuberculosis* (sputum cultured on agar plate and observed for 6 to 8 weeks for colonies of *M. tuberculosis*)

Antibiotic Therapy

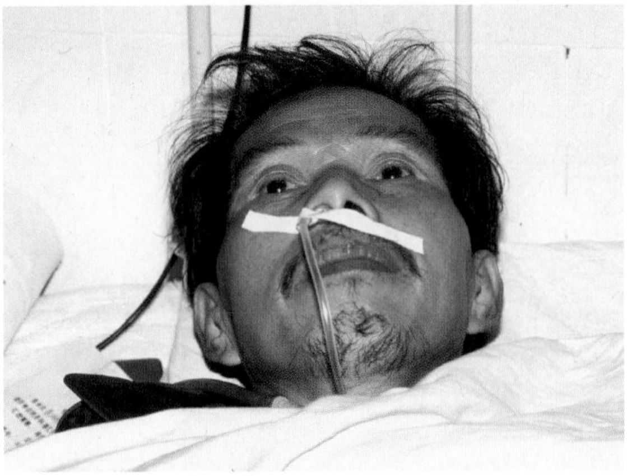

Antibiotic therapy is available for both TB infection and TB disease. For those with TB infection, the antibiotic isoniazid is completely effective if taken for six months as a preventive measure. Treatment for active TB disease requires taking a combination of antibiotics for six to nine months along with supportive measures of adequate rest, a good diet, and oxygen therapy (FIGURE 16.26). Five antibiotics are particularly effective against the tubercle bacilli: isoniazid, ethambutol, rifampin, streptomycin, and pyrazinamide. Toxicity and antibiotic resistance need to be considered in choosing the appropriate combination of antibiotics. The length of time needed for treatment can create a serious problem of noncompliance; further, the antibiotics are expensive. These factors severely hamper the treatment of TB in both developing and developed countries. This is particularly true in the population of homeless people—many of whom have TB and/or AIDS as well as drug or alcohol addiction. At least 25% of the homeless population of London and of San Francisco are infected with TB. The infection rates in prisons and homeless shelters are staggering. Even if the antibiotics are provided free of charge and are readily available, noncompliance remains a major issue. Attempts to control TB by supplying infected individuals with take-home medicines have been unsuccessful because the medications were taken in a haphazard fashion and fostered the development of multiple drug-resistant strains.

FIGURE 16.26 TB patient receiving antibiotic and oxygen therapy. Courtesy of WHO/TDR/Andy Crump. Used with permission.

FIGURE 16.27 Patient in a DOTS program taking her TB medicine. Courtesy of WHO/TDR/Andy Crump. Used with permission.

DOTS

DOTS, or **direct observational therapy short course**, is a simple and effective method adopted by the WHO in 1992 to combat noncompliance and complacency. It has saved thousands of lives and minimized the emergence of drug-resistant strains.

The DOTS strategy combines the five elements of political commitment, microscopy services, drug supplies, surveillance and monitoring, and direct observation. Once an individual's sputum shows tubercle bacilli, healthcare workers must watch the patient swallow the full course of the prescribed anti-TB drugs on a daily basis (FIGURE 16.27). The sputum-smear test is repeated after two months and at the end of the six- to eight-month treatment schedule.

The drugs used in the DOTS system are not new and when used correctly have a cure rate approaching 100%. Within the first two to four weeks of treatment, patients become noninfectious. The DOTS program is designed for large, poor populations. DOTS does not require hospitalization, a luxury not available for the countries hardest hit by TB. Further, the program is cost-effective. The World Bank rates DOTS "one of the most cost-effective of all health interventions."

The DOTS success stories are numerous and almost too good to be true, but they are true; DOTS implementation revolutionized TB control in those countries of the world with massive rates of TB. In China the old TB proverb, "ten get it, nine die," has been replaced by "ten detect it, nine cured." In Bangladesh, a poverty-stricken country with a minimal public health infrastructure, an 85% cure rate has occurred. In Peru DOTS has resulted in successful treatment of about 90% of the TB cases. The strategy has been used in New York City and in other cities of the developed world with significant cure rates. President Clinton got into the act in India when he marked World Tuberculosis Day on March 24, 2000 by administering the final dose of medicine to patients undergoing DOTS therapy. He noted as follows:

> "Today is World Tuberculosis Day. It marks the day the bacteria [sic] which causes TB was discovered 118 years ago. And, yet, even though this is 118-year-old knowledge, in the year 2000, TB kills more people than ever before, including one almost every minute here in India. These are human tragedies, economic calamities, and far more than crises for you, they are crises for the world. The spread of disease is the one global problem from which no nation is immune."

Factors Contributing to Reemergence

The statistics presented earlier are frightening, particularly because TB was on the brink of elimination 30 years ago in the United States and in other industrialized countries (FIGURE 16.28). The public health infrastructure was caught flat-footed, and TB skyrocketed, with much of the world's population becoming infected. What happened? You can probably guess. In the United States the increase was first noted in 1986 in New York City. The nation experienced a 20% increase between 1985 and 1992 as a consequence of TB and HIV coinfection, multiple-drug resistance, complacency, and travel and immigration. Increased attention halted the increase to a low in 2010, but attention to causative factors can further decrease the incidence of tuberculosis.

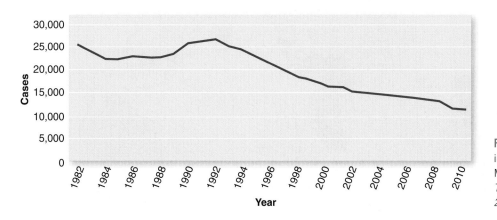

FIGURE 16.28 Reported Cases of TB in the United States, 1982–2010. Modified from the CDC *Reported Tuberculosis in the United States, 2010.*

FIGURE 16.29 HIV-infected patient with TB. Courtesy of WHO/TDR/Andy Crump. Used with permission.

TB and HIV Coinfection

TB and HIV are a dynamic and deadly duo; each accelerates the other's progress. HIV weakens the immune system, fostering an 800 times higher probability that an HIV-infected individual will develop active TB (FIGURE 16.29). TB is the leading cause of death in HIV-infected populations, accounting for about 15% of deaths. It stands to reason that those countries hardest hit by AIDS are those with an increased incidence of TB.

Multiple Drug-Resistant TB and Extremely Drug-Resistant TB

The tuberculosis antibiotic success story was tarnished by the emergence of **multiple drug-resistant tuberculosis (MDR TB)** and **extremely drug-resistant tuberculosis (XDR TB)**. MDR TB tubercle bacilli are resistant to at least two of the best anti-TB drugs, isoniazid and rifampin, both first-line drugs. XDR is a relatively rare type of MDR TB and is resistant to isoniazid and rifampin and at least three of the second-line drugs. Misuse of antibiotics is the major cause. The regimen of TB therapy is long and expensive, frequently resulting in noncompliance, which fosters the development of antibiotic resistance. Progress in global coverage on TB drug resistance is shown in FIGURE 16.30.

FIGURE 16.30 Progress in global coverage of drug resistance, 1994–2010. Modified from WHO Report: *Global Tuberculosis Control 2011.*

Complacency

Human progress frequently suffers because of a "why bother" or "it's too much trouble" attitude. Complacency was a significant factor for the increased infection

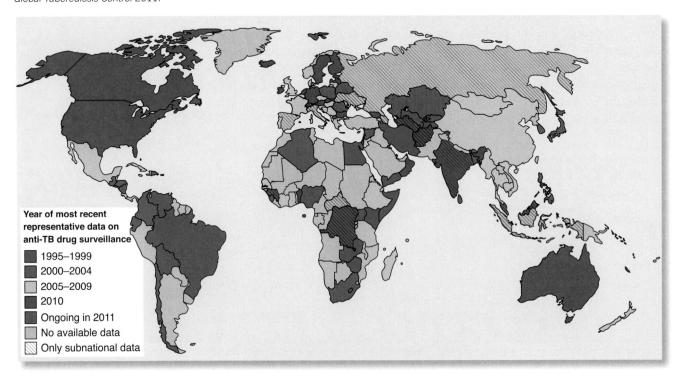

Year of most recent representative data on anti-TB drug surveillance

- 1995–1999
- 2000–2004
- 2005–2009
- 2010
- Ongoing in 2011
- No available data
- Only subnational data

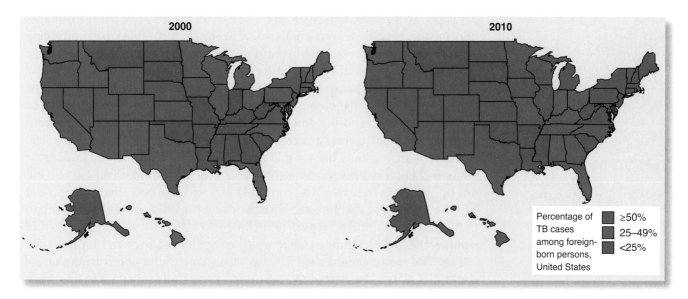

FIGURE 16.31 TB cases among foreign-born people in the U.S. in 2000 and 2010. Modified from the CDC *Reported Tuberculosis in the United States, 2010.*

rates in the 1980s, when the more developed countries of the world relaxed their TB control programs.

Technological Advances

Although the benefits of technological advances cannot be overstated, there is a downside. Consider, for example, organ transplantation, which can transmit TB. In April 2007 an organ donor was found to have TB three weeks after his death; two of the three recipients developed the disease and one recipient died.

Travel and Immigration

As with many microbial diseases, the period of incubation for TB is longer than the time it takes to travel from one part of the world to another. Hence, tourists may return from countries that have a high incidence of TB laden with an armful of gifts and souvenirs but also with a chest full of tubercle bacilli. Health experts warn that immigrants from countries with high rates of TB negate the gains made in reducing the disease in the United States (FIGURES 16.31 and 16.32). The Institute of Medicine recommends that immigrants from Mexico, the Philippines, Vietnam, and other countries with high TB rates be tested for the latent as well as the active form of the disease. If positive, these immigrants would not be denied entry but would be issued special visas requiring them to be treated before becoming permanent residents. Millions of illegal residents from Mexico and Central and South America live in the United States, contributing to the health burden created by TB. Children adopted from some countries are a potential source of infection. As an example, a boy from the Marshall Islands in the western Pacific spread TB to 56 people, including many of his playmates, after moving to rural North Dakota.

Concern exists about the risk of transmission of TB during long airplane flights. The CDC reported an investigation in 1995 involving transmission of the tubercle bacillus from a passenger with active TB to others on the plane; four passengers were infected during the eight and a half hour flight, but none of them

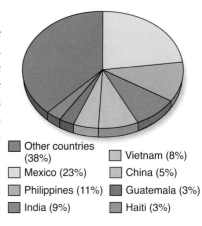

Other countries (38%)
Mexico (23%)
Philippines (11%)
India (9%)
Vietnam (8%)
China (5%)
Guatemala (3%)
Haiti (3%)

FIGURE 16.32 Countries of birth of foreign-born persons reported with TB in the U.S., 2010. Modified from the CDC *Reported Tuberculosis in the United States, 2010.*

developed active disease one year later. The CDC concluded that the risk of transmission on airplanes is no greater than that in any other confined space, including other forms of transportation, but the agency recommended passengers and flight crews be advised when they have been exposed to TB.

An international scare occurred in May 2007 when a man from Atlanta, Georgia previously diagnosed with XDR TB flew to Europe for his wedding and honeymoon, exposing about 450 fellow passengers to TB. He had been diagnosed with the disease as a result of an abnormal chest x-ray taken for a rib injury. The man claims he was told by CDC officials that "he was not contagious and not a danger to anyone, but that officials would prefer that he did not fly." While in Italy the CDC contacted him with the news that not only did he have TB, he was also infected with the dangerous XDR strain. According to the CDC, he was told to report to a clinic there and not to take any further commercial flights, although there were no legalities to prevent him from doing so. The man believed the CDC had abandoned him, and fearing that he could die in Italy, he elected to be treated in the United States. He flew from Prague, Czech Republic to Montreal, Canada and then rented a car to drive to Plattsburgh, New York on May 24. The account is further complicated by the fact that the CDC notified the border control to isolate and quarantine him and to notify public health officials. For reasons that are not known, the border control agent released him. Upon seeking treatment, he was tested again and the new laboratory test revealed that he had TB, but not the XDR strain. The CDC also retested the man's original sputum sample and confirmed the new findings.

Despite the (relatively) happy ending for the man and others that had been exposed, the case pointed out weaknesses and flaws in the public health surveillance system and opened up ethical questions concerning violation of civil rights by quarantine versus the right of the public. The case of Typhoid Mary is but one example.

Prevention

Vaccine Development

Public health strategies focus on prevention of disease. The high worldwide incidence of TB affords top priority to the development of new and effective vaccines against the disease. The **bacillus Calmette-Guérin (BCG) vaccine** for tuberculosis is available, but its use is controversial. The vaccine consists of attenuated (weakened) live tubercle bacilli. The protection rate is about 80% in children and less than 50% in adults, with the duration of protection ranging from five to fifteen years. The vaccine is not used in the United States because the incidence of TB does not justify its use, except among high-risk groups, including health professionals charged with the care of patients with TB and military personnel in areas with a high rate of TB. Further, those receiving the BCG vaccine will have a positive tuberculin reaction for life, necessitating repeated chest x-rays for monitoring purposes. New approaches using the tools of molecular biology, including plasmid DNA vector-based vaccines, recombinant DNA vaccines, mutant BCG vaccines, and subunit vaccines, are encouraging.

DOTS Implementation

More countries need to adopt the DOTS strategy or other measures to treat those infected with TB on an uninterrupted basis with the right combination of antibiotics.

Improved Social Conditions

Many of the world's citizens live in abject poverty and suffer from a lack of clean water, malnutrition, and inadequate housing conditions, particularly in the developing world. These factors favor the development and spread of all microbial diseases. TB is as much a social disease as it is a microbial disease, and alleviation of poor social conditions will alleviate the incidence of TB.

The number of TB cases in the United States has decreased in recent years (Figure 16.28), and the time is ripe to eliminate TB in this country, an attainable goal. Care must be taken to ensure that complacency and neglect, the price of success, will not once again increase the nation's vulnerability. In other nations, particularly in the developing world, the WHO and its partners can only continue to chip away at the problem and to assist in DOTS implementation or in alternative strategies to ensure treatment of those with TB.

Overview

Plagues continue to occur throughout history as circumstances dictate humans and microbes battle for survival. Smallpox, bubonic plague, and other plagues of the Middle Ages as well as the possibility of those caused by new, emerging, and remerging microbes and the rapidity of transmission stand as reasons for continued surveillance.

Influenza viruses have been afflicting humans for centuries. They can spread at alarming speeds because of its short incubation, airborne transmission, and the fact that many symptomatic people continue normal work and social activities, spreading influenza to many contacts. The influenza outbreak of 1918 remains as a benchmark of influenza for over almost a century; various epidemics and threats of pandemics continue to occur via droplet transmission. Wild birds, chickens, and pigs are a major source. Human-to-human transmission on a sustainable basis has not taken place but antigenic shift and antigenic drift does not rule out the possibility. Recently scientists have altered an influenza strain rendering it airborne and increasing anxiety in the population.

AIDS, a plague that surfaced in the last two decades of the twentieth century and continues into the twenty-first century, has taken a terrible toll throughout the world, particularly in Africa. Treatment and prevention of this disease remains elusive, although considerable progress has been made in improving the quality and life span of those infected. In more than thirty-three countries, the HIV epidemic is stabilizing or declining in new cases. A few regions of the world do not fit the overall trend. Countries in Eastern Europe and Central Asia are HIV hotspots. The Russian Federation and Ukraine are among Eastern European Countries with the fastest growing numbers of HIV cases. The primary route of transmission is intravenous drug use.

TB is an ancient disease, but it is not a disease of the past. It has reemerged with a vengeance and has infected one-third of the world's population. Treatment of the disease is seriously complicated by the development of multiple drug-resistant strains, coinfection with HIV, and immigration from countries with a high prevalence of tuberculosis. According to the WHO, the number of TB cases has been falling since 2006. The majority of new cases occur in Asia and Africa. DOTS have been effective, but its implementation is costly in terms of trained manpower and dollars.

■ Self-Evaluation

PART I: Choose the single best answer.

1. Identification of HIV as the cause of AIDS is credited to
   a. Duesberg b. Montagnier c. Gallo d. more than one of the above

2. Which variety of HIV is most common in West Africa?
   a. HIV-1 b. HIV-2 c. HIV-3 d. about the same for HIV-1 and HIV-2

3. Which is correct regarding the BCG vaccine?
   a. killed microbes b. attenuated (weakened) bacteria c. DNA vaccine
   d. attenuated (weakened) virus

4. Which term is associated with TB?
   a. H1N1 b. granuloma c. iatrogenic d. prodromal

5. Isoniazid is the drug of choice for
   a. Kaposi's sarcoma b. AIDS c. influenza d. tuberculosis

6. Some individuals with AIDS develop blue-purplish lesions on their skin known as
   a. thrush b. Kaposi's sarcoma c. ringworm d. dermatitis

7. This unapproved FDA drug was given emergency use approval and administered to individuals with severe H1N1 infections that did not respond to treatment in 2009.
   a. zanamivir b. oseltamivir c. peramivir d. amantadine

8. Location of the world that does not fit the overall trend of HIV epidemic decline:
   a. U.S. b. Eastern Europe c. South America d. Australia

9. Ferrets are used as experimental animals in research involving
   a. AIDS b. tuberculosis c. HIV d. influenza

10. Historically, it was thought that influenza was caused by this microbe:
    a. *Yersinia pestis* b. *Mycobacterium tuberculosis* c. *Bacillus anthracis*
    d. *Haemophilus influenzae*

PART II: Fill in the blank.

1. The _____ period of stage 2 HIV infection can last from two to fifteen years.

2. Some individuals with AIDS develop blue-purplish lesions on their skin known as _____.

3. The first stage of HIV infection is called _____.
4. DOTS stands for _____.
5. TB is frequently found as a coinfection with _____.
6. AZT is used in the treatment of _____ infections.
7. _____ and _____ are the two surface antigens of influenza A viruses that play an important role in attachment or release of the virus to or from infected cells.
8. _____ is an example of an anti-influenza drug.
9. _____ is the practice of prechewing food for infants by which HIV can be transmitted.
10. _____ is the strain of influenza that infects birds and humans and faces a world threat as a potential pandemic in the human population.

PART III: Answer the following.

1. A positive tuberculosis test does not necessarily mean that an individual has the disease; explain this.
2. The course of AIDS can be followed by measuring the patient's CD4$^+$ T-cell level. Explain this.
3. List arguments that supports Duesberg's belief that HIV does not cause AIDS.
4. Distinguish between antigenic shift and antigenic drift of influenza viruses.
5. Distinguish between HIV and AIDS.
6. Compare and contrast the 1918 and 2009 influenza pandemics in regarding to the population infected, mortalities, and information about the influenza strains that caused the pandemics.
7. Explain why the same influenza vaccine cannot be used from year to year to prevent influenza infections.
8. Discuss why is it difficult to fulfill Koch's postulates to prove HIV is the cause of AIDS.
9. Discuss reasons why there is a resurgence of TB in the United States.
10. Explain why researchers were motivated to create an airborne H5N1 influenza virus in the laboratory.

The Unfinished Agenda

Article 25. (1) Everyone has the right to a standard of living adequate for the health and well-being of himself and his family, including food, clothing, housing and medical care and necessary social services, and the right to security in the event of unemployment, sickness, disability, widowhood, old age or other lack of livelihood in circumstances beyond his control.

—Universal Declaration of Human Rights, 1948

■ Preview

This chapter looks into the future. Surgeon General William H. Stewart stated in 1967 with good reason, it was "time to close the book on infectious diseases." Sanitation, vaccines, and antibiotics had made enormous strides against pathogenic microbes, and there was no reason to suggest that this wave of optimism would not continue.

But the forces of natural selection and the ability of microbes to adapt, along with urbanization, complacency, technological advances, ecological disruption, and social evolution reflected in human behavior, proved otherwise. One scientist remarked, "The war has been won, but, by the other side"; this facetious statement reflects the theme of this book, namely, the challenge of new, emerging, and re-emerging infections. Outbreaks of microbial diseases are frequent subjects of current news and indicate the need for continued surveillance and avoidance of complacency.

Text Review

The full text presents the challenges and challengers—the biology of microbes and the diseases they cause. The beneficial aspects of microbes (the other side of the coin) and their role in the cycles of nature, without which life could not exist, and their exploitation in the production of pharmaceuticals, foods, and industrial products as agents for bioremediation are discussed to counter the notion that all microbes are disease producers; the reality is that most microbes are not pathogens. Harnessing microbes for beneficial purposes to the fullest extent possible is in itself, a part of the microbial challenge.

The next part of the text focuses on strategies of meeting the microbial challenge. The human immune system, a system adept at recognizing pathogens, and the body's response to eliminate pathogens by countering with antibodies and cell-mediated immunity is described. Public health measures and partnerships to meet the challenge are considered, with an emphasis on sanitation, vaccines, and antibiotics.

The final part of the text emphasizes the current challenges posed by microbes. Biological warfare continues to pose a threat to the world and is within the capability of many nations; biological terrorism (real or threatened) strikes fear into populations. The potential use of biological weapons indicates the need for research to develop rapid methods of microbial identification. The world is still faced with plagues and disease outbreaks; the spectrum varies from nation to nation and is in a constant state of flux. Tuberculosis and AIDS are rampant throughout the world. Possible pandemics of influenza, Nipah virus outbreaks, and 30 years of AIDS require constant vigilance.

An Ongoing Battle

Microbes have challenged their hosts throughout history and have influenced the course of civilization. Despite major advances in sanitation and the availability of clean water, antibiotics, and immunization, microbial diseases remain a significant cause of death, disability, and socioeconomic burden worldwide. These diseases cause approximately one-third of the world's deaths. Are microbes deliberately challenging humans? Is it in a microbe's best interest to annihilate its human (or other) host, and is it in humans' best interest to annihilate microbes? The answer to each of these questions is a definite "no." The biotic component of ecosystems is complex and varied and characterized by constant adaptation in the community. Adaptation and not annihilation is the key.

Host–parasite interactions are dynamic. There are no scientific equations that can predict the outcome of specific interactions; the outcome depends on a variety of circumstances reflecting the time and place when microbes and humans are on a collision course. History reveals that epidemics come and go as a result of shifts in the seesaw of host–parasite relationships. On the surface it might seem that microbes have the upper hand. Microbes reproduce asexually and quickly without being hampered by requiring suitable partners (sexual reproduction) with which to share their genetic material. There are two advantages of asexual reproduction. First, when large numbers of the population are killed, asexual reproduction, coupled with generation times that can, in many cases, be expressed in minutes, leading to a rapid recovery of the population. Second, rapid reproduction results in enormous genetic variability that, in turn, allows for genetic adaptation.

Antibiotic resistance remains prevalent as a global challenge in the microbial community. Outbreaks of measles, chickenpox, and whooping cough still occur, along with a variety of other infectious diseases, indicating the need for strict measures of surveillance and reassessment of immunization schedules. An ongoing outbreak of cholera began raging in Zimbabwe in December 2008. Currently, there is strong concern about the potential for a pandemic of avian (bird) flu. Many public health officials believe it will happen and the question is "when" and not "if." In April 2009, an outbreak of H1N1 swine flu occurred in Mexico and rapidly spread to many areas of the world. It resembled the 1918 swine flu pandemic and was a cause of great concern as a potential pandemic during the "regular" flu season starting in the fall months of the year.

Chicken flu (H5N1) continues as a threat that may increase particularly now that in 2012 biologists created an airborne strain of the virus. Foodborne diseases continue with a high profile in the United States and around the world. Malaria continues and, as a result of international travel, has increased in the United States. Ecotourism, "living with the natives," and "going on safari" is a growing trend for those seeking *Tarzan of the Apes* adventures. These travelers may get more than they bargained for, including a belly full of worms and malaria. And what about the effect on animal populations as human diseases are introduced into their ecosystems? A group of mountain gorillas in Rwanda's Volcanoes National Park became ill with severe respiratory problems and at least one succumbed. In 2012, bats were found to carry an H17 influenza A virus. A new virus spread by midges (*Culicoides* sp.) that causes birth defects and stillborn births in cows, sheep, and goats was reported in Schmallenberg, Germany in 2011; it quickly spread to the Netherlands, Belgium, France, Luxembourg, Italy, and the United Kingdom. Will it "jump species" into humans as happened in mad cow disease and in Nipah virus (BOX 17.1)?"

Advancements in technology have dramatically improved the life span and health of humans. Some of the biggest medical breakthroughs have not come without prior unanticipated complications. Progress in immunosuppressive treatment has resulted in a dramatic increase in the number of living immunocompromised patients, many of whom are organ recipients. In the past two decades, organ transplantation has become the treatment of choice for many diseases that in the past have led to patient death due to the failure of one or more vital organs. Each year, 40,000 organ transplants are performed worldwide with high success rates (90% of recipients survive at least one year).

Microbial infections are the most frequent complications in transplant recipients and are the main cause of death during their first year after transplantation (FIGURE 17.1). Immunosuppressive drugs are necessary to prevent organ rejection but the downside is that they weaken host immune responses. The

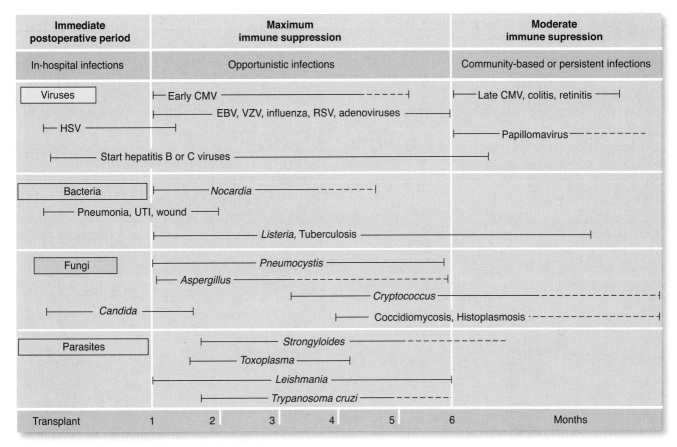

Immediate postoperative period	Maximum immune suppression	Moderate immune supression
In-hospital infections	Opportunistic infections	Community-based or persistent infections

Viruses
├── Early CMV ──────────────── ─ ─ ─ ─ ┤   ├── Late CMV, colitis, retinitis ──┤
                    ├──── EBV, VZV, influenza, RSV, adenoviruses ────┤
├─ HSV ─────────┤   ├────── Papillomavirus ─ ─ ─ ─ ─ ─ ─
├──── Start hepatitis B or C viruses ────────────────┤

Bacteria
    ├──── Nocardia ─────── ─ ─ ─ ─ ─ ┤
├──── Pneumonia, UTI, wound ────┤
        ├────────── Listeria, Tuberculosis ──────────┤

Fungi
    ├────── Pneumocystis ──────┤
    ├─ Aspergillus ─ ─ ─ ─ ─ ─ ─ ─ ─ ─ ┤
                ├──── Cryptococcus ──── ─ ─ ─ ─ ─ ─ ─ ─ ┤
├──── Candida ────┤   ├── Coccidiomycosis, Histoplasmosis ─ ─ ─ ─ ─ ─ ─ ┤

Parasites
    ├────── Strongyloides ──── ─ ─ ─ ─ ─ ─ ─
    ├──── Toxoplasma ────┤
├──────── Leishmania ────────┤
    ├──── Trypanosoma cruzi ──── ─ ─ ─ ─ ┤

Transplant   1   2   3   4   5   6   Months

FIGURE 17.1 Increased susceptibility of immunocompromised persons to infection. CMV: cytomegalovirus, HSV: herpes simplex virus, VZV, varicella zoster virus, RSV: respiratory syncytial virus, UTI: urinary tract infection. Adapted from Salavert, M. et al., 2011. Role of viral infections in immunosuppressed patients. *Medicina Intensiva* 35(2):117–125.

source of infection may come from the donor, the reactivation of latent viruses of the donor or recipient, community-acquired infections, or travel-associated diseases. Organ recipients encounter potentially drug-resistant pathogens during hospital treatments and are frequently given antimicrobials prophylactically to prevent common nosocomial opportunistic infections.

Emerging pathogens pose a unique challenge for physicians to recognize, diagnose, and treat. Immune compromised are high-risk individuals that, in some ways, are like the canary in the coal mines that miners once used to warn of a level of dangerously low oxygen. High-risk individuals are the first ones to become infected making it imperative to monitor these "human canaries" for the emergence of new pathogens that challenge the world's population.

Natural catastrophic events will continue to occur and in their aftermath will be an array of microbial-related public health problems (**FIGURE 17.2**). These events may be nature's way of "cleaning house," but they demand attention to minimize human suffering.

FIGURE 17.2 Catastrophic events will continue to occur and so will microbial diseases in their wake. Author's photo (RIK).

BOX 17.1 A-Pork-Alypse Now

In September of 1998, a small Malaysian village, Nipah, was turned upside down when pigs began suffering from a respiratory illness and paralysis. Farmers who came in close contact through respiratory droplets of the sick pigs developed fevers, headaches, drowsiness, and confusion and lapsed into a coma two days after the onset of symptoms and died from what appeared to be viral encephalitis. Cats and dogs in close contact to pigs also became infected.

Initially, government authorities thought this "unknown ailment" in pigs was thought to be classical swine fever or "hog cholera." The human deaths were speculated to be caused by the Japanese encephalitis virus, an endemic mosquito-borne illness. Various insecticide control measures were implemented but nevertheless, the disease in both pigs and humans spread to the biggest pig-farming region in Malaysia and to Singapore. Transmission was rapid and the fatality rate was high (40% in Malaysia) causing more than 265 human encephalitis cases, including 105 deaths in Malaysia and 11 cases of encephalitis cases and 1 death in Singapore over the next 18 months.

Blood sera and cerebral spinal fluid samples from sick patients sent to the CDC revealed a new paramyxovirus, named Nipah virus after the village in which the first cases were identified. More than 1 million pigs of the 2.4 million pig population were disposed of during outbreaks in Malaysia. This was economically devastating to the pork industry; some referred to it as an *a-pork-alypse*.

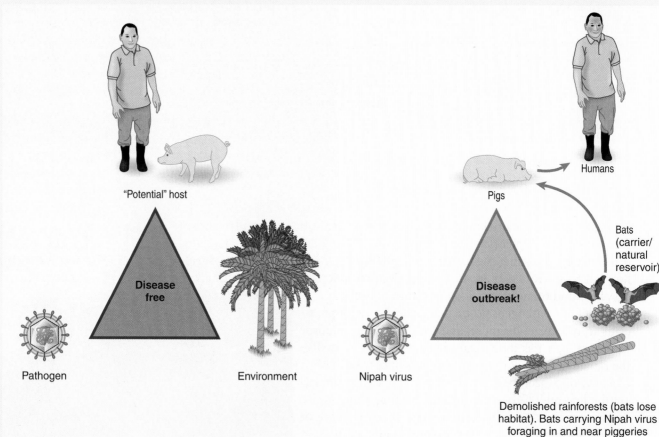

(a)

(b)

FIGURE B17.1a **(left)** Disease triangle model. **(right)** Disease triangle and the Malaysian Nipah virus outbreak.

The government suffered a loss of approximately $450 million, about 618 homes, 111 shops, schools and banks were evacuated, and 36,000 people lost employment due to the closure of farms.

Why did this happen? Where did this emerging zoonotic Nipah virus come from? Why was the outbreak hotspot a tiny village in Malaysia? There are no clear answers, but whatever the answers, the outbreak is a glaring example of how a change in the environment led to the outbreak. The environment (favorable for disease development and transmission) a susceptible host, and a pathogen are the "big three" components of disease causation. Conversely, the disease may be prevented or halted by elimination of any one of these three components referred to as the "disease triangle" (FIGURE B17.1a).

Prior environmental changes led to the Nipah virus outbreak. Approximately 5 million hectares (12,355,269 acres!) of tropical rainforest were cleared to allow room for piggeries (FIGURE B17.1b). Using the "slash and burn" strategy resulted in a blanket haze over the region. Consequently, a number of flowering and fruit trees required by bats (flying foxes), the natural reservoir of Nipah virus, as a food source was destroyed; their fruit and nectar are the primary foods (FIGURE B17.1c). Ultimately, Nipah virus jumped species from bats into human hosts primarily via infected pigs.

Nipah virus is present in the saliva and urine of the wild bats and in partially eaten fruit. Scientists suspect that fruit trees close to the pig pens were foraged by the bats and the virus was spread by this close proximity of pigs with bat saliva or urine, contaminated water, or dead bats. Humans can also be infected by consuming contaminated fruit or juice.

In 2012, a more virulent Nipah virus outbreak with 100% mortality in northern Bangledesh was traced to individuals drinking raw date (fruit tree) sap or juice causing widespread panic throughout the country. Nipah virus is a recurring threat to human health in Southeast Asia. Continued education of healthcare workers, veterinarians, and other responders about infection control procedures, surveillance, and rapid diagnostic tests is essential in high-risk areas.

FIGURE B17.1b   These are several of the hog confinement barns that were affected during the Malaysia Nipah virus outbreak. The reservoir fruit bats live in these caves and feed on the fruit trees that are in close proximity to the hog confinement barns. Courtesy of James Roth, Iowa State University.

FIGURE B17.1c   Flying fox (*Pteropus* bat), natural reservoir of Nipah virus. © iStockphoto/Thinkstock.

Just as there are developments that are worrisome and threatening, there are numerous examples of progress that are decidedly beneficial. For the first time plant-produced HIV antibodies are undergoing clinical trials in the United Kingdom. They were produced in the leaves of tobacco plants grown in a greenhouse in Germany and are anywhere from 10 to 100 times less expensive to produce than by using conventional methods. There may be a vaccine for humans on the horizon against the Ebola virus. Ebola is a potential candidate to be used as a biological weapon prompting the United States government to allocate increasing funds for Ebola vaccine research.

So why have humans (and other hosts) been able to survive the periodic epidemics and pandemics? Nobel Laureate Joshua Lederberg stated, in reference to the 1918 influenza epidemic, "If the mortality rate had been another order of magnitude higher, our species might not have survived." This statement makes it clear that the battle between microbes and humans has been a close call at times.

A notable point to be made is the ability of microbes, based on their genetic variability, to be transmitted cross-species such as from animal hosts to human hosts. Many biologists believe that tuberculosis and plague jumped from animal reservoirs to human hosts centuries ago. More recently, the emergence of AIDS, Nipah virus, and mad cow disease and their human variants support this fact.

The ongoing saga of microbial diseases is a reflection of the ongoing evolutionary dance. The practices of personal and public health can be brought to bear on the prevention and control of outbreaks (BOXES 17.2 and 17.3). Humans and microbes coexist; thus individuals and public health officials need to continue surveillance on a worldwide basis to nip new threats in the bud as they emerge (BOX 17.4).

BOX 17.2 — Good Personal Health Practices

- Wash hands frequently (but not obsessively).
- Cook foods properly and refrigerate them promptly.
- Stay up to date on immunizations.
- Use antibiotics prudently.
- Practice abstinence or safe sex.
- Check on immunization requirements and recommended medications before traveling, particularly if traveling to a developing country.
- Avoid complacency.
- Use common sense in health matters.

BOX 17.3 — Important Public Health Strategies to Contain Infectious Disease

- Develop new partnerships between the public and private sectors.
- Continue the war on poverty in the developing and developed world.
- Maintain and improve vector control programs.
- Continue and increase monitoring and surveillance on a worldwide basis.
- Reduce disparity in standards of living between developed and developing countries.
- Increase research in public health as related to epidemiology.
- Educate the public and promote behavioral changes related to public health.
- Develop and maintain the public health infrastructure.
- Be prepared to handle outbreaks.
- Avoid complacency.

On May 16th, 2011, Dr. Ali S. Khan, Director of the CDC's Office of Public Health Preparedness and Response posted a Public Health Matters Blog titled "Preparedness 101: Zombie Apocalypse" containing information on preparation for a zombie apocalypse. This may seem far-fetched, but according to a CDC spokesman, the idea for the post came after the word "zombie" caused internet traffic to spike. Zombies are fictional, undead mindless creatures that must eat humans (often brains) or animal flesh to survive. They are now a part of pop culture in Hollywood blockbuster movies (e.g., *Resident Evil*, *I Am Legend*, *Night of the Living Dead*), in a popular American Movie Channel TV series called *The Walking Dead*, and in gaming (e.g., *Plants vs. Zombies* and so many other gaming applications).

The purpose of the blog is to prepare the public for real emergencies including tsunamis, hurricanes, tornadoes, or pandemics. Emergency supply kits, plans for families to regroup following zombie invasion, evacuation routes, and isolation and quarantine of those who become infected are a part of the blog. The site includes downloadable posters, badges, (FIGURE B17.4) and a graphic novel, teaching moments from *The Walking Dead* TV series, and other disaster preparedness information.

FIGURE B17.4 CDC badge promoting emergency preparedness using a zombie as a hook to engage individuals to prepare before a disaster strikes. Courtesy of the CDC.

Is there an infectious agent that could turn humans into zombies? Could a "zombie microbe" be engineered in the laboratory? Probably the closest known infectious agents to cause "crazed" symptoms are the rabies virus, prions that cause kuru, variant CJD, or mad cow disease, or the parasite *Toxoplasma gondii*, which can infect the brains of humans, and brain-eating amoebas. A "zombie microbe" is just fiction so relax but the symbolism of zombies with the word "preparedness" is a catchy way to educate the public on how to respond to natural disasters and pandemics.

Research

Continued scientific research leads to continued advancement in society's understanding of its environment and establishes direction for future investigation. The advancements in human welfare that have been made over the centuries, particularly in the past century, are the result of rigorous research. Microbiological research is directed toward an understanding of the microbial world and its interactions with humans and other species.

The scientific and medical community once found it difficult to accept that stomach and intestinal ulcers are caused by the bacterium *Helicobacter pylori,* an organism present in the acidic environment of the stomach. Scientists are still finding links indicating that microbes may play a role in other chronic diseases in which no one suspected their involvement, including stomach cancer, lung cancer, cervical cancer, strokes, asthma, heart disease, and juvenile diabetes (TABLE 17.1). Research is necessary to uncover the causal relationships. Microbes may act as causative agents, as cofactors, or as **triggers** that cause damage through autoimmune reactions. The list of diseases is formidable and includes a multitude of bacteria and viruses. In fact, some biologists believe that many chronic illnesses

TABLE 17.1 Chronic Diseases Possibly Caused By Microbes	
Juvenile diabetes	Cervical cancer
Atherosclerosis	Arthritis
Strokes	Asthma
Heart disease	Chronic lung disease
Stomach cancer	Hypertensive renal disease
Lung cancer	Crohn's disease

are microbial in nature. These diseases might then be amenable to prevention and treatment with antibiotics, antiviral agents, and vaccines.

In a sense, microbes are our partners in life. Total adaptation to a symbiotic or even mutualistic coexistence is out of the question, necessitating continued surveillance and research possibly opening a Pandora's box as more and deeper knowledge into life processes and "life" itself, garnering increasing number of ethical issues. Are we opening a cauldron of trouble? Should we be conducting research to make bigger and better biological weapons, tinkering with ways to make airborne H5N1, or delving into synthetic biology to create new microbes that may make the species leap? Is research for the sake of research without concern for its significance and implementation justifiable and not used as a status symbol and career-fostering strategy linked to the number of published papers? A risk-benefit analysis needs to be in place. Are scientists immune to the consequences of their research, or do they share a responsibility along with nonscientific groups? Well-intended and seemingly justifiable research may fall under a dark cloud as the products may turn out to be unpredictable—that is what research is about—to explore the unknown. Accidents (including unintended release of microorganisms), implementation of knowledge to create bioweapons, and failure by countries and individuals to adhere to established guidelines are ethical considerations that become more pronounced as "life" continues to be explored at deeper and deeper levels. Research is vital for continued survival, but the knowledge/products gained need to be handled with wisdom.

Microbes are essential for the maintenance of life on the planet, and others have been harnessed for a better quality of life. TABLE 17.2 lists major areas of research in microbiology. Genomics, the study of the functions and interactions of all genes in a genome, is a relatively new and rapidly developing area. Microbial genome sequencing projects have led to the sequencing of numerous microbes (and other organisms), an accomplishment with a strong positive impact on an understanding of virulence factors, vaccine design, countermeasures to antibiotic resistance, and microbial diversity. In early January 2009 the National Aeronautics

TABLE 17.2 Key Research Areas in Microbiology and Public Health
Microbial diversity and versatility
Microbial ecology and physiology
Pathogenic microbiology and immune responses
Microbial control of environmental pollution and recycling
Biotechnology
Microbial genomics

and Space Administration (NASA) reported that Mars was venting atmospheric methane, a simple molecule of four hydrogen atoms bound to a carbon atom, indicating the possibility of life on Mars. Life is responsible for 90% of the Earth's atmospheric methane; however, it should be emphasized that methane can also come from minerals.

■ Health for All

Over half a century has passed since the proclamation of the Universal Declaration of Human Rights, and over a quarter of a century has transpired since the 1978 Alma-Ata Conference. Article 25 of the Declaration speaks of the highest attainment of health for all people as a fundamental right, and Alma-Ata defined its goal as "the attainment for all people of the world by the year 2000 of a level of health that will permit them to lead a socially and economically productive life." Substantial progress has been made toward these goals, as witnessed by impressive gains in healthy life expectancy for citizens fortunate enough to live in the industrialized world. But an estimated 85% of the world's population live in the developing countries and will not reap the full benefits of modern health care. Their lives are spent in abject poverty with little access to decent housing, clean water, sanitary waste disposal systems, appropriate nutrition, and education. Their future is bleak, and they live life on the edge (FIGURE 17.3). The reality is that the Declaration of Human Rights and the goal of Alma-Ata remain as empty words with little promise. There is an approximate thirty-year disparity in life expectancy between industrialized countries and Swaziland, where the life expectancy is forty years. It is true that the successes achieved in the twentieth century were remarkable, but developing countries still bear the brunt of the burden of disease (FIGURE 17.4). Balancing the scale and reducing the disparity between the developed industrialized nations and developing nations are priorities in international public health. New strategies and updated public health infrastructures must be implemented (FIGURE 17.5). This is the single most significant item of an unfinished agenda.

FIGURE 17.3 Life on the edge.
© K. Thorsen/ShutterStock, Inc.

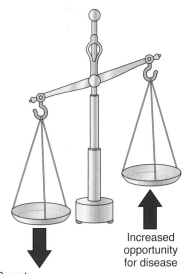

FIGURE 17.4 The scales of (in)justice.

- Poverty
- Inadequate shelter
- Lack of safe water
- Poor public health infrastructure
- Illiteracy

Increased opportunity for disease

FIGURE 17.5 The old must give way to the new to improve the public health infrastructure in less-developed countries. © Stanislav Komogorov/ShutterStock, Inc.

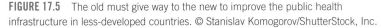

Poverty is considered to be the root of all evils; poverty lies at the heart of the burden of inequality and needs to be the target of social, economic, and public health intervention. Poverty, not microbes, is the fundamental cause of disadvantaged populations suffering more than their fair share of microbial (and other) diseases. Microbes are agents of disease that flourish in the midst of poverty. The war against malaria and tuberculosis, as well as numerous other diseases, is a war against poverty and encompasses social and economic issues as well as public health issues.

The poor exist in pockets of poverty in developed and developing countries. In response the World Health Organization (WHO) launched new initiatives for health that could save millions of lives in the twenty-first century. A former director-general of the World Health Organization stated the following:

The world could end the first decade of the twenty-first century with notable accomplishments. Most of the world's poor people would no longer suffer today's burden of premature death and excessive disability, and poverty itself would thereby be much reduced. Healthy life expectancy would increase for all. . . . The financial burdens of medical needs would be more fairly shared, leaving no household without access to care or exposed to economic ruin as a result of health expenditure. And health systems would respond with greater compassion, quality, and efficiency to the increasingly diverse demands they face.

But the responsibility of addressing poverty does not rest only with the World Health Organization; partnerships at the international level are vital. It is obvious that new initiatives targeted at addressing the disparity that exists in healthcare delivery are necessary for us to realize the intent of the Universal Declaration of Human Rights and the Alma-Ata Conference. The cycle of disease is complex and decidedly multifaceted (FIGURE 17.6).

The Harvard University School of Public Health bears the following

FIGURE 17.6 Relationships among determinants of emerging infections. Reprinted from D. B. Louria, in *Emerging Infections I*, W. M. Scheld, D. Armstrong, and J. M. Hughes (ed.), 1998, ASM Press, Washington, D.C., with permission.

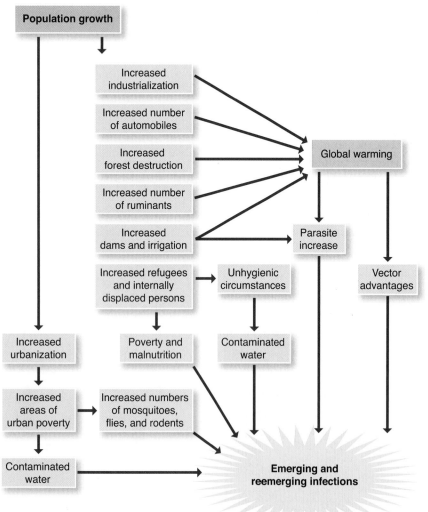

inscription, repeated in several languages, on an outside wall: "The highest attainable standard of health is one of the fundamental rights of every human being."

To conclude this text the Red Queen's Hypothesis, proposed by the evolutionary biologist Leigh van Valen in 1973 based on the novel *Through the Looking Glass* authored by Lewis Carroll in 1871, is appropriate. The Red Queen, a character in the novel said, "It takes all the running you can do to stay in the same place." According to the principle, "for an evolutionary system, continuing development is needed just to maintain its fitness relative to the systems it is co-evolving with."

AUTHOR'S NOTE (RIK)
I have walked by this inscription at the Harvard School of Public Health at least a few hundred times, but each time is like the first time. My eyes, my mind, and my emotions focus on these words, and they continue to inspire me.

■ Self-Evaluation

Part I: Choose the single best answer.

1. Despite major advances in sanitation and the availability of clean water, antibiotics, and immunization, microbial diseases remain
   a. low **b.** an ongoing battle **c.** a mystery **d.** unchallenging

2. Which of the following are the most susceptible to opportunist infections?
   a. healthy people **b.** healthcare workers **c.** organ recipients **d.** zombies

3. Which of these will continue to occur along with microbial diseases in their wake?
   a. catastrophic events **b.** surveillance activities **c.** good sanitation practices **d.** Western medical practices

4. Which of these is used in the CDC blog to educate the public about emergency preparedness?
   a. goats **b.** vampires **c.** Sponge Bob **d.** zombies

5. Some microbial diseases can be prevented by
   a. autoclaves **b.** vaccines **c.** pitch forks **d.** viruses

6. Continued advancement in society's understanding of its environment and establishes direction for future investigation is accomplished by
   a. evolution **b.** sanitation **c.** research **d.** quarantine

7. Travelers may get more than they bargained for including a belly full of worms and malaria by
   a. ecotourism **b.** cruise ship vacations **c.** gardening **d.** farming

8. The fundamental cause of disadvantaged populations suffering more than their fair share of microbial (and other) diseases is
   a. poverty **b.** financial security **c.** politicians **d.** global change

9. Microbes may play a greater role in which of these types of diseases than previously suspected?
   a. acute **b.** respiratory **c.** gastrointestinal **d.** chronic

10. In a sense, our partners in life are
    a. zombies **b.** nurses **c.** microbes **d.** animals

Part II: Fill in the blank.

1. Progress in _____ has resulted in a dramatic increase in the number of living immune compromised patients.
2. Research is vital for continued _____, but the knowledge/products gained need to be handled with wisdom.
3. _____ pose a unique challenge for physicians to recognize, diagnose, and treat.
4. A _____ is defined by efficient spread by a pathogen within a new host species that was not previously susceptible to that pathogen.
5. _____ are essential for the maintenance of life on the planet, and others have been harnessed for a better quality of life.
6. Sanitation, vaccines, and _____ have made enormous strides against pathogenic microbes.
7. The responsibility of addressing poverty does not rest only with the World Health Organization; _____ at the international level are vital.
8. Microbes may act as causative agents, as cofactors, or as _____ that cause damage through autoimmune reactions.
9. The practices of personal and public health can be brought to bear on the prevention and control of _____.
10. The scientific and medical community once found it difficult to accept that stomach and intestinal ulcers are caused by the bacterium _____, an organism present in the acidic environment of the stomach.

Part III: Answer the following.

1. Explain the following statement: "The ongoing saga of microbial diseases is a reflection of the ongoing evolutionary dance."
2. Discuss why chicken flu (H5N1) continues as a threat?
3. Create a list of items that should be included in an emergency supply kit.
4. List at least five ways in which microbes have challenged their hosts throughout history and have influenced the course of civilization.
5. Discuss why microbial-related health problems occur after natural catastrophic events.
6. List at least five examples of advancements in technology that have dramatically improved the life span and health of humans.
7. Explain the importance of continued microbial surveillance on a world-wide basis.
8. List common microbial infections in transplant recipients during their first year after transplantation.
9. Why does the spectrum of disease outbreaks vary from nation to nation?
10. What are the pros and cons of researching the mechanism(s) of microbial virulence or pathogenicity?

Glossary

AB model   An explanation of the mode of toxin activity involving A and B toxin fragments.

Abiotic   Nonliving.

Acquired immune deficiency syndrome   *See* AIDS.

Active carriers   Individuals who have a microbial disease that can be transmitted to others.

Acute leukemia   One of two classifications of leukemia, in which the symptoms appear suddenly and progress rapidly and death occurs within a few months.

Adenoids   Tissues located in the back of the throat that are associated with the secondary immune structures; their lymphocytes play a role in protection from microbes entering through the nose and mouth.

Adenosine triphosphate   *See* ATP.

Adhesions   Molecules on the surface of some bacteria that enable them to stick to host receptor molecules in a Velcro-like manner.

Adsorption   The first stage in the replication cycle of viruses during which viruses attach to the surface of host cells.

Aedes aegypti   A species of mosquito that is a vector for dengue fever.

Aedes albopictus   A species of mosquito that is a vector for dengue fever; these mosquitoes are commonly known as Asian tiger mosquitoes.

Aerobes   Organisms that require oxygen for their metabolic activities.

Aerosols   Suspensions of airborne particles ranging from 1 to 5 micrometers that are a means of transmission of microbes.

African trypanosomiasis   *See* Sleeping sickness.

AIDS   Acquired immune deficiency syndrome, a disease caused by the human immunodeficiency virus in which the immune system becomes severely compromised.

AIDS cocktail   A combination of drugs used to block the replication of human immunodeficiency virus, including two reverse transcriptase inhibitors and one protease inhibitor.

Airport malaria   Malaria that is the result of the survival of malaria-infected mosquitoes from other countries and is transmitted to people within the vicinity of an airport.

Albendazole   A single-dose drug used in the treatment of filariasis.

Algae   Photosynthetic organisms, some of which are unicellular and considered microbes.

Allergy   An adverse immune response to molecules associated with pollen, dust, foods, mites, antibiotics, or bee stings.

Alpha interferon   A drug used in the treatment of hepatitis C.

Amebiasis   A disease caused by the protozoan *Entamoeba histolytica;* it is prevalent in areas with poor sanitation.

Amino acid   Nitrogen-containing molecule that is the building block of protein.

Anaerobes   Organisms that do not require oxygen for their metabolic activities.

Anthrax   A potentially fatal disease caused by *Bacillus anthracis;* the organism has been used in bioterrorism and is a potential agent of biological warfare.

Antibiotics   Metabolic products of bacteria and fungi that inhibit the growth of other microbes.

Antibodies   Protein molecules produced in response to antigens; they react specifically with the antigen that triggered their production and are important defense mechanisms.

Antibody-mediated (humoral) immunity   One of two categories of specific immunity; it is the body's defense strategy for dealing with extracellular microbes.

Anticodon   A series of three nucleotides on tRNA that relate to codons on mRNA.

Antigen-presenting cells   Cells that phagocytize microbes and present antigenic components to other cells of the immune system.

Antigenic drift   A minor change in the H and N spikes of the influenza virus occurring over a period of years.

Antigenic shift   A major and abrupt antigenic change in the H or N spikes of the influenza virus resulting in a new strain of the virus.

Antigens   Components of microbes, usually protein structures, that are recognized as foreign by the immune system and are targeted for destruction.

Antiseptic   An agent used to minimize and destroy the growth of microbes in a wound.

APCs   *See* Antigen-presenting cells.

API 20E   An acronym for analytical profile index; a test strip containing twenty mini-test tubes (cupules) with dehydrated media used in the identification of bacteria.

Arboviruses   Viruses that are borne by arthropods.

Archaea   One of three domains in the Woese system of classification, to which the archaebacteria are assigned.

Arthropoda   A biological phylum characterized by jointed appendages and divided into three classes; fleas, ticks, mosquitoes, and lobsters are examples.

Asexual stage   Part of the malaria parasite's life cycle that occurs in the liver and red blood cells of infected individuals in which sporozoites multiply without sexual reproduction taking place.

Asian tiger mosquito   *See Aedes albopictus.*

Assembly   A stage in the life cycle of viruses in which the viral components are assembled.

ATP   Adenosine triphosphate, a high-energy molecule that drives most cellular processes.

Autoclave   A device that generates steam under pressure used for the sterilization of materials.

Autoimmune disease   A disease in which the immune system fails to distinguish "self" from "nonself."

Autoinfection   A process in which infection is perpetuated within the body; pinworm and strongyloidiasis are examples.

Autotrophs   Microorganisms and plants that are capable of utilizing the energy of the sun or derive energy from the metabolism of inorganic compounds.

Azidothymidine   *See AZT.*

AZT   Azidothymidine, the first clinically safe and effective drug used for the treatment of AIDS; it acts as an inhibitor of the reverse transcriptase enzyme. Also called zidovudine.

B lymphocytes   White blood cells that produce antibodies.

Babesiosis   A tickborne protozoan disease caused by species of *Babesia.*

Bacilli   Rod-shaped bacteria.

Bacillus Calmette-Guérin vaccine   A vaccine against tuberculosis with limited effectiveness.

Bacteria   Unicellular microbes with distinct properties; one of the six distinct types of microbes.

Bacteria   One of three domains in the Woese system of classification.

Basal bodies   The structures by which flagella are anchored to the cell wall and cell membrane.

Basophils   White blood cells that are rich in granules of histamine.

BCG   *See* Bacillus Calmette-Guérin vaccine.

Beaver fever   *See* Giardiasis.

Binary fission   An asexual mode of reproduction in which a cell splits into two new cells.

Binomial system   A system of nomenclature established by Carolus Linnaeus in 1735 in which the genus and species are identified.

Bioaugmentation   Spraying of nutrients on beaches or on other microbe-contaminated sites to foster the growth of microbes indigenous to the area in order to accelerate degradation of pollutants.

Biocrime   A bioterrorist act that targets a specific individual or group rather than the masses.

Biogeochemical cycles   Processes such as the carbon and nitrogen cycles in which bacteria play a critical role.

Biological amplification   The mechanism of activity of the complement system that promotes immune system function.

Biological vectors   Organisms, including mosquitoes, ticks, lice, and flies, that transmit microbial disease.

Biological warfare   The use of microbes, such as the anthrax bacillus and smallpox virus, as weapons in the conduct of war.

Biological weapons   Microbes that are deployed with the intention of producing clinical disease in an attempt to incapacitate or kill large numbers of individuals.

Bioremediation   A process utilizing the enzymatic activity of microbes to break down pollutants such as oil, paper, and concrete.

Biosphere   All the organisms and environments found on Earth.

Bioterrorism   Employment of biological weapons by nonstate governments, religious cults, militants, or crazed individuals.

Biotic   Of or relating to life; refers to the living components of an ecosystem.

Biting plates   Mouth parts found in some worms that allow them to attach to and suck blood from their host.

Black Bain   Term used for anthrax in the 1600s when anthrax killed 60,000 cattle.

Black Death   *See* Plague.

Blender   An organism that facilitates the creation of a new hybrid strain of a disease; an example is that of a pig that becomes infected with a human strain and a bird strain of the flu, resulting in a new hybrid form of the virus.

Blood flukes   Common name for *Schistosoma* species and some other worms.

Boils   Localized skin infections frequently caused by staphylococci.

Bone marrow   Source of all blood cells.

Botox   Minute doses of botulinum toxin used to reduce wrinkles and to treat common disorders associated with muscle overactivity.

Botulism   A form of food poisoning caused by *Clostridium botulinum;* a neurotoxin that causes flaccid paralysis.

Bovine spongiform encephalopathy   A neurological condition in cattle resulting from abnormally folded prions that convert normal prions to the defective form.

Breakbone fever   *See* Dengue fever.

Breath test   A means of diagnosing peptic ulcers based on the production of urease by *Helicobacter pylori.*

Broad-spectrum antibiotic   An antibiotic that is inhibitory to a wide range of bacteria.

Bronchitis   Inflammation of the bronchioles.

BSE   *See* Bovine spongiform encephalopathy.

Buboes   Enlarged lymph nodes that occur when bacteria localize in the lymph nodes, such as in bubonic plague.

Bubonic plague   A form of *Yersinia pestis* infection in which the bacteria localize in lymph nodes, causing them to swell to the size of hens' eggs.

Budding   A mechanism by which some viruses are released from the host cell.

Bulbar poliomyelitis   An extremely serious form of polio in which individuals have difficulty swallowing and breathing because of muscle paralysis.

Burkitt's lymphoma   A malignant cancer of the jaw and abdomen that occurs often in children in central and western Africa.

Campylobacter   A bacterium that is an important cause of intestinal disease in humans and is associated with abortion and enteritis in sheep and cattle.

CA-MRSA   Community associated methicillin-resistant *Staphylococcus aureus.*

Capsid   The protein coat surrounding viruses.

Capsomeres   Protein units that make up the capsid; they confer helical, polyhedral, or complex shapes.

Capsule   A component of the bacterial envelope that may contribute to virulence; it is not present in all species of bacteria.

Carbohydrates   Organic molecules that function as an energy source and are found in some cellular structures.

Carbuncles   Localized skin infections that are larger and deeper than boils and can reach baseball size.

Causative agent: *See* Etiological agent.

Cauterization   An early method of treating wounds inflicted by rabid animals in which long, sharp, hot needles were inserted deeply into the wounds.

CD   *See* Cluster-of-differentiation molecules.

CD4 receptor molecules   Receptor molecules found on some T cells that identify them as T-helper cells.

CD8 receptor molecules   Receptor molecules found on some T cells that identify them as cytotoxic T cells.

Cell   The basic unit of life.

Cell culture   A technique for growing cells that can be used to culture viruses.

Cell-mediated immunity   A type of specific immunity that results in the production of sensitized T lymphocytes directed against a particular antigen.

Cell membrane   A component of the bacterial envelope that regulates the passage of molecules between the bacterial cell and its external environment.

Cell wall   A component of the bacterial envelope that gives the cell its characteristic shape and confers structural integrity.

Cercariae   Freely-swimming immature forms of *Schistosoma* that are responsible for "swimmer's itch."

Cercarial dermatitis   A condition characterized by itching and a rash caused by penetration of the skin by cercariae.

Cerebral malaria   A deadly form of malaria that affects the brain.

Cervix   The opening to the uterus.

Chagas disease   A protozoan disease caused by *Trypanosoma cruzi* that leads to widespread tissue damage, particularly to the heart, causing it to enlarge and impairing its function.

Chancres   Sores on the penis or on the cervix that are characteristic of the primary stage of syphilis.

Chemosynthetic autotrophs   Organisms that derive energy from the metabolism of inorganic compounds.

Chemotaxis   The process of moving toward or away from a chemical stimulus.

Chicken pox   A disease characterized by blisterlike lesions on the body and typically occurring in childhood.

Chimeric (hybrid)   A single microbe or virus composed of two or more different genetically distinct ones.

Chlamydiae   Coccoid bacteria that are obligate intracellular parasites; they are responsible for a variety of diseases, including urethritis, trachoma, and chlamydia.

Chlorella   A photosynthetic alga found on the surface of ocean water.

Cholera   A disease caused by *Vibrio cholerae* that is characterized by severe diarrhea and dehydration.

Chromosome   The structure into which DNA is organized.

Chronic carriers   Individuals who harbor a pathogen for long periods without becoming ill with the disease but who may spread the disease to others.

Chronic fatigue syndrome   A disease that was incorrectly believed to have been caused by Epstein-Barr virus.

Chronic granulomatous disease   An inherited disorder of phagocytes characterized by their inability to kill bacteria.

Chronic leukemia   A form of leukemia, cancer of the white blood cells, that can remain for many months.

Cilia   Hairlike projections that line certain areas of the respiratory system and assist in the removal of bacteria.

Ciliata   A group of protozoans characterized by the presence of cilia.

Clonal selection theory   An explanation of antibody production: A population of antibody-producing B cells exists for every possible antigen and is triggered to clone and produce antibodies upon contact with a specific antigen.

Cluster-of-differentiation molecules   Distinctive molecules found on T cells that identify their specific role.

Coagulase   An enzyme that forms a network of fibers around bacteria, affording protection against phagocytosis.

Cocci   Spherical bacteria.

Codon   A series of three nucleotides on MRNA that relates to anticodons.

Collagen   A protein found in connective tissues.

Collagenase   An enzyme that breaks down collagen.

Colonies   Visible masses of bacterial cells growing on an agar surface, each presumably derived from a single cell.

Colonization   Growth and reproduction of a microbe in a particular niche, resulting in large numbers of cells.

Colorado tick fever   A tickborne viral fever caused by an arbovirus.

Commensalism   A symbiotic relationship between two species in which one benefits and the other is neither harmed nor benefited.

Common cold   A mild illness caused by a variety of viral groups.

Common-source epidemics   Outbreaks of disease arising from contact with a single contaminated source, typically associated with fecally-contaminated food or water.

Common warts   Benign, painless, elevated growths caused by human papillomaviruses that occur most frequently on the fingers.

Competent   Cells that are able to take up DNA from the environment resulting in gene transformation.

Complement   A series of blood proteins that constitute a significant defense mechanism against disease-causing microbes.

Complex   An arrangement of capsomeres in some viruses.

Congenital syphilis   Syphilis resulting from the passage of spirochetes across the placenta from mother to baby.

Conjugation   A recombinational process resulting in the transfer of DNA from donor to recipient during physical contact.

Constitutive enzymes   Enzymes that are constantly produced because the gene switch is always in the "on" position.

Consumers   Organisms that take in oxygen and release carbon dioxide.

Convalescence stage   The time in which recovery from an illness takes place, strength is regained, repair of damaged tissue occurs, and rashes disappear.

Copepods   Water fleas that can harbor infective larvae of guinea worms.

Coronaviruses   A group of viruses that are a major cause of the common cold.

CPE   *See* Cytopathic effect.

Creutzfeldt-Jakob disease   A human transmissible spongiform encephalopathy.

Cryptosporidiosis   A protozoan disease transmitted by drinking fecally-contaminated water.

Culture   Growth of an organism in a laboratory for the purposes of propagation and study.

Cutaneous anthrax   A form of anthrax that is acquired by contact with *Bacillus anthracis* or its spores via wool, hides, leather, or hair products.

Cutaneous leishmaniasis   A form of leishmaniasis that results in skin lesions.

Cyanobacteria   A group of bacteria that is photosynthetic.

Cyclosporiasis   A protozoan disease.

Cyst   A small sac, frequently filled with pus.

Cysticerci   Tapeworm larvae that are enclosed within a membranous sac.

Cysticercosis   A disease resulting from the ingestion of pork tapeworm eggs.

Cytokines   Products that are released by lymphocytes in response to stimuli and that trigger responses in other cells.

Cytopathic effect   A pathological change that occurs in cells as a result of viral replication.

Cytoplasm   The area of the cell enclosed by the cell membrane containing organelles that function in cell metabolism and multiplication.

Cytotoxic T cells   Members of the T-cell subset designated CD8.

Cytotoxins   Exotoxins that damage or kill host cells.

$D = nV/R$ An equation representing the struggle between disease-producing microbes and host resistance. D is the severity of infection; n is the number of organisms; V is for virulence factors; R is for resistance factors.

DALYs   *See* Disability-adjusted life years.

DaPT vaccine   A newer DPT vaccine that utilizes acellular pertussis.

Debridement   Removal of necrotic (dead) tissue in an attempt to halt an infection.

Decomposers   Microbes that break down compounds into simpler constituents.

Defensive strategies   Adaptations that allow microbes to escape destruction by the host immune system.

Deletion   Removal of one (or more) nucleotide leading to a frame-shift change.

Dengue fever   A mosquito-borne disease that is typically self-limiting, with recovery occurring within about ten days.

Dengue hemorrhagic fever   A potentially fatal disease that can result from infection with a dengue virus strain different from the one causing the initial infection.

Denitrifying   Returning nitrogen to the atmosphere.

Deoxyribonucleic acid   *See* DNA.

Dermotropic viruses   Viruses that have an affinity for and cause diseases in the skin and subcutaneous tissues.

Diatom   A type of unicellular alga.

Differential count   A reflection of the ratio of the white blood cell categories.

Diffusion   The passive movement of a substance from an area of high concentration to an area of low concentration.

DiGeorge syndrome   An immune disorder resulting from abnormal development of the thymus gland.

Dinoflagellate   A type of unicellular alga classified as a microbe; dinoflagellates are the primary source of food in the oceans.

Dioecious   Having distinct male and female forms in a species.

Diphtheria   An upper respiratory tract infection caused by *Corynebacterium diphtheriae*.

Diplococci   Groups of cocci that occur in pairs.

Direct fluorescent antibody test   A diagnostic test used for the detection of antigens using fluorescent tagged antibodies.

Direct observational therapy short course   *See* DOTS.

Direct transmission   Person-to-person contact in which the infectious agent is directly transferred from a portal of exit to a portal of entry.

Disability-adjusted life years   A statistic based on healthy years of life (i.e., not including years lost to disability).

Disinfection A process of killing or minimizing microbes usually on surfaces.

Disease A possible outcome of infection in which health is impaired in some fashion.

Disinfectant An agent used to accomplish disinfection.

Distilled spirits Alcoholic beverages resulting from bacterial fermentation and having a high alcoholic content, e.g., brandy, rum, and whiskey.

DNA Molecules that store the genetic information.

Dose The number of microorganisms to which a host has been exposed.

DOTS Direct observational therapy short course, a strategy to ensure that individuals infected with tuberculosis take their prescribed medicines.

DPT vaccine A triple vaccine against diphtheria, pertussis, and tetanus.

Dracunculiasis A disease caused by the parasitic guinea worm, *Dracunculus medinensis*.

Dual-use dilemma A dilemma in which well-intentioned scientific research has the potential to be misused for nefarious purposes (e.g., laboratory-modified bird flu).

Duodenum The first segment of the small intestine.

Eastern equine encephalitis virus A common arbovirus found in the United States.

Ebola hemorrhagic fever A severe viral infection characterized by extreme hemorrhaging with a high fatality rate.

Ecosystem A population of organisms in a particular physical and chemical environment.

Ehrlichiosis A bacterial tickborne infection similar to Lyme disease.

Elephantiasis A disease caused by filarial worms that results in blocked lymphatic vessels and the accumulation of large amounts of lymph fluid in the tissues.

Embryonated (fertile) chicken eggs Chicken eggs containing live embryos; they are used to grow viruses.

Encephalopathy A condition involving brain pathology as in transmissible spongiform encephalopathy.

Encystation A process that allows protozoans to survive outside a host; the parasite is surrounded by a thick capsule.

Endemic disease A disease that is continually present at a steady level in a population and poses little public threat.

Endemic relapsing fever A tickborne bacterial disease caused by *Borrelia recurrentis*.

Endemic typhus A disease caused by *Rickettsia typhi*; it is transmitted to humans from the bite of rat fleas.

Endocytosis A process of engulfment of material, including viruses, displayed by some cells.

Endotoxin A toxin produced by some gram-negative bacteria; it is usually released not during cell growth but upon death and disintegration of the microbe.

Enology The science of wine making.

Enterotest A method to test for the presence of *Giardia* trophozoites; the patient swallows a gelatin capsule attached to a string, and after four hours, the capsule is withdrawn and examined for the presence of trophozoites.

Enterotoxigenic A bacterial strain that produces an enterotoxin.

Escherichia coli strains Strains of *E. coli* that are a common cause of traveler's diarrhea.

Enterotoxin A toxin that affects the intestinal tract.

Envelope A bacterial structure consisting of a capsule, cell wall, and cell membrane.

Enveloped viruses A category of viruses characterized by a membrane surrounding the capsid.

Enzymes Substances that act as catalysts on specific substrates and influence the rate of a chemical reaction; some bacterial enzymes are virulence factors.

Eosinophils White blood cells associated with helminthic infections and allergies.

Epidemic A disease that has a sudden increase in morbidity and mortality in a particular population.

Epidemic relapsing fever A spirochete-caused disease transmitted from person to person by the bite of infected body lice.

Epidemic typhus A disease caused by *Rickettsia typhi*; it is transmitted directly from human to human by the bites of body lice.

Epidemiology The study of the sources, causes, and distribution of diseases and disorders that produce illness and death in humans.

Epstein-Barr virus A DNA virus that is the primary cause of infectious mononucleosis.

Erythrocytes Red blood cells.

Erythrogenic toxin A secretion, produced by some strains of streptococci, that causes the red rash of scarlet fever.

Espundia See Mucocutaneous leishmaniasis.

ETEC See Enterotoxigenic *E. coli* strains.

Etiological agent The microbe responsible for a specific disease. *See* Causative agent.

Eucarya One of three domains in the Woese system of classification.

Eucaryotic cells Cells possessing a nuclear membrane and other membrane-bound organelles.

Excystation The process by which a cyst comes out of dormancy resulting in an active stage.

Exfoliative toxin A substance produced by staphylococci that causes the skin to become blistery and peel away.

Exotoxins Protein molecules that are released by microbes during their growth and metabolism.

Extracellular microbes Microbes that colonize on cell surfaces.

Extremely drug-resistant tuberculosis Extreme drug-resistant strains of tuberculosis.

Extremophiles   Microbes that grow under harsh environmental conditions.

Extrusion   A process whereby mature virus particles are released from the host cell; also called budding.

Facultative anaerobes   Organisms that grow best in the presence of oxygen but are capable of survival in its absence.

Fermentation   A metabolic reaction regulated by enzymes that break down sugars to acids and carbon dioxide.

Filarial worms   Tiny adult threadlike worms that can block the lymphatic vessels, possibly resulting in elephantiasis.

Filariasis   *See* Elephantiasis.

Fixation   The process through which atmospheric nitrogen is converted to a product (typically ammonia) that plants are capable of using in metabolic processes.

Flagella   Structures composed of the protein flagellin that provide motility to certain species of bacilli and cocci.

Flagellin   The protein of which flagella are composed.

Flatworms   A morphological category of parasitic worms.

Flesh-eating strep   A streptococcus causing a severe form of streptococcal infection, necrotizing fasciitis.

Fomites   Inanimate objects that serve as means of transmission of infectious material.

Food infection   The result of ingestion of bacteria in contaminated foods and their subsequent growth in the intestinal tract accompanied by their secretion of toxin.

Food intoxication   The result of ingestion of bacterial toxins.

Foreign   Not normally present in the body and capable of triggering an immune response.

Fungi   A category of eucaryotic organisms including mushrooms and yeasts.

Fusion   A mechanism of penetration employed by some viruses in which there is contact between the viral envelope and the host cell membrane.

Gametes   Male or female sex cells.

Gamma globulin   A fraction of the globulin component of blood plasma in which most of the antibodies are present.

Ganglia   Collections of nerve cells.

Gas gangrene   A condition brought about through contamination of a wound with particles of soil containing *Clostridium perfringens* or its spores.

Gastrointestinal anthrax   A form of anthrax resulting from the ingestion of inadequately cooked meat contaminated with *Bacillus anthracis*.

Generation time   The length of time between rounds of binary fission.

Genes   Segments of the DNA molecule involved in heredity, which encode specific polypeptide, protein, or RNA molecules.

Genetic engineering   A means of manipulating genes to bring about a desired function.

Genital warts   One of the most common sexually transmitted diseases, caused by human papillomaviruses.

Genome   The complete set of chromosomes in a cell.

Genus   A category in the binomial system of nomenclature above the species level.

Giardiasis   An intestinal disease caused by the protozoan *Giardia lamblia* and acquired by drinking contaminated water.

Glomerulonephritis   A kidney disease that is a potential result of streptococcal infection.

Gonorrhea   A sexually transmitted disease caused by *Neisseria gonorrhoeae.*

gp120   A molecule found on the surface of HIV that "docks" with CD4 receptor molecules on T lymphocytes.

Gram-negative   A cell wall-staining property of some bacteria; they do not retain the crystal violet stain after decolorizing with alcohol and appear pink to red after staining with safranin.

Gram-positive   A cell wall-staining property of some bacteria that causes them to retain the crystal violet stain after decolorizing with alcohol; they appear purple.

Granuloma   A pocket-like arrangement of cells associated with chronic inflammation that walls off the inflammatory agent.

Growth   An increase in the size of individual cells or an increase in a population of cells.

Guillain-Barré syndrome   A potentially fatal and rare nervous system disease of unknown cause; it sometimes occurs after administration of a vaccine.

Gummas   Tumorlike lesions that characterize tertiary syphilis.

H5N1 influenza   A strain of the influenza virus that mainly occurs in birds and is highly contagious among birds. Human infections are rare but usually severe in symptoms.

H spike   *See* Hemagglutinin.

HAART   Highly active antiretroviral therapy; a cocktail of medications used to treat AIDS.

Halophiles   Bacteria that live in environments containing extreme salt concentrations.

Hanging drop   A method of observing bacterial motility under a microscope by suspending a drop of culture on a slide.

Hansen's disease   A disease caused by *Mycobacterium leprae;* formerly known as leprosy.

Hantavirus   A virus that causes hantavirus pulmonary syndrome.

Hantavirus pulmonary syndrome   A disease caused by a hantavirus that produces severe influenzalike respiratory problems and can result in death.

HBV   *See* Hepatitis B virus.

HCV   *See* Hepatitis C virus.

Healthy carriers   Individuals who have no symptoms of a particular microbial disease but harbor the microbes and may unwittingly pass the disease on to others.

Helical   Arranged in a continuous tube containing a virus's nucleic acid. Refers to arrangement of capsomeres.

Hemagglutinin   A protein containing spikes on the outside of influenza viruses that aids in the attachment/fusion to and entry of cells by the virus.

Hemolysins   Secretions produced by some microbes that destroy red blood cells through the destruction of cell membranes.

Hemolytic uremic syndrome   Kidney damage occurring primarily in young children as a result of bacterial toxins.

Hepatitis   Inflammation of the liver.

Hepatitis A virus   A virus found in feces and transmitted through contaminated drinking water and food. It is the most common hepatitis virus and usually causes only mild disease.

Hepatitis B virus   A virus transmitted in body fluids. It causes subclinical to severe disease.

Hepatitis C virus   A virus transmitted mainly by intravenous drug use; a major reason for liver transplants in the United States.

Hepatitis D virus   An incomplete virus that requires hepatitis B virus to replicate. It causes severe disease with a high mortality rate.

Hepatitis E virus   A virus transmitted by the fecal–oral route. It causes a disease that is usually of moderate severity.

Herbivorous   Refers to organisms that feed on plants.

Herd immunity   Immunity of enough members of a population to protect against an epidemic.

Hermaphroditic   Producing both sperm and eggs.

Herpes simplex virus   Commonly referred to as HSV; HSV-1 is the main cause of cold sores and fever blisters around the mouth while HSV-2 is the main cause of genital herpes.

Heterotrophs   Organisms that require organic compounds as an energy source.

HGE   *See* Human granulocytic ehrlichiosis.

Highly active antiretroviral therapy   *See* HAART.

Hiker's diarrhea   *See* Giardiasis.

HIV   *See* Human immunodeficiency virus.

HIV seropositive   Having antibodies against human immunodeficiency virus.

HME   *See* Human monocytic ehrlichiosis.

Homologous   Refers to stretches of DNA with identical or closely related sequences.

Hookworm disease   A disease that is caused by the roundworms *Ancylostoma duodenale* and *Necator americanus*.

Horizontal gene transfer   A process by which genes are transferred from one organism to another.

Horizontal transmission   A means by which a disease spreads from one person to another.

HPV   *See* Human papillomaviruses.

HSV   *See* Herpes simplex virus.

Human Genome Project   The mapping of the genes located on the 23 pairs of human chromosomes.

Human granulocytic ehrlichiosis   A tickborne bacterial infection caused by an unknown species of *Ehrlichia*.

Human growth hormone   A hormone that is used for the treatment of dwarfism and is now produced by genetic engineering.

Human immunodeficiency virus   The virus that causes AIDS. Human immunodeficiency virus type 1 is the most common cause of AIDS worldwide, and human immunodeficiency virus type 2 is the most common cause of AIDS in West Africa.

Human insulin   A hormone that regulates the amount of glucose in the bloodstream. Much of the insulin now used to treat diabetes is a product of genetic engineering.

Human monocytic ehrlichiosis   A tickborne bacterial infection caused by *Ehrlichia chaffeensis*.

Human papillomaviruses   Viruses that cause warts in humans.

Hyaluronidase   An enzyme produced by some bacteria that breaks down hyaluronic acid found in connective tissue.

Hydrophobia   Fear of water; an old term for rabies.

Hyperthermophiles   Bacteria that live in extremely hot environments.

Hyphae   Intertwined filaments characteristic of molds.

Iatrogenic infection   Infection induced in a patient by a medical procedure.

Icosahedrons   Three-dimensional, 20-sided structures with triangular sides; they confer a geodesic shape to a virus.

ID   *See* Infectious dose.

Ig   *See* Immunoglobulins.

Illness stage   The time in which disease develops to the most severe stage, as evidenced by typical signs and symptoms.

Immunocompromised   Characterized by a weakened immune system as a result of AIDS or other conditions.

Immunoglobulins   Categories of antibodies (immunoglobulins A, D, E, G, and M), each possessing specific properties.

Impetigo   A superficial infection of the skin that is typically manifested by blisters around the mouth and is usually caused by staphylococci.

Incubation stage   The time between a pathogen's access into the body and the appearance of signs and symptoms.

Indirect transmission   The passage of infectious material from a reservoir to an intermediate host and then to a final host.

Inducible enzymes   An enzyme that is the result of genes that can be turned on or off depending upon the circumstances.

Infantile paralysis   A form of poliomyelitis in infants and young children, characterized by muscle paralysis.

Infection   The presence of microbes in the body without definitive symptoms.

Infectious agents   Plasmids capable of exchanging genetic material from one microbe to another; some confer antibiotic resistance.

Infectious dose   The minimum amount of bacteria needed to establish an infection.

Infectious mononucleosis   A disease caused by Epstein-Barr virus in which the salivary glands are infected by the virus; frequently referred to as "mono."

Inflammation   Swelling, pain, redness, and heat in an area of tissue.

Influenza   Exaggerated coldlike symptoms caused by influenza virus.

Influenza A virus   A strain of influenza virus that causes epidemics and occasionally pandemics.

Influenza B virus   A strain of influenza virus that causes epidemics and does not have an animal reservoir.

Influenza C virus   A strain of influenza virus that does not produce epidemics and causes mild respiratory illness.

Inhalation anthrax   The most severe form of anthrax, resulting from the intake of anthrax spores.

Inorganic compound   Generally, a chemical compound that does not contain carbon (although carbon dioxide and several other carbon-containing compounds are considered inorganic).

Insects   A large class of arthropods characterized by three body segments and six legs.

Insertion   Addition of one (or more) nucleotides leading to a frame-shift change.

Interferon   A component of blood that interferes with viral replication.

Internally displaced persons   Persons or groups of persons for whom social disruptions have forced to move but have not crossed international borders.

International Committee on Taxonomy of Viruses   A group charged with the indentification of new viruses and their classification.

Internationally displaced persons   Persons or groups of persons for whom social disruptions have forced to move across international borders.

Intracellular bacteria   Bacteria that penetrate and grow inside host cells.

Iron lung   A barrel-like device formerly used for persons who had difficulty breathing on their own because of muscle paralysis as a result of polio.

Ivermectin   A drug used to treat onchocerciasis.

Kala-azar   A severe and usually fatal form of leishmaniasis in which the parasites invade the liver and other organs.

Kaposi's sarcoma   A type of vascular cancer that occurs in people suffering from AIDS.

Keratitis   An aggressive eye infection that can be caused by microbial contamination of a contact lens solution.

Kinases   Enzymes that break down clots in the blood.

Kissing bug   An insect that is the vector for Chagas disease.

Koch's postulates   A set of rules used to establish that a particular organism is the cause of a particular disease.

Koplik's spots   Spots that appear in the mouth in the early stage of measles.

Kuru   A transmissible spongiform disease formerly found among members of the Fore people of New Guinea.

Larva   A stage in arthropod development that occurs after hatching and before maturity is reached.

Laryngitis   Inflammation of the larynx.

Latency   The period in which a microbe is in the host without displaying any visible symptoms.

LD50   50% lethal dose; a laboratory measurement of virulence to determine the dose that kills 50% of the test animals in a given time.

Legionnaires' disease   An airborne pneumonialike disease caused by *Legionella pneumophila*.

Leguminous plant   A type of plant with swellings or nodules along the root system containing *Rhizobium* and other nitrogen-fixing bacteria.

Leishmaniasis   A disease caused by parasites and transmitted by the bite of female sand flies.

Lepromas   Tumorlike skin lesions seen in leprosy.

Leprosy   *See* Hansen's disease.

Leptospirosis   A disease caused by the spirochete *Leptospira interrogans*.

Leukemia   A type of cancer characterized by uncontrolled reproduction of white blood cells.

Leukocidins   Bacterial enzymes that destroy white blood cells.

Leukocytes   White blood cells.

Lipopolysaccharide   A molecule with both a lipid and a polysaccharide component found in the outer wall of gram-negative bacteria; it acts as an endotoxin.

Lockjaw   A symptom of tetanus characterized by contraction of the muscles in the jaw; also an alternative name for tetanus.

Lower respiratory tract infections   Infections that occur in the trachea, larynx, bronchi, bronchioles, or lungs.

Lymph   A tissue fluid derived from blood that is returned to the blood by lymphatic vessels.

Lymph nodes   Structures along lymphatic vessels that act as microbe filters.

Lymphadenopathy   Swelling of lymph nodes that occurs in AIDS and other infections.

Lymphatic system   A system in which lymph is transported in lymphatic vessels.

Lymphatic vessels   Structures that transport lymph fluid.

Lymphocytes   A category of white blood cells that play a key role in immunity.

Lymphocytic leukemia   A form of leukemia characterized by overproduction of lymphocytes, leading to abnormally large numbers of immature and nonfunctional lymphocytes.

Lymphoid path   The development of blood cells that leads to the production of lymphocytes.

Lysis   Bursting of cells.

Lysogenized   Refers to bacterial cells that contain phage nucleic acid.

Lysozyme   An enzyme present in tears that contributes to the nonspecific immune system by disrupting bacterial cell walls.

M protein   A protein found in the cell walls of streptococci that confers resistance to phagocytosis.

Macrophages   Phagocytic cells that function in the immune system.

Macroscopic   Visible without the aid of a microscope.

Mad cow disease   A disease of cattle caused by prions that results in spongy degeneration of the brain accompanied by severe and fatal neurological damage.

Major histocompatibility molecules   Molecules associated with presentation of antigenic material to receptor molecules on cytotoxic T cells.

Malaria   A tropical disease caused by the bite of female *Anopheles* mosquitoes infected with the *Plasmodium* protozoan parasite.

Mantoux test   *See* Tuberculin skin test.

Mastigophora   One of the four groups of protozoans, characterized by the presence of flagella.

MDR TB   *See* Multiple drug-resistant tuberculosis.

Mechanical vectors   Arthropod vectors that transmit microbes passively on their body parts; the microbes do not invade, multiply, or develop in the vector.

Memory cells   B cells that do not progress to antibody-producing plasma cells but are retained for a quick immune response in the event of future exposure to the same antigen.

Meninges   Membranes around the spinal cord and the brain.

Meningitis   Inflammation of the meninges that can be caused by microbes.

Merozoites   Forms in the malaria life cycle resulting from the asexual multiplication of sporozoites.

Metastasis   The spreading of cancer cells from one region or tissue to another.

Miasma   A term, used before microbes were identified as agents of disease, meaning "bad air" or "swamp air."

Microfilariae   Mature filarial worms in the blood that can be taken up by mosquitoes to infect other people with elephantiasis.

Micrometer   A unit of measurement equal to one-millionth of a meter.

Microscopic   Visible only with the aid of a microscope.

Missense mutation   A change in a codon resulting in incorporation of a different amino acid.

MMR vaccine   Measles, mumps, and rubella vaccine.

Monocytes   A category of phagocytic white blood cells.

Mononucleosis   *See* Infectious mononucleosis.

Morbidity   A measure of the rate of illness of a particular disease.

Mortality   A measure of the rate of death resulting from a particular disease.

MRSA   Methicillin-resistant strains of *Staphylococcus aureus.*

Mucocutaneous leishmaniasis   A form of leishmaniasis characterized by invasion of the parasite into the skin and mucous membranes, causing destruction of the nose, mouth, and pharynx; also called espundia.

Multicellular   Having more than one cell.

Multiple drug resistant   Resistant to more than one antibiotic.

Multiple drug-resistant tuberculosis (MDR TB)   A strain of the tubercle *Bacillus* that is resistant to the two most anti-TB antibiotics, rifampicin and isoniazid.

Multiplication   A process resulting in an increase in the total number of cells.

Murine typhus   A variety of typhus fever caused by *Rickettsia typhi* that is transmitted by fleas.

Mutagens   Physical or chemical agents that cause mutation.

Mutation   A change in DNA that is transferred to subsequent generations.

Mutualism   A symbiotic relationship in which both organisms benefit from the association.

Mycoplasmas   Bacteria that have no cell walls.

Mycoses   Diseases caused by fungi.

Myeloid leukemia   A type of cancer resulting from overproduction of monocytes and granular leukocytes.

Myeloid path   The lineage of blood cell maturation that leads to the production of platelets, red blood cells, monocytes, neutrophils, eosinophils, and basophils.

N spike   *See* Neuraminidase.

Naked viruses   Viruses that are not enclosed by an envelope.

Nanometer   A unit of measurement equal to one-billionth of a meter.

Narrow spectrum antibiotic   An antibiotic that is inhibitory to limited range of bacteria.

Necrotic   Dead; used in reference to tissue.

Necrotizing fasciitis   A condition caused by highly invasive streptococci in which the subcutaneous tissue is infected; the streptococci are sometimes referred to as "flesh-eating."

Negri bodies   Cytopathic structures found in rabies virus-infected cells.

Neonatal (newborn) tetanus   A manifestation of tetanus in newborn children that results from unsanitary conditions during delivery.

Neuraminidase   An enzyme containing spikes on the outside of surface of some influenza viruses, aiding in the release of new virions.

Neurocysticercosis   The presence of cysticerci in the brain or spinal cord resulting in seizures, headaches, and possibly death.

Neurosyphilis   A late stage of syphilis involving neurological damage possibly characterized by paralysis and insanity.

Neurotoxins   Toxins released by microbes that interfere with the transmission of neural impulses.

Neutrophils   White blood cells that are phagocytic and characterized by a multilobed nucleus.

Nitrification   A stage in the nitrogen cycle in which ammonia is converted into nitrogen.

Nitrogen cycle   A cycle in nature in which atmospheric nitrogen is recycled through a pathway involving bacteria.

Nonenveloped viruses   Viruses that do not have an envelope around them; also referred to as "naked."

Nonsense mutation   A change in a codon leading to a stop codon.

Nonspecific immunity   Physiological defenses that prevent microbes from gaining access into the body or eliminate those that have penetrated the body.

Norwalk virus   A virus that is a frequent cause of gastroenteritis in older children and adults.

Norwalk-like viruses   A group of viruses related to the Norwalk virus that frequently cause gastroenteritis in older children and adults.

Nosocomial infections   Infections acquired by patients during hospitalization or during stays in long-term healthcare facilities.

Nuclein   An early term for DNA.

Nucleocapsid   A viral structure consisting of the viral nucleic acid and the protein coat.

Nucleoid   The DNA-rich area in procaryotic cells; it is not surrounded by a membrane.

Nutrient agar   A semisolid growth medium used to grow bacteria in the laboratory.

Nymph   A preadult stage in tick development.

Obligate intracellular parasites   Microbes that can only replicate within cells.

Offal   A ground-up mixture of organs and trimmings of dead animals used to feed other animals.

Offensive strategies   Adaptations by microbes that result in their ability to damage the host and establish disease.

Oligodynamic effect   Killing or toxic effect of heavy metals on bacteria.

Onchocerciasis   A parasitic disease caused by *Onchocerca volvulus* that frequently results in blindness.

Oocyst   Infectious form of *Cryptosporidium parvum*.

Operons   A group of functionally related genes that act as "on and off" switches.

Ophthalmia neonatorum   A condition in which the corneas are damaged as a result of the transmission of *Neisseria* into the eyes of newborns during delivery.

Opisthotonos   A body position in which the back is severely arched; characteristic of tetanus.

Opportunistic pathogens   Organisms that cause disease when the host's immune system is weakened, as in AIDS.

Oral rehydration therapy   A method of cholera treatment designed to replace lost body fluids; an alternative to intravenous rehydration.

Orchitis   Inflammation of the testes that sometimes occurs in males infected with mumps virus.

Organ system   A collection of organs that contribute to an overall function.

Organic compound   Generally, a chemical compound that contains carbon (although carbon dioxide and several other carbon-containing compounds are considered inorganic).

Organisms   Living entities composed of cells.

Organs   Structures composed of more than one tissue type.

Oriental sore   A form of leishmaniasis that results in skin lesions primarily on exposed parts of the body; also known as cutaneous leishmaniasis.

Oseltamivir   A relatively new drug that shortens the duration of influenza caused by influenza A and B viruses. The trade name of this drug is Tamiflu®.

Osmosis   The movement of a solvent from an area of low solute (high solvent) concentration to an area of high solute (low solvent) concentration.

Outer membrane   An additional layer external to the peptidoglycan layer in gram-negative bacteria that is associated with virulence.

Pandemic   A worldwide outbreak of disease.

Paralytic poliomyelitis   A form of polio caused by replication of poliovirus in nerve cells, sometimes resulting in severely deformed limbs and paralysis.

Parasitic cycle   A chain of events, sometimes quite complex, by which a parasite exits from a host and gains access to a new host.

Parasitism   A form of symbiosis in which one biological agent lives at the expense of another.

Parotid glands   One of the three pairs of salivary glands; these glands may become infected and cause swelling on one or both sides of the face, a characteristic of mumps.

Parthenogenesis   A process by which females produce eggs without fertilization by males.

Pasteurization   A process of disinfection by which liquids are heated to reduce the number of pathogens.

Pathogens   Microbes capable of producing disease.

PCR   An acronym for polymerase chain reaction.

Pellicle   A protective, rigid cover outside the cell membrane found in some protozoans.

Pelvic inflammatory disease   A condition that occurs in about 50% of untreated female gonorrhea patients characterized by abdominal pain and sometimes sterility.

Penetration   A stage in the replication of viruses in which the virus enters the host cell.

Peptidoglycan   A compound found in gram-positive and gram-negative bacterial cell walls that confers rigidity and tensile strength.

Peramivir   An antiviral drug authorized by the FDA for the treatment of severe H1N1 cases during the 2009 influenza pandemic.

Perforin   A membrane-penetrating protein released by cytotoxic T cells that leads to lysis of cells harboring viruses.

Peritonitis   A potentially fatal condition caused by leakage of intestinal fluids into the abdomen.

Petri dish   A round, platelike container in which bacteria are grown on agar in the laboratory.

Phage conversion   A process by which bacteriophage DNA is incorporated into the bacterial chromosome and confers new properties on the bacterial host.

Phagocytic cells   Cells that engulf microbes and bring about their destruction by enzymatic activity; these cells play a vital role in the nonspecific immune system.

Phagocytosis   A nonspecific body defense mechanism by which bacteria are ingested and killed by cells.

Photosynthetic autotrophs   Organisms that utilize the energy of the sun and that use carbon dioxide as a carbon source.

PID   *See* Pelvic inflammatory disease.

Pili   Bacterial appendages in gram-negative bacteria that act as adhesins; some serve as a bridge allowing genetic exchange.

Pilin   The protein that composes pili.

Pinworm   A common helminthic disease caused by the worm *Enterobius vermicularis.*

Plague   A bacterial disease caused by *Yersinia pestis,* a highly virulent bacterium; also known as the Black Death.

Plankton   The primary food source for many aquatic organisms, consisting mainly of protozoans.

Plantar warts   Deep, painful warts found on the soles of the feet; caused by human papillomavirus.

Plasma cells   Antibody-producing cells derived from B cells.

Plasmids   Small molecules of nonchromosomal DNA found in some bacteria.

Platelets   Blood cells that initiate the clotting of blood.

Pleurisy   Inflammation of the pleural lining of the lungs.

Plug drugs   Drugs that act to both prevent and treat influenza by interfering with the N spikes of the virus.

Pneumonia   Inflammation of the lungs; can be caused by a variety of microbes.

Pneumonic plague   A form of bubonic plague that develops into pneumonia.

Pneumotropic viruses   Viruses that have an affinity for the respiratory tract.

Point mutation   A genetic process in which one nucleotide is replaced with another nucleotide; simplest mutation.

Polio   Poliomyelitis, a highly infectious viral disease that can lead to muscle paralysis.

Polyhedral   Consisting of icosahedra; a term used to describe an arrangement of viral capsomeres.

Polymerase chain reaction   A molecular biology technique used to amplify fragments of DNA.

Polypeptide   A series of amino acids chemically held together by peptide bonds.

Pontiac fever   A pneumonialike illness caused by *Legionella pneumophila.*

Portal of entry   The site at which microbes enter a host.

Portal of exit   The site from which microbes leave a host and may infect another host.

PPD   *See* Purified protein derivative.

Premastication   The practice of prechewing food for infants by which HIV can be transmitted.

Primary immune structures   The bone marrow and the thymus gland in which maturation of B and T lymphocytes takes place.

Primary producers   Photosynthetic organisms that utilize the energy of the sun and produce organic compounds and oxygen.

Primary syphilis   An early stage of syphilis characterized by the formation of chancres on the genitals.

Prions   Infectious, highly stable, misfolded proteins that can cause neurological disease; they do not contain DNA or RNA.

Procaryotic   Not having the genetic material contained within a membrane in the cell.

Prodromal stage   An early stage in a microbial disease characterized by headache, tiredness, and muscle aches. Also the first stage of human immunodeficiency virus infection characterized by fever, diarrhea, rash, aches, fatigue, and lymphadenopathy.

Proglottids   Compartmentlike segments of tapeworms containing both testes and ovaries.

Promoter site   A site on a gene strand that marks the beginning of transcription.

Propagated epidemics   Epidemics resulting from direct person-to-person transmission.

Prophage   A segment of phage integrated into a bacterial chromosome.

Protease inhibitors   A class of drugs used in AIDS therapy that prevent viral replication.

Protozoans   A category of unicellular eucaryotic microbes.

PrP gene   Gene located on chromosome 20 that is responsible for prion proteins.

Psychrophiles   Cold-loving bacteria that grow best at about 15°C.

Purified protein derivative   A product derived from *Mycobacterium tuberculosis* that is used in the tuberculin skin test.

R (resistance) factors   Genes carried on plasmids that confer antibiotic resistance.

Rabies   A viral disease transmitted by the bite of a rabid animal, resulting in damage to the nervous system and eventual death if not treated promptly by vaccination; formerly called hydrophobia.

Rapid Strep Antigen Test   A rapid test used in diagnosis of streptococcal sore throat.

Recombinant DNA technology   *See* Genetic engineering.

Red blood cells   Cells of the circulatory system that carry oxygen throughout the body; also called erythrocytes.

Release   The last stage in the viral replication cycle during which mature viruses are released from host cells.

Relenza   *See* Zanamivir.

Rendering process   The process by which animal parts are turned into animal feed.

Replication   The process by which viruses multiply in host cells.

Reservoir   A site in nature where microbes survive and multiply and from which they may be transmitted.

Resistance   A defensive function of the immune system affording protection against microbial invasion; also, the ability of microbes to counter the effects of antimicrobial agents.

Respiratory syncytial virus   A prevalent cause of respiratory illness, most commonly found in infants, that produces nonspecific symptoms, including fever, runny nose, ear infection, and pharyngitis.

Rheumatic fever   A condition involving the heart and joints resulting from repeated bouts of streptococcal infection.

Rhinoviruses   A large group of viruses that are responsible for many common colds.

Ribavirin   A drug used in the treatment of hepatitis C and in the treatment of pneumonialike illnesses caused by respiratory syncytial virus.

Ribosomal ribonucleic acid   A type of ribonucleic acid (RNA) that is associated with ribosomes.

Rickettsiae   Rod-shaped bacteria that are (with a single exception) transmitted through the bite of arthropods; they are intracellular obligate parasites.

River blindness   A common type of blindness resulting from the migration of larval forms of the worm *Onchocerca volvulus* into the eye; also called onchocerciasis.

R nought   An epidemiological term used as a measure of the potential for transmission; also called R_0.

Rocky Mountain spotted fever   A tickborne disease common in the southeastern United States that is caused by *Rickettsia rickettsii*.

Ropy milk   A condition that occurs when *Alcaligenes viscolactis* sheds its slime layer into milk.

RotaShield   An oral vaccine against rotavirus; the vaccine was withdrawn about a year after its approval because of adverse reactions.

Rotaviruses   A group of viruses that are a common cause of viral gastroenteritis in children under the age of five years.

Roundworms   A morphological category of worms.

Rubella   A viral disease characterized by a rash on the face that spreads to the trunk and the extremities.

Saber shins   A condition sometimes seen in syphilis patients in which the shinbone develops abnormally.

Sabin vaccine   A polio vaccine developed by Albert Sabin consisting of attenuated polioviruses.

Salk vaccine   The first vaccine against polio developed by Jonas Salk containing inactivated poliovirus.

Salmonellosis   A condition caused by ingestion of salmonella bacteria resulting in gastroenteritis manifested by nausea, vomiting, abdominal cramps, and diarrhea.

Sand flies   Insects that transmit *Leishmania* species.

Sarcodina   A group of protozoans that exhibit a "creeping" movement.

Scalded skin syndrome   A condition caused by a staphylococcal toxin that results in the skin's becoming blistery with a tendency to peel.

Scarlet fever   A disease caused by a strain of streptococcus that produces an erythrogenic toxin leading to the development of a red rash and a strawberry-colored tongue.

Scavengers   A synonym for decomposers.

Schistosomiasis   A parasitic disease caused by blood flukes that enter the body through the skin.

SCID   *See* Severe combined immunodeficiency.

Scolex   The head of a tapeworm; it attaches to the host intestinal wall by suckerlike projections.

Secondary syphilis   A stage of syphilis characterized by a rash on the palms and soles; during this stage, the spirochetes multiply and spread throughout the body.

Secondary bacterial infections   Infections that result from bacteria in individuals suffering from the flu or other conditions that lower immune resistance.

Secondary immune structures   The spleen, tonsils, adenoids, lymph nodes, and patches of tissue associated with the intestinal tract that are seeded with mature B and T cells and phagocytic cells.

Selectively permeable   Able to be permeated by some but not all types of molecules; a property of cell membranes.

Septicemic plague   A variety of plague resulting from the spread of infection from the lungs to other parts of the body.

Serum sickness   A condition characterized by the formation of antigen–antibody complexes that are deposited in the skin, kidney, and other sites, as might occur in treatment with immunoglobulin.

Severe combined immunodeficiency   A disease in which individuals lack both functional T and B cells and cannot mount either an antibody- or a cell-mediated immune response.

Sex pili   Structures that function in exchange of DNA between bacteria by forming a bridge between cells.

Sexual stage   A stage of malaria in which sporozoites are produced from *Plasmodium* parasites in the blood.

Sexually transmitted diseases   Diseases caused by transmission of microbes from the warm, moist mucous membranes of one individual to the mucous membranes of another individual during sexual contact; formerly called venereal diseases.

Shigellosis   A bacterial gastrointestinal illness caused by the ingestion of *Shigella*-contaminated foods and water resulting in symptoms of diarrhea, abdominal cramping, and, in some cases, dysentery.

Shingles   A disease caused by varicella-zoster virus that occurs in some individuals with a history of chicken pox; the virus infects nerve fibers, creating intense pain.

Sleeping sickness   A protozoan disease caused by the bite of a tsetse fly carrying the parasite. Two types (West and East African trypanosomiasis) are known.

Slime layer   A heavy mucuslike material that accumulates around some bacteria.

Sludge   The product of the conversion of waste material that may be used as fertilizer; also called biosolid.

Small animalcules   The name for microbes first described by Antony van Leeuwenhoek during his examination of tooth scrapings with a primitive microscope.

Species   The fundamental rank in the binomial system of classification of organisms; in the name *Escherichia coli,* "coli" refers to the species.

Specific immunity   Immunity that is acquired after birth and responds to the presence of foreign and potentially harmful microbes that have breached the external and nonspecific defense mechanisms.

Spikes   Projections that extrude through the surface of some viruses.

Spirilla   Spiral-shaped bacteria.

Spirochetes   Flexible, corkscrew-shaped, motile bacteria.

Spleen   An organ that is part of the secondary immune system; it contains phagocytic cells and both mature B and T cells.

Spontaneous generation   A false but once popular theory that nonliving structures could give rise to living organisms. Louis Pasteur played a major role in disproving this theory.

Sporadic   Occurring occasionally and at irregular intervals in a random and unpredictable fashion.

Spores   Structures contained within dormant bacterial cells that are highly resistant to heat, drying, radiation, and a variety of chemical compounds. The genera *Bacillus* and *Clostridium* are sporeformers.

Sporozoites   Forms of *Cryptosporidium parvum* that penetrate the intestinal cells; also, infective forms of malaria parasites.

St. Louis encephalitis virus   A common mosquito-borne arbovirus found in the United States that causes encephalitis.

Stage of decline   The time following an illness in which symptoms begin to disappear and the body returns to normal.

Staph food poisoning   An illness caused by the secretion of an enterotoxin from *Staphylococcus aureus;* characteristic symptoms are abdominal cramps, nausea, vomiting, and diarrhea.

Staphylococci   Bacteria of the genus *Staphylococcus* that group together in clusters resembling bunches of grapes.

STDs   *See* Sexually transmitted diseases.

Stem cells   Undifferentiated cells capable of differentiating into specialized cells.

Sterilization   A process of disinfection used to destroy all live microbes, spores, and viruses.

Stomach flu   A common but meaningless term that is associated with gastroenteritis.

Strep throat   A mild airborne infection caused by *Streptococcus pyogenes* with characteristic symptoms of red or sore throat, fever, and headache.

Streptococci   Bacteria of the genus *Streptococcus* that group together in chains resembling strings of pearls.

Streptokinase   A product of streptococci that dissolves blood clots.

Stromatolites   Fossilized mats formed from microorganisms and dating as far back as 3.5 billion to 3.8 billion years.

Strongyloidiasis   Illness caused by the nematode *Strongyloides stercoralis;* symptoms are nausea, vomiting, anemia, weight loss, and chronic bloody diarrhea.

Subclinical   Asymptomatic and not diagnosed.

Sucking disks   Structures by which *Giardia lamblia* trophozoites attach to the lining of the intestine.

Surfactant   A surface-active agent used for cleaning (e.g., soaps and detergents).

Swamp fever   *See* Leptospirosis.

Swimmer's itch   *See* Cercarial dermatitis.

Symbiosis   A relationship between two or more organisms that live together; mutualism, commensalism, and parasitism are all forms of symbiosis.

Synthetic biology   A discipline that produces unnatural biological molecules from natural biological molecules.

Syphilis   A sexually transmitted disease caused by *Treponema pallidum;* the symptoms are sores on the penis or cervix, rash, and degeneration of organs and tissues.

Systemic infections   Bloodborne infections that spread throughout the body.

T-cell subset   A collective term for the categories of T cells differentiated by clusters of differentiation.

T-helper cells   Members of the T-cell subset designated CD4.

T lymphocytes   A subcategory of lymphocytes that function in the immune system in several ways.

Tamiflu   *See* Oseltamivir.

Taq polymerase   An enzyme produced by *Thermus aquaticus* that is essential to the PCR technique.

Taxonomy   The science of classification of organisms.

Terminator sequence   A site on DNA marking the end point of transcription.

Tertiary syphilis   The third stage of syphilis during which organs and tissues undergo degenerative changes.

Tetanospasmin   A neurotoxin produced by *Clostridium tetani* that results in rigid paralysis.

Tetanus A bacterial disease acquired by exposure to *Clostridium tetani* or its spores. Symptoms include stiffness in the jaw and contraction of muscles in the limbs, stomach, and neck. Also called lockjaw.

Tetrads Groupings of cocci in clusters of four.

Thermus aquaticus A hyperthermophile isolated from hot springs.

Thymus gland The organ in which T-cell maturation is completed; it is located behind the sternum and just above the heart.

Tissue A group or collection of cells that are all of the same type.

Tonsils Structures of the secondary immune system located at the back of the throat that aid in protection from microbes entering through the nose and throat.

Toxemia An illness resulting from the presence of an exotoxin in the body.

Toxic shock syndrome A condition caused by certain strains of toxin-producing staphylococci; primarily associated with the use of highly absorbent tampons.

Toxigenicity The ability of microbes to produce toxins.

Toxins Major virulence factors that are harmful to the body and are produced by many pathogenic microbes; tetanus, botulinum, and erythrogenic toxins are examples.

Toxoplasmosis A protozoan disease caused by *Toxoplasma gondii* and acquired by the ingestion of oocysts present in cat feces. Symptoms include sore throat, low-grade fever, and lymph node enlargement.

Tracheitis Inflammation of the trachea.

Tracheotomy The cutting of a hole in the throat to facilitate breathing.

Transcription A genetic process in which DNA is "read" into mRNA.

Transduction A recombinational process characterized by bacteriophage-mediated transfer of DNA.

Transfer RNA (tRNA) A clover leaf-shaped molecule that transfers a single specific amino acid to the ribosome to build a polypeptide bond.

Transformation A recombinational process characterized by the uptake of "naked" DNA into competent cells.

Transforming principle A term used by Frederick Griffith to describe the transfer of virulence from virulent to nonvirulent bacteria; the active component was later discovered to be DNA.

Translation A genetic process in which the mRNA nucleotide sequence is converted into an amino acid sequence, which is assembled into a protein.

Transmissible spongiform encephalopathies Degenerative brain diseases that are thought to be the result of abnormally folded prions that latch onto normal prions and convert them into an altered, defective form.

Transmission A link in the cycle of microbial disease between reservoir and portal of entry.

Transovarial transmission The passage of microbes from adult ticks to their eggs.

Transposons Also called "jumping genes," genetic elements that move from one site on a chromosome to another or from a chromosome to a plasmid or from a plasmid to a chromosome.

Traveler's diarrhea An illness caused by enterotoxigenic *Escherichia coli* strains.

Treponemes Spirochetes released by chancres in individuals suffering from syphilis.

Triatomid insect The vector for Chagas disease, commonly referred to as the kissing bug.

Trichinellosis A roundworm disease caused by *Trichinella spiralis* that is transmitted in undercooked pork. Symptoms include nausea, diarrhea, vomiting, fatigue, fever, headaches, chills, aching joints, and itchy skin.

Trichomoniasis A sexually transmitted disease caused by *Trichomonas vaginalis*. Symptoms include intense itching, urinary frequency, pain during urination, and vaginal discharge in females. In males, symptoms include pain during urination, inflammation of the urethra, and a thin, milky discharge.

Trophozoite The reproductive and feeding stage of parasitic amoebae and other protozoan parasites.

Tuberculin skin test A method for diagnosis of tuberculosis; purified protein derived from *Mycobacterium tuberculosis* is injected into the arm of a patient, and the presence of an induration between 48 and 72 hours after injection indicates past or present exposure to the tubercle bacillus.

Tuberculosis A contagious lower respiratory tract disease caused by *Mycobacterium tuberculosis*. Symptoms include fever, night sweats, weight loss, fatigue, and coughing up of blood-tinged sputum.

Typhoid fever A disease caused by *Salmonella typhi* that is transmitted by flies and fomites.

Typhoid Mary A nickname for Mary Malone (1869–1938), a notorious carrier of typhoid fever.

Ulcer A lesion (sore) on a soft part of the body including the skin, duodenum, or stomach.

Unicellular Having only one cell.

Uncoating A process following the penetration stage in the viral replication cycle in which nucleic acid is released from the surrounding protein coat.

Upper respiratory tract infections Infections occurring in the tonsils or pharynx.

Urea A waste product of protein metabolism present in urine and other body fluids.

Urease An enzyme that digests urea.

Urinary tract infection (UTI) Bacterial infection that occurs in any part of the urinary tract.

Variant Creutzfeldt-Jakob disease A fatal human transmissible spongiform encephalopathic disease caused by prions. It is similar to mad cow disease in that it causes degeneration of brain tissue and severe neurological damage.

Varicella-zoster virus The DNA virus that causes shingles and chicken pox.

VD *See* Venereal diseases.

Vector-borne Carried by a particular organism.

Vegetative cells Spore-forming bacteria without spores.

Venereal diseases A former name for sexually transmitted diseases, commonly referred to as VD.

Vertical transmission A method of transmission characterized by passage of pathogens from parent to offspring across the placenta, in breast milk, or in the birth canal.

Vesicle A membrane-bound structure in the cytoplasm; also, a fluid-filled lesion that appears on the skin.

Vibrios Spirillar or S-shaped bacteria in the shape of comma-curved rods; *Vibrio cholerae*, the causative agent of cholera, is an example.

Virion A complete viral particle.

Virulence The capacity of microbes to produce disease as a result of defensive and offensive strategies.

Viruses One of the categories of microbes; characterized as subcellular obligate intracellular parasites.

Visceral leishmaniasis The most severe form of leishmaniasis; symptoms are fever, weakness, weight loss, anemia, and protrusion of the abdomen. Also known as kala-azar.

Viscerotropic viruses Viruses that affect internal organs (viscera) such as the liver, spleen, and intestines.

Wandering macrophages Macrophages derived from monocytes of the blood that are phagocytic and move freely about the tissues.

WBC See White blood cells.

West Nile virus An arbovirus transmitted by the bite of a mosquito that causes encephalitis; it emerged in New York in 1999.

Western equine encephalitis virus A common arbovirus in the United States.

White blood cells (WBCs) Group of cell types found in the bloodstream of which there are five categories; they play a role in the immune system.

Whoop The characteristic sound of whooping cough; deep and rapid inspirations in the partially obstructed passages are responsible for the sound.

Whooping cough A highly infectious disease caused by *Bordetella pertussis*. Symptoms include spasms of violent hacking and persistent, recurrent coughing with a whooplike noise. Also called pertussis.

XDR TB *See* Extremely drug-resistant tuberculosis.

Xenodiagnosis A diagnostic method used in Chagas disease. "Kissing bugs" free of trypanosomes are allowed to feed on individuals suspected of having the disease. A few weeks later, the insects are examined for the presence of the parasite.

Yellow fever A mosquito-borne viral disease; symptoms include fever, bloody nose, headache, nausea, muscle pain, vomiting, and jaundice. Also known as yellow jack.

Zanamivir An antiflu drug that is effective against both A and B influenza viruses. It reduces the severity of the flu if taken within forty-eight hours of the appearance of symptoms. The trade name of this drug is Relenza.

Zidovudine *See* AZT.

Zoonoses Diseases for which domestic and/or wild animals are the reservoirs and which can be transmitted to humans.

Appendix: Suggested Readings

Note: The most current books are at the top of the list.

Each book is followed by a letter or letters to indicate something about its content.
A: antibiotics; B: bacteria; F: fungi; H: history; P: protozoans; Pr: prions; V: viruses.

Encounters in Virology by T. Shors (2012) V

The Graves Are Walking: The Great Famine and the Saga of the Irish People by J. Kelly (2012) F

The Mirage Man: Bruce Ivins, the Anthrax Attacks, and America's Rush to War by David Willman (2011) B, H

The Making of a Tropical Disease: A Short History of Malaria by Randall M. Packard (2011) P, H

The Viral Storm: The Dawn of a New Pandemic Age by Nathan Wolfe (2011) V

The Black Death in London by Barney Sloane (2011) B, H

Pox: An American History by Michael Willrich (2011) H, V

Superbug: The Fatal Menace of MRSA by Maryn McKenna (2011) A

The Immortal Life of Henrietta Lacks by Rebecca Skloot (2010) H, V

A Child of Sanitariums: A Memoir of Tuberculosis Survival and Lifelong Disability by G. Paris (2010) B

Panic in Level 4: Cannibals, Killer Viruses, and Other Journeys to the Edge of Science by R. Preston (2009) V

The Illustrious Dead: The Terrifying Story of How Typhus Killed Napeleon's Greatest Army by Stephan Talty (2009) B, H

Deadly Companions: How Microbes Shaped Our History by D. H. Crawford (2009) H

Virus: The Co-Discoverer of HIV Tracks Its Rampage and Charts the Future by Luc Montagnier (2008) H, V

20th Century Microbe Hunters by Robert I. Krasner (2007) H

The Making of a Tropical Disease: The History of Malaria by R. M. Packard (2007) H, P

Twelve Diseases that Changed Our World by I. W. Sherman (2007) H

The Ghost Map: The Story of London's Most Terrifying Epidemic—and How it Changed Science, Cities and the Modern World by S. Johnson (2006) B, H

The Power of Plagues by I. W. Sherman (2006) H

Splendid Solution by Jeffrey Kluger (2006) V

Rats: Observations on the History and Habitat of the City's Most Unwanted Inhabitants by Robert Sullivan (2005) H

Typhoid Mary by A. Bourdain (2005) B, H

Guns, Germs and Steel: The Fates of Human Societies by Jared Steel (2005) H

Polio: An American Story, The Crusade That Mobilized the Nation Against the 20th Century's Most Feared Disease by D. M. Oshinsky (2005). H, V

Louis Pasteur: Revolutionary Scientist (Great Life Stories) by A. Lassieur (2005) H

Penicillin Man: Alexander Fleming and the Antibiotic Revolution by Kevin Brown (2005) A, H

The Mold in Dr. Florey's Coat: The Story of the Penicillin Miracle by Eric Lax (2004) A, H

The Great Influenza: The Epic Story of the Deadliest Plague in History by John M. Barry (2004) H, V

Lab 57: The Disturbing Story of the Government's Secret Plum Island Germ Laboratory by M. C. Carrol (2004) V

Beating Back the Devil: On the Front Lines with Disease Detectives of the Epidemic Intelligence Service by M. McKenna (2004) V

The Greatest Killer: Smallpox in History by D. R. Hopkins (2002) H, V

Plague Time: The New Germ Theory of Disease by P. W. Ewald (2002) H

The Demon in the Freezer by R. Preston (2002) V

Year of Wonders by Geraldine Brooks (2002) B

The Trembling Mountain: A Personal Account of Kuru, Cannibals, and Mad Cow Disease by R. Klitzman (2001) Pr

Flu: The Story of the Great Influenza Pandemic of 1918 and the Search for the Virus that Caused It by G. Kolata (1999) H, V

Influenza, 1918: The Worst Epidemic in America by L. Iezzoni and forward by D. McCullough (1999) H, V

The River: A Journey to the Source of HIV and AIDS by E. Hooper (1999) H, V

Biohazard: The Chilling True Story of the Largest Covert Biological Weapons Program in the World—Told from the Inside Man Who Ran It by K. Alibek and S. Handelman (1999) B, H, V

A Paralyzing Fear: Triumph Over Polio in America by N. G. Seavey, J. S. Smith, and P. Wagner (1998) H, V

Virus Hunter: Thirty Years Battling Hot Viruses Around the World by C. J. Peters and M. Olshaker (1997) V

Deadly Feasts: Tracking the Secrets of a Terrifying New Plague by Richard Rhodes (1997) Pr

The Evolution of Infectious Disease by Paul W. Ewald (1996) H

Yellow Fever, Black Goddess: The Coevolution of People and Plagues by C. Wills (1996) V

Level 4: Virus Hunters of the CDC by J. B. McCormick and S. Fisher-Hoch with L. A. Horvitz (1996) V

The Coming Plague by L. Garrett (1995) B, V

The Hot Zone by Richard Preston (1995) V

The Path Between the Seas: The Creation of the Panama Canal, 1870–1914 by David McCullough (1978) H, P, V

The Malaria Capers: Tales of Parasites and People by Robert S. Desowitz (1991) P

America's Forgotten Pandemic. The Influenza of 1918 by A. W. Crosby (1989) H, V

And the Band Played On: Politics, People, and the AIDS Epidemic by Randy Shilts (1988) H, V

Princes and Peasants: Smallpox in History by D. R. Hopkins (1983) H, V

Fever! The Hunt for a New Killer Virus J. G. Fuller (1974) V

The Andromeda Strain by M. Crichton (1969, reprinted in 1992) V

The Day of St. Anthony's Fire by J. G. Fuller (1969) F

Microbe Hunters by P. de Kruif (1926; reprinted in 2002) H

Arrowsmith, with a new afterward by E. L. Doctarow, revised edition, by S. Lewis (1925; reprinted in 1998) V

The Jungle by Upton Sinclair (1906; reprinted in 2012) H

Index

Note: Italicized page locators indicate figures/photos; tables are noted with *t*.

anaerobic bacteria, culturing, *31*
anal cancer, 274
Analytical Profile Index (API), 89, *90*
Analyzing a Bioterror Attack on the Food Supply: The Case of Botulinum Toxin in Milk, 465
anaphylactic shock, 348
Anaplasma phagocytophilum, 247
anatomy
    genitourinary system, *227*
    human digestive system, *203*
    immune system, *349,* 349–356
    lymphatic and circulatory system, 353, *353*
    respiratory system, *217*
ancient warriors, 433
Ancylostoma duodenale, 328
Anderson, W. French, 61
Animalia kingdom, 36, *36*
animals
    antibiotic-resistant pathogens in, 406
    deforestation and, 13
    fungal diseases of, 337
    helminthic diseases from, 324
    leishmaniasis from, 308
    rabies and, 278, 279, 281
    as reservoirs of infection, 183, 184
    ringworm in, 339, 340
    salmonellosis carried by, 208
    T. cruzi and, 306
    toxoplasmosis from, 304, *304*
    West Nile fever and, 289
Anisakis simplex, 325
Annals of Medicine, 399
Anopheles mosquitoes, 312
Anopheles pupa, *315*
anthrax, 239–240, 239*t,* 441*t, 442*
    attacks timeline, post-9/11, 448–449
    as biological weapon, 239, 435, 440, 441*t,* 442–444, 442*t,* 452
    British experiments with, 434, 442
    cutaneous, 240, *240,* 442, 447, 448, 449
    gastrointestinal, 240, 442
    hoaxes, 450
    inhalation, 240, 436, 442, 447, 448, 449
    military vaccination program, 240, 443–444, 452
    spores of, 240
    threats, U.S., 443
    varieties of, 193
antibiotics, 398–406, 408.
        See also penicillin
    adaptation to, 22
    broad-spectrum, 400
    H. pylori infections, 233

history of, 399–400
mechanisms of antimicrobial activity and, 400–402
misuse of, *404,* 404–406, 490
narrow-spectrum, 400
nosocomial infections and, 195, 197*t*
resistance to, 145
    acquisition of, 402–403. *See also* drug resistance
    mechanisms of, 403–404
    phage therapy and, 114
    solution for, 406
sensitivity, *89*
for tuberculosis, 282, 488
types of, 400
viruses and, 108
antibodies, 111, 350
    antigenic variation and, 459
    in gamma globulin, 350
    in immunity, 360–364, *362*
    influenza, 459
antibody-mediated immunity, 360–364, *362*
antibody molecules, *360, 362*
anticodons, 134
antigenic drift, 459
antigenic shift, 459
antigenic variation
    in influenza, 459
    in trypanosomes, *162,* 162–163
antigen-presenting cells, 364
antigens, 111, 350, 359
antimetabolites, 402
antimicrobial activity, mechanisms of, 400–402, *401*
antimicrobials, 399, 400
antiseptics, 397
Anti-Terrorism and Effective Death Penalty Act (1996), 437, 437*t*
antitoxins, 369
antiviral activity, *407*
antiviral agents, 407–408, 407*t*
API-20E test strip, 89
Apophysomyces trapeziformis, 20
appendages, bacterial, *76,* 83
appendix, *349*
    normal *vs.* ruptured, *156*
arboviral encephalitides, 285*t*
arboviruses, 188, 285
Archaea, 35*t,* 37, *37,* 37*t,* 38, 42*t*
arenaviruses, as biological weapon, 441
arginine, 135*t*
"Armageddon virus," 456, 464–465
Armed Forces Institute of Pathology, 461
arsenic, 392

Arthropoda phylum, 187, 187*t,* 320
arthropodborne diseases, 187–191, 189*t*–190*t,* 243–251, 244*t*
    bacterial, 243–251
    ehrlichiosis, 246–247
    helminthic, 317*t,* 324–327
    Lyme disease, 247–249
    plague, 244–246
    protozoan, 300*t,* 304–315
    relapsing fever, 249–250
    Rocky Mountain spotted fever, 250
    typhus fever, 250–251
    viral, 285–290, 285*t*
artificial immunity, 366
ascariasis, 317–319, *318*
Ascaris lumbricoides, 316, 317, 318, 319
Ascaris worms, 168
asexual reproduction, 497–498
asexual stage, of mosquitoes, 312
Asia, sanitation facilities in, *387*
asparagine, 135*t*
aspartate, 135*t*
assembly, 103, 103*t,* 121, 407
association, in disease causation, 158
asthma, 503
Aswan Dam, schistosomiasis and, 333
athlete's foot, 339
atmosphere, 37–38
atomic bomb, 465
atypical bacteria, 91–92, 91*t*
Aum Shinrikyo ("Supreme Truth") cult, 437
autism, 264, 372, 373
autoimmune disease, 348. *See also* AIDS (acquired immune deficiency syndrome)
autoinfection, 330, *330*
autotrophs, 31, *31*
Avery, Oswald, 129, 130*t*
avian flu, 96, 109, 197, 462–466, *463,* 498
azidothymidine (AZT), 408, 475, 477, 478
azithromycin, 405

B

Babesia divergens, 315
Babesia microti, 315
babesiosis, 188, 315
bacilli, 74, *74,* 92
Bacillus anthracis, 75, 77, 239, 240, *240*
    as biological weapon, 434, 440, 441, 442, 447, 448, 449
    spores of, 81, *81,* 82, 436
bacillus Calmette-Guérin (BCG) vaccine, 368, 492

botulinum, soil as reservoir of, 184
botulinum toxin, 166, *166*, 436
botulism, 82, 165, 165*t*, 204–207, 392, 441*t*
　as biological weapon, 440, 441
　incubation stage, 169
bovine spongiform encephalopathy (BSE),
　119, 120, 290, 291
　atypical, 293
　Creutzfeldt-Jakob disease and, 291
Boy in the Plastic Bubble, The (film), 375
Bradford cocci, 100
Brazil, AIDS program in, 480–481
bread, 55
breakbone fever, 285
breath test, 233
Brown, Paul, 294
Brucella species, as biological weapon, 441
brucellosis, 441
Brugia malayi, 325
Brundtland, Gro Harlem, 310
BSE. *See* bovine spongiform
　encephalopathy (BSE)
Bt, 63
bubble boy disease, 375
buboes, 245, *245*
bubonic plague, 157, 170, *245,* 245–246,
　456, 493
budding, of viruses, 107, *108*
bulbar poliomyelitis, 258
Burkholderia mallei, as biological
　weapon, 441
Burkitt's lymphoma, 276
Bush, George W., 419, 446, 447
buttermilk, 56

C
Cairns, John, 13, 382
Campylobacter, 214, 215
campylobacteriosis, 214–215
Campylobacter jejuni, 214
CA-MRSA. *See* Community acquired
　MRSA infections
cancer, *7*
　anal, 274
　Burkitt's lymphoma, 276
　cervical, 274, 503
　gastric, 233
　　H. pylori in, 233
　Kaposi's sarcoma, 466, *466*
　leukemia, 375
　liver, *276,* 276–277
　lung, 503
　penile, 274
　prostate, 276
　stomach, 503

throat, 274
　virotherapy and, 116–117, 121
　vulvar, 274
Candida albicans, 339–340
candidiasis, 297, 339–341, 374
capsid, 101, 121
capsomeres, 101
capsules, 75, *76,* 77, 160*t,* 161–162, *162*
carbohydrates, 38
carbon cycle, 51, *51*
carbon dioxide
　in carbon cycle, 51, *51*
　in fermentation, 54
　in stationary phase of growth, 86
carbuncles, 238
cardiovascular disease, *7*
Caribbean region, AIDS in, 481
carriers of disease. *See also* reservoirs
　of infection
　gonorrhea, 230
Carter, Jimmy, 322, 420
Carter Center, 327, 412, 425
Category A/B/C agents, 440, 441, 441*t*
cats, toxoplasmosis from, 304, *304*
cattle plague, 412
causation, in disease, 158
causative agent, 339
cauterization, 280–281
CD4+ cell count, 466, 467
CD4 receptor molecules, 359
CD8 receptor molecules, 359
cell cultures, 108–110, *109, 110*
　bacterial, 87–91, *88*
　viral, 102–103
cell-mediated immunity (CMI), 363–364
cell membrane, 402
　bacterial, 75, 79–80
cells, 30, *34*
　differentiation and specialization in,
　　33–34
　envelope of, 75, *76*
　shapes and patterns of bacteria in, 74, *74*
cell theory, 30
cellulose, 51
cell walls
　antibiotics and, 400–402
　bacterial, 75, 77–79
Centers for Disease Control (CDC), 5, 7,
　174, 181, 191, 209, 254
　on AIDS, 466, 467, 478, 482
　on bacterial diseases, 201
　biological agents/diseases regulated by,
　　437–438, 438*t*
　biological weapon monitoring by, 431,
　　437, 447, 452

biosafety levels in, 285, *285*
bioterrorism and plan of readiness
　by, 447, 452
building 15, 417–418
on Chagas disease, 307
on chickenpox, 266–267
childhood immunization schedule
　by, *374*
on chlamydia, 231
on *Clostridium difficile,* 216
Committee on Immunization
　Practices, 372
on dengue fever, 286
on Ebola virus, 285
Emergency Response Team, 462
on foodborne disease, 202, 204, 255,
　393, 394, *394*
framework for preventing infectious
　diseases, 418
functions of, 416
Green Sheet, 418
headquarters, Atlanta, *416*
on hepatitis, 257
on HIV infection stages, 472–473
on human papillomavirus, 274
on influenza, 460
on leptospirosis, 243
on mad cow disease, 291
on measles, 263, 264
on MRSA, 405
notification of disease outbreaks to, 415
on onchocerciasis, 327
partnerships with, 424
on pertussis, 221
pharmaceutical stockpiles developed
　by, 446
on plague, 245
on public health achievements, 364,
　380, 380*t*
on reptiles and risk of salmonellosis, 208
research labs, 416–417
on rotavirus infections, 256
on rubella, 266
on smallpox vaccine, 445
surveillance by, 426
traveler's health information, 419
on trichinellosis, 320
on tuberculosis, 492
on vectorborne diseases, 190
water surveillance by, 391
on West Nile fever, 289
zombie apocalypse campaign, 503, *503*
Central Asia, AIDS in, 481
Central Intelligence Agency (CIA), 380, 449
cephalosporins, 401

Watson, James, 123, 128, 130, 130*t*
Weller, Thomas, 109, 110, 112, 113
western equine encephalitis, 287, *288,* 290
West Nile fever, 287, 288–289
West Nile virus, 3, 23, 96, *96,* 188, 191, 197, 294
Wheelis, Mark L., 36
whipworm infection, 319
White, Owen, 40
White, Ryan, 466
white blood cells, 33, 350, *351,* 352*t*
Whittaker, Robert, 36, 37
whoop, 220
whooping cough. *See* pertussis
wildfires, 19
William J. Clinton Foundation, 425
Wimmer, Eckard, 146
wines, 58, 59–60, *60*
wine tasting, 59
Winfrey, Oprah, 290
Woese, Carl, 36, 37
World Bank, 314, 385, 424, *424*
Worldfact Book, 380
World Health Assembly, 412
World Health Organization (WHO), 7, 96, 181, 493
  on clean water, 389, 391
  Constitution of, 6
  on Ebola virus, 284
  Emergency Response Team, 462
  on filariasis, 326
  on foodborne disease, 394

  on food safety, 392
  on global disease outbreaks, 426
  on guinea worm disease, 322
  on Hansen's disease, 237
  health initiatives of, 412, 506
  on hepatitis B, 277
  on hookworm, 330
  on influenza, 460, 463
  on leptospirosis, 243
  on malaria, 5, 310–311, 314
  on measles, 263
  on onchocerciasis, 327
  partnerships with, 422
  on plague, 245
  on sanitation and clean water, 382, 386
  on SARS, 269
  on shigellosis, 209
  on smallpox, 157
  on trypanosomiasis, 306
  on tuberculosis, 225, 483–484, 488, 494
World Malaria Day, 314
World Malaria Report, 5, 310
World Organization for Animal Health, 464
World Toilet Organization, 385
World Trade Center, terrorist attack on, 447
World Tuberculosis Day, 483, 489
World Urbanization Prospects, 10
World War I, 432, 433, 457
World War II, 398, 399, 433–434

worms. *See* helminths
Wuchereria bancrofti, 325

X
xenodiagnosis, 307

Y
yeasts, 32, 54–55, *55*
yellow fever, 285*t,* 286–287, 412, 419, 441, 456
  vaccine, 368
  vector for, 184
Yersinia pestis, 191, 193, 244, *244,* 245, 434, 437, 441, 465. *See also* plague
yogurt, 56–57, *57*

Z
zanamivir, 408, 460
zidovudine, 408, 475
Zigas, Vincent, 118
Zinsser, Hans, 6, 156, 455
Zolinza, 482
zones of inhibition, 89
Zoogloea ramigera, 69
zoonoses, 183
zoonotic diseases, 9, 183–184, 183*t*
  Lyme disease, 247–249
  plague, 244–246
  rabies, 278
  toxoplasmosis, 304
  tularemia, 226
Zostavax, 268
Zygomycetes, 336